# PHARMACEUTICAL BIOTECHNOLOGY

S. S. PUROHIT
H. N. KAKRANI
A. K. SALUJA

AGROBIOS (INDIA)

**Dr. S. S. Purohit**
*Ex-Head*, P.G. Department of Botany
Dungar College, Bikaner
**Dr. H. N. Kakrani**
*Principal*, Tolani Institute of Pharmacy
Opposite Hotel Mid Town, Ward 2A, Adipur (Kutch)
**Dr. A. K. Saluja**
*Principal*, A. R. College of Pharmacy & G. H. Patel Institute of Pharmacy
Vallabh Vidyanagar

*Published by:*
**AGROBIOS (INDIA)**
Agro House, Behind Nasrani Cinema
Chopasani Road, Jodhpur 342 003
Phone: 91-0291-2643993, 2642319
E. Mail: agrobiosindia@gmail.com, Website: agrobiosonline.com

AGROBIOS (INDIA)

**AGROBIOS INDIA**
First Published, 2003
**Print: 2026**



ISBN: 978-81-7754-194-6

***Published by:*** Dr. Updesh Purohit for Agrobios (India), Jodhpur
***Lasertypeset at:*** Yashee Computers, Jodhpur
***Cover Design by:*** Shyam
***Printed at:*** Agrobios Digitals, Jodhpur

# Preface

*One should respect the power of biotechnology and apply it for the benefit of humankind.*

Biotechnology in pharmaceutical industry employs techniques of modern molecular biology to help humankind. Pharmaceutical biotechnology works to make manufacturing processes more efficient in industrial production of pharmaceutical products. Biotechnology will transform the industrial sector in much the same way that it has changed the pharmaceutical sector.

Pharmaceutical Industrial Biotechnology companies develop biocatalysts, such as enzymes, to be used in chemical synthesis. Enzymes are proteins produced by all living organisms. In humans, enzymes help digest food, signal cells to turn on and off, and perform other complex functions. Enzymes are characterized according to the compounds they act upon. Some of the most common enzymes are proteases, which break down protein; cellulases, which break down cellulose; lipases, which act on fatty acids and oils; and oils; and amylases, which break starch down into simple sugars. To improve the productivity-to-cost ratio, scientists are modifying genes to increase enzyme productivity in microorganisms commonly used in manufacturing. They also use genetic engineering to make other microbial enzymes that are too expensive or even impossible to cultivate. The biotechnology techniques of protein engineering and directed protein evolution have been used to modify the substrate specificity of enzymes, improve catalytic properties or broaden the reaction conditions under which enzymes can function so that they are more compatible with existing industrial processes. Companies involved in industrial biotechnology are constantly striving to discover and develop high-value enzymes and bioactive compounds that will enhance current industrial processes.

The present book is an attempt to provide overall information to the readers interested in Pharmaceutical Biotechnology.

Authors

August, 15th 2003

# Contents

# 1. Pharmaceutical Biotechnology: An Introduction

## ORIGIN AND DEFINITION: OLD VS NEW BIOTECHNOLOGY

Biotechnology (*bio* = the use of biological processes, and *technology* = to solve problems or make useful products), in the recent years, has created unprecedented opportunities, not only for the manipulation of biological systems for the benefit of logical systems for the benefit of mankind but also for undertaking studies to understand the fundamental life processes. Consequently, it has become the world's fastest growing and most rapidly changing technology.

The story of the use of biological systems for the fulfillment of human needs perhaps started in 6000 B.C. when Sumerians and Babylonians fermented a kind of beer. Beginning with fermentation, use of biological processes then experienced many changes over the centuries. But greatest revolution till now commenced with 1970's and 1980's, when a product of interaction between the science of biology and technology came into wider existence. This relationship got the name *Biotechnology*. Fermentation, antibiotic production, baking and brewing are included in old biotechnology, whereas, techniques related with cell culture, fusion, bioprocessing, genetic engineering etc. were named as new biotechnology. Besides the several educational institutions, many commercial companies also engaged themselves in biotechnology research for potential gains.

Biotechnology is the ***third wave*** in biological sciences and represents such an interface of basic and applied sciences, where gradual and subtle transformation of science into technology can be witnessed. The prokaryotes, eukaryotic algae, glycophytes and halophytes all are likely to contribute in commercially viable future propositions. Biotechnology is the application of biological organisms, systems or processes to manufacture and service industries, comprises a number of technologies based upon increasing understanding of biology at the cellular and molecular level. The technique includes recombinant DNA manipulations, monoclonal antibody preparation, tissue culture, protoplast fusion, protein engineering, immobilized enzyme, cell catalysis, sensing with the aid of biological molecules, etc. The term *biotechnology* gained several definitions from different groups as follows:

### Definitions

There are many definitions of biotechnology and yet there is *more confusion* as to what it is, and what is involved in the practice activity; is practiced in University and Polytechnic Laboratories, Government and private research institutes and large and small industries, who are each involved to some extent in research, development, testing, evaluation and production exercises. Clearly what goes on in industry, is better defined as **Process Biotechnology.** This is simply described as a discipline which enables its exponents to convert raw materials to final products when either the raw material and/or a stage in the production process involves biological entities. Biotechnology in the broad sense, however, involves the cloners, the hybridisers and the molecular and cell biologists. It involves the conjoint interaction of two identifiable sub-components; bio**Science** and bio**Technology**.

1. Biotechnology is also defined as the applications of scientific and engineering principles to the processing of material by biological agents to provide goods and services.
2. The Spinks Report (1980) defined biotechnology as the application of biological organisms, systems or processes to the manufacturing and service industries.
3. **United States Congress's Office of Technology Assessment** defined biotechnology as any technique that used living organisms to make or modify a product, to improve plants or animals or to develop microorganisms for specific uses. The document focuses on the development and application of modern biotechnology based on new enabling techniques of recombinant-DNA technology, often referred to as genetic engineering.
4. **British Biotechnologist:** Application of biological organisms, system or processes to manufacturing and service industries'
5. **European Federation of Biotechnology:** The integrated use of biochemistry, microbiology and engineering sciences in order to achieve technological (industrial) application of the capabilities of micro-organisms, cultured tissue cells and parts thereof'
6. **Japanese Biotechnologists:** 'A technology using biological phenomena for copying and manufacturing various kinds of useful substances
7. **US National Science Foundation:** The controlled use of biological agents, such as microorganisms or cellular components for beneficial use.

Unfortunately, biotechnologies, are still in the process of early development, do not possess a sharp and easily defined form. Only in last fifteen years progress have been made by microbiologists and genetic engineers, and we are hopeful to solve many fold problems of the present day, specially energy and food crisis to cater the need of growing population of the world. Mineral ore deposits are also becoming more scarce and expensive to recover from earth's crust. Microorganisms can be used to enhance the recovery of metals from low-grade ores and from effluents containing undesirable quantities of heavy metals or other toxins.

When these technologies are applied at industrial level, they constitute bio-industry (**TABLE 1.1**) which include, on the one hand, industrial activities where biotechnologies can replace technologies normally or currently in use and, on the other hand, industrial activities where biotechnologies play an important driving role. There are several areas in which these technologies are being used, the important are chemical industry, food industry (mass production of yeast, algae and bacteria with a view of providing proteins, amino acids, vitamins, and use of enzymes), agricultural productivity, pharmaceutical industry, environmental protection and abatement of pollution, etc.

**TABLE 1.1:** Schematic distribution of the principal products of bio-industry

| **Fermentation Technology** | |
|---|---|
| Health | Antibiotics, Vitamins, Enzymes, Amino acids, Polysaccharides (dextran), Nucleotides, Steroids, Alkaloids, Diagnostic reagents, etc. |
| Food & Agricultural | Citric acid, Lactic acid, Gluconic acid, Malic acid, Amino acids, Nucleotides, Enzymes, Biopolymers, etc. |
| Agriculture | Biopesticides, improvement of varieties, etc. |
| Energy | Ethanol, Acetone, Butanol, Biogas, etc. |
| Chemical Industries | Chemistry of ethanol, Ethylene, Acetaldehyde, Acetone, Butanol, etc. |
| **Enzymatic Engineering** | |
| Food & Agricultural | Isoglucose, Glucose syrup, etc. |
| Energy | Ethanol |
| **Genetic Engineering and Cell Cultures** | |
| Food and Agriculture | Single cell protein, clones, etc. |
| Health | Interferons, Hormones, Vaccines, Blood products, Monoclonal antibiotics, etc. |

## Historical Background

The oldest biotechnological processes are found in microbial fermentations, as born out by a Babylonian tablet dated circa 6,000 B.C., unearthed in 1881 and explaining the preparation of beer. The Sumerians were able to brew as many as twenty types of beer in the third millennium BC. In about 4000 BC. leavened bread was produced with the aid of yeast. **TABLE 1.2** presents chronological history of biotechnology. The term biotechnology was described in a Bulletin of the Bureau of Biotechnology published in July, 1920 from the office of the same name in Leeds in Yorkshire. The articles in this bulletin described the varied roles of microbes in leather industry to pest control.

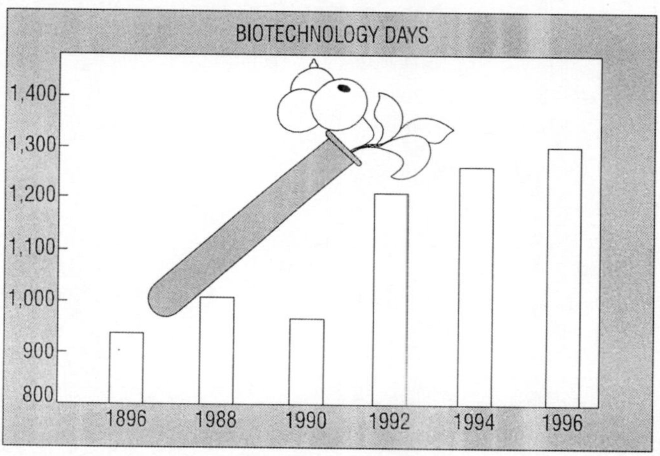


**FIGURE 1.1:** Biotechnological Days. Despite the potential gains, the number of companies engaged worldwide in biotechnology research remains stagnant.

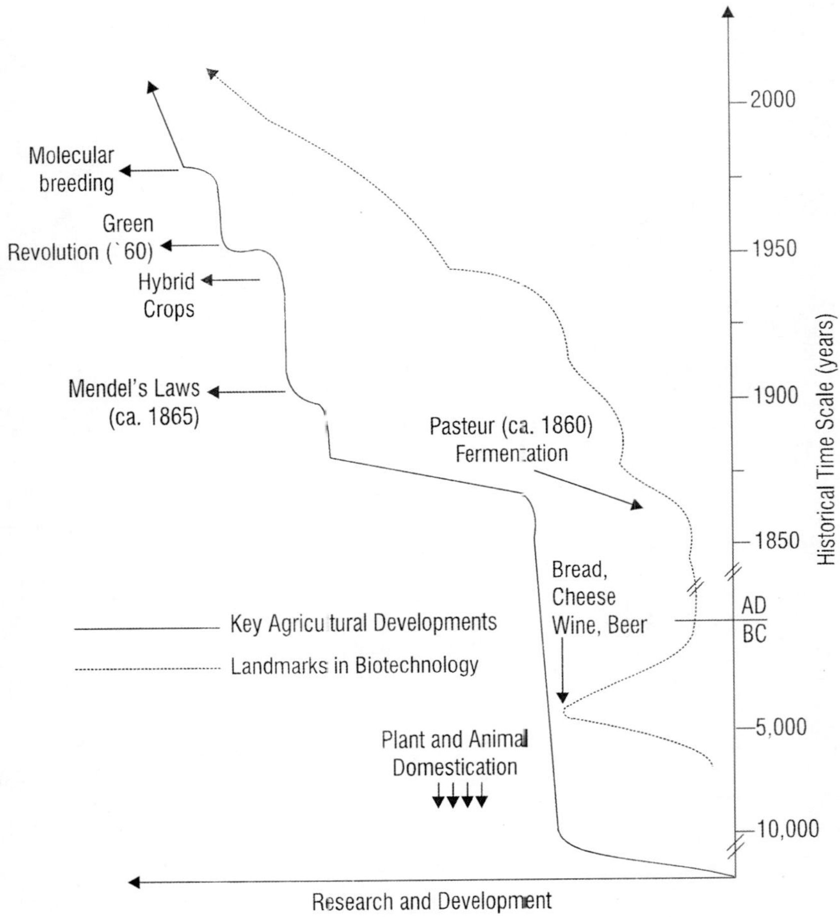


**FIGURE 1.2:** Research and Development in biotechnology through ages.

**TABLE 1.2:** Biotechnology through ages

| Time Period | Work |
|---|---|
| 8000 B.C. | Humans domesticate crops and livestock.<br>Potatoes first cultivated for food. |
| 4000-2000 B.C. | Biotechnology first used to leaven bread and ferment beer, using yeast (Egypt).<br>Production of cheese and fermentation of wine (Susmeria, China and Egypt).<br>Babylonians control date palm breeding by selectively pollinating female trees with pollen from certain male trees. |
| 500 B.C. | First antibiotic-Moldy soybean curds used to treat boils (China). |
| A.D. 100 | First insecticide-Powdered chrysanthemums (China). |
| 1322 | An Arab chieftain first uses artificial insemination to produce superios horses. |
| 1590-1875 | It took more than 300 years to discover and understand cells:<br>1590-Janssen invents the microscope. |

| Time Period | Work |
|---|---|
| | 1663-Hooke discovers existence of the cell.<br>1675-Leeuwenhoek discovers bacteria. |
| 1761 | Koelreuter reports successful crossbreeding of crop plants in different species. |
| 1797 | Jenner inoculates a child with a viral vaccine to protect him from smallpox. |
| 1830-1833 | 1830-Proteins discovered.<br>1833-First enzyme discovered and isolated. |
| 1835-1855 | Schleiden and Schwann propose that all organisms are composed of cells, and Viichow declares, "Every cell arises from a cell." |
| 1857 | Pasteur proposes microbes cause fermentation. |
| 1859 | Charles Darwin publishes the theory of evolution by natural selection. The concept of carefully selecting parents and culling the variable progeny greatly influences plant and animal breeders in the late 1800s despite their ignorance of genetics. |
| 1865 | Science of genetics begins: Austrian monk Gregor Mendel studies garden peas and discovers that genetic traits are passed from parents to offspring in a predictable way-the laws of heredity. |
| 1870-1890 | Using Darwin's theory, plant breeders cross-breed cotton, developing hundreds of varieties with superior qualities.<br>Farmers first inoculate fields with nitrogen-fixation bacteria to improve yields.<br>William James Beal produces first experimental corn hybrid in the laboratory.<br>1877-A technique for staining and identifying bacteria is developed by Koch.<br>1878-The first centrifuge is developed by Laval.<br>1879-Fleming discovers chromatin, the rod-like structures inside the cell nucleus that later came to be called chromosomes. |
| 1900 | *Drosophila* (fruit flies) used in early studies of genes. |
| 1902 | The term "immunology" first appears. |
| 1906 | The term "genetics" in introduced. |
| 1911 | The first cancer-causing virus is discovered by Rous. |
| 1914 | Bacteria are used to treat sewage for the first time in Manchester, England. |
| 1915 | Phages, or bacterial viruses, are discovered. |
| 1919 | First use of the word "biotechnology" in print. |
| 1920 | The human growth hormone is discovered by Evans and Long. |
| 1928 | Penicillin discovered as an antibiotic by Alexander Fleming.<br>A small-scale test of formulated *Bacillus thuringiensis* (Bt) for corn borer control begins in Europe. Commercial production of this biopesticide begins in France in 1938.<br>Karpechenko crosses radishes and cabbages, creating fertile offspring between plans in different genera.<br>Laibach first uses embryo rescue to obtain hybrids from wide crosses in crop plants-known today as hybridization. |
| 1930 | U.S. Congress passes the Plant Patent Act, enabling the products of plant breeding to be patented. |
| 1933 | Hybrid corn, developed by Henry Wallace in the 1920s, is commercialized. Growing hybrid corn eliminates the option of saving seeds. The remarkable yields outweigh the increased costs of annual seed purchases, |

| Time Period | Work |
|---|---|
| | and by 1945, hybrid corn accounts for 78 percent of U.S.-grown corn. |
| 1938 | The term "molecular biology" is confined. |
| 1941 | The term "genetic engineering" is first used by Danish microbiologist A. Jost in a lecture on reproduction in yeast at the technical institute in Lwow, Poland. |
| 1942 | The electron microscope is used to identify and characterize a bacteriophage-a virus that infects bacteria<br>Penicillin mass-produced in microbes. |
| 1944 | DNA is proven to carry genetic information-Avery et al.<br>Waksman isolates streptomycin, an effective antibiotic for tuberculosis. |
| 1946 | Discovery that genetic material from different viruses can be combined to form a new type of virus, an example of genetic recombination.<br>Recognizing the threat posed by loss of genetic diversity, the U.S. Congress provides funds for systematic and extensive plant collection, preservation and introduction that U.S. agriculturalists had pursued on a limited and random basis for 200 years. |
| 1947 | McClintock discovers transposable elements, or "jumping genes," in corn. |
| 1949 | Pauling shows that sickle cell anemia is a "molecular disease" resulting from a mutation in the protein molecule hemoglobin. |
| 1951 | Artificial insemination of livestock using frozen semen (a longtime dream of farmers) is successfully accomplished. |
| 1953 | The scientific journal *Nature* publishes James Watson and Francis Crick's manuscript describing the double helical structure of DNA, which marks the beginning of the modern era of genetics. |
| 1954 | Cell-culturing techniques are developed. |
| 1955 | An enzyme involved in the synthesis of a nucleic acid is isolated for the first time. |
| 1956 | The fermentation process is perfected in Japan.<br>Korenberg discovers the enzymes DNA polymerase, I leading to an understanding of how DNA is replicated. |
| 1958 | Sickle cell anemia is shown to occur due to a change of single amino acid.<br>DNA is made in a test tube for the first time. |
| 1959 | Systemic fungicides are developed. The steps in protein biosynthesis are delineated. |
| Also in the 1950s | Discovery of interferons.<br>First synthetic antibiotic. |
| 1960 | Exploiting base pairing, hybrid DNA-RNA molecules are created.<br>Messenger RNA is discovered. |
| 1961 | USDA registers first biopesticide; *Bacillus thuringiensis*, or Bt. |
| 1963 | New wheat varieties increase yields by 70 percent by Norman Borlaug. |
| 1964 | The international Rice Research Institute in the Philippines starts the Green Revolution with new strains rice that double the yield of previous strains if given sufficient fertilizers. |
| 1965 | Harris and Watkins successfully fuse mouse and human cells. |
| 1966 | The genetic code is cracked, demonstrating that a sequence of three nucleotide bases (a codon) determines each of 20 amino acids. |
| 1967 | The first automatic protein sequencer is perfected. |
| 1969 | An enzyme is synthesized in vitro for the first time. |
| 1970 | Specific restriction nucleases are identified, opening the way for gene cloning. |

| Time Period | Work |
|---|---|
| | Norman Borlaug receives the Nobel Peace Prize.<br>Discovery of restriction enzymes that cut and splice genetic material. |

## SCOPE AND IMPORTANCE OF BIOTECHNOLOGY

Biotechnology is the applied science and has made advances in two major areas, viz., molecular biology and production of industrially important biochemicals (including enzymes).

- **Biotechnology of plants and specific crops,** including genetic engineering for yield and quality traits, metabolite production, rapid clonal propagation, and germplasm conservation.
- **Microbial agro-biotechnology,** including new approaches to plant pest control, biofertilization, bioremediation, and fermentation of agricultural products.
- **Livestock biotechnology,** including genetic manipulations for improving fertility and reproduction, growth, milk quality, egg production, disease control, and genepharming.
- **Marine biotechnology,** including the use and improvement of fish, macro- and micro-algae for food and biochemical production and environmental uses.
- **Pharmaceutical Biotechnology:** This science is based on the production of therapeutic proteins and hormones, fermentation products like antibiotics, vaccines or drugs (using receptor hypothesis), gene correction, drug delivery to specific tissue (targeted delivery), production control using biosensors, artificial bioprostheses, standardization of chemotherapeutic agents and the diagnostic aids using the gene cloning technology, recombinant DNA technology, enzyme immobilization, monoclonal antibodies and mutagenesis have been exploited and attempted for possible use. Furthermore, all the fields of medical biotechnology have to be linked with the pharmaceutical biotechnology in order to attain the approvals from FDA.

The last few decades have witnessed some revolutionary advancement in biological sciences, which have dramatically affected the future of design and development of pharmaceuticals in an unprecedented manner. The concept of molecular biology and biotechnology has altogether changed the scenario of pharmaceutical developments. Emergence of DNA cloning technology, known commonly as recombinant DNA (rDNA) technology and the methodologies for production of monoclonal antibodies called hybridoma technology, are the counter pieces of revolution, which are the outcome of decades of continual basic research directed towards molecular biology, cell biology and immunology. Similar developments in the field of biological sciences, commonly referred as 'biotechnology', have facilitated the development, designing discovery, as well as production of drug(s). Not only drugs but also the technology for better therapeutics and diagnostics have been put forth.

Some major advancements in the field of pharmaceuticals in the last few decades have come forth as a result of extensive research and advancement in pharmacology, analytical chemistry, synthetic organic chemistry, biochemistry and drug delivery. The field of biotechnology has thus an attractive opportunity for maximum developments. It is 10 years now after FDA approval was granted for the first biotechnological product, recombinant human insulin; over 100 products derived from biotechnological principles have progressed to the stage of clinical use or human trials. Over the next 10 years, the global sale for biotechnology-based products is expected to approximate $30 billion per annum. It is therefore abundantly clear that the practice of pharmacy has been and will continue to be, strongly influenced by the field of biotechnology and the practicing pharmacist should surely understand the concepts now being created. Some milestones in the history of pharmaceutical biotechnology are listed in **TABLE 1.3**

**TABLE 1.3:** Milestones in biotechnology

| S N | Milestone | Scientist and Year |
|---|---|---|
| 1. | Proposed double helix model for 3 dimensional structure of DNA after X-ray diffraction data | R.E. Franklin and M.H. Wilkins; J.D. Watson and F. H. Crick, 1952-1953 |
| 2. | Cleavage of DNA by restriction endonucleases | W. Arber, 1962; M. Meselson and R. Yuan, 1968 |
| 3. | Determination of genetic code | M. Nirenberg, S. Ochoa and P. Heder, 1966; H. G. Khorana, 1966 |
| 4. | Identification of DNA ligase | M. Gellert, 1967 |
| 5. | DNA cloning techniques | H.W. Boyer, S. Cohen and P. Berg 1971-1972 |
| 6. | Emergence of rDNA technology | Bordon Conference on Nucleic acids, June 1973 |
| 7. | Hybridoma creation | C. Milstein and G. Kohler, 1976 |
| 8. | Guidelines issued by Recombinant Advisory Committee (RAC) | 1976 |
| 9. | DNA sequencing technology | F. Sangar, 1977; W. Gilbert, 1977 |
| 10. | US approval of first diagnostic kit using monoclonal antibody technology (MAb) | Anti-C3d Bioclone: Orthro diagnostics 1981 |
| 11. | US approval for first pharmaceutical product derived from DNA technology [Human Insulin (Humulin)] | Genetech and Eli Lilly and Co. 1982 |

*Huber, B.E., FASEB, 3(1), 1989, 5

The production of enzyme or hormone based pharmaceuticals has been revolutionized by the recombinant DNA technology. Consider the case of insulin; since its discovery in 1921 by Banting and Best, many animals have been sacrificed so as to achieve very small, quantities of the protein. However, since 1986 the recombinant production process has overtaken the markets. The human genetic coding for proinsulin is inserted in *Escherichia coli* cells, which were then grown by fermentation to produce proinsulin. The connecting peptide was further cleaved and converted enzymatically from proinsulin to human insulin. Currently Eli-Lilly and company is manufacturing the product. The recombinant DNA technology offered many added advantages like production of non-immunogenic, more effective, cheaper products, sparing animal life too.

It has been a subject of debate for years as to how to define pharmaceutical biotechnology as well as how to justify the nomenclature. It was evolved out of vigorous discussion, that as pharmaceutical aspects of chemistry and analysis were respectively accepted as pharmaceutical chemistry or pharmaceutical analysis, it appears to be appropriate to accept the name for the field that encompasses pharmaceutical aspects of biotechnology as pharmaceutical biotechnology. Obviously the products which occur naturally and of microbiological or biological origin having applicable potential in pharmaceutical industry in human therapeutics, in disease diagnosis as well as clinical monitoring of patients may largely be covered under the discipline of pharmaceutical biotechnology. Obviously, the major areas which could be considered include antibodies as microbial secondary metabolites, monoclonal antibodies, genetic engineering and related products, enzyme products, microbial steroid conversion, recombinant vaccine, single cell proteins, animal and plant cell cultures for production of pharmaceuticals, immunomodulators, blood products, tissue banks, protein hydrolysates, and glandular products. Furthermore, other products, tissue banks, protein hydrolysates, and glandular products. Furthermore, other products which too belong to biologicals may include sera, diagnostic agents, organic acids, vitamins, nucleotides, oligonucleotides, antisense, plasma expanders, alkaloids, sutures and ligatures and other microbiological products used in diagnostic and biological assays.

*The unprecedented development and growth of biotechnology has been an outcome of some of the following mentioned successive, discrete milestone discoveries and events in basic biological research**

A concept central to pharmaceutical development is the postulation of the presence of 'receptor substances' in biological systems to which drugs would bind with specificity, thus eliciting their responses. Ehrlich has propounded receptor concept from his studies on antigen-antibody interaction and binding of dye to cells. Earlier developments in drug discovery were based upon hypothetical receptor mediated drug action since methods for isolation of receptors were no established.

The new technological advancements of recombinant DNA and monoclonal antibody technology address the receptor site of the drug selectively and enable scientists to have real understanding of mode of the drug action mechanism. The ability to obtain detailed structure of receptors and the molecular legends using biological analysis of the gene sequence, production of receptors in large quantities for structural analysis and development of specific antibody reagents to analyze their presence in organs and tissues have opened new vistas for newer drug development and for modification of old drug entities.

The recombinant DNA technology may be appreciated as a milestone, which has led to a qualitative and quantitative growth in research in the field during this century. Some of the products originating from genetic engineering are clotting factor, colony stimulating factors, dismutases, erythropoietins, MAbs, recombinant soluble CD4, tumor necrosis factor and tissue pasminogen. Using monoclonal antibodies a variety of diagnostic kits have been developed. Some of the classical advents include: ovarian cancer detecting test, and kits for detection of tumor in stomach and intestine as well as kits for pregnancy testing. The diagnostic kits developed are of greater precision, accuracy and reproducibility. Similarly, immunoradiometric assay kits for measurement of interferon or other autocoids are interesting examples of biotechnological development, which have significantly contributed towards diagnostic pharmaceuticals. The hybridoma technology, which provided pivotal base for monoclonal antibody production, is one of the important inventions in biotechnology on which what we call modern pharmaceutical biotechnology mainly depends upon. The most important utilization of the hybridoma-based antibodies is their specific utilization in the target-oriented therapy so as to attain cell specific delivery of the drugs like anticancer and anti-HIV drugs etc.

Enzyme related products, which nearly count two-dozen is employed in therapeutics and preventive medicine. The therapeutic values are mainly classified to be as clotting factor, digestive aids, antithrombolytic agents, diffusive enzymes, anti-inflammatory enzymes and some specifically designed or aimed enzymes (e.g., urokinase, chymotrypsin, penicillinase, glucocerbrosidase, streptokinase, etc.). One of the various functions of an enzyme is biochemical conversion, which has been greatly exploited for the conversion of racemic DL-amino acids to biologically active α-amino acids tyrosine to L-dopa, etc. Similarly the steroidal conversion where the conversion products are of pharmaceutical importance, have been reported using free and immobilized enzyme systems as well as immobilized microbial cells.

Effect of most of the drugs is through interaction with molecular constituents of cells. Amongst these nucleic acids, mainly DNAs are receptors for a variety of chemotherapeutic agents used in the treatment of cancer. In the wake of the detailed understanding of gene expression provided by molecular biological tools, newer therapeutic approaches of targeting to nucleic acids are under development. An analysis of cellular carbohydrates, which are often linked to proteins, promises newer opportunities for drug development. This area named as '**glycobiology**' could be identified as one of the recent developments in biotechnology. In a very similar direction, an in depth knowledge of cell biology has revealed the importance of lipids and lipid conjugates in cellular functioning. These molecules play a vital and structural role in anchoring proteins to membranes. Lipids and lipid derivatives are involved as intracellular effectors in the pathways of transduction of extracellular signals.

The most beneficial part of biotechnology is the ability to produce therapeutically active proteins and peptides. Use of therapeutically active proteins and peptides predates the developments in pharmaceutical biotechnology. For instance antihaemophilic factor VIII, immunoglobulin fraction isolated from human sources was well established prior to the advent of recombinant DNA techniques. But due to reasons of safety and low production rates they could not be adopted and exploited as therapeutics. Recombinant materials revolutionized the production of peptides. Now they can be produced in large quantities.

Proteins and peptide based drugs have been extensively utilized in replacement therapies in cases where the patient is unable to produce the required protein in sufficient quantities as in the case of insulin in Type I diabetics. The case of subunit and peptide vaccines represent the class, which has been utilized for their improved therapeutic potential due to their ability to produce large quantities of proteins with, directed changes in primary sequences. The development of the recombinant based protein products is increasing day by day like growth hormone, interferon, interleukines, and cytokines. It could well be concluded that the field of professional pharmacy is being affected significantly by biotechnology.

Contribution of above cited scientists including many Nobel Laureates is landmark in this field. The scientists are now diverting themselves toward biotechnological companies; this has caused the development of many biotechnological industries. In USA alone about 225 companies have been established and successfully working, like Biogen, Cetus, Genetech, Hybritech, etc. In world, USA, Japan, and many countries of Europe are leaders in biotechnological researchers encouraged by industrialists. These companies are working for human welfare and opted following areas for research and development.

- Automated bioscreening.
- Bioprocessing alkenes to valuable oxides and glycols.
- Developing immobilized cell and enzyme systems for chemical process industries.
- Engineering of a series of organisms for specific industrial use.
- Genetical improvement of microorganisms for production of pharmaceutical products.
- Human gene therapy.
- Improved production of Vitamin $B_{12}$.
- Large-scale production of fructose from inexpensive forms of glucose.
- Manufacturing ethanol by continuous fermentation.
- Microbiological based production of human insulin and interferons.
- Microbiologically upgradation of hydrocarbons.
- Production and development of vaccine to prevent calibacillosis (a disease develops in newborn calves and piglets).
- Production of biopesticide and biofertilizers.
- Production of diagnostic kits for toxoplasmosis identification.
- Production of monoclonal antibodies for organ transplant tissue typing.
- Production of photosynthetically efficient plants.
- Production of transgenic plants and animals.
- Production of xanthan gum in oil fields for recovery of crude mineral oils.

## Genetic Engineering

Genetic engineering is the most fundamental mechanics of biotechnologies and is a recent offshoot of biotechnological research. It involves gene splicing, recombinant DNA cloning and tissue culture technology. The technique overall involves into two steps:

1. The *in vitro* incorporation of the gene or segment of DNA of interest into a small, self-replicating chromosomes and,
2. The introduction of the recombinant *minichromosome* into a host cell where it will replicate. Step one involves synthesis of **recombinant DNA** and step two is the **Gene cloning**.

These two, **recombinant DNA** and **gene cloning** technologies are the most powerful tools developed in field of biology. Genetic Engineering involves manipulation of the genetic material of an organism to give an altered expression of our choice. It deals with identification and isolation of desired gene and then joining this gene of interest into another organism. The desired gene expresses itself in that organism by the gene product. Various steps of genetic engineering are:

1. **Gene isolation:** Desired gene is identified, isolated and purified. This DNA of interest is also called donor DNA or target DNA.
2. **Selection of vector:** A vector is a self-replicating molecule of DNA or replicon to which desired gene is linked. Vector molecule with foreign DNA inserted is known as chimeric DNA. Vector acts as carrier and transports the gene into the host cell. Thus, it is also known as a cloning vehicle or a carrier molecule. Suitable vector is identified for a system; commonly used vectors

are plasmids and viral DNA molecule. However, recent techniques involve physical delivery of DNA by various methods.

3. **Cloning of desired gene:** Multiple copies of desired gene can be obtained by placing them in host cell with the help of vectors. Here, the desired gene along with the vector is amplified. Large number of identical copies of gene of interest is produced for subsequent gene transfer into target cell. Now, gene cloning is a fast and mechanized process, using polymerase chain reaction (PCR) machines.
4. **Specific gene transfer:** The gene of interest is finally transferred to host cells. Transformed cells are selected, multiplied and produce transgenic plants by tissue culture technique.
5. **Expression of desired gene:** The desired gene produce, the product in new environment of host, thus the desired traits.

Genetic Engineering includes the propagation of chimeric DNA in a different host organism. The ability to cross natural species barriers and place genes from any organism in an unrelated host organism is one important feature of gene manipulation. A second important feature is the fact that a defined and relatively small piece of DNA is propagated in host organism. Thus genetic engineering opened the door to a range of molecular biological opportunities including nucleotide sequence determination, site-directed mutagenesis, and manipulation of gene sequences to ensure very high-level expression of an encoded polypeptide in a host organism. If used wisely genetic engineering promises to enhance the quality of human life. However, only future will determine the scope and final outcome of this technology.

## Tissue Culture

Plant tissue culture is a field of many facets. The term *tissue culture* is actually a misnomer borrowed from the field of animal tissue culture. It is a misnomer because plant micropropagation is concerned with the whole plantlet and not just isolated tissues, though the explant may be a particular tissue. The terms *plantlet culture* or *micropropagation*, therefore, are more accurate. Plant tissue culture is the technique of growing plant cells, tissues and organs in an artificial prepared nutrient medium static or liquid, under aseptic conditions. Better control of environmental conditions like light, temperature, gas mixtures and nutrients can be achieved for plant tissues growing *in vitro*. But whether we call it cloning, tissue culture, micropropagation, or growing *in vitro*, the process remains the same; it is a vegetative method for multiplying plants.

Tissue culture is the ever-ready tool for specialists who hybridize plants by either sexual or asexual means. It is a clean and rapid way for genetic engineers to grow material for identifying and manipulating genes or to transfer individual characteristics from one plant to another. It plays a role in a wide array of fields, such as botany, chemistry, physics, genetic engineering, molecular biology, hybrid development, pesticide testing, and food science.

A piece of a plant, which can be anything from a piece of stem, root, leaf, or bud to a single cell, is placed in that tiniest of greenhouses, a test tube. In an environment free from microorganisms and in the presence of a balanced diet of chemicals, that bit of plant, called an *explant*, can produce plantlets that, in turn, will multiply indefinitely, if given proper care. The *medium* (plural, *media*) is the substrate for plant growth, and in the context of plant tissue culture it refers to the mixture of certain chemical compounds to form a nutrient-rich gel or liquid for growing cultures, whether cells, organs, or plantlets. The process of tissue culturing plants from the explant stage to the final stage of transferring a mature plant to field or greenhouse conditions involves 4 basic stages. The 4 stages of culture growth are:

- Stage I: Explant establishment or initiation
- Stage II: Multiplication
- Stage III: Rooting; and
- Stage IV: Acclimatization or hardening off

These stages can overlap in certain cases, and the requirements of each stage vary widely from plant to plant.

Tissue culture can serve a number of purposes, and growers have started their own commercial production laboratories for a variety of reasons. Growing plants from seeds or cuttings can be unacceptable or impractical due to some of the following (and other) factors **TABLE 1.4**:

**TABLE 1.4:** Unacceptable factors that influence growing plants from seeds or cutting.

| **Seed-grown products lack uniformity** | **Seed-grown products are not true to type** |
|---|---|
| Seeds take too long to grow to mature plants | Seeds are difficult to handle |
| Seeds are not available | Cuttings grow too slowly |
| Cuttings have poor survival rate | Cuttings require too much care |
| Cuttings are too vulnerable to disease. Tissue culture is often the only practical way to produce the large numbers required. | There is a shortage of stock plants from which to take cuttings because there is:<br>only one hybrid,<br>only one virus-free plant and<br>only one desirable mutant |
| There is insufficient room for stock plants | It is not cost effective to maintain the stock plants |

If an ample supply of seeds is available, or if plants from cuttings are acceptable and cutting stock is available and not a problem to maintain, then tissue culture may not be the most practical option because it can be an expensive, labor-intensive process, especially if less than a thousand plants are needed. If plants from seeds are acceptable but there are not enough seeds, then seeds or excised embryos can be used as starting material for tissue culture.

Every year excessive amounts of growers' time, labor, and space are spent on unproductive seeds, cuttings, and grafts. Significant numbers of young plants are lost to pests, diseases, or other environmental factors. Tissue cultured plantlets are less subject to such attacks and disasters because in the sterile environment of the laboratory they are not exposed to the pathogens or extreme conditions that afflict many plants grown in the field or greenhouse. Material usually comes out of culture as well-started plantlets or microcuttings with a stockpile of nutrients and vigor often superior to that of conventional cuttings. It is not a secret that healthy plants are the first line of defense against disease.

Enormous amount of the daily care is required with cuttings and seedlings. Cultures usually need to be divided and transferred to a fresh medium every 2 to 6 weeks, but between transfers there is no need to water or tend to the cultures, other than casual surveillance. How different this is from the daily watering and weeding requirements accompanying greenhouse growing!

The simplest tissue culture hobby is the multiplication of easy, fast-growing plant material, such as *Kalanchoe*, Boston ferns (*Nephrolepis*), African violets (*Saintpaulia*), or *Begonia*; next in order of complexity are carnations (*Dianthus*), strawberries (*Fragaria*), or *Syngonium*. The chemist looking for a tissue culture hobby may be challenged to explore the field of plant by products; dyes, flavorings, medicinal, and oils are just some of the by-products of certain plants.

The earlier researchers explored problems of academic interest, then the emphasis shifted to applied aspect such as haploid from pollen, triploids from endosperms, somatic hybridization, tissue culture of cereals, legumes and oil crops, and clonal multiplication of elite species of plants. Our knowledge of cell and tissue culture is developing with a fast speed specially in many areas, like totipotency, differentiation, cell division, cell nutrition, metabolism, radiobiology, cell preservation, etc. We are now in position to cultivate cells in quantity, or as clones from single cell, to grow whole plant from isolated meristems, to induce callus or even single cell to develop into complete plant either by organogenesis or directly by embryogenesis *in vitro*. The production of haploid through tissue culture from anthers or isolated microspores and of protoplasts from higher plant cells has served as the basic tools for genetic engineering and somatic hybridization. Protoplasts can also be used as genetic material present in nuclei and chloroplasts as well as isolated DNA molecules. This technique provides the opportunity to combine by fusion of genotype of species, which are sexually incompatible and to introduce foreign genetic material such as organelles or DNA into the genome.

## Protoplast Culture and Micropropagations

The hybridization programme is hampered in some cases because of sexual incompatibility. Synthetic production of hybrids has, however becomes possible through the novel technique of protoplast culture and their fusion. The isolation, culture and fusion of protoplast are one of the most fascinating fields of research, though still in developing stage. The protoplast culture technique can be suitably used for microinjection and other genetic engineering experiments. The techniques are important, specially, because of their far-reaching effects on crop improvement by somatic hybridization and cell modification. The protoplast culture can be regenerated into an entire plant. It is imperative that the technique in future will be one of the most frequently used research tools for tissue culturists, physiologists, pathologists, molecular biologists, cytogeneticists, and biotechnologists. The discovery of enzymes, which could separate the cells and remove the cell wall, revolutionized the concept of plant tissue culture. It has provided a tool for isolating protoplasts and exploring the possibilities of genetic engineering.

## Biosensors, Biochips, Biofilms and Biosurfactents

The natural biosensors are the *sense organs* primarily the chemical sensors of smell and taste. Biosensors technically in their various forms share a reliance of biological materials as sensing elements. Technical biosensors have been under intense development since the middle of 1960 with the prospects of commercial potential offered by biotechnology. Biosensors are the combinations of biologically active material displaying characteristic specifically with chemical or electronic sensor to convert the response into electrical signals. Biosensors have their roots in military research, as means of detecting nerve gases and other chemical warfare toxins. Their applications have branched out to include simple-to-use *alternate site* diagnostic devices for home, doctor's office, or drug use screening; medical and surgical monitors (small enough to fit inside a blood vessel) and environmental quality monitors.

**Biochips** are made from different biological materials and can control the computer by replacing silicon chips. Biomolecular computers thus made, promise to be ten to thousand times smaller than the best super computers with much faster switching times and extremely low power dissipation. They are made up of **semi-conducing molecules** inserted into the protein framework and fix the whole on to a protein support. They can be used in implanting of several sorts in human body, like regulation of heart beats, responses to nerve impulses by artificial limbs (bearing such computer device), overcoming of blindness and deafness, etc. Biochips may be infected by microbes since they are made up of proteins.

A **biofilm** is an accumulation of microbial cells and inorganic components held together in a polymeric matrix and firmly attached to a substratum. Accumulation of biofilms is encountered in many natural and modulated environments. It may be fundamental to process performance i.e., for fixed-film biological waste water treatment, sudden deterioration of water quality and deterioration of substrata.

**Biosurfactents** are surface active compounds produced by microorganisms. Biological surfactants possess a number of potential advantages over their chemically manufactured counterparts, including lower toxicity, biodegradability, a wide variety of possible structures, and ease of synthesis from inexpensive, renewable feed stocks. Consequently, Biosurfactents may have applications in numerous areas, particularly for enhance oil recovery and in foods, beverages, cosmetics, pharmaceutical preparations, etc.

## THE BIOTECHNOLOGIES AND THEIR APPLICATIONS

Here are a few of the new biotechnologies that use cells and biological molecules and examples of their applications in medicine, agriculture and environmental management.

### Monoclonal Antibody Technologies

Monoclonal antibody technology uses immune system cells that make proteins called antibodies. We have all experienced the extraordinary specificity of antibodies: those that attack a flu virus one winter do nothing to protect us from a slightly different flu virus the next year. The specificity of antibodies makes them powerful tools for locating substances that occur in minuscule amounts and measuring them with great accuracy. For example, we use monoclonal antibodies to:

- Distinguish cancer cells from normal cells
- Locate environmental pollutants
- Detect harmful microorganisms in food and to diagnose infectious diseases in humans, animals and plants more quickly and more accurately than ever before.

### Cell Culture Technology

Cell culture technology is the growing of cells outside of living organisms.

With mammalian cell culture, we can replace animal testing with cell testing when evaluating the safety and efficacy of medicines. Cellular therapeutics is one the horizon as well. New, healthy cells, produced with embryonic stem cell culture, could replace the malfunctioning cells in people with Parkinson's disease and restore function in stroke and heart attack victims.

Plant cell culture provides us with an environmentally sound and economically feasible option for obtaining naturally occurring products with therapeutic value, such as the chemotherapeutic agent paclitaxel, a compound found in yew trees and marketed under the name Taxol.

By using cell culture to grow microorganisms that infect insects, we will be able to kill insect pests, such as mosquitoes and caterpillars. Biocontrol agents infect only certain insects without harming beneficial insects, such as honeybees.

## Cloning Technology

Cloning technology allows us to generate a population of genetically identical molecules, cells, plants or animals. In molecular cloning, the word "clone" refers to a gene or DNA fragment and also to the collection of cells or organisms, such as bacteria, containing the cloned piece of DNA. Cellular cloning, a type of cell culture, produces "cell lines' of identical cells. In Monoclonal antibody technology, scientists isolate one cell from an array of antibody-producing cells, and then generate a clonal cell line from that cell. We can produce plant clones by rooting small pieces of fully developed plants, but animal cloning requires cells specialized for reproduction: either eggs or embryonic cells at the very earliest stages of development.

Although the debut of the cloned sheep Dolly in 1997 brought animal cloning technology into the spotlight, biologists first cloned animals in 1952 by transferring genetic material from the nucleus of frog embryonic cells to frog eggs, which then developed into genetically identical tadpoles. Using similar nuclear transfer techniques, scientists first cloned cattle in the mid-1980s and have now produced hundreds of cattle, sheep and swine embryos that have been carried to term by surrogate females. For two decades, animal breeders have also used another cloning technology, embryo twinning, to improve livestock herds. They artificially inseminate a genetically superior cow, then divide a 3 to 4-day-old embryo into two and implant the twin embryos into surrogate females.

Dolly was considered a scientific breakthrough not because she was a clone, but because the source of her genetic material was an adult animal cell, not an embryonic one. Before Dolly, scientists assumed genetic material from adult animal cells could not direct the full development of an animal because the cells had become specialized into certain types and could not "remember" how to give rise to a complete organism.

Because cloning technology can be used to produce molecules, cells, plants, some animals, and mammalian embryos, the applications of cloning technology are extraordinarily broad.

***Molecular cloning*** provides the foundation of the molecular biology revolution and is a fundamental and essential tool of biotechnology research, development and commercialization. Virtually all applications of recombinant DNA technology described below, from basic research to pharmaceutical production, depend on molecular cloning. This application of molecular cloning is so pervasive that in scientific circles "to clone" can mean to identify and map a gene, indicating the importance of molecular cloning to these applications. Both the ever-expanding scientific definition of "clone" and the essential role molecular cloning plays in biotechnology research and production necessitate that any laws or regulations covering other types of cloning be carefully crafted.

***Cellular cloning*** is also a fundamental and essential tool of biotechnology research, development and commercialization. All applications of monoclonal antibody technology; the regeneration of transgenic plants from single cells; pharmaceutical manufacturing based on mammalian cell culture; and generation of therapeutic cells and tissues depend on cellular cloning. These and other potential applications must also be taken into account in any legislative actions directed at cloning.

***Animal cloning*** has helped us rapidly incorporate improvements into livestock herds for two decades and has been an important tool for scientific researchers since the 1950s. Genetic engineering, in conjunction with animal cloning, is providing us with excellent animal models for studying genetic diseases, aging and cancer, and in the future, will help us discover drugs and evaluate other forms of therapy, such as gene and cell therapy. Finally, in the future, zoo researchers may help save endangered species with animal cloning. In August 1998, a rare breed of cow was successfully cloned.

## Genetic Modification Technology

Genetic modification technology is often referred to as recombinant DNA technology. Humans make Recombinant DNA, both in nature and, by combining genetic material from two different sources. Humans began to preferentially combine the genetic material of domesticated plants and animals thousands of years ago by selecting, which individuals would reproduce. By selectively breeding, individuals with valuable genetic traits while excluding others from reproduction, we intentionally changed the genetic makeup of the plants and animals we domesticated. Techniques for making selective breeding more predictable and precise have been continually evolving, especially since we discovered the genetic basis of heredity in the early 1900s.

Now, in addition to using selective breeding to combine valuable genetic material from different organisms, we combine genes at the molecular level using the more precise techniques of genetic modification. Selective breeding and genetic modification fundamentally resemble each other, but there are important differences.

In genetic modification, we move single genes whose functions we know from one organism to another; in selective breeding, large sets of genes of unknown function are transferred. By making our manipulations more precise and our outcomes more certain, we decrease the risk of producing organisms with unexpected traits and avoid the time-consuming trial-and-error approach of selective breeding.

**Currently, we are using genetic modification techniques to:**

- Produce new medicines and safer vaccines
- Treat some genetic diseases
- Enhance biocontrol agents in agriculture
- Increase agricultural yields and decrease production costs
- Decrease allergy-producing characteristics of some foods
- Improve food's nutritional value

- Develop biodegradable plastics, and
- Decrease water and air pollution.

A special type of genetic modification technology, antisense technology, uses small nucleic acids to block the protein production of specific genes. Currently, researchers use antisense to (1) slow food spoilage, (2) control viral diseases, (3) inhibit inflammation.

## Protein Engineering Technology

Protein engineering technology will often be used in conjunction with genetic modification to improve existing proteins, usually enzymes, and to create proteins not found in nature. These new improved proteins will encourage the development of ecologically sustainable industrial processes because they are renewable, biodegradable resources.

Unlike other catalysts used in industrial manufacturing processes, enzymes, as biocatalysts, dissolve in water and work best at neutral pH and comparatively low temperatures. Because biocatalysts are more specific than chemical catalysts, they also produce fewer unwanted by products.

The chemical, textile, pharmaceutical, pulp and paper, food and feed, and energy industries are all benefiting from cleaner, more energy-efficient production made possible by incorporating biocatalysts into their production processes.

The traits that make biocatalysts environmentally advantageous may, however, become detrimental in certain industrial processes. Most enzymes fall apart at temperatures above 100°F. Scientists are circumventing these limitations by using protein engineering to increase enzyme stability under harsh manufacturing conditions.

## Hybrid Technologies

The biotechnologies described above, which rely almost exclusively on knowledge of cells and biological molecules, have provided us with an extraordinary array of new options. We are also combining our understanding of biological processes with scientific advances and technological innovations in other disciplines, which gives birth to a synergistic set of new technologies.

## Biosensor Technology

Biosensor technology couples our knowledge of biology with advances in microelectronics. A biosensor is composed of a biological component, such as a cell or antibody, linked to a tiny transducer. Biosensors are detecting devices that rely on the specificity of cells and molecules to identify and measure substances at extremely low concentrations. When the substance of interest collides with the biological component, the transducer produces a digital electronic signal proportional to the concentration of the substance. Biosensors can measure the nutritional value, freshness and safety of food, provide emergency room physicians with bedside measures of vital blood components and locate and measure environmental pollutants.

## Tissue Engineering Technology

Tissue engineering technology combines advances in cell biology and materials science, allowing us to create semi-synthetic tissues and organs in the lab. These tissues consist of biodegradable scaffolding material plus living cells produced through cell culture.

The most basic forms of tissue engineering use natural biological materials, such as collagen, for scaffolding. For example, infiltrating a collagen gel with connective tissue cells, then creating the outer skin with a layer of tougher protective cells make two-layer skin. In other methods, the scaffolding, made of synthetic polymers, is shaped and then placed in the body where new tissue is needed. Adjacent cells invade the scaffolding, which eventually degrades and is absorbed. At other times, the biodegradable implant is spiked with cells grown in the laboratory.

Simple tissues, such as skin and cartilage, were the first to be engineered successfully. Ultimately the goal is to create complex organs, consisting of a number of tissue types, that can replace diseased or injured organs.

## DNA Chip Technology

DNA chip technology, a marriage of the semi-conductor manufacturing industry and molecular genetics, will transform genetic analysis because it allows us to analyze tens of thousands of genes simultaneously on a single chip. The manufacturing process of microchips and DNA chips is similar, in principle; but instead of shining light through a series of masks to etch circuits into silicon, automated DNA chipmakers use a series of masks to lay down an array of DNA fragments on a glass slide.

**DNA chip technology is being used in:**

- Detect mutations in disease-causing genes
- Monitor gene activity
- Diagnose infectious diseases and identify the best antibiotic treatment
- Identify genes important to crop productivity and
- Improve screening for microbes used in bioremediation.

DNA chips will be essential for converting the raw genetic data provided by the Human Genome Project into useful products. Sequencing the human genome, while a remarkable achievement, provides only the first milestone in the upcoming medical revolution. The gene sequence and mapping data mean little until we determine what those genes do. This field of study, known as functional genomics, helps us translate gene identification and DNA sequence data into biological functions.

Any study of gene function is, at its core, a study of proteins. Each cell produces thousands of proteins, each with a specific function. This collection of proteins in a cell is known as the proteome, and, unlike the genome, which is constant irrespective of cell type, the proteome varies from one cell type to the next. The science of proteomics attempts to identify the protein profile of each cell type, assess protein differences between healthy and diseased cells, and uncover not only a protein's specific function but also how it interacts with other proteins.

Neither functional genomics nor proteomics is an end in itself. Their medical value will be in identifying specific therapeutic targets and helping us understand the complex biochemistry of disease processes. For more information on the exciting advances spawned by genomics and proteomics, see the "Health Care" section.

## Bioinformatics Technology

Bioinformatics technology uses computational tools provided by the information technology revolution, such as statistical software, graphics simulation and database management, to organize and analyze information about biological systems, which, for biotechnology, information about cells and biological molecules. Using another product of the information revolution, the Internet, scientists broadcast this information around the world.

The technologies described above - from monoclonal antibody technology to DNA chips - have provided us with massive amounts of information, in addition to useful products. Without methods for organizing and analyzing the raw data, however, we will not be able to turn it into knowledge, understanding and, ultimately, products. Bioinformatics technology provides us with tools and methods for consistently organizing, accessing, processing and integrating data from different sources. This uniformity, in conjunction with the universal language of life at the molecular level, enables international collaboration among scientists studying any plant, animal or microbe. Bioinformatics technology helps us to map genomes and identify genes, determine protein structure and simulate protein interactions, discover new therapeutic targets and design medicines aimed at the targets and to assess the effects of virtual mutations on gene function.

## RESEARCH PROMOTION AND PRIORITIES IN INDIA

Recognizing the importance of biotechnology in national development, the Govt. of India set up the National Biotechnology Board (NBTB) in 1982 to coordinate and oversee priority areas of biotechnology. The board was converted into a full-fledged Department of Biotechnology (DBT) in 1986, under the Ministry of Science and Technology. The DBT has initiated several programmes for the promotion of biotechnology in diverse fields, out of which six R&D projects are viz., tissue culture propagation of bamboo, cardamom, coconut, oil palm, natural rubber and sugarcane. The Department has also initiated model systems of post-graduate teaching in Agricultural Biotechnology at four institutions, namely the IARI, New Delhi, G.B. Pant University, Pantnagar and Tamil Nadu Agricultural University (TNAU), Coimbatore.

Under the ICAR, the NBPGR was established in New Delhi during the late 70s and a Division of plant Genetic Resources (PGR) was started at the IIHR in the early 80s to initiate research on cryopreservation of germplasm. An International Centre for Genetic Engineering and Biotechnology (ICGEB) has also been established for developing countries under the auspicies of United Nations. This centre has two locations, one in New Delhi and and other in Triesta, Italy. This centre in India started working in 1987.

The NFPTCR has been established recently at New Delhi and is founded by DBT.

Tissue culture laboratories and molecular biology laboratories at ICAR Institutes such, IARI, New Delhi, IIHR, Bangalore, CPCRI, Kasargod, CTCRI, Trivandrum, CPRI, Shimla, etc. are engaged in biotechnological work. A major programme for large-scale fermentor based production of *Rhizobia* inoculants has been taken up by the DBT with the objective of making available *Rhizobium* fertilizer to some major crops. The Department has also set up a National Facility for Blue green algal collection at the IARI, which facilitates the collection, screening and identification of agroclimatic regions.

A programme has also been launched to use naturally occurring biological agents for management of insects, helping the poor farmer, also help preserve the environment. Pilot plants are underway at the Tata Energy Research Institute (TERI), New Delhi and the National Chemical Laboratory (NCL), Pune, to produce certified planting material.

The important biotechnological centres in India are:

1. Indian Agricultural Research Institute, New Delhi (the centre was named as Lab Bahadur Shastri Centre for Advanced Research in Biotechnology)
2. National Dairy Research Institute (NDRI), Karnal.
3. Indian Veterinary Research Institute (IVRI), Izatnagar (U.P.)

Many other centres are also in action, which is by one way, or other related with biotechnological researches.

### Bioinformatics

In India, nine Distributed Information Centres (DICs) have been established along with 14 user centres. The nine DCIs located at the following institutions: (1) Indian Institute of Science (IISc), Bangalore, (2) Madurai Kamraj University (MKU), Madurai, (3) Bose Institute, Calcutta, (4) Jawahar Lal Nehru University, New Delhi, (5) Poona University, Poona, (6) Centre for Cellular and Molecular Biology (CCMB), Hyderabad, (7) National Institute of Immunology, New Delhi, (8) Institute of Microbiological Technology, Chandigarh and (9) Indian Agricultural Research Institute, New Delhi. These DICs are linked with each other and with the Central Information Network at DBT in New Delhi which access to International Databases in biotechnology.

### Plant Molecular Biological Researches

In India, six plant molecular biology centres have been established by DBT at the following institutions. (1) Jawahar Lal Nehru University, New Delhi, (2) Madurai Kamraj University, Maduri, (3) Tamil Nadu Agricultural University, Coimbatore, (4) Bose Instotute, Calcutta, (5) Osmania University, Hyderabad and (6) National Botanical Research Institute (NBRI), Lucknow.

### Service Oriented Infrastructural Facilities

For advancement and strong Research and Development (R&D) the efforts of DBT are remarkable. DBT has set up service oriented infrastructural facilities in the following fields:

Animal Houses, (2) Antibody development consortium, (3) Bioprocess optimization and pilot plants, (4) Carbohydrate cell surface and cellular transport, (5) Centre for reproductive biology and molecular endocrinology, (6) Genetic Engineering R&D units, (7) Germplasm banks for plant, animals, algae and microbes, (8) Marine cyanobacterial germplasm collection, (9) NMR facility, (10 Oligonucleotide synthesis (DNA synthesizers), (11) Production, import and distribution of enzymes, reagents and radio-labelled compounds, (12) Protein and peptide sequencing.

## BIOTECHNOLOGY INDUSTRY STATISTICS: SOME FACTS

- More than 250 million people worldwide have been helped by the more than 117 biotechnology drug products and vaccines approved by the U.S. Food and

Drug Administration (FDA). Of the biotech medicines on the market, 75 percent were approved in the last six years.

- There are more than 350 biotech drug products and vaccines currently in clinical trials targeting more than 200 diseases, including various cancers, Alzheimer's disease, heart disease, diabetes, multiple sclerosis, AIDS and arthritis.
- Biotechnology is responsible for hundreds of medical diagnostic tests that keep the blood supply safe from the AIDS virus and detect other conditions early enough to be successfully treated. Home pregnancy tests are also biotechnology diagnostic products.
- Consumers are already enjoying biotechnology foods such as papaya, soybeans and corn. Hundreds of biopesticides and other agricultural products are also being used to improve our food supply and to reduce our dependence on conventional chemical pesticides.
- Environmental biotechnology products make it possible to clean up hazardous waste more efficiently by harnessing pollution-eating microbes without the use of caustic chemicals.
- Industrial biotechnology applications have led to cleaner processes that produce less waste and use less energy and water in such industrial sectors as chemicals, pulp and paper, textiles, food, energy, and metals and minerals. For example, most laundry detergents produced in the United States contain biotechnology-engineered enzymes.
- DNA fingerprinting, a biotech process, has dramatically improved criminal investigation and forensic medicine, as well as afforded significant advances in anthropology and wildlife management.
- There are 1,273 biotechnology companies in the United States, of which 300 are publicly held.
- Market capitalization, the amount of money invested in the U.S. biotechnology industry, increased 156 percent in one year, soaring from $137.9 billion in 1999 to $353.5 billion in 2000.

**TABLE 1.5:** Industry Statistics (1993-2000)

| Year | 2000 | 1999** | 1998 | 1997 | 1996 | 1995 | 1994 | 1993 |
|---|---|---|---|---|---|---|---|---|
| Sales* | 16.1 | 14.5 | 13 | 10.8 | 9.3 | 7.7 | 7 | 5.9 |
| Revenues* | 22.3 | 20.2 | 17.4 | 14.6 | 12.7 | 11.2 | 10 | 8.1 |
| R&D Expense* | 10.7 | 10.6 | 9 | 7.9 | 7.7 | 7 | 5.7 | 4.9 |
| Net Loss* | 5.6 | 4.4 | 4.1 | 4.5 | 4.6 | 4.1 | 3.6 | 3.4 |
| Market Capitalization* | 353.5 | 137.9 | 93 | 83 | 52 | 41 | 45 | n/a |
| Number of Public Companies | 300 | 316 | 317 | 294 | 260 | 265 | 235 | 225 |
| Number of Companies | 1273 | 1311 | 1274 | 1287 | 1308 | 1311 | 1272 | 1231 |
| Employees | 162000 | 155000 | 141000 | 118000 | 108000 | 103000 | 97000 | 79000 |

*U.S. dollars; **The 1999 column shows numbers updated since the chart appeared on the BIO Web site in 2000.

- The biotechnology industry more than doubled in size since 1993, with revenues increasing from $8 billion in 1993 to $22.3 billion in 2000.
- The U.S. biotechnology industry currently employs 150,800 people; that's more than all the people employed by the toy and sporting goods industries.
- Biotechnology is one of the most research-intensive industries in the world.

The U.S. biotech industry spent $10.7 billion on research and development in 2000 (**TABLE 1.5**).

- The top five-biotech companies spent an average of $89,400 per employee on R&D in 2000. This compares with an average of $37,200 per employee spent by the top pharmaceutical companies.
- The Food and Drug Administration (FDA), the Environmental Protection Agency (EPA) and the Department of Agriculture (USDA) regulate the biotech industry.

# 2. Biological Genetics and Variations in Characters of Microbes

The term *genetics* came from a Greek word meaning *to generate* and the *gene* from the last syllable of Darwin's term *pengene*. Genetic is the fundamental biological science for without genes, there is no life. Today genetics is growing at an unprecedented rate. Genetic Engineering is now a reality, and all these recent developments are related with the hereditary unit i.e., Deoxyribonucleic acid (DNA).

## IMPORTANCE OF MICROBIAL GENETICS

The importance of microbial genetics is due to the following reasons:

1. The function of the gene is a base for the function of the cell. It is, therefore, indicative of the functioning of microorganisms.
2. Microorganisms are simple to study through their genetic phenomena.
3. Microorganisms can be exploited for isolation and duplication of specific genes from other organisms, which is referred to as molecular cloning.
4. In molecular cloning genes are manipulated and are incorporated into microorganisms where they can be induced to increase in number.
5. The use of genetically improved microorganisms for large-scale industrial production comprises an important part of biotechnology.
6. By understanding the genetics of disease causing microorganisms we can control the diseases caused by them.
7. By studying processes such as conjugation we can understand how the genes are transferred from one organism to another.

### Important Terms used in Microbial Genetics

- **Auxotroph:** Auxotroph is an organism, which has developed a nutritional requirement through mutation.
- **Bacteriocins:** They are ribosomally synthesized peptides. Many of them require posttranslational modification for activity. The bacteriocins are named accordance with the species of organisms that produce them. Thus *Escherichia coli* they are colicins coded by Col plasmids and *Bacillus subtilis* produces subtilisin.
- **Conjugation**: Conjugation is the transfer of genes from on prokaryotic cell to another by a mechanism involving cell to contract and plasmid.
- **Diploid:** A eukaryotic cell or organism containing two sets of chromosomes.
- **Electroporation**: The use of an electric pulse to induce cells to accept free DNA or plasmid. **Method of DNA can be transferred by electorporation:** Small pores are formed on the membranes of bacterial cells exposed to

pulsed electric fields. The DNA molecules present outside the cells during electric pulse can enter the cells through these pores. By this method plasmid can be directly transferred from one cell to another if both the types of cells are present during electorporation.
- **Gamete:** The haploid germ cell resulting from meiosis in eukaryotes.
- **Genetic map:** Genetic map is the arrangement of genes on a chromosome.
- **Genotype:** The precise genetic make up of an organism.
- **Haploid:** It is a cell or organism that has only one set of chromosomes
- **Histidine auxotroph:** *Escherichia coli* or any other microorganisms with a His phenotype called histidine auxotroph.
- **Phenotype:** The observable characteristics of an organism comprise phenotype.
- **Plasmid:** An extra chromosomal genetic element, which has no extracellular form.
- **Prototrophs:** A technical term for the wild type parent from which an auxotroph is obtained.
- **Recombination**: Recombination is the process by which parts of or all of the DNA molecules from two separate sources are exchanged or brought together into a single unit.
- **Replica plating:** It is the technique by which nutritional mutants can be detected.
- **Selection:** It is placing of organisms under conditions where the growth of those with a particular genotype will be favoured.
- **Selection:** It is the isolation of a single mutant from population containing millions or billions of parental organisms. **Penicillin selection method:** It is to isolate mutants that need amino acids or other growth factors. The penicillin selection method is based on the concept that penicillin destroys only the growing cells. If penicillin is added to a population with wild parents and mutants growing in a medium which lacks growth factor, the parent cells would be destroyed while the non-growing mutants would survive. Thus after preliminary incubation in the absence of growth factor the penicillin is washed away leaving the mutant population free from penicillin which can be transferred to plates containing growth factor. The colonies that finally grow are growth factor mutants.
- **Transformation:** Transformation is transfer of bacterial genes involving free DNA.
- **Transposable element:** It is the genetic element, which has the ability to move (transpose) from on site on a chromosome to another.
- **Transposon:** It is type of transposable element, which transfers genes in addition to those involved in transposition.

## DNA AS GENETIC MATERIAL

In the mid–1940s Avery, Macleod and McCarty identified DNA as the molecular material that made up genes, and by the mid-1950s Watson and Crick, using the data from Franklin and Wilkins, had determined the structure of DNA. These workers have convincingly demonstrated that DNA, not protein, was the *transforming principle*, which produces heritable changes. Avery and his associates demonstrated that DNA isolated from virulent strains of pneumonia-causing bacterium *Diplococcus pneumoniae* was able to produce heritable variations or transform a nonvirulent *D. pneumoniae* strain. The progeny of the transformed strain were virulent, causing pneumonia upon injection into mice. In the early 1950s, Chargaff demonstrated that DNA always contained dual proportions of some of its chemical constituents. That is the amount of two sets of nucleotide bases- adenine and thymine, and guanine and cytosine. The amount always remains equal. Watson and Crick (1953) using X-ray diffraction patterns elucidated the helical, symmetrical structure of DNA and earned Nobel Prize.

According the Watson and Crick, a DNA molecule has double helix, i.e., it is composed of two long and parallel polynucleotide chains or strands helically coiled around the same axis like spiral staircase or ladder (**FIGURE 2.1**).

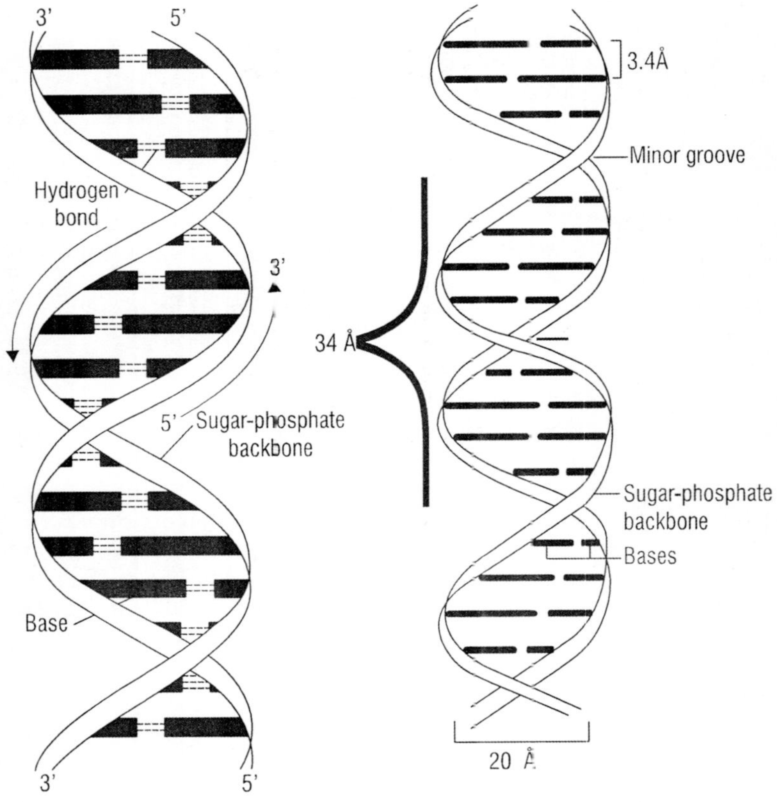


**FIGURE 2.1:** DNA molecule -the basic tool for genetic engineering.

The simplest way of looking at the structure of DNA is as two strings of molecular units, almost like beads on a string. These wind around each other in what has been called the double helix. The structure of an individual string of beads or DNA strand is made of repeating units, the **beads**. However, each of these is one of only four types, and each of these types is related and is, in fact, very similar. Each of the simplified units shown is referred to as a **monomer**, and many of these will join together, or polymerize, to make the **polymer**. **This is called DNA**.

The **sugar** (five-carbon sugar or pentose sugar) is also involved in making DNA structure. Sugar is of two forms, either as a **ring** structure or as a **long**, five-carbon chain. In DNA ring form of sugar is found. One of the important points to see here is that the 2' carbon (pronounced **two prime carbon**) has no OH group attached to it, unlike the 1' and 3' carbons around the ring, and hence this sugar is called **2'-deoxyribose**. In RNA, the 2' carbon does attach to an OH group. In this way DNA differ from RNA.

**Sugar:** A sugar unit is attached to a second unit known as a **base** (a nitrogenous base due to its nitrogen atom content). Together, the sugar and the base are collectively called a **nucleoside**. In DNA four types of base, and so four types of monomer units are found. The information, which is stored in the DNA molecule is found according to base arrangement that is it is the order of these bases in the polymer, which ultimately determines the protein, which is being encoded by a particular stretch of DNA polymer.

**Bases:** The four bases can be separated into two groups, each group having two members. The simplest bases are the **pyrimidines**, which have a single ring, structure. The two forms found in DNA are *thymine* (T) and *cytosine* (C). The other bases are the **purines**, where there are two rings found in the structure. The forms found in DNA are *adenine* (A) and *guanine* (G).

**The phosphates:** To complete the monomer structure the addition of a phosphate unit to the nucleoside. This whole monomer unit is called a **nucleotide.** In a nucleotide three phosphate groups are joined in a long chain. The phosphate groups are labelled $\alpha$, $\beta$ and $\gamma$.

Therefore DNA is a polymer of nucleotide units. The four bases, which are incorporated into the nucleotide are:

- 2'–deoxythymidine 5'–triphosphate, using thymine as the base (dTTP)
- 2'–deoxyguanosine 5'–triphosphate, using guanine as the base (dGTP)
- 2'–deoxyadenosine 5'–triphosphate, using adenine as the base (dATP)
- 2'–deoxycytidine 5'–triphosphate, using cytosine as the base (dCTP)

where *d* denotes deoxy sugar while the first letter refer to the base incorporated.

The above building blocks when polymerize they form a long chain, or **polynucleotide**. Here, the nucleotides are held together by the bonding of the **3' carbon** of the deoxyribose sugar to the first or $\alpha$ phosphate group of the next nucleotide. This bond is called **phosphodiester** bond, and results in the $\beta$ and $\gamma$ phosphate groups of the bonding phosphate chain.

The polynucleotide has a direction to it, that is, the ends are not the same. At one end there is still a triphosphate unit attached to the **5' carbon** of the sugar, and therefore this end of the chain is called the 5' end (or **five prime**). At the other end is an unused 3' (**three prime**) carbon still attached to an untouched OH group. This is the 3' end of the chain. Further polymerization could still take place but, even if it did, we still end up with a 5' end and a 3' end.

The phosphate groups contain a negative charge. It develops due to the dissociation of a proton from on OH group, leaving $O^-$. Therefore the molecule is both acidic and negatively charged.

DNA consisted of two polynucleotide strands. They run in opposite directions. These DNA strands wound around each other to form a double helix. The sugars and phosphate groups effectively form the staging of the helix and make two continuous strands from one end to the other. These bases are like the steps on a spiral staircase. Each step of the staircase is actually two bases, holding each other's 'hands' by way of **hydrogen bonding.**

Thymine on one strand always hydrogen bonds to adenine on the other strand while cytosine always hydrogen bonds to guanine. These two strands are called **complementary** to each other, and the phrase **complement strand** is often used. There are **three** hydrogen bonds between guanine and cytosine, while there are only **two** hydrogen bonds between adenine and thymine.

Model of DNA molecule shows that DNA is in helix and has a regular pattern. The helix is approximately 20 Å across, and has two grooves running around and along it. One groove is referred to as the **wide groove**, while the other is the **narrow groove**. One total rotation of the molecule takes place every 34 Å along, while the distance between the bases, the step height so to speak, is 3.4 Å.

**Z form DNA:** The helix of this kind of DNA is left-handed. The nucleotide sequence of the DNA appears to be influential in the structure adopted and areas of naturally occurring DNA have been identified, which take up the Z form structure. The significance of this is as yet unknown.

**Supercoil DNA:** When DNA is twisted it into a structure known as a **supercoil**. This can only be achieved if the ends of the DNA are prevented from rotating, for example if the DNA is circular s in a plasmid. Enzymes known as **topoisomerases** can introduce or remove turns in the DNA double helix and, if this is done, the DNA twists up into a helical structure, that is the double helix as described is then further twisted into a helix. This super-coiling may be **positive super-coiling**, resulting from the addition of turns to the double helix, or alternatively **negative super-coiling** as a result of the removal of turns from the helix (**FIGURE 2.2**).

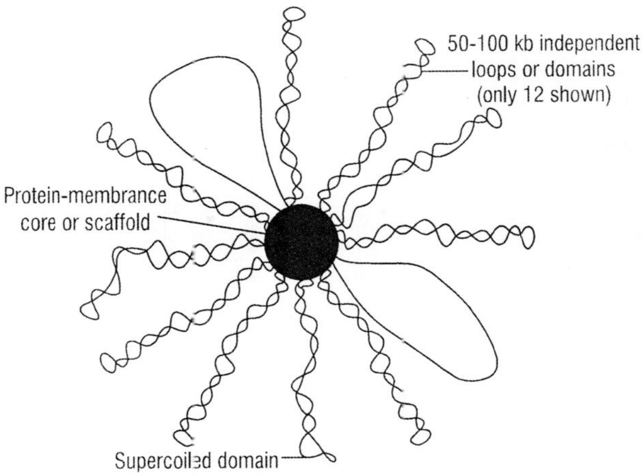


**FIGURE 2.2:** Presentation of *E. coli* chromosome (4600 kb). The thin line present double halix.

**Noncoding DNA:** A large proportion of the DNA in eukaryotic cells does not code for protein; most genes have large intron sequences, and genes or gene clusters are separated by large stretches of sequence of unknown function. Much of this DNA consists of multiple repeats, sometimes thousands or hundreds of thousands of copies, of a few relatively short sequence elements.

**Unique sequence DNA:** The fraction of DNA which reassociates most slowly is unique sequence , essentially comprising single - copy genes, or those repeated a few times, *Escherichia coli* DNA is essentially all unique sequence.

**Dispersed repetitive DNA:** Much of the moderately repetitive DNA comprises a number of DNA sequences of a few hundred base pairs (SINES or short interspersed elements) or between one and five thousand base pairs (LINES, or long interspersed elements), each repeated more than 100000 times and scattered throughout the genome. The most prominent examples in humans are the *Alu* and the *L1* elements, which may be parasitic DNA sequences, replicating themselves by transposition.

**Satellite DNA:** Satellite DNA, which occurs mostly near the centromeres of chromosomes, and may be involved in attachment of the mitotic spindle, consists of huge numbers of tandem repeats of short (up to 30 bp) sequences. Hypervariability in satellite DNA is the basis of the DNA fingerprinting techniques.

**Genetic polymorphism:** Base changes (mutations) in a gene or a chromosomal locus can create multiple forms (polymorphs) of that locus which is then said to show genetic polymorphism. The term can describe different alleles of a single copy gene in a single individual as well as the different sequences present in different individual in a population. Common types are single nucleotide polymorphisms (SNPs) and simple sequences length polymorphisms (SSLPs). Where SNPs create or destroy the sequence recognized by a restriction enzyme, restriction fragment length polymorphism (RFLP) will result.

The structure of many biological molecules, including that of DNA and many proteins, can be elucidated by an experimental method known as **X-ray diffraction**. Erwin Chargaff determine the exact composition of the bases within a DNA molecule. The method he used was **paper chromatography.**

## DNA REPLICATION

Every cell in a body synthesizes proteins and contains the appropriate DNA encoding for that information. All the cells are derived from a single cell in the very early stages of development, then DNA has to be replicated many millions of times in order for all the cells to receive their complement. The whole molecule may be replicated into a new molecule in which the **old** DNA remains the same and the **new** DNA is made of all new components. This is called **conservative replication**. The DNA may be broken up and new bits are added in a random pattern in which both strands of both resulting DNA molecules contain old and new components. This is called **dispersive replication**. DNA could be unzipped into two single strands and use each one of these strands to form the basis of the new DNA molecules. This type of replication is called **semiconservative replication.** The direction of replication is always in 5'–3' direction.

### How Replication is achieved?

Firstly the DNA is in a double helical structure and therefore, for each of the new DNA molecules to receive an old strand, this helical structure has to be **unzipped** (unwinding) and then the new DNA is synthesized. The unwinding of the DNA helix is carried out by a group of enzymes called **topoisomerases**. These enzymes fall into two groups, Type I and Type II. Type I enzymes. These enzymes (Type I) break one of the DNA strands, while Type II enzymes break both DNA strands. **DNA gyrase** is common examples is an enzyme.

Once the DNA has been unwound the two strands have to be separated. This starts at a point known as the **replication origin**. The DNA strands are separated by breakage of the **hydrogen bonding** between the bases. This process should remain continue to enable the replication. This process is catalyzed by an enzyme called a **helicase** and the break points are refereed to as the **replication forks**. However, DNA would naturally reanneal back to the double helix with the reformation of the double bonds is allowed to do so. Therefore, proteins known as **single-strand binding proteins** (SSBs) bind to the single-stranded DNA formed by the helicase and stop the rennealing process from taking place.

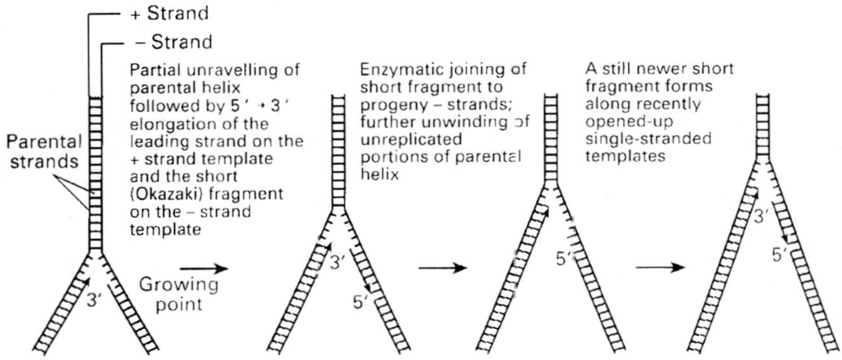


**FIGURE 2.3: (A):** Replication of a double-stranded DNA molecule. At the replication fork the two template strands become unwound one from another. The DNA strand, which is extending in the same direction as the replication fork grows in an uninterrupted fashion and is called the leading strand. The lagging strand grows in the opposite direction to the fork by a series of initiation and elongation reactions. At the end of the whole process the segments of newly replicated DNA on the lagging strand are joined together enzymatically

Once the DNA has been opened up into a single-stranded structure, then the synthesis of new strand starts. However, how does this process start, as the first bases have to be there for strand extension to take place? The process starts by using a **primer**. This is a short length of RNA synthesized by an enzyme called **RNA polymerase,** or **primase**. The RNA primer may be of 6 to 30 bases length. **DNA polymerase** catalyze synthesis of the DNA strand. In *E. coli* there are three DNA polymerases, but types I and III seem to be most essential for DNA replication, the function of type II remaining more obscure. Most of the DNA synthesis is carried out by polymerase III, with polymerase I removing the RNA primers and basically tidying up the missing bits.

The polymerases catalyse the formation of new **phosphodiester bonds**. The newly formed strand has a **free 3'** OH group on the ribose sugar, and to this the $\alpha$ phosphate of the next nucleotide can be attached, with the formation of the phosphodiester bond and the release of the $\beta$–$\gamma$ phosphates as an orthophosphate unit. However, such a process means that the DNA synthesis always takes place in the 5'–3' direction. This is fine for synthesis along one strand of the DNA molecule, known as the **leading strand**, as continuous synthesis is obtained.

The other strand (**lagging strand)** is being effectively unteased backwards, synthesis along the second strand cannot follow the direction of the replication fork. Therefore, on the second strand the synthesis of the new DNA is **discontinuous**, a process was confirmed by the work of **Okazaki** (1968). Hence the short stretches of DNA are known as **Okazaki fragments**. Each fragment has to be started by its own RNA primer and so, along this strand, you would find short DNA stretches following short RNA stretches and gaps between. An enzyme called **DNA ligase** carries out the final covalent bonding, which joins all the resulting Okazaki fragments together.

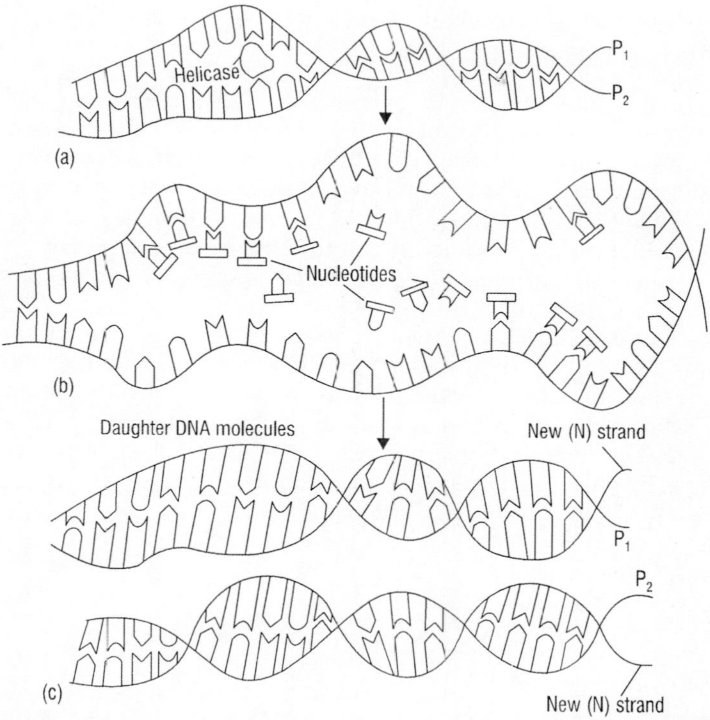


**FIGURE 2.4: (B):** Simplified steps in semiconservative replication of DNA. (*a*) A helicase unwinds the double helix into two parent strands ($P_1$ and $P_2$). (*b*) The replication of new complementary strands proceeds through the action of an enzyme that attaches nucleotides, using the exposed strands as templates. (*c*) Completed daughter molecules each contain one strand that is newly synthesized and one of the original parent strands.

## Enzymes Involved in Replication

The initiation of the replication process requires a **primer**. It is a short RNA molecule, which is complementary to one of the DNA strands. Once the RNA molecule is formed by the enzyme RNA polymerase, DNA synthesis begins. Next, the RNA primer is removed and replaced by newly formed DNA. The replication process continues with the enzyme DNA polymerase adding nucleotides only to the 3' hydroxyl (OH) end of the replicating DNA strand or strands in a 5 to 3' direction. The fact that this polymerase can only add nucleotides to the 3' hydroxyl group of the strand raises the question of how DNA synthesis can proceed in proper order (sequentially) from the single starting point along *both* DNA strands. As mentioned earlier, the two strands of DNA are antiparallel, since on strand runs from the 5' OH to the 3' OH free end and the complementary strand runs from the 3' OH to the 5' OH free end. Therefore, the formation of the complementary strands must proceed in opposite directions. One of the new strands is formed *continuously* as the replication fork moves, since it runs in the appropriate direction for the addition of nucleotides to the free 3' OH of the primer. The other new DNA strand can start only after unwinding of the DNA molecule has begun, it will leg behind the continuous strand.

## How Bacteria Replicate their DNA

The genetic material of a bacterium is contained within a single double-stranded molecule that is generally organized into one circular chromosome. Unlike eukaryotic chromosomes, the bacterial chromosome is not enclosed by a nuclear envelope and may be attached to one or several points along a bacterium cell membrane. While the replication process in bacteria is complex, and involves the enzymes described earlier, it is not difficult to visualize overall process.

The DNA replication process begins at single points of origin, or nick site, and moves in two opposite directions (*bi-directional*) around the chromosome, resulting in the formation of two replicating forks. The newly created replication forks move away from the original starting site and around the circular chromosome to form a newly replicated DNA molecule. Since the chromosome is a circular structure, the forks eventually meet when the process is completed. DNA replication is relatively rapid process in bacteria at their respective optimal temperatures. Bacterial cells regulate the initiation of the process to conserve energy and limit the formation of unwanted products.

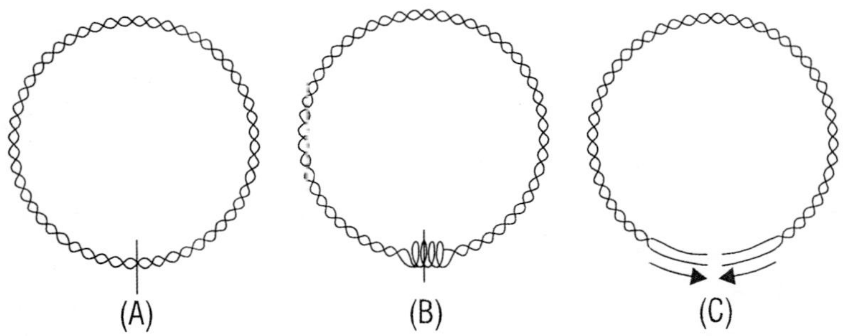


**FIGURE 2.5:** Generation of supercoiled DNA. If a closed circle (a) of DNA is cut, rotated in the opposite direction to the twist of the helix (b), and then cut ends rejoined, (c), the DNA becomes negatively supercoiled.

Although the functions of the three major kinds of RNA are different, the mechanism for their formation is identical. The enzyme *DNA-dependent RNA polymerase* (also called *RNA polymerase*) catalyzes their polymerization from the ribonucleotides: adenosine triphosphate (ATP), cytodine triphosphate (CTP), guanosine triphosphate (GTP) or uridine triphosphate (UTP). One strand of DNA serves as a template or pattern.

RNA molecule is lengthened by the addition of one ribonucleotide. The specific nucleotide added does the DNA template specify difected by the hydrogen bonding. The same hydrogen bondings occur in RNA formation as in the case of DNA replication with *exception that thymine is replaced by uracil.* A pyrophosphate (P–P) molecule that is formed with each step subsequently is split into two inorganic phosphate molecules.

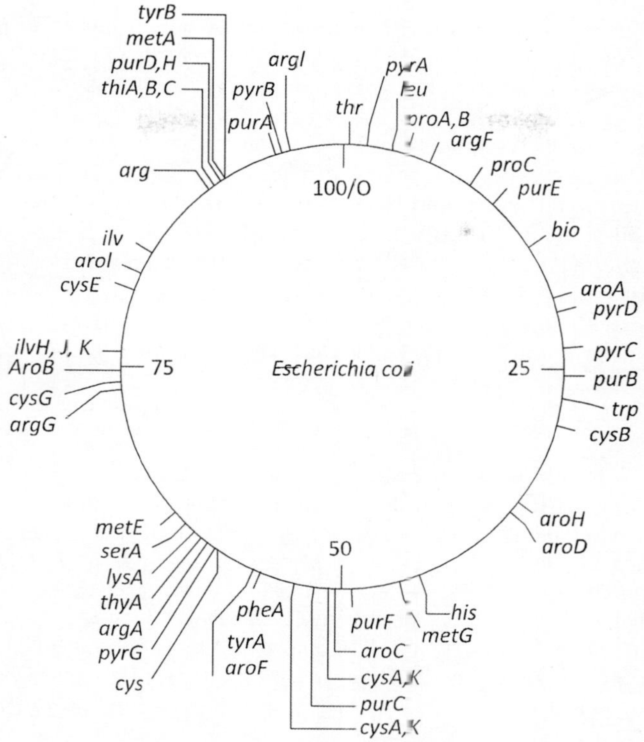


**FIGURE 2.6:** A map of the *E. coli* chromosome showing a few of the genetically mapped loci. *E.coli* is by far the best-characterized living organism. Approximately 1000 genes have been mapped to its chromosome, there is a complete restriction map of the genome and a large fraction of its DNA sequence is known.

## RIBONUCLEIC ACID (RNA)

Like DNA, RNA also consists of a sequence of nucleotides, but unlike DNA, it usually exists as a single-stranded molecule. There are two other basic differences: in RNA the pentose ribose is substituted for deoxyribose, and the pyrimidines base, uracil, is substituted for thymine. RNA participates in several different ways to bring about the expression of the genetic information enclosed within DNA molecules. The genetic or hereditary information specifies the particular type and position of amino acids in a protein molecule (amino acid sequence). Based on their respective functions and roles, three different forms of RNA are recognized: *messenger RNA (mRNA), transfer RNA (tRNA), and ribosomal RNA (rRNA)*. All RNA is synthesized as a complementary copy of a DNA sequence.

**Messenger RNA (mRNA):** This form of RNA is synthesized as a complementary copy of a DNA nucleotide sequence. The function of mRNA is to obtain and carry the genetic information encoded in the RNA to the location where proteins are to be synthesized. Messenger RNA molecules are very unstable and survive for only a short time in microbial cells.

**Transfer RNA (tRNA):** Transfer RNA is the shortest of the ribonucleic acid molecules. It assumes a three dimensional structure, which is usually shown in the shape of a cloverleaf. Each tRNA molecule, of which there are at least 20, is equipped to transport a single amino acid to the site of protein synthesis. Transfer RNA makes up approximately 15 per cent of a cell's total RNA.

**Ribosomal RNA (rRNA):** Ribosomal RNA is combined with specific ribosomal proteins to form the structural units known as *ribosomes*. Ribosomes serve as the sites or workbenches where the information in mRNA is read and proteins are formed. Ribosomal RNA comprises the bulk of the RNA in cells.

The expression of the information encoded in the base sequence of DNA begins with the synthesis of an RNA copy of DNA sequence making up a gene. This is called **transcription** because the DNA base sequence is being copied into an RNA base sequence. The RNA that carries information from DNA and directs protein synthesis is mRNA. The last phase of gene expression is **translation** or protein synthesis.

## GENE

The genetic information is stored in packages, on **genes**, where one gene holds the information for the production of one polypeptide sequence. This idea was first hypothesized by Beadle the Tatum in 1941. *Gene is a presumed unit of heredity.* They are defined *as discrete sequences of DNA which code for given biological molecules such as structural proteins and enzymes. Thus a gene might be defined as a polynucleotide, tRNA, or rRNA.*

As the information is in the form of a series of bases along the DNA molecule, a gene is simply a **stretch of DNA**, with a defined start site and a defined end. It can then be processed as a unit and the information passed on to the machinery, which will make the protein.

Not all the DNA in a cell are used for the storage of information therefore not all the DNA is used as part of a gene. The stretches of DNA that lie between genes are known as **intergenic DNA**. Any stretch of DNA has two strands and therefore the capacity for holding two different sequences of bases, depending on which strand is read. However, a gene will only use the information held by one of the strands. That is not to say though that both strands are not used. Some genes will be on one strand while others will be of the other strand. Therefore, varied terms have been coined to refer to the strands, which hold the important sequence information and those which do not, such as **sense** and **antisense**, **coding** and **non-coding.** A gene sequence is always read from the 3' end to the 5' end, and therefore genes, which reside on different strands are read in opposite directions; again a fact that needs considering when a researcher is trying to decipher the gene sequence.

3' end ⟶ to the 5' end

**The arrangement of genes:** In most of the eukaryotic organisms, each cell may contain more than one DNA molecule and these together are known as chromosomes. The DNA is packaged and each DNA molecule may carry thousands of genes. Therefore, on each DNA molecule thousands of genes are found in apparently random manner. The genes may be grouped into clusters of related genes. While in prokaryotes (bacteria), it is not unusual to have the need to express several genes, which are not the same but are related, in that the proteins, which are encoded by the genes are required along a common metabolic pathway. These clusters of genes are known as **operons.**

**Operons:** Operons do not occur in higher organisms. The most studied operon is known as the lactose operon in *E. coli*. This operon contains three genes, which are required for the utilization of lactose as a metabolic fuel in the cell. The operon encodes for a **permease** to allow entry of the lactose into the cell along with a transacetylase and β-galactosidase used in the metabolic pathway using lactose. The operon also contain all the control sequences needed to ensure efficient expression of the genes together.

The clusters of genes, which either encodes for the same protein or very similar proteins are called **homologous.** Their nucleotide sequences are very similar. Clusters of similar genes are called **multigene families.** These are of two kinds:

1. **Simple multigene families:** These are clusters of identical or extremely homologous genes, e.g., the genes for 5S rRNA.
2. **Globin genes:** These are made up of related but different genes. The two clusters are found one for the β-like subunits. The β clusters are located on chromosomes 11 in humans. They contains six globin genes, referred to as $\varepsilon$. ${}^{G}\gamma$ ${}^{\psi}\beta_1$, ${}^{A}\gamma$ $\delta$ and $\beta$. Different genes from the cluster are expressed at different times during the development of the animal.

The genes clusters come about because of **gene duplication** events in the DNA. When DNA contains a single copy of the gene, which is duplicated. Therefore the DNA will have two copies of the gene. The areas within the base sequence of any one gene may not be very important. The regions within the protein encoded by that gene may be fairly redundant, and these can be changed without major changes to the proteins function. Therefore, nucleotide changes may take place, which can alter the protein sequence quite markedly, but the areas of importance within the protein remain the same, as alterations here would be catastrophic.

- **Pseudogenes**: Not all genes that can be identified on a chromosome are functional. Such genes are called **pseudogenes**.
- **Tandem gene clusters** - Regions of the genome where certain gene or gene clusters are tandemly repeated up to a few hundred times comprise part of the moderately repetitive DNA. Examples include r RNA genes and histone gene clusters.
- **Genes may not be continuous:** Although a gene is sequences of bases along a length of DNA, this sequence is not necessarily continuous and may be interrupted by non-coding regions.
- **Exons:** The stretches of coding DNA are known as exons.
- **Intervening sequences (introns)**: The stretches interrupting the coding area are known as **intervening sequences** (IVSs). A gene may have no introns or several. When the DNA sequence is uncoded, the intron areas have to be ignored.

A simplicified rule is that the sequence GT is found at the start of the intron with AG at the end of the intron, giving the gene sequence.

exon/GT–intron–AG/exon

However, four nucleotides seem hardly sufficient to ensure that such an important event takes place properly, and by the use of the technique known as

**site-directed mutagenesis**, that is where individual bases are changed in a sequence and the effects studied, **consensus sequences** for the splice sites in genes have been identified. These bases in the consensus sequence have been studied in natural mutations, where the splicing of a gene's introns is defective and therefore a non-functioning product is formed. The process of splicing involves other RNA molecular. It is called **U-RNA**, or **small nuclear RNAs** (snRNAs).

**Procaryotic and viral gene structure** differs greatly from that of eukaryotes. In bacterial and viral systems, the coding information within a cistron normally is continuous (some bacterial genes do contain introns); however, in eukaryotic organisms, many if not most genes contain coding information (**exons**) interrupted periodically by noncoding sequences (**introns**). An interesting exception to this rule in eukaryotic histone genes, which lack introns.

## Functional Genes

**Genes encoding RNA molecule:** The genes encoding 80S rRNA are located in nucleolus. Both subunits as well as 5s rRNA (genes not in nucleolus) are transcribed from tandemly arranged repeat sequence. Coding sequences are separated by spaces sequences of differing sizes.

**Genes encoding proteins:** Protein-encoding genes are distributed in the euchromatic regions of chromosomes.

**Multigene families:** These are structurally related and their products may have similar physical and biochemical properties, and include storage proteins of cereals, isoenzymes etc.

## Genes of Structure and Genes of Regulation

A gene that determines the structure of a particular protein (e.g., an enzyme) is called a **structural gene**. The activity of a structural gene, in terms of production of messenger RNA for enzyme synthesis, is strictly regulated in the cell. It has been shown that many structural genes lie adjacent to specific sites concerned with the regulation of structural gene activity. Such a site is called an **operator**. Under certain conditions, the cell produces cytoplasmic proteins called **repressors**; when an operator gene binds its specific repressor, the structural gene adjacent to it is prevented from producing mRNA and is thus inactivated. In many cases, a series of structural genes determining a series of coordinated enzymes (e.g., the enzymes of a particular metabolic pathway) form a continuous segment of DNA under the control of a single adjacent operator gene. A gene sequence under the coordinated control of a single operator is called an **operon**.

Each specific type of repressor molecule of the cell must, of course, be formed by its own structural gene. A gene concerned with the production of a repressor is called a **regulator gene**. Both operator genes and regulator genes can be detected when they occur in mutant form. For example, the operator can mutate to a state in which it is unable to bind repressor. The operon now functions under all conditions and is said to be **derepressed**.

Mutations of the regulator gene produce phenotypes similar to those produced by mutations of the corresponding operator gene. For example, a mutated regulator gene may fail to make repressor, giving rise to the derepressed phenotype. Regulator genes can be distinguished from operator genes, however, by the behaviour of diploid cells carrying one normal gene and one mutated gene.

In such diploids the derepressed state is dominant if the mutation has altered the operator gene but recessive if the mutation has inactivated the regulator gene.

Note that the set of structural genes, which make up an operon is transcribed as a unit, forming a polygenic mRNA molecule. Transcription begins at the operator end of the operon; a nonsense mutation in one of the genes proximal to the operator may stop translation not only of that gene but also of all genes in the operon distal to it. Such mutations are called **polar mutations**; the polarity of nonsense mutations reflects the inability of a ribosome, once discharged from the mRNA, to reinitiate translation further downstream.

## *Prokaryotic Genes*

The genes of prokaryotic organisms and their regulation were the first to be studied. In bacteria, genes for proteins catalyzing related activities are often arranged tandem so that they are transcribed on a single polycistronic mRNA in which each cistronic component is the messenger template for a single protein. Such a collection of genes is referred to as an operon. Bacteria must be able to adapt to rapid environmental changes in a similar manner as plants hence their genes are constructed to enable them to function at highly variable rates. When required mRNA molecules can be produced at high levels, but only when their genes receive appropriate signals. The compounds responsible for these signals are called **inducers**.

The classical study of Jacob and Monod in 1961 on the lactose (lac) operon in *Escherichia coli* paved the way to an understanding of the regulatory components of prokaryotic genes. The lac operon contains the structural genes for three proteins β–galactosidase, which hydrolyzes lactose into glucose and galactose, lactose permease that is the transporter for the uptake of lactose into the cell and thiogalactosidase transacetylase, an enzyme that transfers an acetyl group from acetyl–CoA to β–galactosides. The last enzyme appears to be involved in the detoxification of non–metabolizable β– galactosides, which can then be secreted from the cell.

### Structure of Cyanobacterial Genes

The available knowledge of cyanobacterial gene organization is not as extensive as that for bacterial genes. There is reason to believe that they are very similar and probably regulated identically. The genes of the nitrogenase complex of the bacterium *Klebsiella* are sufficiently homologous to those from the cyanobacterium– *Anabaena*, that they can be used as high stringency probes to detect these genes in the blue–green alga (Haselkorn, 1986). In the bacterium the genes are arranged in the *nif*HDK operon which contains three open reading frames, each for a polypeptide component of the complex. This is also the case in heterocysts in *Anabaena*. During the differentiation of heterocysts from normal vegetative cells an interesting gene rearrangement occurs which results in a novel means of gene regulation.

*Anabaena* genes sometimes possess more than one promoter. This is the case with *gln*A, a gene involved in ammonium assimilation. When ammonia is plentiful, promoter sequences similar to those of *E. coli* are regulatory. However, under conditions where anaerobic nitrogen fixation occurs and involves the *nif* gene products, a promoter region similar to those of the *nif* genes controls transcription.

### Gene Arrangement in the Cauliflower Mosaic Virus

The cauliflower mosaic virus (CaMV) genome is a circular double stranded DNA molecule approximately 8 kbp in length. In the host two transcripts are produced. One is from the entire viral genome (35S) and the other (19S) is the messenger for the portion of the genome responsible for the inclusion body protein involved in viral assembly.

### *Eukaryotic Genes*

Eukaryotic genes differ in a number of ways from those of lower organisms. For example, *they are not arranged in operons* and, therefore, each coding region is associated with its own regulatory sequence. Transcription of mRNA is performed by RNA polymerase II, one of three eukaryotic RNA polymerases. RNA polymerase I and III are responsible for the synthesis of ribosomal and transfer RNA, respectively. In prokaryotes all these functions are performed by a single RNA polymerase.

### *Organelle Genes*

Mitochondrial and plastid genomes code for some, but not all, proteins found in these organelles. They produce their own ribosomal and transfer RNAs, and a small number of mRNAs (i.e., around 90 in chloroplasts and 8 or more in mitochondria). The nuclear genome codes for most of the organellar proteins, which are imported from the cytoplast. In some cases separate protein subunits are encoded in the different genomes (e.g., nuclear *rbcS* and chloroplast *rbcL* genes of ribulose biphosphate carboxylase), which poses interesting questions regarding regulatory interactions.

*Organelle genes are essentially prokaryotic in nature.* They possess Shine–Dalgarno promoter sequences, which have been recognized by *E. coli* RNA polymerase. They are usually polycistronic, but they do occasionally contain introns, characteristics very rarely seen in prokaryotic genes.

## Gene that Code for Proteins

During transcription DNA is double stranded, only strand contains coded information and directs RNA synthesis. This strand is called the **sense strand**, and the complementing strand is known as the **natisense strand**. Since the mRNA is made from the 5' tc the 3' end, the polarity of the DNA sense strand is therefore 3' to 5'. Therefore, the beginning of the gene is at the 3' end of the sense strand (also the 5' end of the netisense strand). An RNA polymerase recognition/binding the regulatory site known as the **promoter** is located at the start of the gene.

**Promoters** are sequences of DNA on the sense strand that are usually upstream from the actual coding or transcribed region – that is, the promoter is located before or upstream from the coding region in relationship to the direction of transcription (*the direction of transcription is referred to as downstream*). In *E. coli* the promoter has two important functions, and these relate to two specific segments within the promoter. Although these two segments do vary slightly in sequence between bacterial strains, they are fairly constant and may be represented by consensus sequences. These are idealized sequences composed of the bases most often found at each position when the sequences from different bacteria are compared. The RNA polymerase recognition site, with a consensus sequence of 5'TTGACA3' on the antisense strand, is located about 35 base pairs before (the– 35 region) the transcriptional start point (labeled as +1) of RNA synthesis. This sequence seems to be the site of the initial association of the RNA polymerase with the DNA.

The **RNA polymerase binding site**, also known as the **Pribnow box**, is located at the –10 region and has a consensus sequence 5'TATAAT3' (*a sequence that favors the localized unwinding of DNA*). This is where the RNA polymerase begins to unwind the DNA for eventual transcription. The initially transcribed portion of the gene is not necessarily sense or coding material. Rather, a **leader sequence** may be synthesized first. The leader is usually a nontranslated sequence that is important in the initiation of translation and sometimes is involved regulation of transcription.

**Shine–Dalgarno sequence**: The leader in prokaryotes generally contains a consensus sequence known as the Shine–Dalgarno sequence, 5'AGGA3', the transcript of which complements a sequence on the 16S rRNA in the small subunit of the ribosome. The binding of mRNA leader with 16S rRNA properly orients the mRNA on the ribosome. The leader also sometimes regulates transcription by attenuation. Downstream and next to the leader is the most important part of the structural gene, the coding region.

The **coding region** of genes that direct the synthesis of proteins typically begins with the sense DNA sequence 3'TAC5'. This produces the RNA translation initiation codon 5'AUG3', which codes for N–formylmethionine. This modified form of methionine. is the first amino acid incorporated in most procaryotic proteins. The reminder of the coding region of the gene consists of a sequence of codons that specifies the sequence of amino acids for the particular protein. Transcription does not stop the translation stop codon but rather at a **terminator sequence**. The terminator often lies after a nontranslated **trailer sequence** located downstream from the coding region. The trailer sequence, like the leader, is needed for the proper expression of the sense portion of the gene.

## Gene that Code for tRNA and tRNA

The DNA segments that code for tRNA and rRNA also are considered genes, although they give rise to structurally important RNA rather than protein. Genes coding for tRNA may code for more than a single tRNA molecule or type of tRNA. The segments coding for tRNAs are separated by short spacer sequences that are removed after transcription by special ribonucleases, at least one of which contains catalytic RNA. Thus the synthesis of tRNA and rRNA involve post–transcriptional modification, a relatively rate process in prokaryotes.

## Eukaryotic Gene Organization and Expression

The major difference between prokaryotic and eukaryotic gene expression is that, in a eukaryotic cell, mRNA must pass from the nucleus to the cytoplasm for translation. There is a complex processing pathway for mRNA precursors in the nucleus, and the cytoplasmic mRNA differs in structure from prokaryotic mRNA.

In eukaryotes, there are three different RNA polymerases, I, II, and III. RNA polymerase I is located in the nucleolus and catalyzes the synthesis of rRNA. RNA polymerases II and III are present in the nucleoplasm (outside the nucleolus). RNA polymerase III transcribes the genes for small nuclear RNAs and tRNAs. Its binding sites are located within these genes, rather than on the 5' or "upstream" sides of the genes.

RNA polymerase II transcribes the majority of the nuclear structural genes; it is responsible for pre– mRNA synthesis. Comparisons of the sequences of the promoters or RNA polymerase II–binding sites of over 200 different eukaryotic genes reveal consensus sequences located at about 25 and 75 nucleotide–pairs, respectively, before the transcription–start site. The consensus sequences for the **–25 sequence** (also called the **Hogness box** and the **TATA box**) and the **–75 sequence** (also called the **CAAT box**) are **TATAAAA** and **GGCCAATCT** respectively.

While in prokaryotes, all well–characterized genes consist of continuous sequences of nucleotide pairs, in case of eukaryotes, coding sequences (called **exons** for **expressed** sequences) of these genes (sequences specifying the functional gene–products RNA or protein) are often, but not always interrupted by noncoding **intervening sequences** or **introns**. This mRNA transcript with both introns & exon sequences of gene is called heterogeneous nuclear RNA (hn RNA). The mRNAs of eukaryotes, derived from primary gene transcript then undergoes processing, which includes:

1. addition of 7–methyl guanosine caps to 5'end of mRNA.
2. addition of poly adenylated nucleotide laid to the 3'end of mRNA
3. mRNA splicing for removal of introns. Spliced mRNA is then translated to form proteins.

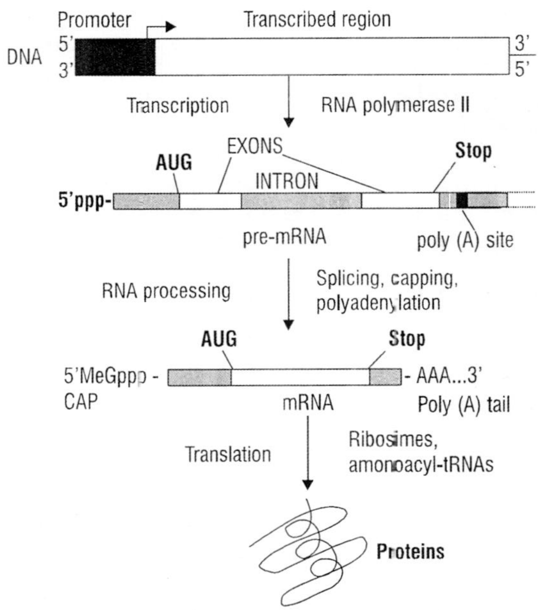


**FIGURE 2.7:** Presentation of eukaryotic gene expression.

Eukaryotic genes are regulated by the **promoter elements** located just upstream (5' terminus) from the transcription initiation sites in a manner similar to prokaryotic gene regulation. However, these eukaryotic promoters may be very complex with binding sites for many different regulatory proteins. In addition to the nearby promoters, many eukaryotic genes are also regulated by more distant cis–acting elements called enhancers and silencers.

## Prokaryotic Gene Organization and Expression

The primary difference between cells of a eukaryote and a prokaryote is the presence or absence of nuclear membrane, which acts as barrier, separating the genetic information contained within the chromosomes from the cytoplasm. Genetic Engineering makes use of eukaryotic as well as prokaryotic cell and its application concern gene organization and their expression.

The mechanism of gene expression was first understood in *E. coli.* The genetic information of this bacterium is contained within a circular chromosome of approximately three million base pairs. Most genes are tightly packed together, with very little superfluous DNA between them. As there is no nuclear membrane, the transcription and translation of prokaryotic gene is co–temporaneous.

In each gene in the **operon** of prokaryotes, there is a short sequence just upstream from the start codon (AUG), known as **Shine–Dalgarno sequence** (after its discoverer). It is the binding site for ribosome and is essential for efficient translation.

**RNA polymerase enzyme** transcribes the gene. This enzyme contains five distinct polypeptides (2∝, ß, ß' and ω together make core enzyme) and one subunit (sigma=σ factor) which is involved in initiation of transcription. Holoenzyme (core enzyme with sigma factor) initiates synthesis of RNA chain followed by release of σ factor and chain elongation by core enzyme. The function of sigma factor is to recognize and bind RNA polymerase to the correct initiation sites, the *promoter* sites, on DNA. Hundreds of promoters have now been sequenced in *E.coli*, and these sequences have surprisingly little in common. Two short sequences within these promoters are sufficiently conserved to be recognized, but even these are seldom identical in two different promoters. The midpoints of the two conserved sequences occur at about 10 and 35 nucleotide–pairs, respectively, before the transcription–initiation site. Thus, they are often called the **–10 sequence** and the **–35 sequence**, respectively. The most common or consensus –10 sequence (also called the **Pribnow box**) is **TATAAT.** The consensus –35 sequence (also called the **recognition sequence**) is **TTAGACA.**

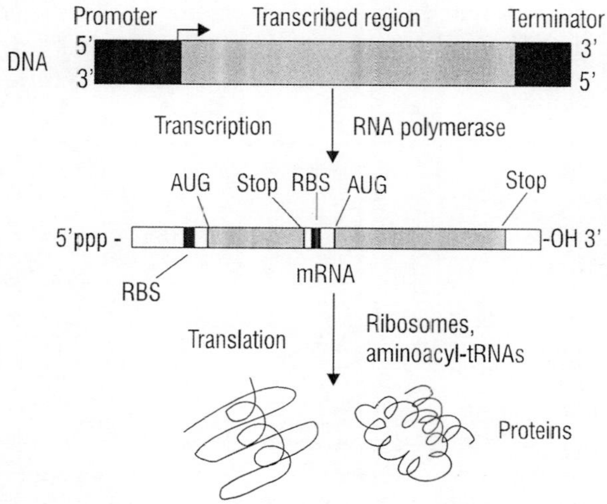


**FIGURE 2.8:** Presentation of prokaryotic gene expression.

If RNA polymerase is arrested during elongation (typically by omission of precursor nucleotides during *in vitro* reaction) then two proteins (Gre A and Gre B proteins) of *E.coli* release the RNA polymerase from elongation arrest. Gre A can act only if it is present before the polymerase arrest.

The termination of transcription occurs at specific **terminator sequences** in DNA. Most prokaryotic mRNAs terminate with the sequence 5'UUUUUUA–3' suggesting that the sequence 3'–AAAAAAT–5' in the sense strand of DNA is at least part of the transcription terminator sequence. Some transcription–termination signals require the presence of a protein called **rho**.

Although only one of the two DNA strands is transcribed in any given region, both strands of DNA in a chromosome usually participate in transcription, with some mRNAs being transcribed from one strand and other mRNAs (from different genes) being transcribed from the other strand. Because the two strands of a DNA double helix have opposite polarity, transcription events using opposite strands, as templates will proceed in opposite direction along the DNA molecule. In T4 viruses, transcription occurs off both DNA strands (but only rarely in the same region). It is important to study the basic features of gene organization and expression in *E.coli* and other prokaryotes as they are of central importance in genetic engineering. *E.coli* has been mapped and the information is useful for cloning techniques.

In eukaryotic cells comprise three interacting genetic systems viz. nuclear genome and chloroplast and mitochondria are semiautonomous bodies with their own organizational and functional properties but they do not synthesize all their proteins. The nuclear genome plays an important role in organelle biogenesis.

A genome is the complete set of chromosomes found in each nucleus of a given species, which contain entire genetic material (measured in picograms).

## GENETIC CODE

The genetic code is the message sent by the genetic material (DNA) of the cell through RNA to translate the synthesis of proteins. There are 64 possible triplets for the genetic code. Each triplet (i.e., 3 units or bases) specifies the synthesis of one amino acid after translation. The code appears to be universal in the sense that any organism uses the same code triplets to synthesize amino acid and proteins through the aid of m RNA.

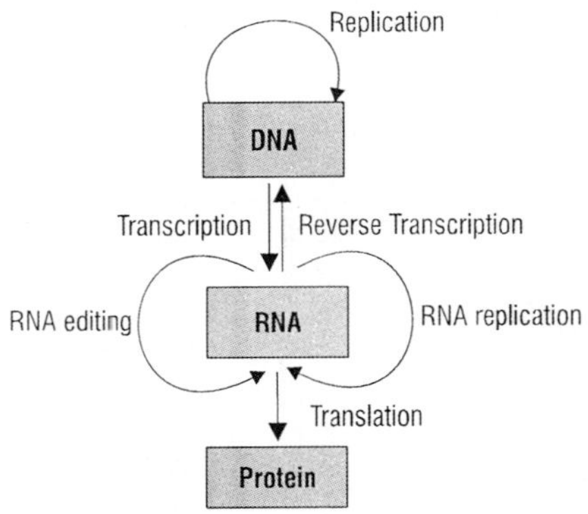


**FIGURE 2.9:** Flow of genetic information from DNA to RNA up to formation of protein.

The code was determined by the use of two kinds of '*in vitro*' systems:

1. Synthetic polymers, each containing a mixture of two different nucleotides were used as message for the synthesis of polypeptides 'in vitro', by analyzing the ability of different polymers to stimulate the incorporation of different amino acids, many codons could be described.
2. Synthesis nucleotide triplets were used to stimulate the binding of specific t RNA molecules to ribosomes 'in vitro'.

The genetic code specifies amino acid codons in different ways. More than one codon can specify the same amino acid, e.g., leucine is specified by 6 different codons (UUA, UUG, CUU, CUC, CUA, and CUG). This is made possible by the existence of several different RNA species, all of which are charged with leucine but each of which has a different anticodon. In some cases a given anticodon can match either of two codons, which differ only in the 3rd nucleotide (e.g., UUU, and UUC). The smallest number of code words is 1 for methionine (AUG) and there are two for aspartic acid (GAU and GAC), 3 for isoleucine (AUU, AUC and AUA) and 4 for serine (UCU, UCC, UCA and UCG). If there is a change in the bases of the codon and this change does not affect the code words of the amino acid, then the product will not be altered or affected, e.g., if the code words from DNA came for lysine as AAG and AAG instead of AAA and AAG than lysine will be synthesized normally because AAG is still another triplet for lysine.

Research work done by Nirenberg, Holley, and Khorana established the basis for deciphering the so–called **genetic code**. Just as DNA from very diverse biological sources usually contains only the four nucleotides A, T, G and C, proteins from diverse sources are commonly made up of only 20 amino acids (the building blocks of proteins). The question arises, how only four nucleotides code for 20 amino acids. That is how long a sequence of nucleotides, and which arrangements of the nucleotides, was necessary to code for 20 amino acids?

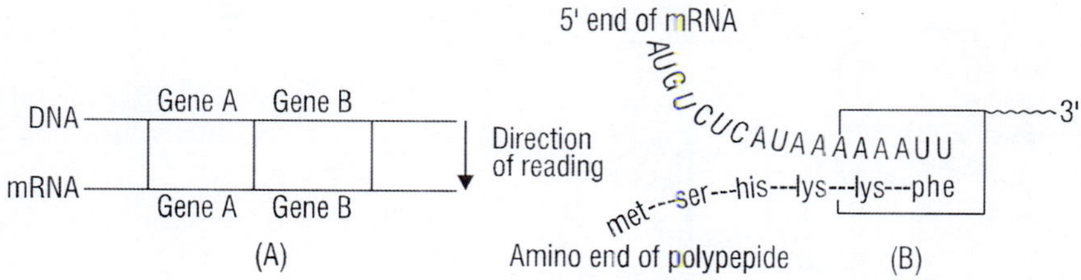


**FIGURE 2.10:** (A) In prokaryotes mRNA is collinear with the regions of the DNA that is transcribed. (B) Protein synthesis (translation is directional process. The mRNA is read in the 5'P to 3'OH direction, and the peptide chain is synthesized from amino acid end to the carboxylic acid end.

Mathematically, that all possible combinations of four bases, taken one or two at a time, could account for only four or sixteen amino acids respectively; taken as triplets (**codons**), ($4^3$) or 64 arrangements could be possible. This is more than enough to code for the 20 amino acid. However, the triplet code also provided for more than one code for amino acid. There are few codons, which cannot code for amino acids and are called **missense codons**. They are considered as the punctuation's for the genetic alpha or code. The finding that more than one codon could code for a given amino acid is known as **degeneracy** in the genetic code.

## Concepts of Degeneracy

The foundation of genetic engineering is based on the concepts of degeneracy, missense and nonsense codons. Since they provide an understanding of how the DNA of various organisms may code for the same amino acids. They also provide numerous ways to code not only for the composition but also for the unique arrangements or sequences of amino acids in the diverse proteins of those organisms. The genetic engineering researches have the potential, specifically rearranging, inserting, deleting, synthesizing, or transferring DNA. Such studies also include the knowledge now, where, and when genes function? **Genes** are defined as discrete sequences of DNA, which code for given biological molecules such as structural proteins and enzymes.

## DNA Encodes Genetic Information

DNA usually encodes genetic information in all living cells. It has complementary strands of opposite polarity, which are wrapped around each other to form a double helix. Each strand is composed of a phosphate–sugar backbone with one of the four different bases, adenine, guanine, cytosine or thymine, covalently attached to sugar residue. A complex of base, sugar and phosphate is called a **nucleotide;** the polynucleotides constitute a strand. The bases pair by non–covalent hydrogen bonds. Adenine forms two hydrogen bonds with thymine, and cytosine forms three hydrogen bonds with guanine. This specificity of base pairing allows the precise duplication of DNA, and enables it to act as the repository of genetic information.

Second Letter

| First Letter | U | C | A | G | Third Letter |
|---|---|---|---|---|---|
| U | UUU Phe | UCU Ser | UAU Tyr | UGU Cys | U |
| | UUC Phe | UCC Ser | UAC Tyr | UGC Cys | C |
| | UUA Leu | UCA Ser | UAA Stop | UGA Stop | A |
| | UUG Leu | UCG Ser | UAG Stop | UGG Trp | G |
| C | CUU Leu | CCU Pro | CAU His | CGU Arg | U |
| | CUC Leu | CCC Pro | CAC His | CGC Arg | C |
| | CUA Leu | CCA Pro | CAA Gln | CGA Arg | A |
| | CUG Leu | CCG Pro | CAG Gln | CGG Arg | G |
| A | AUU Ile | ACU Thr | AAU Asn | AGU Ser | U |
| | AUC Ile | ACC Thr | AAC Asn | AGC Ser | C |
| | AUA Ile | ACA Thr | AAA Lys | AGA Arg | A |
| | AUG Met | AGC Thr | AAG Lys | AGG Arg | G |
| G | GUU Val | GCU Ala | GAU Asp | GGU Gly | U |
| | GUC Val | GCC Ala | GAC Asp | GGC Gly | C |
| | GUA Val | GCA Ala | GAA Glu | GGA Gly | A |
| | GUG Val | GCG Ala | GAG Glu | GGG Gly | G |

**FIGURE 2.11:** Presentation of 64 codons of mRNA.

During **DNA replication,** the helix is unwound and each strand acts as a template for the synthesis of a complementary strand. New strand is formed in 5' to 3' direction. This reaction is enzymatically catalyzed by DNA polymerase. In opposite polarity (3' to 5' direction), a series of discontinuous short pieces (**Okazaki fragments**) are formed, which are subsequently joined up by the enzyme DNA ligase.

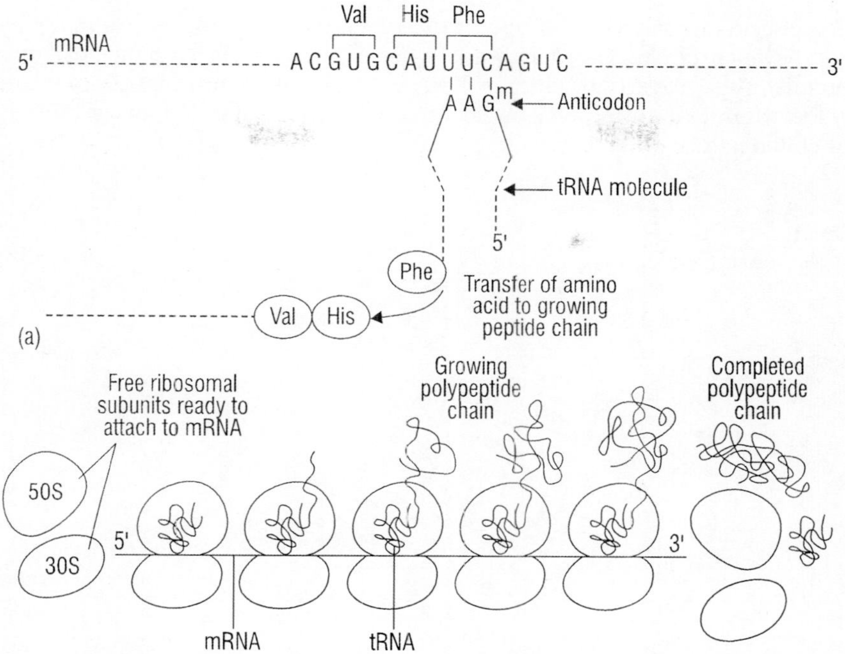


**FIGURE 2.12:** Protein synthesis: The translation process occurs on ribosomes complex macromolecular structures located in the cytoplasm.

RNA differs from DNA in containing a ribose, rather than a deoxyribose sugar and uracil base in place of thymine. There are three different classes of RNA, messenger RNA, ribosomal RNA and transfer RNA, all of which are transcribed from DNA templates. mRNA is the carrier of genetic information and forms only a few per cent of the total cellular RNA.

DNA can be linear or circular. Viruses, unlike cellular organisms, have a variety of genome structures, linear or circular, single–stranded or double–stranded, and DNA or RNA. Some viruses, such as the SV40 DNA tumour virus, have a genome of double–stranded DNA.

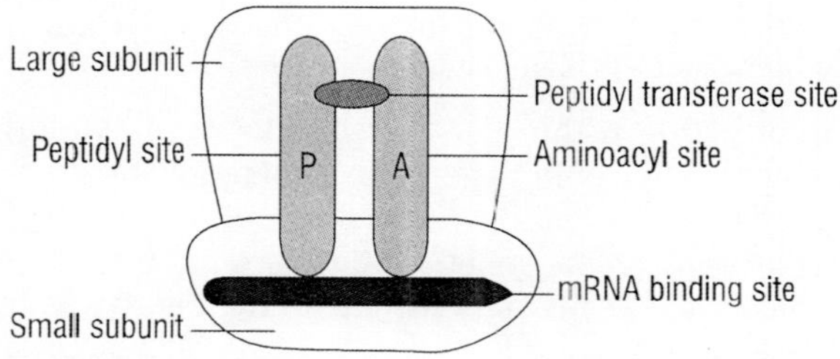


**FIGURE 2.13:** Structure of ribosome.

The genome of a prokaryote cell, such as *E. coli* consists of a gigantic circular molecule of DNA. The DNA molecule found within eukaryotic mitochondria is also circular, although the normal chromosomal DNA is linear.

A circular double stranded DNA molecule in which each strand is intact is termed closed circular DNA. If one of the two strands is broken, the molecule is termed as open circle. The open ends when join to regain circular structure, it supercoils. This property of circular DNA is technically important, because open and closed circular forms can be separated from one another by virtue of the super coiled nature of the latter molecule.

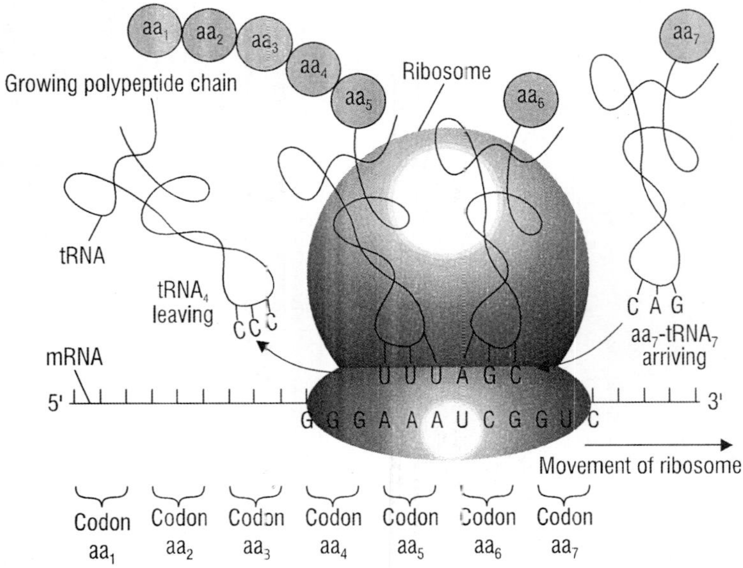


**FIGURE 2.14:** The translation of mRNA into protein. (a) Insertion of a single amino acid into the polypeptide chain. This is a highly diagrammatic representation of the process of translation. The codons in the mRNA are recognized by the anticodon loop on the tRNA. In this case a UUU codon is interacting with a phenylalanine tRNA. In all such interactions the last two nucleotides of the anticodon loop in the tRNA are the direct complement of the cognate codon in the mRNA. The third residue in the anticodon loop is often a 'non–standard' base, in this case, a methylated guanosine. The base in the third position is called the wobble base because it has the ability to pair with more than one base in the third position of the codon. Thus this phenylalanine tRNA is able to pair with either a UUU or a UUU phenylalanine codon. The enables a cell to utilize fewer tRNAs. (b) Orientation of a tRNA molecule relative to the large and small ribosomal subunits in the 3 main binding states designated A/A, P/P, and E.

The genetic information within DNA directs the production of the structural proteins and enzymes, which build the cell. Production of proteins (or protein synthesis) is a two step process viz. transcription and translation. Transcription is production of RNA by copying the genetic information in DNA. It is fundamentally similar to DNA replication, except that only one of the two DNA strands is copied.

The nucleotide sequence of mRNA is read as a series of triplets (Genetic code). Each codon specifies the insertion of a particular amino acid into the growing peptide chain. This is then translated into the language of protein with the help of 20 naturally occurring amino acids.

Prokaryotes and eukaryotes use essentially similar mechanisms to decode the information contained within their DNA; however, they differ greatly in cell structure and in the size and organization of their genome.

## MOLECULAR CLONING OF DNA

Only part of the DNA of any genome is directly involved in coding for amino acid sequences in proteins. Other noncoding regions function in such processes as DNA replication, recombination, the control of gene expression, and the structural organization of the chromosome. In many cases, this is accomplished by short base pair sequences in the DNA that serve as binding sites for specific proteins or RNA molecules; in other cases, it seems likely that recognition sites are provided by secondary structures in the DNA, formed by internal base–pairing in regions of repeated base sequences.

To obtain definitive information concerning such processes, it is necessary to perform a **functional analysis of DNA structure**: Specific segments of DNA of known base pair sequence must be altered and rearranged in known ways, then reintroduced into host cells or put into in vitro systems to determine the effects of the changes on specific functions. Such functional analysis has been made possible by 3 major technical advances: (1) the ability to determine the complete base–pair sequence of DNA molecules (**DNA sequencing**); (2) the ability to synthesize DNA sequences 20 or more base sequences in length (performed by automatic machines); and (3) the ability to carry out the **molecular cloning of DNA**. These techniques not only permit the functional analysis of DNA structure but also make possible the production in large quantities of almost any desired gene product, e.g., otherwise scarce hormones, enzymes, and antigens.

### Restriction Enzymes

The name restriction enzyme is derived from the genus and species of bacterium from which it is isolated. The first letter of the genus name and first two letters of the species are combined to form the enzyme name. This is followed by a strain designation if needed. In many instances, a bacterial strain contains more than one restriction endonuclease. When this occurs, each enzyme is assigned a Roman numeral. Thus, *Bam* HI was the first enzyme activity reported from *Bacillus amyloliquefaciens* strain H.

Structurally, most restriction enzymes reported in purified form are composed of two equal subunits with molecular weights of 20,000– 25,000 or single polypeptides with molecular weights of 30,000–35,000.

Enzyme activities can be differentiated from each other by their characteristic patterns of digestion of small viral DNAs. The DNA from bacteriophage Lambda is the most widely used substrate for screening restriction enzymes. Because it's often difficult to determine a characteristic pattern from a Lambda digest, smaller DNAs, such as the replicative form of bacteriophage ϕX174 and of SV40 are also used as substrates. The resulting DNA digest is displayed on agarose or polyacrylamide gels and visualized by staining with Methylene Blue Plus™ or ethidium bromide.

Sequence–specific, or Type II, endonucleases are commonly described as restriction enzymes. In contrast with nonspecific or Type I enzymes, Type II endonucleases generate producible nucleotide fragments from specific DNAs. They cleave double–stranded DNA by hydrolyzing two phosphodiester bonds (one per strand) within defined nucleotide sequences. Over 1500 H.O. Smith and collaborators have discovered enzymes since the first report, with a variety of bacterial cells as their source.

A given sequence in DNA is recognized and can be cleaved by more than one restriction enzyme. The term *isoschizomers* describes a group of enzymes that recognize the same sequence in DNA. In such cases, the representative enzyme that is used most commonly is selected on the basis of ease of purification and stability in purified form. The sequences recognized by these enzymes are for the most part centrosymmetric *palindromic* sequences that are usually hexamers, pentamers, or tetramers. Several Type II enzymes recognize DNA at a specific site and hydrolyze phosphodiester bonds at a defined distance from that site. An example of this group of enzymes is Bgll, which recognizes a sequence containing two groups of specified residues separated by completely unspecified residues – GCCNNNNNGGC; it, therefore, generates DNA fragments with

**Endonucleases** are enzymes that hydrolyze phosphodiester bonds at internal locations within a polynucleotide chain. Exonucleases, by comparison, break phosphodiester bonds sequentially, starting at one free end of a polynucleotide and working towards the other. Restriction endonucleases cleave duplex DNA only at specific sites, producing double strand breaks. They generally cut to the 5' side of the phosphate, resulting in breaks, which are amenable to the DNA ligase repair process, suggesting that such accidental nicking and repair may happen in nature.

## DNA Fingerprinting

DNA typing (also called DNA profile analysis or DNA fingerprinting is a recently developed method that allows for the unambiguous identification of the source of unknown DNA samples. The method has become very important in forensic biochemical laboratories where it has been used to provide evidence in paternity and criming cases. In contrast to the more conventional methodologies, such a blood typing, which can only exclude a suspect, DNA fingerprinting, can provide positive identification with great accuracy.

DNA fingerprinting involves the electrophoretic analysis of DNA fragment sizes generated by restriction enzymes.

## Plasmids

Jacob Schaeffer and Wollman (1958, 1960) coined the term *episome* for supernumerary element, which can exist either independently of bacterial chromosome or integrated into it. Thereafter, various types of other conjugation factors or sex factors were discovered. These factors determine the conjugational ability and transfer the genetic material to recipient cell in *E. coli* K–12 strain. Lederberg (1952) used the word plasmids for these extracellular and autonomously reproducing genetic structure. They are circular, double stranded molecules of DNA and exist independently of chromosomal DNA in bacterial cell. The genes associated with plasmid determine various biological characters in bacterial cell. The number of genes in plasmids varies from 3–4. They have no role in viability or growth of bacteria. Their function is additional to the functions of chromosomal genes. Two types of plasmids have been identified:

**Conjugative plasmid:** It carries genes that promote the transfer of plasmids from host cell to a recipient cell by conjugation.

**Non–conjugative plasmid:** It cannot promote its own transfer by conjugation.

**Structure of plasmids:** They are circular dsDNA molecules and appear as super helical coil. Each super helical turn consists of 400–600 base pairs. The twisted arrangement is known as covalently closed circular (CCC) DNA. When the closed circular DNA opens, it appears like cleavage or nicking DNA strands.

## Types of Plasmids

Bacteria are hosts to small, extra chromosomal genetic elements called plasmids. Plasmids are dispensable to the cell under ordinary conditions of growth. Their presence is detectable when the genes they carry confer new properties on the host, and they are generally named for these properties. About 9 types of plasmids are known, they are: (1) Transfer and sex factor, (2) R–plasmids, (3) Plasmids conferring pathogenicity to mammals, (4) Col plasmids, (5) Degradative plasmids, (6) Mercury's–resistance plasmids, (7) Tumour inducing plasmids in *Agrobacterium tumefaciens*, (8) Cryptic plasmids, (9) Staphylococcal plasmids.

Transfer and sex factors: They are conjugative plasmids and determine their own transfer in bacteria and have the character to promote transfer of other genes from the donor to recipient cell at the time of conjugation. Those conjugative plasmids, which cause transfer of chromosomal genes during conjugation, are termed as sex factors. It is extrachromosomal genetic element and was known as sex factor F (for fertility in *E. coli* K–12). The chromosomal fragment or genes, which are transferred to the recipient cell, then joins with its chromosome.

Sometimes exchange between DNA segments of plasmids and chromosome takes place after a break on the both strands where the nucleotide sequences are similar. This results in insertion of plasmid into the chromosome. A bacterial cell which contains chromosomal DNA + plasmid DNA rings is known as *Hfr* (high frequency of recombination) cells.

All plasmids discovered to date are circular, double–stranded DNA molecules. Their molecular weights range from 3 x $10^6$ to 1 x $10^8$, which is sufficient to dose for 5–160 average polypeptides. Plasmids contain 3 to 4 genes. Most of the plasmids used as vector are derived from psc 101 to $EcolR_1$ plasmids by addition or deletions. $P^{BR\ 322}$ is one such popular derived plasmid vector having unique sites for restriction enzymes, which are available for insertion of DNA fragment. Shuttle vectors are generally employed which can replicate in more than one types of hosts. Generally, plasmids with 2 genes for different antibiotic resistance are used

- to identify bacteria that carry plasmid, and
- to distinguish chimeric plasmid (recombinant plasmid) from the parental vector.

Thus, chimeric plasmid is resistant to one antibiotic while parental plasmid is resistant to two antibiotics. Multicopy plasmids are more suitable since 10 to 30 copies of one plasmid in a cell along with inserted gene will give an amplified response. Addition of chloramphenicol often affects new rounds of chromosome replication while plasmid replication is unaffected. This makes isolation of plasmid easier.

**Sex factors**: These mediate chromosome transfer by the mechanism described below. The most widely studied is F, the sex factor found in *E. coli* D12.

**Col factors**: These carry genes that cause their hosts to produce **colonies**, proteins that are lethal toxins for coliforms bacteria. Most members of this group are small (less than 10 megadaltons of DNA), are present in the cell in many copies (20–100), and are nonconjugative– lacking the gene system mediating self–transfer.

**Resistance (R) factors**: These carry genes conferring on the host cell resistance to various antimicrobial agents, such as antibiotics. A single plasmid, for example, may carry separate genes for resistance to streptomycin, chloramphenicol, tetracyclines, and sulfonamides. In some cases it has been possible to dissociate the plasmid into several smaller ones: an element called resistance transfer factor (RTF), carrying the genes governing the intracellular transfer process; and separate elements called **R determinants**, carrying the resistance genes. In other cases, the various elements remain tightly linked, and the entire plasmid is then referred to as an **R factor**.

**Penicillinase plasmids of staphylococci**: These plasmids carry a gene that causes the cell to produce a potent penicillinase, thus rendering it resistant to penicillin. They differ from R factors in that they are not capable of transfer by conjugation. They can be carried from cell to cell, however, by phage mediated transduction.

**Degradative plasmids of *Pseudomonas***: These plasmids carry sets of genes determining the enzymes of catabolic pathways, such as the enzymes for the degradation of camphor, toluene, octane, or salicylic acid.

**Virulence plasmids**: A number of plasmids carry genes whose products contribute to the relative pathogenicity of the host cell. These plasmids include the enterotoxin and colonization antigen plasmids of enteropathogenic *E coli* strains and a staphylococcal plasmid carrying a gene for exfoliative toxin.

As will be discussed below, plasmids share many properties with bacterial viruses (phages). They are of major clinical significance, not only because they may carry genes for resistance to therapeutic drugs and for virulence factors but also because some of them mediate gene transfer– a process that leads to the emergence of strains with new combinations of antigenic and virulence properties.

## Replication of Plasmids

Plasmid DNA replicates while maintaining a supercoiled, circular state. Replicating plasmid DNA can be isolated in the form of **relaxation complexes**, in which the DNA is bound to proteins that appear to provide the following functions:

- binding of the DNA to the membrane replicator site;
- nicking and unwinding of the strands to permit replication; and
- resealing of the nicks. Replication is bi–directional in some plasmids and unidirectional in others.

Plasmids regulate their own replication, such that each plasmid exhibits a typical copy–number ratio (ratio of plasmid copies to chromosome copies in the cell). In general, the small nonconjugative plasmids have high copy–number ratios whereas the large conjugative plasmids have copy number ratios closer to 1. Plasmid replication is unusually sensitive to inhibition by such agents as acridine dyes and ultraviolet light; by using these agents at threshold doses, cells can be cured of their plasmids.

Even in plasmids having a copy–number ratio of 1, partition between daughter cells following replication is highly efficient. This is affected by a plasmid DNA site called *par*, which is the functional equivalent of the eukaryotic chromosomal centromeres. It may act by binding to a cell membrane site or, alternatively, by associating with the host chromosome.

**Incompatibility of Plasmids:** Plasmids can be classified into a number of incompatibility groups; 2 members of the same group cannot coexist in the same cell. This phenomenon may involve competition for a specific attachment site in the cell membrane or may reflect the action of group–specific repressors. Members of the same incompatibility group are closely related as indicated by the extensive hybridization of their DNAs.

### *Self–Transfer of Plasmids*

**Gram–negative bacteria**: Most of the plasmids of gram–negative bacteria are conjugative: they carry the genes (called the ***tra* genes**) mediating their own transfer by the process of cell conjugation. There are 12 or more *tra* genes arranged in an operon (see below); some of these code for the production of the **sex pilus**.

Plasmid transfer in gram–negative bacteria begins with the extrusion of a sex pilus, a protein thread several times the length of the cell. The tip of the sex pilus adheres to gram–negative cell walls.

Binding of the recipient cell to the donor pilus requires the presence of specific receptor sites on the recipient cell's surface. In certain conjugation deficient mutants, one of the major proteins of the outer membrane is missing and the lipopolysaccharides is altered, indicating a role for both of these components in pair formation.

Following specific pair formation, the plasmid undergoes a special type of replication called **transfer replication**, one parental strand passing into the recipient and the other remaining in the donor cell. Complementary strands are synthesized in the donor and recipient simultaneously with transfer. The daughter molecules are circularized by ligase action immediately after transfer replication is complete.

**Gram–positive bacteria**: Plasmid genes have determined several antibiotics produced by members of the genus *Streptomyces*, and some of these plasmids are transferable from one streptomycetes to another. The process resembles conjugation in that direct hyphal contact is necessary, but the molecular details of the process are not understood. The best known such plasmid is SCPI, which carries the genes for the biosynthetic pathway producing the antibiotic methylenomycin. SCPI also carries the genes responsible for transfer by a mechanism, which has some similarities to that of plasmid–mediated chromosome transfer in gram–negative bacteria.

Plasmid–mediated conjugation also occurs in members of the gram–positive genus *Streptococcus*. In *Streptococcus lactis*, the ability to ferment lactose is conjugally transmitted as a plasmid–borne β–galactosidase (*lac*) gene. A series of conjugative plasmids have been analyzed in *Streptococcus faecalis*, carrying genes for hemolysin, bacteriocin, and single or multiple drug resistance. These plasmids can mobilize other (nonconjugative) plasmids as well as chromosomal markers.

**Sex Pheromones**: An unusual feature of the *S faecalis* mating system is the production by recipient cells of specific **sex pheromones**– diffusible proteins that stimulate donor cell aggregation (with each other and with recipient cells) and, if the donor cells are pre induced by the pheromone, increase the frequency of donor plasmid transfer as much as a million–fold. When a recipient cell acquires a given plasmid, it ceases to produce the pheromone specific for that plasmid but continues to excrete pheromones that plasmid but continues to excrete pheromones that act on donor cells harbouring other conjugative plasmids.

### *Recombination*

Plasmids undergo crossing over with each other and with the host chromosome, depending on the extent of their base sequence homologies or on the presence of transposable elements. Since both plasmids and chromosome are circular, an odd number of crossovers serves to integrate the 2 DNA structures, which then replicate as a single unit. An even number of crossovers, on the other hand, brings about an exchange of segments.

A plasmid capable of integrating with the bacterial chromosome is called an **episome**. When integration occurs, a double replicon is formed bearing both a chromosomal and a plasmid replicator site. In some cases, replication continues to start at the chromosomal replicator; in others (e.g., at nonpermissive temperatures in mutants temperature–sensitive for DNA synthesis), replication of the entire structure is taken over by the plasmid replicator, a phenomenon known as **integrative suppression**.

## Cell Properties Determined by Plasmid Genes

**Drug resistance**: In gram–negative bacteria, genes governing resistance to such drugs as neomycin, kanamycin, streptomycin, chloramphenicol, tetracyclines, penicillins, and sulfonamides are found on one or another plasmid in various combinations. In the gram–positive staphylococci, genes governing resistance to such agents as penicillin, erythromycin, and heavy metals (e.g., $Hg^{2+}$ and $Co^{2+}$) are found on plasmids. Genes conferring resistance to the same agents may also be found on the chromosome, but in such cases they do so by different mechanisms. Most plasmid–governed resistance is mediated by enzymatic inactivation of the drug (e.g., by acetylation or phosphorylation), whereas chromosome–mediated resistance usually reflects a lowered affinity of its target molecule for the drug. These findings are compatible with the fact that plasmids are (by definition) dispensable to the cell: Only chromosomal genes can confer resistance by structurally altering the binding site for the drug, since the genes determining the structures of indispensable cell components must themselves be indispensable.

**Virulence:** Enteropathogenic strains of *E coli* of porcine or human origin cause diarrhoeal illnesses by the liberation of enterotoxins. Two types have been recognized: a heat–liable toxin (LT) and a heat–stable toxin (ST). Plasmid genes determine both. Other plasmids of *E coli* carry genes determining surface hairs (pili) that serve as "colonization antigens" only strains possessing these pili can adhere to host epithelial cells and establish themselves in sufficient numbers to cause disease.

Plasmid–borne genes can contribute to host cell virulence in other ways. The colicinogenic plasmid Col V, for example, determines enzymes that synthesize cell–associated, hydroxamate–containing iron sequestering compounds which are induced in media containing low concentrations of iron. It is believed that invasiveness of symbiotic bacteria depends heavily on the ability of the bacteria to compete with their mammalian host cells for iron; most *E coli* strains isolated from bacteremias carry Col V, suggesting a role for the Col V –determined iron–sequestering system in host cell virulence. The invasiveness of the human pathogen *Yersinia enterocolitica* has also been shown to be plasmid–determined, although in this case the mechanism is not known.

**Production of antimicrobial**: Anti–microbial agents include antibiotics, which are the products of metabolic pathways, and polypeptide toxins. Plasmid genes encode a number of these agents, including some antibiotics of *Streptomyces*, bacteriocins (such as colonies), and microcins– oligopeptides produced by certain strains of *E coli*.

**Other properties**: As discussed above, plasmids are known that encode enzymes of special catabolic pathways (in *Pseudomonas*), enzymes of nitrogen fixation (in *Klebsiella*), and tumorigenicity for host plants (in *Agrobacterium tumefaciens*).

**Relation to Viruses**: Bacterial viruses possess all of the properties described above for plasmids; a *Pseudomonas* phage has even been found to promote conjugation. The major difference thus appears to be the ability of phages to form mature, protein–coated virions that can be liberated and passed to other cells through the medium. Their many similarities suggest a close evolutionary relationship between phages and plasmids.

**Clinical Significance**: The ease with which plasmids can transfer from cell to cell and the strong selection that chemotherapy has exerted for drug resistance have combined to produce striking results. In Japan, for example, the frequency of R–governed multiple–resistance *Shigella* strains rose from 10–20% in 1955 to 80% in 1968. In England, multi–resistant *Salmonella typhimurium* first appeared in 1961 and rose to 21% of all *S typhimurium* isolates by 1964.

In general, 60–90% of resistance markers in the gram–negative pathogens are carried on transferable plasmids. Although the experimentally determined rate of transfer of resistance in vivo is very low, epidemiological evidence indicates that such transfer accounts for the rapid rise in incidence of drug–resistant strains of gram–negative pathogens.

A common method for cloning DNA molecules is to cleave both plasmid and insert DNA with the same restriction endonuclease. The sticky ends created by enzyme can be sealed easily with phage T4 DNA ligase. T4 DNA ligase can also be used to join the blunt–ended DNA molecules together. Sticky end or cohesive end joining is more efficient and advantageous because foreign DNA can be retrieved easily as restriction sites (say, for $EcoR_I$ enzymes) are created on both sides of inserted DNA.

Bacteriophages can also be used as vectors. λ–Phase of *E. Coli* is normally used. The amount of DNA that can be packed in phage head has to be 80% to 105% of its normal size for encapsulation to occur, hence there is limitation for quantity of external DNA that can be inserted in the phase vector. About 20% of non–essential portion of original phase can be removed to make room for foreign DNA to be inserted. In insertion and transfer to the host the procedure comprises: (i) allow the virus to undergo lytic cycle first; (ii) get more viruses; (iii) prepare chimeric phages; (iv) allow it to undergo lytic cycle; (v) more chimeric phages obtained.

As pBR322 plasmid or phages are vector for *E. coli* cells, similarly, a range of vectors is now available for *Bacillus subtilis*, *Streptomyces* spp. and *Saccharomyces cerevisiae*.

### Transfer of Foreign Gene through Vector to Appropriate Host (Cloning Organism)

When foreign DNA is ligated (joined) to the vector, the recombinant vector is to be transferred to a suitable host. This is done by transformation of competent cells by treating it with cold $CaCl_2$ in the case of plasmid vectors and transfection in the case of phage vectors. This cloning organism should be deficient in the major pathway of DNA recombination (i.e. rec A–minus). This will reduce the possibility of undesirable recombinations. The cloning organism should preferably lack restriction enzymes that can degrade foreign DNA. Mini cells lacking in major chromosome but having all other cell machinery are suitable for testing expression and there will be no interference from the products of chromosomal expression in such cells. In eukaryotic cells, primary and secondary tissue cultures of African green monkey kidney cells have been used for recombinant DNA experiments.

*E. coli* was used and is still very useful as a cloning organism. *Saccharomyces cerevisiae* is also used, as it is safer, nonpathogenic, not living in human being and is close to human being (eukaryotic). *Bacillus subtilis* is a competent cloning organism when it is starved of nutrients. Many other bacteria and *Streptomyces* spp. can only be transformed when they are converted into protoplasts by removing their cell walls. So is the case with yeasts also, in which $CaCl_2$ and PEG treatment is given to facilitate protoplasts for the uptake of DNA.

## GENOME

Understanding the information contained in genome sequence will be the major challenge of the early 21st century. The genome is the entire DNA content of a cell, including all of the genes and all of the intergenic regions. Prokaryotic genomes are small with very little space between genes. The yeast genome contains 6000 genes and has a more compact organization.

Every organism is specified by genomes, which contain the biological information needed to construct and maintain a living example of that organism. Most genomes, including those for all cellular life forms, are made of DNA (deoxyribonucleic acid) but a few viruses have RNA (ribonucleic acid) genomes. DNA and RNA are polymeric molecules made up of linear, unbranched chains of monomeric subunits called nucleotides (see structure of DNA and RNA).

The biological information contained in a genome is encoded in the nucleotide sequence of its DNA or RNA molecules and is divided into discrete units called **genes**. The information in a gene is read by proteins that attach to the genome at the appropriate positions and initiate a series of biochemical reactions referred to as gene expression.

For organisms with DNA genomes, this process was originally looked on as comprising following two stages:

1. **Transcription:** First producing an RNA copy of the gene.
2. **Translation:** Resulting in synthesis of protein whose amino acid sequence is determined, via the genetic code, by the nucleotide sequence of the RNA transcript.

This is still an accurate description of gene expression in simple organisms such as bacteria, but it gives an incomplete picture of the events involved in conversion of genomic.

**Complete copy of the genome:** A complete copy of the genome must be made every time a cell divides. DNA replication has to be extremely accurate in order to avoid the introduction of mutations into the genome copies. Some mutations do, however, occur, either as errors in replication. It may be due to the effects of chemical and physical mutagens that directly alter the chemical structure of DNA. DNA repair enzymes. These enzymes correct many of these errors. Those that escape the repair processes become permanent features of the lineage descending from the original mutated genome. These events, along with genome rearrangements resulting from recombination, underlie molecular evolution, the driving force behind the evolution of living organism.

## Genomes of Microorganisms

Those cells, which contain membrane–bound compartments, including a nucleus and organelles such as mitochondria and, in the case of plant cells, chloroplasts, called **eukaryotes**. They include animals, plants, fungi and protozoa. Those cells, which lack extensive internal compartments, called **prokaryotes**. Eukaryotes and prokaryotes have quite different types of genome and we must therefore consider them separately. There are two very different groups of prokaryotes, distinguished from one another by characteristic genetic and biochemical features:

**TABLE 2.1:** Characteristic genetic and biochemical features of different groups of prokaryotes

| Bacteria | Archaea |
|---|---|
| The bacteria, which include most of the commonly encountered prokaryotes such as the gram–negatives (e.g. *Escherichia coli*), the gram–positives (e.g. *Bacillus subtilis*), the cyanobacteria (e.g. *Anabaena*) and many more. | The archaea, which are less well studied, and have mostly been found in extreme environments such as hot springs, brine pools and anaerobic lake bottoms. |

## Genomes of Eukaryotic Microorganisms

All of the eukaryotic nuclear genomes are divided into two or more linear DNA molecules. Each contained in a different chromosome. All eukaryotes have smaller, usually circular, Mitochondrial genomes. In plants and other photosynthetic organisms a third genome is found, located in the chloroplasts.

Although the basic physical structures of all eukaryotic nuclear genomes are similar, one important feature is very different in different organisms. This is genome size. The smallest eukaryotic genomes being less than 10Mb in length while the largest over 100 000 Mb (18.1).

**TABLE 2.2:** Sizes of genomes in microorganisms`

| Organism | Genome size (Mb) |
|---|---|
| **Prokaryotes** | |
| *Mycoplasma genitolium* | 0.58 |
| *Escherichia coli* | 4.64 |
| *Bacillus megaterium* | 30 |
| **Eukaryotes** | |
| *Fungi* | |
| *Saccharomyces cerevisiae (yeast)* | 12.1 |
| *Aspergillus nidulans* | 25.4 |
| *Protozoa* | |
| *Tetrahymena pyriformis* | 190 |

The variability in size coincides to a certain extent with the complexity of the organism, the simplest eukaryotes such as fungi having the smallest genomes, and higher eukaryotes such as vertebrates and flowering plants having the largest ones.

This makes sense as one would expect the complexity of an organism to be related to the number of an organism to be related to the number of genes in its genome – higher eukaryotes need larger genomes to accommodate the extra genes. However, the correlation is not precise: if it was, then the nuclear genome of the yeast *Saccharomyces cerevisiae*, which at 12Mb is 0.004 times the size of the human nuclear genome, would be expected to contain 0.004 × 80 000 genes, which is just 320. The *S. cerevisiae* genome contains about 6000 genes.

A correlation between the complexity of an organism and the size of its genome was a bit of a puzzle or so–called **C–value paradox**. The space is saved in the genomes of less complex organisms because the genes are more closely packed together.

### *Yeast (Saccharomyces cerevisiae)*

For example, in *S. cerevisiae* genome, the chromosome III contains 26 genes thought to code for proteins and two coding for transfer RNAs (tRNAs), short molecules involved in reading the genetic code during translation. These 28 genes take up 66.4% of the 50–kb sequence.

1. **Relatively few of the yeast genes are discontinuous**. In this segment of chromosome III none of the genes are discontinuous. In the entire yeast genome there are only 239 introns, which is a remarkably small number compared with individual genes can have more than 100 introns.
2. **There are fewer genome–wide repeats.** This part of chromosome III contains a single LTR element, called *Ty*2, and four truncated LTR elements called *delta sequences*. These five genome–wide repeats make up 13.5% of the 50–kb segment. However, this figure is not entirely typical of the yeast genome as a whole. When all 16 yeast chromosomes are considered the total amount of sequence taken up by genome–wide repeats is only 3.4% of the total.

Yeast genome is much more economical and compact. It has fewer introns. The spaces between the genes are relatively short with much less space taken up by genome–wide repeats and other noncoding sequences.

### Genomes of Prokaryotic Microorganisms

Prokaryotic genomes are very different from eukaryotic ones. Prokaryotic genomes are much smaller. There is some overlap between the largest prokaryotic and smallest eukaryotic genomes.

The *E. coli* genome, for example, is just 4639 kb, two–fifths the size of the yeast genome. The physical organizations of prokaryotes are also different. In prokaryotes, most if not all of, the genome is contained in a single DNA molecule, and this molecule is circular rather than linear. In addition to their single *chromosome*, prokaryotes may also have additional genes on independent, smaller, circular or linear DNA molecules called plasmids.

Genes carried by plasmids are useful. They code for properties such as antibiotic resistance or the ability to utilize complex compounds such as toluene as a carbon source. The plasmids appear to be dispensable – *a prokaryote can exist quite effectively without them.* For this reason, a bacterial or archael genome is usually defined as just the main DNA molecule, with plasmids looked on as ancillary components, not parts of the genome itself.

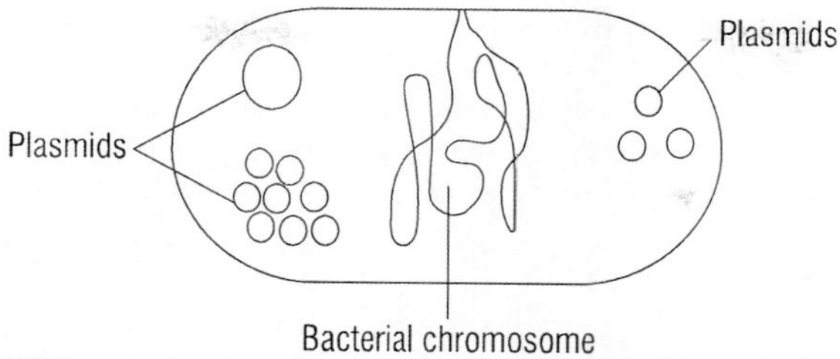


**FIGURE 2.15:** Bacterium showing chromosome and plasmids. Plasmids are small circular DNA molecules that are found inside some prokaryotic cells.

Prokaryotes have smaller genomes and fewer genes than eukaryotes. *E. coli* has just 4397. A 50–kb segment of the *E. coli* genome have more genes and less space between them, with 43 genes taking up 85.9% of the segment. Some genes have virtually no space between them: *thr* A and *thr* B, for example, are separated by a single nucleotide, and *thrC* begins at the nucleotide immediately following the last nucleotide of *thrB*. These three genes are an example of an **operon** (*a group of genes involved in a single biochemical pathway*).

### Important Features of Prokaryotic Genomes

1. There are no introns in the genes present in this segment of the *E. coli* genome. In fact *E. coli* has no discontinuous genes at all and it is generally believed that this type of gene structure is absent in prokaryotes with a few exceptions, mainly among the archaea.
2. The infrequency of repetitive sequences. Prokaryotic genomes do not have anything equivalent to the high–copy –number, genome–wide repeat families found in eukaryotic genomes. They do, however, possess certain sequences that might be repeated elsewhere in the genome, examples being the **insertion sequences** ISl and IS186 that can be seen in the 50–kb segment. These are examples of **transposable elements** (*sequences that have the ability to move around the genome and to transfer from one organism to another, even sometimes between two different species*)

## Importance of Genome in Research

Genome sequencing represents the frontier of molecular biology. Its territory was inaccessible just a few years ago. It has innovative approaches and a lot of sheer hard work. The major goal of molecular biology is to obtain the complete sequences of a many genomes as possible. Considerable effort is being devoted to a wide range of organisms. Following are the reasons for developing activity devoted to genome sequences:

1. The genome sequences are the key to the continued development of our knowledge of molecular biology, genetics, biochemistry, cell biology and physiology, etc.
2. A catalog containing a description of the sequence of every gene in a genome is immensely valuable, if at first the functions of many of the genes are unknown.
3. They help to catalog sequences (which they contain) of the coding parts of every gene, it will also include the regulatory regions for these genes.

4. The genome sequence therefore opens the way to a comprehensive description of the molecular activities of living cells and the ways in which these activities are controlled.
5. Gene catalogs also aid the isolation and utilization of important genes, such as those human genes responsible for inherited disease, or bacterial genes whose protein products have industrial value.
6. The genes can be isolated from a genome even if the complete genome sequence is not known.
7. It is easy to obtain a copy of the desired gene if the sequence of its genome is already known, so that the gene can simply be withdrawn from the catalog.
8. The genome of all higher eukaryotes contains extensive amounts of noncoding DNA. It is assumed that most of the noncoding DNA has no function, but perhaps this is because we do not know enough about the role of noncoding DNA. This is a matter of further research.

## Mapping Genomes by Genetic Techniques

Genetic mapping uses genetic techniques such as crossbreeding and pedigree analysis to construct genome maps. The basis of genetic mapping is the calculation of recombination frequency by linkage analysis. Genetic analysis of bacteria requires gene transfer from one cell to another. Genes were originally used as markers but these are now supplemented with DNA markers such as RELPs, SSLPs and SNPs. These techniques involve much more than just methods for sequencing DNA molecules.

### *Limitations*

1. Even with the most sophisticated technology it is rarely possible to obtain a sequence of more than about 750 bp in a single experiment. This means that the sequence of a long DNA molecule has to be constructed from a series of shorter sequences. One approach is to break the molecule into fragments, determine the sequence of each one, and use a computer to search for overlaps and build up the master sequence. This is done by shotgun method in prokaryotic genomes (The shotgun approach on its own is inappropriate for larger and more complex genomes).
2. The shotgun method may lead to errors when repetitive regions of a genome of a genome are analyzed. When a repetitive sequence is broken into fragments many of the resulting pieces contain the same, or very similar, sequence motifs. It would be very easy to reassemble these sequences so that a portion of the repetitive regions was left our or even so that two pieces of the same or different chromosomes were mistakenly connected together.

### *Shotgun Technique*

The DNA molecule is broken into small fragments, each of which is sequenced. Searching for overlap between the sequences of individual fragments assembles the master sequence. In practice, an overlap of several tens of base pairs would be needed to establish that two sequences should be linked together. The DNA molecule contains two segments of repetitive DNA. Each made up of many copies of the motif GATTA. When the sequences are examined, two fragments from different parts of the DNA appear to overlap. If the error is not recognized then the central segment of the DNA molecule will be left out of the master sequence. If the two repetitive DNA segments were on different chromosomes, then the sequences of these chromosomes might be mistakenly assembled.

Once a genome map is available, the sequencing phase of the project can proceed in either of two ways.

1. **By the clone containing approach:** The genome is broken into manageable segments, each a few hundred kb or a few Mb in length, which are then sequenced individually, usually by the shotgun method. Once the sequence of a segment has been completed, it can be positioned at its correct location on the map.
2. **By a directed shotgun approach:** This strategy uses the distinctive features on a genome map as landmarks to aid assembly of the master sequence from the huge numbers of short sequences obtained by the shotgun approach.

With both approaches, the map provides the framework for carrying out the sequencing phase of the project.

### *Genetic and Physical Maps*

Genetic mapping is based on the use of genetic techniques to construct maps showing the positions of genes and other sequence features on a genome. There are two conventional methods for genome mapping:

The technique, for prokaryotes, includes crossbreeding experiment. Physical mapping uses molecular biology techniques to examine DNA molecules directly in order to construct maps showing the positions of sequence features including genes.

## *Markers for Genetic Maps*

Genes are very useful markers but they are by no means ideal. Mapped features that are not genes are called DNA markers as with gene markers, to be useful a DNA markers. As with gene markers, to be useful a DNA marker must exist in at least two allelic forms. There are three types of DNA sequence feature that satisfy this requirement. Genetic mapping is based on the principles of inheritance as first described by Gregor Mendel in 1865 (Orel, 1995).

Genes are segments of segments of DNA molecules. They are the abstract entities responsible for the transmission of heritable characteristics from parent to offspring. To be useful in genetic analysis, a heritable characteristic had to exist in two alternative forms or phenotypes. This approach was fine in the early days but geneticists soon realized that there are only a limited number of visual phenotypes whose inheritance can be studied, and in many cases the analysis is made less easy because more than one gene affects a single physical feature.

A genome, consisting of a linear DNA molecule of 2.5 Mb. It has been mapped and the positions of eight markers (ABCDEFGH).

**On the left**, the clone counting approach starts with a segment of DNA whose position on the genome map has been identified because it contains markers A and B. The segment is sequenced by the shotgun method. The master sequence placed at its known position on the map.

**On the right**, the directed shotgun approach involves random sequencing of the entire genome. This result in pieces of contiguous sequence, possibly hundreds of kb in length. If a contiguous sequence contains a marker, then it can be positioned on the genome map.

**TABLE 2.3:** Typical biochemical markers used for genetic analysis of *Saccharomyces cerevisiae*

| Marker | Phenotype | Method by which cells carrying the marker are identified |
|---|---|---|
| ADE2 | Requires adenine | Grows only when adenine is present in the medium |
| CAN1 | Resistant to canavanine | Grows in the presence of canavanine |
| CUP1 | Resistant to copper | Grows in the presence of copper |
| CYH1 | Resistant to cyclocheximide | Grows in the presence of cyclocheximide |
| LEU2 | Requires leucine | Grows only when leucine is present in the medium |
| SUC2 | Able to ferment sucrose | Grows if sucrose is the only carbohydrate in the medium |
| URA3 | Requires uracil | Grows only when uracil is present in the medium |

Microbes (bacteria and yeast) have very few visual characteristics. A big advantage of these markers over visual phenotypes is that many of the relevant genes have multiple alleles. The problem is made worse by the fact that only a fraction of the total number of genes exists in allelic forms that can be distinguished conveniently. Gene maps are, therefore, not very comprehensive. We need other types of markers.

**Restriction Fragment Length Polymorphisms (RFLPs)**

RFLPs were the first type of DNA marker to be studied. Restriction fragments are produced when a DNA molecule is treated with a restriction endonuclease, which is a type of enzyme that cuts DNA molecules at defined sequences. For example, the restriction endonuclease *Eco*RI cuts double–stranded DNA at:

5'–G A A T T C – 3'
3'–C T T A A G – 5'

and at no other sequence. This sequence specificity means that treatment of a DNA molecule with a restriction enzyme should always produce the same set of fragments. This is not always the case with genomic DNA molecules because some restriction sites are polymorphic, existing as two following alleles:

1. Allele *displaying the correct sequence for the restriction site* and therefore being cut when the DNA is treated with the enzyme.
2. Allele with a *sequence alteration* so the restriction site is no longer recognized.

In order to score an RELP, it is necessary to determine the size of just one or two individual restriction fragments against a background of many irrelevant fragments. This is not a trivial problem: an enzyme such as *Eco*RI, with a 6–bp recognition sequence, should cut DNA approximately once every $4^6$ = 4096 bp and so would give almost 750 000 fragments when used with human DNA. Fortunately, there are techniques that enable individual fragments to be detected.

**Southern hybridization** was the first technique used to type RELPs. Agarose gel electrophoresis and the ones being studied detected by hybridization probing separate the DNA fragments resulting from restriction from one another.

The **polymerase chain reaction** or PCR is now more commonly used because it s faster than Southern analysis. In PCR, the region of DNA containing the RELP is amplified into multiple copies before the DNA is treated with the restriction enzyme. Now the RELP can be treated with the restriction enzyme and can be visualized by staining an agarose gel of the restriction fragments with ethidium bromide, which is a less sensitive detection method than hybridization probing.

Note that these techniques directly detect both alleles of an RELP: this means that (as with other DNA markers) the alleles are codominant.

**Simple Sequence Length Polymorphisms (SSLPs)**

SSLP are arrays of repeat sequences that display length variations, different alleles containing different numbers of repeat units. Unlike RFLPs, SSLPs can be multiallelic as each SSLP can have a number of different length variations. There are two following types of SSLP:

1. **Minisatellites**, also known as variable number of tandem repeats (VNTRs), in which the repeat unit is a few tens of nucleotides in length.
2. **Microsatellites** or simple tandem repeats (STRs), whose repeats are much shorter, usually di–, tri– or tetra nucleotide units. We met two Microsatellites in Chapter 1 when we looked at a typical 50–kb segment of the human genome.

## Southern Hybridization

In Southern hybridization, a collection of restriction fragments is transferred from an agarose gel to a nylon membrane and the specific ones being studied are detected by hybridizing probing. As well as its use in RELP mapping. It has many applications in other procedures that require identification of specific restriction fragments, for example to check the identity of a cloned fragment.

**Agarose gel electrophoresis** separates linear DNA molecules, such as restriction fragments, according to their sizes. If there are only a few fragments in the digest then each one forms a separate band that can be seen after staining the gel with ethidium bromide. If there are more than about 50 fragments then the bands merge together into a smear and individual fragments cannot be seen. Instead, the fragments are transferred to a nylon membrane, simply by placing the membrane on the gel and allowing buffer to soak through, taking the DNA from the gel to the membrane where it becomes bound. One can buy an expensive machine to do this, but is simpler to use a pile of paper towels.

A **hybridization probe** is a labeled DNA molecule whose sequence is complementary to the target DNA that we wish to detect. Because the probe and target DNAs are complementary they can base pair or hybridize. Detecting the signal given out by the label attached to the probe therefore identifies the position of the target restriction fragment on the membrane. The probe could be a synthetic Oligonucleotide or it could be a cloned DNA fragment. For detecting an RELP, the probe is usually a cloned DNA fragment that spans the polymorphic restriction site.

## DNA Chip and its Technology

The gene expression studies are useful in determining the function of a genes. However, there is a serious limitation of the current technique like monitoring gene expression, namely via Northern blots, because one can analyze expression of only *one gene at a given time*. DNA chips allows simultaneous and *en masse* analysis of hundreds and thousands of genes (DeRisi and Brown, 1997, *Science*, **278**: 68—686; Ferea and Brown, 1999, *Curr. Opin. Genei. Dev.* **9**: 715–722). In fact, it has become possible to analyze expression of all the genes of an organism such as yeast on a single chip. This technology has been given various names like, *lab–on–a–chip*, *array–of–hope*, or *revolution on a square cm*.

### Structure

A DNA chip is a wafer of silicon, 2 $cm^2$ or less in area, carrying many different oligonucleotides in high–density array. The DNA to be tested is labeled with a fluorescent marker and pipetted on to the surface of the chip. Hybridization is detected by examining the chip with a fluorescence microscope; the positions at which the fluorescent signal is emitted indicating which oligonucleotides have hybridized with the test DNA. Many SNPs can therefore be scored in a single experiment (Wang et al., 1998).

A DNA chip carries a large number of DNA probes, each one with a different sequence and each at a defined position on the surface of the chip. The probes can be synthetic oligonucleotides or other short DNA molecules such as complementary DNAs. In the earliest technology, the oligonucleotides or cDNAs were spotted on to a glass microscope slide or a piece of nylon membrane to form a micro array. With this approach only, relatively low density can be achieved, typically 6400 spots as an $80 \times 80$ array in an area of 18 mm by 18 mm. To prepare really high–density arrays a more sophisticated technology was needed. The answer has been to synthesize oligonucleotides *in situ* on the chip surface, which in turn has required an advance in the methodology used for preparing oligonucleotides, as shown in the **FIGURE 2.15**. The normal method involves adding nucleotides one by one to the growing end of oligonucleotides, the sequence determined by the order in which the dNTP substrates are added to the reaction mixture. If used for synthesis on a chip, this method would result in each Oligonucleotide having the same sequence. Instead, modified dNTP substrates are used, ones that have to be light activated before they will attach to the end of a growing Oligonucleotide. The dNTPs are added one after another to the chip surface, photolithography being used to direct pulses of light at individual positions in the array and hence determine which of the growing oligonucleotides will be extended by addition of the particular dNTP added at each step.

DNA chips are designed to allow many hybridization experiments to be performed in parallel. Their main applications have been in the screening of polymorphisms such as SNPs and comparing the RNA populations of different cells. They also have potential in novel DNA sequencing methodologies.

A density of up to one million oligonucleotides per $cm^2$ is possible (Ramsay, 1988), so if used for SNP screening half a million polymorphisms can be typed in a single experiment, presuming there are oligonucleotides for both alleles of each SNP. A DNA chip is not complicated to use. The chip is incubated with labeled target DNA to allow hybridization to take place. Which oligonucleotides have hybridized to the target DNA is determined by scanning the surface of the chip and recording the positions at which the signal emitted by the label is detectable. Radioactive labels can be used with a low–density micro array, signals being detected electronically by phosphor imaging. The does not provide enough resolution for a high–density chip, so with these it is necessary to use a fluorescent label. The fluorescent signal is detected by laser scanning or, more routinely, by fluorescent confocal microscopy.

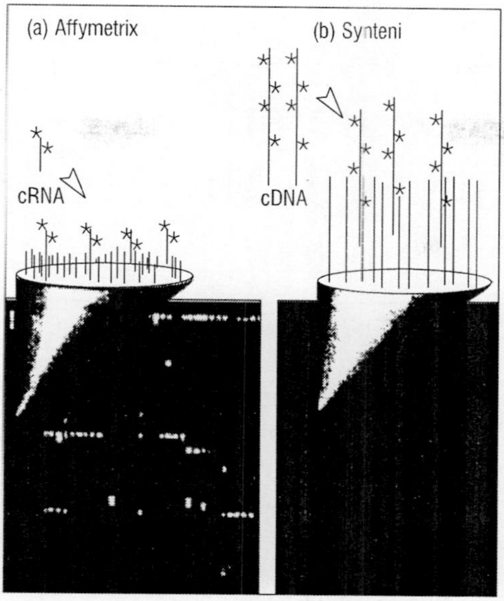


**FIGURE 2.16:** Diagrammatic representation of the types of DNA chips 'micro arrays' that are currently available. On the left is a synthetic chip marketed by Affymetrix Company where short oligos representing unique sequences of various genes of an organism are synthesized *in situ* by a combination of photolithography and organic chemistry techniques. On the right is another kind of DNA chip for example, one pioneered by Synteny/Incyte Company where cDNAs representing various genes of an organism are spotted on glass sides or another appropriate support by a microarrayer. Both types of chips are employed to monitor gene expression by labeling the mRNAs with fluorescent tags and allowing them to hybridize to specific spots. Hybridization is monitored by a scanner attached to a computer (Gerhold, D. et al., May 1999).

Under highly stringent hybridization conditions, a stable hybrid occurs only if the oligonucleotide is able to form a completely base–paired structure with the target DNA. If there is a single mismatch then the hybrid does not form. To achieve this level of stringency the incubation temperature must be just below the melting temperature or $T_m$ of the oligonucleotides. At temperatures above the $T_m$, even the fully base–paired hybrid is unstable. At more than 5°C below the $T_m$, mismatched hybrids might be stable. The $T_m$ for the oligonucleotide shown in the figure would be about 58°C. This is calculated from the formula $T_m$ = (4 × number of G and C nucleotides) + (2 × number of A and T nucleotides) °C.

# 3. Concepts, Perspectives and Techniques of Microbial Genetic and Protein Engineering

Genetic engineering or genetic manipulation may be defined as the extracellular (i.e., *in vitro*) creation of new forms or arrangements of DNA in such a way as to allow the incorporation or continued propagation of the altered genetic condition in nature. In short, genetic engineering means the manipulation of genes under controlled laboratory conditions. Genetic engineering deals with identification and isolation of desired gene and then joining this gene of interest into another organism. The desired gene expresses itself in that organism by the gene product. Various steps of genetic engineering are:

1. **Gene isolation:** Desired gene is identified, isolated and purified. This DNA of interest is also called *donor DNA* or *target DNA*.
2. **Selection of vector:** A vector is a self–replicating molecule of DNA or replicon to which desired gene is linked. Vector molecule with foreign DNA inserted is known as chimeric DNA. Vector acts as carrier and transports the gene into the host cell. Thus, it is also known as a cloning vehicle or a carrier molecule. Suitable vector is identified for a system; commonly used vectors are plasmids and viral DNA molecule. However, recent techniques involve physical delivery of DNA by various methods.
3. **Cloning of desired gene:** Multiple copies of desired gene can be obtained by placing them in host cell with the help of vectors. Here, the desired gene along with the vector is amplified. Large number of identical copies of gene of interest is produced for subsequent gene transfer into target cell. Now, gene cloning is a fast and mechanized process, using polymerase chain reaction (PCR) machines.
4. **Specific gene transfer:** The gene of interest is finally transferred to host cells. Transformed cells are selected, multiplied and produce transgenic plants by tissue culture technique.
5. **Expression of desired gene:** The desired gene produce, the product in new environment of host, thus the desired traits.

Genetic engineering includes the propagation of chimeric DNA in a different host organism. The ability to cross natural species barriers and place genes from any organism in an unrelated host organism is one important feature of gene manipulation. A second important feature is the fact that a defined and relatively small piece of DNA is propagated in host organism. Thus, genetic engineering open the door to a range of molecular biological opportunities including nucleotide sequence determination, site–directed mutagenesis, and manipulation of gene sequences to ensure very high–level expression of an encoded polypeptide in a host organism.

## ADVANTAGES OF GENETIC ENGINEERING

### *Modifying Gene Expression*

- Abolishing expression of a gene
- Quantitatively reducing expression of a gene
- Over–expressing a gene
- Putting gene expression under the control of an external trigger
- Changing the tissue specificity of expression
- Changing the timing of expression during the cell cycle, during development, or in response to a stimulus

### *Modifying a Gene Product*

- Giving it a new biological specificity
- Abolishing one among several biological effects
- Changing the rate at which it is metabolized or degraded
- Changing the circumstances in which it is metabolized or degraded
- Inducing resistance or sensitivity to a drug
- Changing its intracellular location
- Causing or preventing secretion of the product

### Practical Applications

1. **Genetical research**: The genetic engineering techniques are useful tools for genetical research. They can help to gain in the structure, function and regulation of genes. By using restriction enzymes and gene cloning techniques.
2. **Industrial products:** Genetic engineering helps in improving genetic strains of industrial microbes. For example the bacterium *Methylophikls methylotrophus* is a rich source of a protein, which is supplemented in animal feeds. A new strain of this bacterium has been developed by recombinant DNA producing the sufficient protein. In the **pharmaceutical industry** essential amino acids such as tryptophan and phenylalanine are used as nutrients. The **food industry** also uses amino acids in flavouring. The amino acid cysteine, because of its reducing poor, is used in the backing of breads with improved quality and in fruit juices as an antioxidant. In the **agricultural industry** amino acids such as methionine and lysine are added to feeds for more efficient utilization of protein.
3. **Super bug:** The super bug has potential for cleaning up oil spills. It is a newly constructd bacterium which can degrade most of the major hydrocarbon component of petroleum. The different strains of *Pseudomonas putida* are being used for this purpose.
4. **Hormones:** The hormones such as somatostatin, insulin and the human growth hormone are synthesized easily and cheaply by using genetic engineering techniques. Somatostatin (brain hormone originating from hypothalamus and is a peptide chain consisting of 14 amino acid units) is the first human hormone synthesized by genetic engineering. It acts to inhibit the release of the human growth hormone and insulin. It is related to the treatment of diabetes, pancreatic and few other conditions.
5. **Vitamins:** Vitamins are organic substances that are used by organisms in trace amounts as co–factors for enzymatic reactions. They are vital for most

forms of life and their biological importance. Most vitamins are produced by microbial fermentation. Apparently, metabolic routes are involved in the fermentation process and genetic engineering would offer scope for increasing the dosage of the genes. The strains of microorganisms could be produced with enhanced vitamin yield.

6. **Interferon**: The interferon is an anti–viral protein. Interferon is prepared from the mammalian cells by recombinant DNA technology. By cloning cDNA to genes for human interferon, it has been found that there are a large number of interferon differing in amino acids sequences and properties. A large amount of interferons are prepared in yeast cells by fermentation process.
7. **Antibiotics:** They are chemical substances produced by several microorganisms by aerobic fermentation. They are classified under three broad heads namely penicillins. Cephalosporins and tetracycline. Apart from their use as anti–microbial agents, antibiotics find application in the agricultural and food industry. For example, **erythromycin** is used as growth promoter in poultry and farm animals. **Nisin** which is an erythromycin polypeptide, is used as preservative of food. Recombinant DNA technology helps in producing strains capable of increased antibiotic yield
8. **Enzymes:** Enzymes are proteins, which act as catalysts of biochemical reaction and the industry. The recombinant DNA technology can be used in the construction of strains that combine all the desired properties. Both gene doses for the enzyme and repressor and regulator molecules can be altered to achieve high concentration of enzyme with minimal detrimental by–products.
9. **Biochips** would have enormous storing and retrieval power. They could be produced inexpensively because bacteria can be made to synthesize them for us. The prediction is that single biochips (1 cubic centimeter in size) will have a capacity equal to a hundred thousand times that of modern big size computer. Since biochips are very small in size, they can even be implanted into the human body and so programmed that they are able to react with light waves.
10. **Gene therapy:** More than 2000 genetic diseases of man could be cured by gene therapy. The conditions for gene therapy are: (1) the gene that causes the genetic disorder should be responsible for a product that functions are the terminal stage of its formation, (2) the etiology of the disease should be restricted to single tissue so that the replacements is easy, and (3) the clinical symptoms of the disease should result from the expression of the single pair of recessive genes e.g., sickle cell anemia; under this condition, it would be necessary to replace in the target cell only alleles. Not all–genetic disease will meet these conditions, but those that do will include **sickle cell anemia**, the **thalassemias** and the hemophilia. The sickle cell anemia and thalassemias are cancer alterations in the hemoglobin protein molecule. The hemophilias result from deficiencies in the factors responsible for the coagulation of blood.

## IMPORTANT TERMS OF GENETIC ENGINEERING

- **β–Replicase:** It helps in replication of viral RNA.
- **Alkaline phosphatase:** It catalyzes hydrolysis of monophosphate ester groups from compounds, including from the 3'–and 5'–terminal ends of nucleic acids and nucleotides

- **Antisense RNA:** A new class of regulatory genes have been discovered with direct the synthesis of RNA and can inhibit the translation of its target, another RNA molecule, by base pairing with it, prevents the translation of the massage.
- **Capping enzyme:** It attaches 7–methylguanosine to the penultimate nucleoside of RNA via a 5', 5'–triphosphate bridge.
- **Clones:** Identical daughter cell derived from a single mother cell by multiplication
- **DNA ligases:** Enzymes which bind DNA fragments together
- **Enhancer element:** These are short regions of DNA, which have the effect of increasing the levels of transcription, but unlike promoters, they affect genes from chromosomal loci at a long distance.
- **Epigenetic gene control:** This is the new concept. It describes that the control of gene expression in higher eukaryotic organisms is related to the methylation of cytosine residues in gene coding sequence.
- **Glycosylases:** They catalyze hydrolysis of nucleoside bases–gylcosidic bonds, resulting in production of an apurinic/apyrimidinic (AP) site of DNA and a free nucleoside base.
- **Gyrase:** It is Type–II topoisomerase that catalyzes conversion of relaxed DNA to a superhelical form and requires ATP.
- **Insertases:** They catalyze the reinsertion of a missing base into the appropriate apurinic or apyrimidinic AP sites of DNA.
- **Introns:** Sequences in DNAs which are in fact transcribed into mRNAs but are subsequently split out and therefore form no permanent constituents of the mRNAs (only in Eukaryotes)
- **Kinases:** They catalyze the transfer of the $\alpha$–phosphate of a nucleoside 5'–triphosphate to the 5'–OH terminus of a DNA or RNA molecule.
- **Ligases:** They catalyze formation of phosphodiester linkages between DNA chain segments.
- **Ligation:** Direct linkage of nucleic acid fragments
- **Methylases:** They catalyze transfer of methyl groups from donar *S*–adenosyl–L–methionine (AdoMet) to specific bases of acceptor DNA molecules.
- **Nucleases:** They catalyze hydrolysis of the phospho–diester bonds of polynucleotides.
- **Nucleotidyl transferases:** They catalyze attachment of mononucleotide triphosphates to the 3'–OH of the initiator polymer, with release of pyrophosphate.
- **Operator:** Regulatory section in a nucleic acid with binding capacity for a repressor protein for blocking the synthesis of mRNAs.
- **Plasmid:** Annular nucleic acid in cells outside the central chromosome with a lower molecular weight than the central chromosome; bearer of nonessential additional information, e.g., sex factors (F factors) and resistance factors (R factors), plasmids and viral nucleic acids with known functions such as resistance or sequences are denoted by trivial abbreviation derived from their function or origin, e.g., SV 40 for simian virus, pBR322 for a plasmid imparting resistance to antibiotics.
- **Poly(A) adding enzymes:** They attach AMP residues, from ATP, to the 3', 5' phosphodiester bond.

- **Polymerases:** They catalyze addition of nucleoside triphosphates to 3'–OH terminals of single strands of preexisting polynucleotides (primers) with release of pyrophosphate.
- **Priming enzymes:** They provide a free 3'–OH at terminal end of initiator strand to serve as acceptor for transfer of the first nucleotide and formation of first phosphodiester bond in DNA synthesis.
- **Processing of primary RNA:** Through a complex and orderly series of steps primary, or the initial, RNA transcript is modified to become the final, which is functional in the cell. These steps include RNA splicing translating RNA editing and C to U conversion, where cytosine bases are changed to uridine.
- **Promotor:** Regulatory section in a nucleic acid with the capacity for binding RNA polymerase
- Protein stability: Protein, in eukaryotic cells has very diverse half–lives ranging from a few minutes to several hours. The degradation of unnecessary proteins is dependent upon the small protein ubiquitin, which when conjugated to proteins leads to their rapid degradation.
- **RecA protein:** Essential for homologous genetic recombination, DNA repair and other SOS functions as a result of DNA damage.
- **Repressor:** Regulatory protein with binding capacity for operators.
- **Restriction endonucleases:** Enzymes, which specifically cleave DNAs at a definite information–bearing sequence of nucleotides and thereby permit an insertion of foreign sequences. Restriction Endonucleases are named in genetic engineering by trivial cell, e.g., EcoRI from *Escherichia coli* R Y 13, Hind II from *Haemophilus influenza* Hpall from *Haemophilus parainfluenzq* or pits from *providencia stuartii*
- **Restriction endonucleases:** They catalyze hydrolysis of the phospho–diester bonds of polydeoxyribo–nucleotides. Strict specificity for double strandes (dsDNA) and extended nucleotide sequences.
- **Reverse transcriptase:** It catalyzes transcription of RNA nucleotide sequence into complementary DNA (cDNA).
- **Topoisomerases:** They catalyze changes in the topological structure of DNA molecule by breakage and resynthesis of the same phospho–diester bond in DNA strands.
- **Transcript stability:** The mRNAs of those proteins, which expressed occasionally, contain AU rich sequences at their 3'–end. These mRNAs have short half–lives, whereas protein which are required for long durations do not contain these sequences and have mRNAs which are stable for longer times, e.g., several hours.
- **Transcription:** Conversion of a DNA (gene) information sequence into a mRNA sequence
- **Transduction:** Introduction of foreign genetic material into a cell by means of a virus or, in the case of bacteria, by means of phages
- **Transformations:** Uptake of free nucleic acids or plasmids into a cell
- **Translation control:** In certain cases, after a gene has been transcribed, the mRNA can fold back upon itself to prevent its translation. This process is known as attenuation.
- **Translation:** Synthesis of proteins on mRNAs
- **Unwinding enzymes:** They catalyze separation of complementary strands of DNA; require ATP.

- **Vector:** Plasmid or virus that after ligation with a foreign DNA is used to introduce new information into the cell.

## PREREQUISITES IN PLANT GENETIC ENGINEERING METHODS

- Isolation of DNA fragment and availability of gene(s)
- It must be possible to propagate identical (clone) piece of DNA.
- A specified method must be available by means of which the cloned DNA can be integrated into the genome.
- The function of new genes must be controlled (its expression and regulation).
- Selection and screening of recombinant clones.

## PERSPECTIVES OF GENETIC ENGINEERING

The term genetic engineering of plants broadly refers to genetic manipulation of plant cells and organs at molecular level. In the past, Mendelian genetics and conventional plant breeding methods have been the tools for the genetic improvement of crop plants. Recently the development of efficient procedures for the successful culture of somatic cells, pollen, protoplasts and plant regeneration from a large number of plant species and availability of various restriction enzymes, improved DNA vector systems based on Ti and Ri plasmids of *Agrobacterium*, transposable elements, series of promoters, marker genes and a large number of cloned DNA fragments/genes, have made genetic manipulations more precise and directed.

Such development has further widened the scope of genetic engineering to precisely insert foreign genes into crop plants from diverse biological systems (plants, animals, and microorganisms), and in cloning of useful genes and to facilitate molecular marker based selection for complex agronomic traits including quantitative trait loci. Through the use of various molecular probes, characterization of pathogens in the infected plant materials and introgression of specific DNA sequences have become more precise and rapid.

Though dramatic progresses have been made in genetic engineering of crop plants, several factors need to be investigated further before the potential of this technology can be realized on commercial scale for the improvement of crops.

Although plant regeneration has been achieved in a number of plant species, it is often limited to only specific genotypes. Hence, to incorporate desirable genes into crop plants, it is essential to have better plant regeneration systems applicable to a wide range of elite germplasm. In spite of the extensive information available in the molecular biology on the use of organ–specific promoters, yet very little is known about the mechanism of gene expression particularly how to switch the genes on and off. In this respect, the molecular cloning of DNA binding proteins from higher plants may facilitate our knowledge concerning regulatory factors in the transcriptional level.

These days gene isolation and genetic transformation are routine in many laboratories. Transgenic plants have been produced in more than 30 plant species. Besides, selectable marker genes including chloramphenicol acetyltransferase, neomycin phosphotransferase, hygromycin phosphotransferase genes conferring herbicide tolerance, insect tolerance, virus protection, storage protein, and environmentally regulated genes have been transferred to plants from a wide range of living organisms. In majority of the cases, foreign genes show expression in transgenic plants and are stably inherited to the progeny without any detrimental effects on the host plants. More interestingly, transgenic plants under field conditions have also maintained increased level of insect resistance, herbicide resistance and higher level of virus protection. Nevertheless, based on the experiment utilizing a large number of transgenic plants, Deroles and Gardner (1988) pointed out a loose association between high copy number of the T–DNA and a normal or non–Mendelian inheritance of the kanamycin resistance gene. They observed that rearrangements and multiple insertions are relatively frequent events during *Agrobacterium*–mediated transformation of plant cells. Thus, there is a need to have extensive laboratory and field–testing of transgenic plants and their progenies before their usefulness can be realized in crop improvement for commercial application.

Isolation of useful genes governing economic characters such as disease resistance, salinity, drought, and improved nutritional quality is still a major bottleneck for producing transgenic crop varieties with desirable genetic properties. Also, there is a need to improve DNA vectors for genetic transformation of cereals, and develop high frequency plant regeneration systems applicable to a wider range of genotypes. It is expected that the insertion of a few or single (specific) genes into crop varieties would become an important component of genetic improvement.

The **RFLP** and **oligonucleotide polymorphism techniques** are becoming important tools for genetic analysis of traits at DNA level including quantitative trait loci. The linkage of RFLP markers with the agronomic traits would facilitate the marker–based selection for complex agronomic traits, which are largely influenced by the environments and are difficult to select in the conventional crop improvement program.

Insertion of **transposons** into new plant systems would result into the tagging of genes for which mRNA or gene products are not known. The availability of new cloned DNA fragment/genes would be increasingly useful in monitoring alien gene introgression into crop plants and to detect viral and bacterial pathogens infecting crop plants. Further, isolation of new genes and transgenic plants would accelerate our understanding of the mechanism (regulation) of gene expression. Hence, the success of the genetic engineering would depend upon the basic knowledge of the plant genome and the close interaction between the scientists engaged in solving fundamental problems of plant genetics and tissue culture and those aiming at developing improved germplasm with new genetic properties. The genetic engineering would be an important supplement to the conventional crop improvement programmes but not a substitute.

## Other Applications of Genetic Engineering

1. **Seed Proteins:** The genes encoding the major seed storage proteins are available, so in principle the alteration of their coding sequences so as to increase the levels of certain amino acids, followed by their reintroduction into the plant is feasible.
2. **Pest and Disease Resistance:** A plant expressing the insecticidal toxin of *Bacillus thuringiensis* has already been developed and the transfer of other anti metabolite protein genes is well underway.
3. **Herbicide and Pesticide Resistance:** Calgene have recently announced the transfer of a gene conferring resistance to the widely used herbicide glycophosphate (Tumble weed) from bacteria to plants.

4. **Nitrogen Availability:** The agriculture's nitrogen problems can be solved by the manipulation of soil and root bacteria.
5. **Photosynthesis:** Protein engineering may be applied to improve the efficiency of the major rate–limiting step, that is catalyzed by Ribulose–1, 5–bisphosphate carboxylase–oxygenase (**Rubisco**).
6. **Drought and Stress Resistance:** As little is known of the molecular basis of such phenomena, these remain long–term prospects.
7. **Modification of Biochemical Pathways:** Transfer of enzymes capable of alteration of undesirable compounds, e.g., tannins, may extend the range of species acceptable for consumption.
8. **Expression of *High–Value* Products:** The drift from bacteria to yeast and animal cells as expression factories in recent years was necessitated by the special subtleties of gene regulation and protein processing in higher organisms. Plant cells are capable of most of the processes as performed in animal cells, so they may be a viable alternative. However, except for the most expensive products, tissue culture is unlikely to be economic, but the plant in the field is a highly efficient bioreactor using free energy. Thus given the availability of the correct expression cassette, high value products could be produced in seeds, and encapsulated in protein bodies, which protect the merchandise, and minimize the problems of downstream processing.
9. **Ornamentals:** The possibilities for producing unusual ornamental species e.g., blue roses, or black tulips or even luminescent species, are quite staggering.
10. **Industrial Microbes:** Another goal in the genetic engineering is the improvement of genetic strains of industrial microbes. Certain substances like antibiotics and alkaloids are synthesized by pathways, which are controlled by a number of different enzymes. Thus, their production is not under the control of a single gene. It is difficult to manipulate a large number of genes encoding these various enzymes. But attempts have been made to modify the genome of existing strains.
11. **Petroleum Industry:** The *super bug* is a newly constructed bacterium, which can degrade most of the major hydrocarbon component of petroleum and also have the capacity to clean oil spills. The different strains of *Pseudomonas putida* contained a plasmid, which has genes coding for enzymes that digest a single family of hydrocarbons. The plasmid can digest octane, hexane and decane; XYL both xylene and toluene; CAM and camphor; and NAH naphthalene.
12. **Success in Pre–determining the Sex**: It has been possible for some time to diagnose the sex of embryos before birth, but not, until now, to pre–determine sex at the time of conception. It is based on a method to sort the sperms into two groups – those with an X–chromosome, which result in female calves, and those with a Y–chromosome that result in male calves. Sex is then pre–determined by fertilizing eggs in vitro (IVF), with the appropriate sperm. The resulting embryos are used to establish pregnancy. Sexed embryos will be distributed and transferred in future by veterinary surgeons, trained laboratory technicians and artificial insemination organizations
13. **Applications to agriculture:** Modern techniques for isolating and analyzing eukaryotic genes have raised hopes (and fears) that these genes can be used to transform plant and animal cells. *Transformation is the insertion of foreign genes into a cell.* It has been observed in bacterial systems for many

years. The *wish or desired list* of many geneticist whose objectives include the genetic manipulation of an organism for either functional dissection or engineering purposes include (i) Isolation of any gene or DNA sequence to insert; (ii) site–specific mutagenesis of any isolated DNA sequence; (iii) introduction of any gene or DNA sequence, either native or foreign, of interest into the desired host (integrated into random chromosomal sites, integrated into homologous chromosomal sites, not integrated, on autonomously replicating vectors, with control of copy number); (iv) targeted disruption of a gene or DNA sequence of interest; (v) direct retrieval of a mutant gene with its wild–type alleles or replacement of a wild–type gene with a specific mutant allele.

## PROTEIN ENGINEERING: TRENDS AND BASIC CONCEPTS

Recently, a question in the Cambridge University tripods examinations asked students to discuss an extension to the **central dogma**; that DNA makes RNA makes Protein makes Money. This variant dogma has been the driving–force behind the phenomenal growth in biotechnology where, to date its success is largely attributable to advances in obtaining expression of foreign proteins in novel host organisms. Biotechnology is growing on the claim that it can do things better, and protein engineering is now illustrating a further route to substantiating that claim. Catalysis and transduction are processes that are fundamental to life. The central role played by proteins in these two processes is increasingly falling under industrial scrutiny, both to harness the catalytic potential of enzymes and to exploit the specificity of protein–legend interactions when designing novel pharmaceutical and agrochemical products. Protein engineering has the potential to access numerous markets in industrial enzymes, food technology, the pharmaceutical and agrochemical industries and the developing biosensors field.

### Enzyme Characteristics as Target for Protein Engineering

Numerous properties of enzymes have been discussed as candidates for manipulation by protein engineering. Often, the modifications envisaged for engineered proteins segregate according to in vitro or in vivo applications. Thus, stability characteristics such as enhanced thermal stability, activity over a broad pH range, prolonged activity after immobilization onto various process substrata, resistance to proteases and broad range solubility characteristics are advantageous in industrial enzymology. Target features could be Michaelis constants ($K_m$), substrate specificity, susceptibility to product inhibition, regulation by allosteric factors, and cofactor requirements. Engineering to enhance catalytic activity towards a typical substrate is of considerable industrial relevance.

### Single Cell Protein (SCP) Production

Single cell protein (SCP) can be obtained from certain members of algae, cyanobacteria, yeasts, molds and higher fungi. These plant species are harvested and dried for use of protein source. It is estimated that worldwide food protein deficiency was about 12 million tones in 1985, and likely to increase to approximately 24 million tones by 2000. SCP has developed hope to solve this problem. The SCP is characterised by high percentage of protein by weight (sometimes upto 85%). For the production of SCP from plant sources, a variety of raw materials are required which serve as carbon source including solar radiation (for photoautotrophic plant species). The important are $CO_2$, $NaHCO_3$, sunlight (*Spirulina maxima*, protein = 60%), $CO_2$ and sunlight (*Scenedesmus acutus*, protein 55%), Bagasse (*Cellulomonas*, protein 85%), molasses (*Saccharomyces cerevisiae*, protein 50%). The SCP is being manufactured under different commercial names like *Brovile*, *Marmite*, *Pruteen*, etc. Pruteen is manufactured by Imperical Chemical Industries, U.K. from the bacterium *Mythylophilus methulotropous* that is grown on methyl derived from natural gas. The company produces about 50,000 tons of Protein every year as animal feed. Similarly Rank Hovis McDougall Company, U.K. produces SPC under the trade name *Mycoprotein* using potato, wheat, and cassava as carbohydrate source for protein production.

SCP is also produced from paper industry waste, because the industry faces a serious problem in disposing one of the waste products – sulphide liquor. If this pollutant is discharged in the river, it quickly depletes the water's oxygen reserves and causes death of fishes. The sulphide liquor is fed to *Paecilomyces* molds. This process is known as *Pelilo*. This process purifies the waste liquids from paper factories but also yields a rich harvest of microbes, which are sold as animal feed.

*Spirulina* (a cyabobacterium) offers a rich source of SCP. *S. platensis* contains about 60% crude protein and is rich in tryptophan, lysine and vitamin $B_{12}$. In India, NBRI, Lucknow and CFTRI, Mysore has started mass culture and utilization of this cyanobactrium as a source of SCP. *S. platensis* is cultured in sewage water containing 9 g/l of sodium bicarbonate and 1 g/l of sodium nitrate. The average yield of the microorganism varies between 6 g/m$^2$/day and 9 g/m$^2$/day on dry weight basis. Similarly A.M.M. Marugappa Chettiar Research Center, Madras has isolated *Spirulina fusiformis* and produced SCP after filtration from cloth filters (30 mesh/inch).

### Protein Engineering: Achievements

Without any doubt, the successes achieved by protein engineering in the rational modification of protein function have been remarkable both from the point of view of the diversity of systems already investigated and the unambiguous specificity of information generated. Zoller and Smith have summarized many of the early examples of site–directed mutagenesis for protein engineering; candidate proteins.

## TECHNIQUES OF GENETIC ENGINEERING

### Cleaving DNA with Restriction Enzymes

After locating gene of interest in colony or on chromosome, the DNA is extracted and cleaved with enzymes. Most restriction enzymes cleave DNA molecules in a site–specific manner. The genomic DNA from an individual organism is digested with either a single or more restriction enzymes. This results in fragmentation of DNA molecules with variation in length. These fragments are known as restriction fragment length polymorphisms or RFLPs. Since restriction enzymes cut DNA in a sequence specific manner, every homologous DNA molecule will be cut at exactly the same sites. This allows isolating large quantities of specific DNA fragments for sub cloning and isolating the DNA sequence of interest.

The sizes of the restriction fragments can be determined by polyacrylamide or agarose gel electrophoresis. It also helps in separation of various RFLPs. Here few electrophoretic and blotting techniques are described.

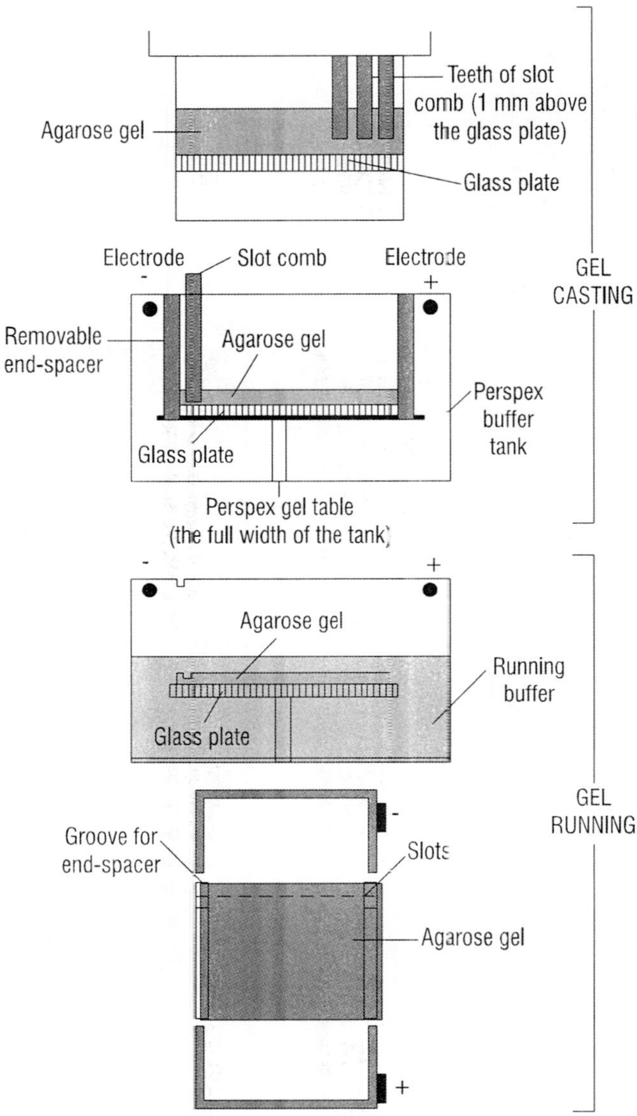


**FIGURE 3.1:** Agarose gel electrophoresis.

## Gel Electrophoresis

Nucleic acid molecules can be distinguished by differences in their chain length, and study of gene organization and expression includes nucleic acid hybridization with a chain length measurement.

DNA and RNA have a net negative charge. At neutral pH therefore, these molecules migrate towards the positive terminal when placed in an electric field. This process of electrophoresis is performed in an agarose or polyacrylamide gel. Nucleic acids are loaded into slots in the gel and allowed to migrate towards the positive terminal. The pores in the gel act to sieve the molecules. So that the mobility of a particular nucleic acid species depends on its length. All the molecules of a particular size move at approximately the same rate through the gel, forming a band, which gradually increases in width during electrophoresis because of diffusion.

Running DNA molecules of known size in a parallel track on the gel, and determining their migration position relative to the species of unknown size can estimate the size of the molecules.

Polyacrylamide has a smaller average pore size than agarose and so is effective at separating molecules of 10–1000 nucleotides in length (1000 nucleotides is termed 1 kilobase or 1 kb), with very high resolution. The larger pores in agarose allow it to resolve much bigger molecules, up to 100 kb in length. However, the separated molecules migrate as a wider band so that the resolution (i.e. the ability to separate molecules which differ only slightly in size) is lower with agarose than with polyacrylamide. In pulsed–field gel electrophoresis (PFGE) the electrical field across the gel is repeatedly switched back and forth between two or more directions. This allows very long DNA molecules, several million base pairs in length, to be separated, since very big molecules change direction less rapidly as they move through the agarose gel.

After gel electrophoresis, the nucleic acid molecules must be detected. They can be visualized directly by including ethidium bromide in the polyacrylamide or agarose gel. This dye binds to double–stranded nucleic acids by intercalating between adjacent base pairs, and emits a red fluorescence when exposed to UV radiation. As little as 5–10 ng of DNA can be detected. If the nucleic acid molecules are radioactively labelled, then individual resolved species could be detected by autoradiography.

## BLOTTING TECHNIQUES

### Analysis of DNA by Southern Blotting

In 1975, E. M. Southern invented a procedure to identify the locations of genes and other DNA sequences on restriction fragments separated by gel electrophoresis. The essential feature of this technique is the transfer of the DNA molecules separated by gel electrophoresis to a nitrocellulose or nylon membrane. The DNA is denatured either prior to or during transfer by placing the gel in an alkaline solution. After transfer is complete, the DNA is immobilized on the membrane by drying or UV–induced cross–linking to the filter. A radioactive DNA (*probe*) containing the sequence of interest is then hybridized or annealed with the immobilized DNA on the membrane. The probe will anneal to form a double helix only with complementary DNA molecules on the membrane.

Nonannealed probe is then washed off the membrane, and the washed membrane is exposed to X–ray film that detects the presence of the radioactivity in the bound probe. After the autoradiogram is developed, the dark bands show the position(s) of DNA sequences that have hybridized with the probe.

The ability to transfer DNA restriction fragments or other DNA molecules that have been separated by gel electrophoresis to nitrocellulose or nylon membranes for hybridization studies and other types of analyses has proven to be extremely useful. Such transfers of DNA to membranes are called Southern blots after E. M. Southern, who developed the technique.

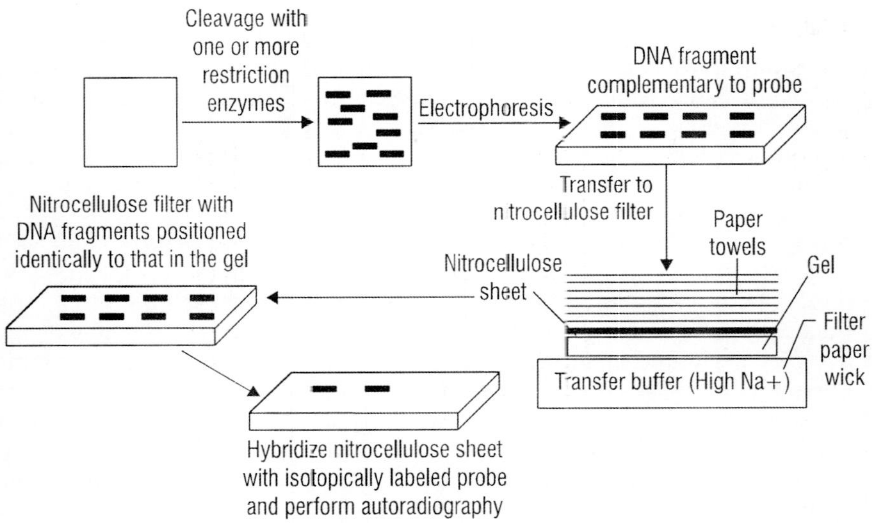


**FIGURE 3.2:** Southern blotting method of analyzing DNA segment that share homology with a nucleic acid probe.

## Analysis of RNAs by Northern Blot Hybridizations

As DNA molecules can be transferred from agarose gels to nitrocellulose or nylon membranes for hybridization studies, RNA molecules can also be separated by agarose gel electrophoresis similarly and transferred for analyses analyzed. Such RNA transfers are used routinely in molecular genetics laboratories and are called *northern blots*, as it is the mirror image of the Southern blotting technique.

The northern blot procedure is essentially identical to that used for Southern blot transfers however, degradation by RNA molecules are more sensitive to degradation by RNases. Thus, gloves must be worn at all times to prevent contamination of solutions with RNases on one's fingers. In addition, all glassware must be baked overnight to inactivate RNases, and all chemical reagents must be baked or be highly purified to ensure that they are RNase–free. RNase molecules are extremely stable enzymes; they are not inactivated by boiling or other treatments that would totally destroy most enzymes. Most RNA molecules contain considerable secondary structure. Therefore, they must be kept denatured during electrophoresis to separate them on the basis of size. Denaturation is accomplished by adding formaldehyde or some other chemical denaturant to the loading buffer and the electrophoresis running buffer. After transfer to an appropriate membrane, the RNA blot is hybridized to either RNA or DNA probes just as with Southern blots.

Northern blot hybridizations are extremely useful in studies of gene expression. They can be used to determine whether a particular gene is transcribed in all tissues of an organism or only certain tissues. They can also be used to study the temporal pattern of expression of individual genes during growth and development. Major disadvantage with northern blot hybridizations is that it only measures the accumulation of RNA transcripts. But explains nothing about why the observed accumulation occurs.

## Analysis of Proteins by Western Blot Techniques

Polyacrylamide gel electrophoresis has been used as an important tool for the separation and characterization of proteins. Since many functional proteins are composed of two or more subunits, individual polypeptides are separated by electrophoresis in the presence of the detergent sodium dodecyl sulfate (SDS), which denatures the proteins.

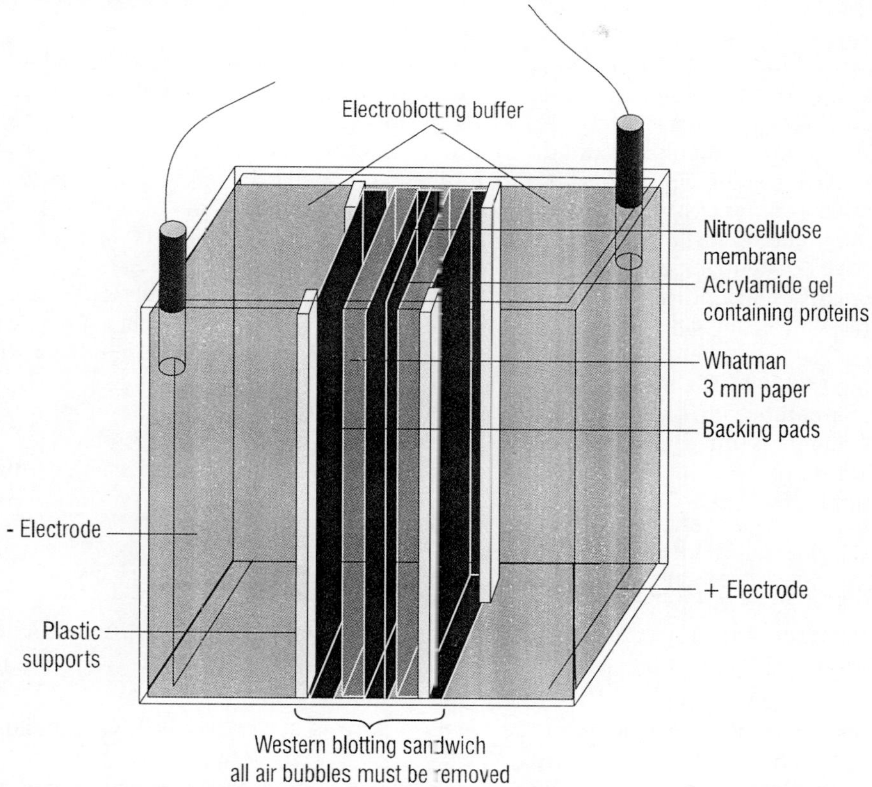


**FIGURE 3.3:** Diagram of a typical Western blotting or electro blotting apparatus. An electric current is used to transfer the proteins from the gel to the nitrocellulose membrane placed next to it in the blotting sandwich. All other components of the sandwich function to provide gentle but firm support; tight contact between the gel and the membrane is essential for good transfer. All bubbles must be carefully removed from the sandwich to assure uniform flow of current and complete transfer of all proteins.

After electrophoresis, the proteins are commonly detected by treating with Coomassie blue or silver stain. However, the separated polypeptides in the gel can also be transferred to a nitrocellulose membrane, and individual proteins can be detected by using specific antibodies. This transfer of proteins from acrylamide gels to nitrocellulose membranes is called western blotting and is performed by using an electric current–so–called **electroblotting**. After transfer, a specific protein of interest is identified by placing the membrane with the immobilized proteins in a solution containing an antibody to the protein. Nonbound antibodies are then washed off the membrane, and the presence of the initial (*primary*) antibody is detected by placing the membrane in a solution containing a *secondary* antibody. This *secondary* antibody reacts with immunoglobulins (the group of protein containing all antibodies) in general.

The secondary antibody is conjugated to either a radioactive isotope (permitting autoradiography) or an enzyme that produces a visible product when the proper substrate is added. The western blot procedure is a very powerful tool for identifying and characterizing specific gene–products.

**Dot blots and slot blots.** Another variety of Southern or Northern blots is dot blots or slot blots. In dot blots, spotting them adjacent to each other on a nitrocellulose membrane tests cloned or purified extract of DNAs. DNA blots thus produced are immobilized and denatured so that they are bound on the membrane as single stranded DNA blots. The membrane is then hybridized with radioactively labelled probe (DNA or RNA). The dot representing sequence related to probe would light up by autoradiography. The intensity of the dot will indicate the relative concentration of a sequence related to the probe in the DNA sample used for a particular dot. No steps involving digestion with enzyme, gel electrophoresis or transfer from gel to membrane is needed in this technique. Therefore, it is considered much more convenient, when no restriction sites are needed to be studied.

Sometimes, an apparatus is used for placing spots on the membrane, through slots made in this equipment. The spots made thus are in the form of oblong slots rather than round blots. These slots are used just like dot blots and are described as slot blots.

## DETECTION OF RFLPS

A RFLP can be demonstrated using the following steps:

- Extract and purify DNA from several individuals, which may differ in some respects among themselves.
- Digest DNA samples with a restriction endonuclease.
- Subject each DNA digest to gel electrophoresis separately, but on the same gel slab.

  The DNA fragments of different lengths resulting from digestion can be separated by gel electrophoresis, but the differences in the distribution patterns of fragments in several DNA samples due to digestion by several Endonucleases cannot be detected directly. This is because the number of fragments is large and the range in size is rather continuous, such that they form a continuous smear on the gel. The following steps therefore, follow electrophoresis.
- Fragments are transferred from gel to nitrocellulose filters using the technique of **Southern blotting**.
- The filters with small fragments, fully or partially homologous to the probe, can be detected by autoradiography after hybridization.

With a sufficient number of digests, it is possible to deduce the origin of each band on the gel, that is, which enzyme or combinations of enzymes have generated it. This yields a restriction enzyme cleavage map or restriction map. Obtaining such a map is very often essential first step in analyzing a piece of DNA of unknown structure. Another method is by labelling with probe followed by autoradiography. Since DNA clones having repetitive DNA may hybridize with many fragments, they will give a smear on autoradiography. Therefore, unique DNA sequences are generally used as probes for detecting RFLPs. The DNA fragments hybridizing with the probe, can be visualized by autoradiography, and will be called RFLPs at the phenotypic level. RFLPs can be studied in a set of related species, using a random or a specific DNA probe. The similarities and differences be used to inter–phylogenetic relationships. This has been done in a number of cases, both in plants and in animals. RFLPs have also been used to prepare chromosome maps in humans, mice, fruitfly, maize, tomato, lettuce and rice.

## DNA SEQUENCING

Restriction enzyme mapping and Southern blotting help establish the gross organization of a piece of DNA, and allow the approximate position of sequences of interest to be determined. The next step in analysis is normally to determine the nucleotide sequence.

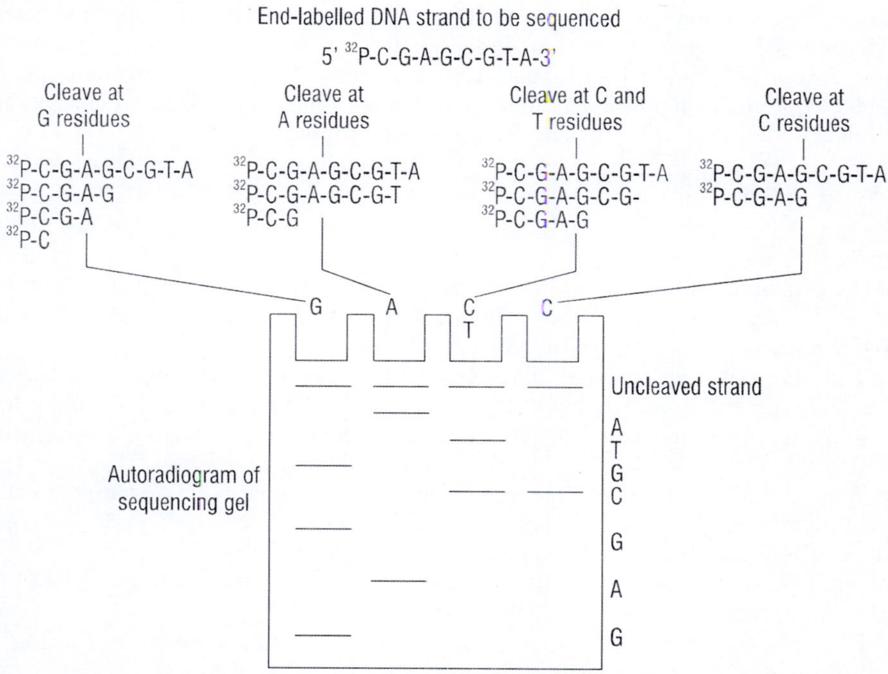


**FIGURE 3.4:** DNA sequence analysis by the chemical degradation (Maxam and Gilbert) technique.

### Methods for Analyzing DNA Sequence

Our present ability to sequence essentially any DNA molecule is the result of four major developments. (1) The most important breakthrough was the discovery of restriction enzymes and their use in preparing homogeneous samples of specific segments of chromosomes. (2) A second major advance was the improvement of gel electrophoresis procedures to the point where DNA fragments that different in length by a single nucleotide could be resolved. (3) The development of gene cloning techniques greatly facilitated the preparation of large quantities of a particular gene or DNA sequence of interest for use in sequencing studies. (4) Finally, two different approaches to the rapid sequencing of DNA molecules (i.e. by chemical degradation and by chain termination) were developed in the period from 1974 to 1977.

## Sequencing by Chemical Degradation

Determination of DNA sequence depends upon the resolution obtained from polyacrylamide gels, which are run in the presence of denaturant e.g. urea. If two single–stranded DNA molecules, identical in sequence over most of their length but differing in just one base on one of the two molecules, are subjected to electrophoresis they will be separated.

In the chemical degradation procedure a DNA restriction fragment is isotopically labelled with P at either its 5' or its 3' termini and each of the two strands is separated and isolated. The DNA is then partially modified with chemical reagents specific for the different bases, and cleaved at the modified nucleotides. This generates a set of molecules differing in length but with the same isotopically labelled terminus. These fragments are subjected to denaturing gel electrophoresis in parallel slots of high–resolution gel, and the sequence of the DNA can be deduced from the resulting autoradiogram.

## Sequencing by Chain Termination

In practice the chemical degradation procedure has been replaced by chain termination method. In this procedure DNA is synthesized rather than degraded, but the principle is similar to the chemical method. DNA polymerases function by adding a primer, a short synthetic oligonucleotide, to pre–existent DNA strand, to provide a fixed 5' terminus for the growing chain.

The reaction is performed in the presence of chain terminators. These nucleotide derivatives act as blocking group at their 3' hydroxyl terming and once there are incorporated into the growing chain, they prevent further addition of nucleotides. For example, dideoxy GTP, is mixed with unmodifies GTP and the other three nucleoside triphosphates terminates at a subset of G residues in the newly synthesized strand. Four parallel reactions are run, with dideoxy radiolabelled nucleoside triphosphates of all four kinds are included. The nucleotide sequence of the template strand can be deduced by comparison of the pattern of autoradiographic bands in the four lanes of the gel.

DNA sequence analysis using chain terminators is best performed using a single–stranded template. This allows the primer to be free accessible to the template. A suitable template for single–strand DNA sequence analysis can be created by inserting the double–stranded template DNA into the genome of a bacterial virus, which naturally packages only one of the two strands into viral particles. Another method is denaturing of double–stranded DNA template by treatment with alkali. An excess of primer is added before re–annealing occurs, so that primer will compete with the complementary strand for annealing to the template. This is a preferable method over insertion of DNA into viral vector.

The major disadvantage of double–stranded DNA sequence analysis is the presence of competing complementary strand. This leads to a higher level of sequencing artifacts like presence of more than one band. In such situation it may be impossible to decide which is the correct residue at that position. Artifacts of this kind are not restricted to double–strand DNA sequence analysis; some level of artifacts is a universal problem in sequence analysis. The DNA sample then re–sequenced using another primer, or by using the other strand (i.e. the strand which is complementary to that used in the original analysis) as template.

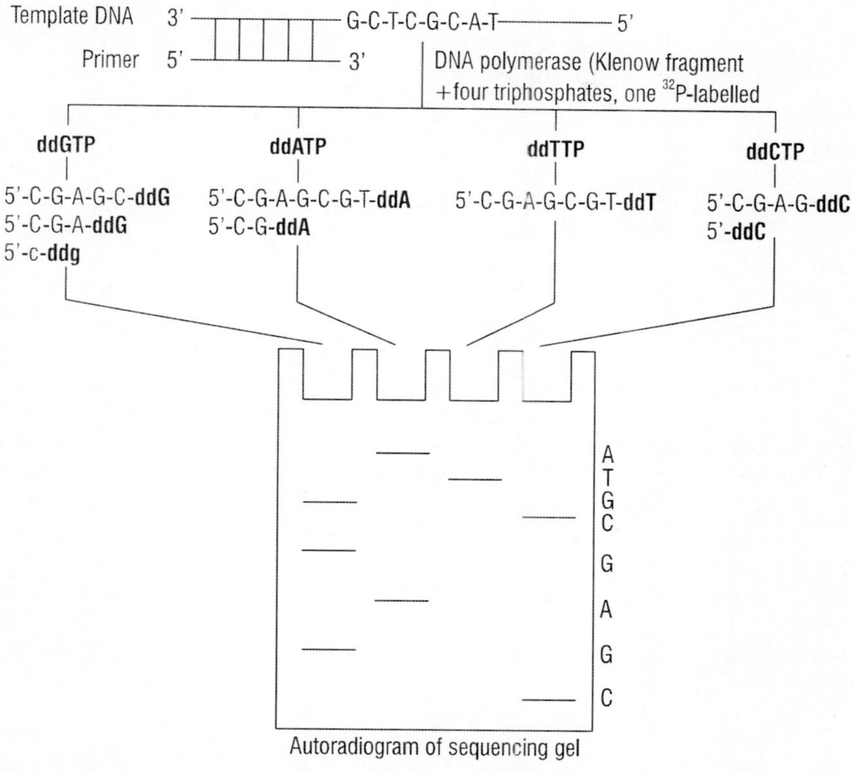


**FIGURE 3.5:** DNA sequence analysis by the chain termination (Sanger) method.

Even when a single–stranded DNA, molecules with more than 500 nucleotides, clearly it is not possible on a polyacrylamide gel to resolve when the molecules are larger than about the nucleotides. This sets the upper limit to the length of sequence to be determined. One way to extend from a known into an unknown region of sequence is to synthesize an oligonucleotide primer, which is complementary to a tract near the end of the region whose sequence has already been established. In this way it is possible to *walk* along a sequence, at each step synthesizing a new primer, which is used to extend the region of known sequence. Alternatively, it is possible to delete sequences enzymatically from the template at a position downstream of the annealing site of the primer, to generate a deletion series. The nucleotide sequence of a set of such molecules can be used to deduce the sequence of the intact template.

By determining the sequence of both strands of a DNA molecule, and confirming that the two deduced sequences are perfectly complementary, it is possible to detect and eliminate artifacts generated in the reactions or during electrophoresis. When this check is performed, DNA sequence can be established with a very high degree of accuracy.

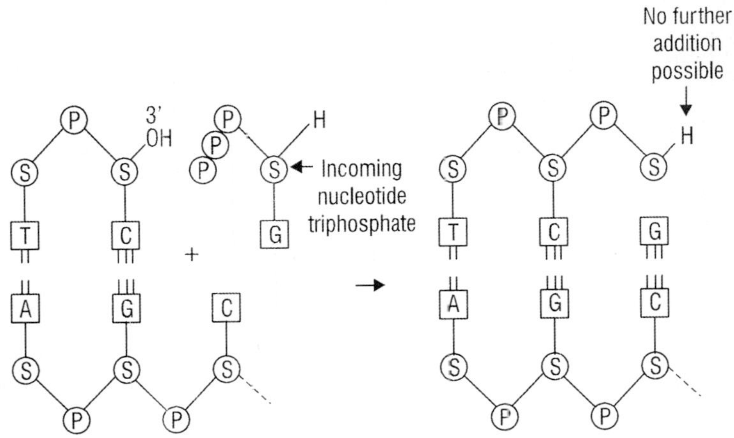


**FIGURE 3.6:** The chemical basis of sequence analysis using chain terminators. This is a representation of a reaction in which a dideoxy G residue incorporated into the growing chain blocks further incorporation.

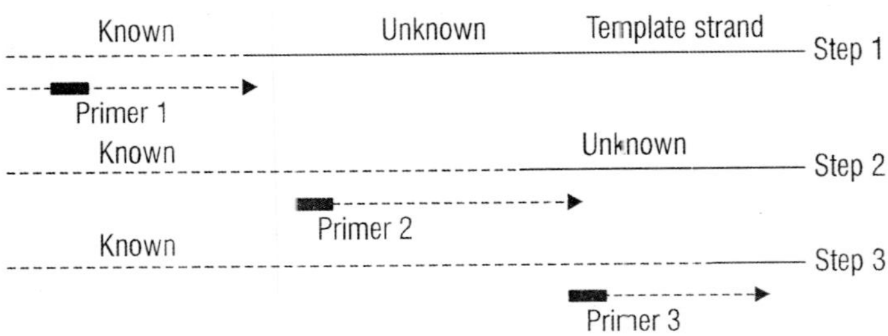


**FIGURE 3.7:** DNA sequence analysis by *primer walking*.

## Direct DNA Sequencing using PCR (also called Ligation Mediated PCR = LMPCR)

Polymerases chain reaction (PCR) has also been used for sequencing the amplified DNA product. This method of DNA sequencing is faster and more reliable and can utilize either the whole genomic DNA or cloned fragments for sequencing a particular DNA segment. The DNA sequencing using PCR involves two steps: (i) generation of sequencing templates (double stranded or single stranded) using PCR and (ii) sequencing of PCR products eight with the thermolabile DNA polymerase or with the thermostable Taq DNA polymerase. Thus the DNA sequencing method using PCR eliminates the need of cloning the DNA in single stranded DNA phage vector i.e. M13.

Although double stranded DNA product of PCR can be utilized for sequencing, this may cause difficulty due to reassociation after denaturation thus preventing the sequencing primer from annealing to its complementary sequence to allow extension. To reduce this problem, either a variant of the standard method for sequencing double stranded DNA is emplcyed or single stranded DNA templates are produced following **asymmetric PCR.**

| - Genomic DNA<br>- Cloned DNA → | DNA amplification (PCR) → | Pipetting robots → | Automated analysis of sequencing reactions |
|---|---|---|---|
| 1. Mutation screening for known gens | 1. Generation of sequencing templats from genomic DNA using PCR | 1. Annealing of sequencing primer | 1. Electrophoretic separation of sequencing reaction |
| 2. Isolation of DNA fragments by cloning | 2. Generation of sequencing templats from Cloned fragments using PCR and primers to vector sites flanking the insert | 2. Labelling reaction<br>3. Termination reaction<br>4. Stop | 2. Computer registration of seqencing data |

**FIGURE 3.8:** Use of PCR in automation of DNA sequencing.

A number of thermolabile DNA polymerases have been used for sequencing of *in vitro* amplified DNA. Alternatively, **Taq polymerase** (thermostable enzyme) used for PCR, can also be used for sequencing reaction. In either case, Sanger's synthetic method involving incorporation of dideozynucleotide (ddNTP) for chain termination is used for sequencing. However, Maxam and Gilbert's chemical degradation method can also be used.

In Sanger's method, as usual, four mixtures ar prepared, each using one of the four ddNTP. The sequencing primer is labelled with $^{32}P$ and the mixtures with amplified DNA, Taq polymerase and appropriate buffer are incubated at 70°C for 5 min. The reaction is stopped by addition of formamide stop solution in each tube and mixtures are run on polyacrylamide sequencing gel to obtain the ladders, which can be read by computer or manually. This method using PCR helped in automation of DNA sequencing.

## Automated DNA Sequence Analysis

Using the chain termination method, DNA sequencing can be mechanized to make the process rapid. In a conventional sequence reaction described above four separate single–strand extension reactions are performed, with each containing a different chain–terminating dideoxynucleoside triphosphate.

The radiolabelled reaction products are separated on four separate tracks of a polyacrylamide gel. In machine–based sequencing a single reaction is run, with each of the four dideoxynucleoside triphosphates containing a different fluorescent label.

When excited by an argon laser, the fluorescent labels emit light at a different wavelength. The same chain–termination reactions are carried out; however, it is no longer necessary to run the products in separate lanes of a gel. The reaction products are mixed and run in a single lane in a gel and as the different length fragments pass the laser each fluorescent label gives a unique signal, which is stored in the computer.

Machine–based DNA sequencing is beginning to change the speed and scope of sequencing projects dramatically. The first complete genomic sequences were those of simple viruses such as the SV40 tumor virus, containing around 500 based of DNA. Recently, the DNA sequence of over 106 000 bases of DNA from human chromosome 19 was determined in a few months using DNA sequencing machines. The sequence of a significant fraction of the *E.coli* genome has been established, the entire sequence of yeast chromosome 3 is known, and there are plans eventually to determine the sequence of the entire human genome.

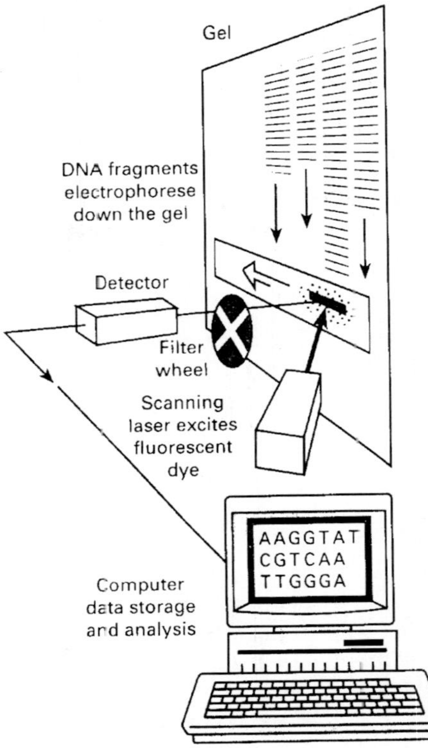


**FIGURE 3.9:** Automated DNA sequence analysis. The chain terminator method of DNA sequencing has been modified to allow DNA sequencing to be automated. Fluorescent dyes have been developed and these are attached to each of the dideoxynucleotides, which replaces the need for radioactive nucleotides. After the sequencing reactions have been completed the products of all four reactions are pooled and electrophoresed in one lane of an acrylamide sequencing gel. As the reaction products pass through the gel the specific DNA bases are detected by a scanning laser that excites each fluorescent dye at different wavelengths, this information is captured on a detector and passed to a computer for data storage and analysis.

## CHEMICAL SYNTHESIS OF DNA

H. Gobind Khorana et al (1972), for the first time synthesized the structural gene for an alanine tRNA from yeast. Since then, chemical synthesis of DNA has found wide applications in genetic engineering as: (1) Partial or total gene synthesis, (2) Primers for DNA and RNA sequencing, (3) Hybridization probes for the screening of RNA, DNA and cDNA genomic libraries. (4) Site directed mutagenesis. (5) Synthesis of Adapters. (6) RNA/DNA structure determination.

### Chemical synthesis of tRNA Genes

The structure of gene could, therefore, be inferred only from its product. For instance, if a gene is responsible for giving rise to a polypeptide chain and the structure of this chain is known, then from the genetic code dictionary, structure of the gene could be easily inferred. Such genes were initially considered to be too long to be synthesized because an average gene contains about 1,500 base pairs. On the other hand, since tRNA molecules are fairly small in size (about 80 nucleotides), a gene responsible for giving rise to a tRNA molecule was within the reach for synthesis.

## INSERT IN DNA SEQUENCING

### Synthesis of Gene for Yeast Alanyl tRNA

The first tRNA whose detailed structure was discovered was yeast alanyl tRNA. The structure of gene yeast alanyl tRNA would obviously be such that one of the two strands of DNA (gene) would be complementary to base sequence of yeast alanyl tRNA. The other strand would then automatically have the same sequence as in tRNA except that in place of uracil of tRNA there would be thymine in DNA.

As synthesis of such a long chain (77 base pairs) of double stranded DNA was rather difficult small oligo deoxyribonucleotides ranging in length from 5 to 20 nucleotides synthesized. Such oligonucleotides were then joined to form double stranded DNA, 77 nucleotide pairs long.

### Synthesis of Gene for a True Precursor tRNA

Before Khorana could complete the synthesis of gene for yeast alanyl tRNA in 1970, as outlined above, it became obvious that tRNA was not the direct product of transcription. Instead a precursor molecule is first synthesized which subsequently, after losing segments of RNA by cleavage, gives rise to tRNA. Obviously therefore, the actual gene for yeast alanyl tRNA will be longer than the DNA duplex synthesized by Khorana. In view of this, Khorana subsequently initiated synthesis of a gene for *E.coli* tyrosine suppressor tRNA precursor. DNA duplex, which will give rise to this tRNA precursor, was synthesized in the form of 26 small oligonucleotide segments. Later, in 1979, Khorana reported completion of the total synthesis of a biologically functional tyrosine suppressor transfer RNA gene carrying all regulatory sequences. This gene was 207 base pairs long.

### Total Synthesis of a Human Leukocyte Interferon Gene

Interferon is a protein, which has a property of inhibiting viral infection and cell proliferation and thus can be used for treatment of viral infection and cancer. These properties have generated great interest in human interferons (IFNs). Atleast three kinds of interferon genes are known in human genome: (i) leukocyte interferon genes (IFN–Â genes), (ii) fibroblast interferon gene (IFN–ß genes), and (iii) immune interferon gene (IFN–Ó genes). Genes for different kinds of interferons differ and details of base sequences are not known for all of them.

Total synthesis of this human leukocyte interferon gene, 514 base pairs long, was achieved and published in 1981. Since there are 514 base pairs in the gene, it involved synthesis of DNA strands with 1028 nucleotides in predetermined way. This was the longest gene synthesized till 1982. The synthesis involved coupling of initial nucleotide onto a polyacrylamide resin to which further nucleotides could be added in pairs. There are no introns sequences present in this gene (introns, common in eukaryotic genes, that are known as split genes).

The plasmid vectors with this newly synthesized gene have been transferred into bacterial host for gene cloning. (*E.coli*) This is being utilized now for synthesis of interferon at industry leveled has greatly reduced the cost of production of this drug (interferon).

## GENE SYNTHESIS MACHINES

The genes can now be synthesized rapidly and in high yields with the help of **Gene Synthesis Machines** (developed in 1980s). The availability of gene synthesis machines become possible, due to (i) development of silica based supports, which are insoluble and provide support for solid phase synthesis of DNA chains and (ii) development of stable protonated deoxyribonucleotides phosphoramidites as synthons, which are remarkably stable towards oxidation and hydrolysis and thus are ideal for DNA synthesis. These two procedures can be used to prepare DNA molecules more than 100 nucleotides long within a few days.

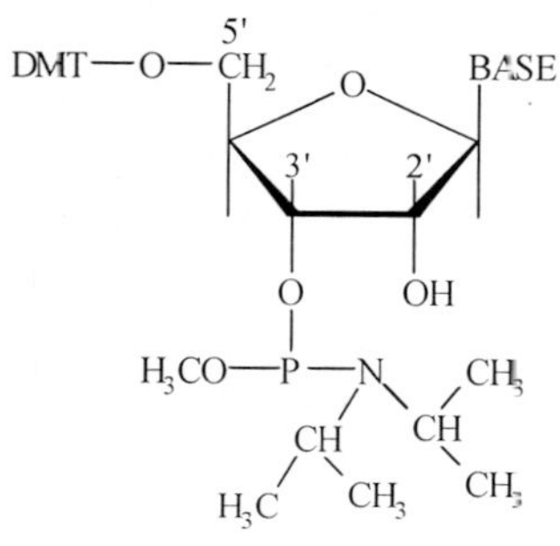


**FIGURE 3.10:** A protonated phosphoramidite (5' hydroxyl is blocked by DMT=dimethoxytrityl group).

Several versions of *gene machines* are now available. These machines, under the control of a microprocessor, synthesize specific short sequences of single stranded DNA automatically. The desired sequence is entered on a keyboard and the microprocessor automatically opens the valves of the containers of successive nucleotides, reagents and solvents needed at each step, into a synthesizer column, which is packed with tiny silica beads. These beads provide support on which DNA molecules are assembled.

## OTHER TECHNIQUES OF GENETIC ENGINEERING

It involves the insertion of DNA into the cell and the incorporation of the DNA into a replicating chromosome. The preparations of pure DNA can be injected into cells with micropipettes or inserted using other methods such as electroporation, large scale, efficient incorporation of new DNA into chromosomes is most often performed using a vector DNA, analogous to the plasmid and viral DNAs used to transform many bacteria.

Following are the three main vectors that are now being investigated:

## Transposable Elements in Prokaryotes

### *Structure*

Transposable elements in prokaryotes were discovered when a series of polar mutations in an *E coli* operons were found to represent the insertions of specific DNA sequences 800–1400 base pairs in length. By heteroduplex analysis and DNA–DNA hybridizations, it was found that copies of a small number of these specific sequences (called **insertion sequences** or **ISs**) are present at numerous sites on the *E coli* chromosome as well as on many plasmids and phage genomes. ISs are transposed as discrete elements and are integrated at new sites by mechanisms that are independent of DNA sequence homology.

Soon after the discovery of ISs, it was found that many of the R determinants of R factors are similarly transposable; in every such case, the R determinant is flanked by repeated sequences. In some cases, these repeated sequences are known ISs; 2 copies of ISI, for example, flank the R determinant for kanamycin resistance. Such elements, consisting of repeated sequences flanking a gene coding for an unrelated function, are called **transposons**.

All ISs and transposons terminate in **inverted repeats** 20–40 base pairs long. These repeats serve as recognition sites for enzymes that mediate a variety of reactions, including not only replication and transposition but also the deletion of DNA segments adjacent to the transposable element. When such an element is transposed from one replicon to another within the same cell, replicon fusion may occur as an intermediate step. Replicon fusion can also occur by normal recombination between replicons carrying the identical element, the 2 copies of the element providing sequence homology.

Transposons code for one or more proteins, including an enzyme (**transposase**) that catalyses their own transposition. The transposon Tn3, for example, encodes for β–lactamase (conferring penicillin resistance), transposase, and a repressor of the transposase. Probably all transposable elements, including ISs, code for their own transposases. If that is the case, the distinction between ISs and transposons becomes a minor one, a transposon being an insertion sequence that carries one or more additional genes coding for functions unrelated to the transposition process itself.

### *Transposition*

Transposition takes place by a process that combines recombination and replication: At the completion of the process, the original transposable element remains in position, and a new copy has been inserted within a **target site** elsewhere in the genome. The new copy may appear within the same replicon or in a different replicon; in the latter case, cointegrate structure– representing the fusion of the 2 replicons– may occur as an intermediate stage in the transposition process. In all cases, the process results in the generation of short (5–11 base pairs long) direct repeat of target DNA on each side of the transposed sequence. Transposable elements show a wide range of specificity for target sites, ranging from elements having no detectable specificity to those having only a single integration site.

**Deletion:** Deletions of DNA segments immediately adjacent to a transposable element occur at rates that are 100 to 1000 fold higher than the spontaneous background for the remainder of the genome. The deletion mechanism appears to be closely related to that of transposition.

**Excision:** When a transposable element inserts within a functional gene, that gene is inactivated for normal expression. Such events appear as loss mutations such as auxotrophy. This elimination is not associated with further transposition but rather appears to be a specific deletion event.

## *Transposable Elements in Eukaryotes*

**Yeast:** A transposable element called Ty1 has been discovered in *Saccharomyces cerevisiae*. Ty1 is a 5.6 kilobase sequence that is present in the yeast genome in about 35 copies. It carries a repeated sequence 338 base pairs long at each end; this sequence, in turn, is present in the genome in about 100 copies. Ty1 generates 5 base pair repeats of target DNA and can inactivate functional genes by insertion.

*S cerevisiae* has a locus controlling mating type, depending on which allele–a or α– is presently. A given strain may switch between the a and α– phenotypes, but this change does not reflect mutation. Rather, the a and α– genes are separately and silently stored at other specific loci; a copy of the a or α– gene is occasionally transposed into the mating–type locus, displacing the pre–existing allele and becoming expressed. The a and genes thus belong to a special class of transposable elements.

**Mammalian Tumour Viruses:** A number of integrated tumour viruses have been found to be flanked by 0.6 kilobase direct repeats. Furthermore, sarcoma, leukemia, and mammary tumour viruses of mice generate 5 to 6 base–pair repeats in their target DNA. These viruses thus share fundamental molecular properties with the transposable elements.

## *Genetic Roles of Transposable Elements*

**Gene Expression:** Transposable elements contain transcriptional promoters, transcriptional termination signals, and nonsense codons. Their insertion adjacent to (or within) a functional gene can thus lead to the switching on or switching of that gene's expression.

**Structural Properties of Transposons:** The isolated plant transposons have the length of 1.4 to 1.7 kb. They carry the same sequences at each end but in different orientation and create characteristic short– sequence duplications upon insertion at a target site. Until recently, most plant transposons isolated belong to two main groups with respect to the sequences of their inverted terminal repeats and the size of target site duplications.

- **First Group:** The Ac/Ds system of maize and the Tam 3 element of *Antirrhinum majus* belong to the first group. These elements have similar inverted termini (7 out of 11 nucleotides are identical) and create duplication of 8 nucleotides upon insertion.
- **Second Group:** The second group comprises the En (Spm) system of maize, the Tam 1 and Tam 2 elements of *Antirrhinum majus* and Tgm 1 sequence of soybean. These elements also have similar inverted termini (13–15 nucleotides) and generate a duplication of three nucleotides upon insertion.

**Transposon Foot Prints:** Unlike bacterial transposable element, the short sequences duplicated at the insertion site of the transposon are not removed after excision of the element. The number of nucleotides left behind after excision can be different from the number of nucleotides duplicated upon insertion of the element. Excision of a transposon from a coding gene region can restore gene activity only if the number of nucleotides left behind is three or multiple three of. Therefore the proteins produced after the excision of a transposon might have slightly different properties than the protein produced from the wild type progenitor allele. The plant transposons not only mutate a gene by insertion, but also create a new type of mutation when they leave their position. Such putative transposons are called **footprints.**

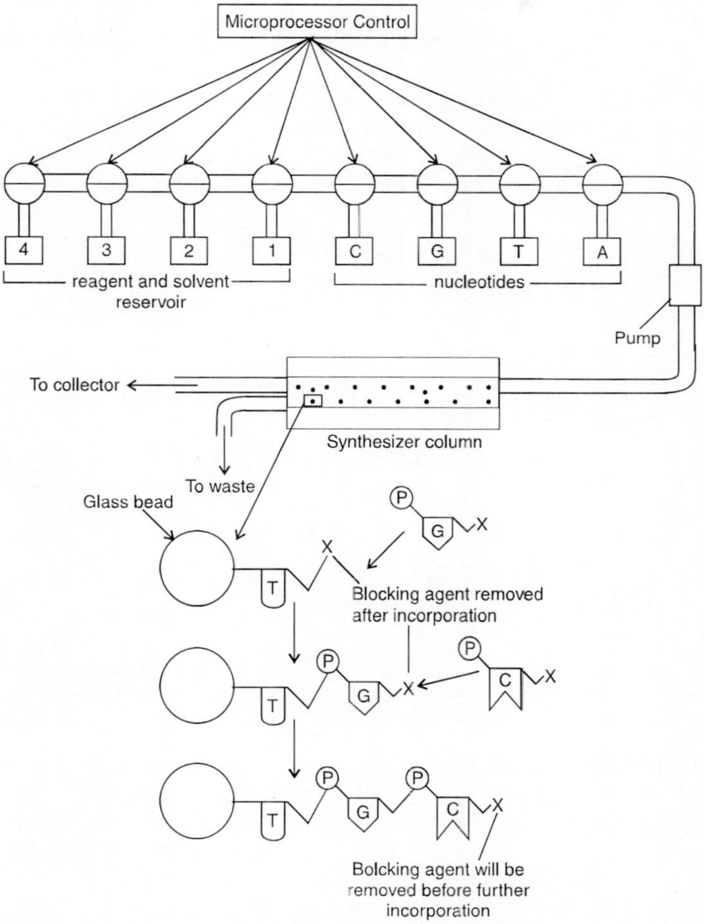


**FIGURE 3.11:** Presentation of the working mechanism of gene machine

**Transposon Tagging:** A transposable element must first be cloned and characterised at the molecular level before it can be used in gene tagging experiment. The first step in transposon tagging is to isolate an interesting mutation, which is the result of the insertion of transposon within a gene. Two following genetic strategies have been used to isolate gene mutations:

i) **Targeted Tagging:** The first is directed towards isolating one specific gene and is often referred to as targeted tagging. This is achieved by crossing a plant containing transposons with one, which is homozygous for a recessive mutant phenotype.

ii) **Non–Targeted Tagging:** The second strategy to obtain transposon–insertion mutations (non–targeted tagging) requires an extra generation. An attempt is made to identify and recover new mutations for a variety of different plant characteristics. This is the more general approach and more likely to recover genes of interest in future studies of this type. It involves isolating plants, which have inherited a germinal transposon excision event.

The major steps in isolating genes by transposon tagging are as follow:

i) **Step I –** Insertion of a transposon within a gene. This can induce a mutation, which may be detected phenotypically.
ii) **Step II –** DNA fragment containing the transposon and gene sequences can then be isolated using the transposon sequences as a probe.
iii) **Step III –** Sequences flanking the transposon can be used to screen a library made from non–mutant plants (e.g. to isolate the intact gene).

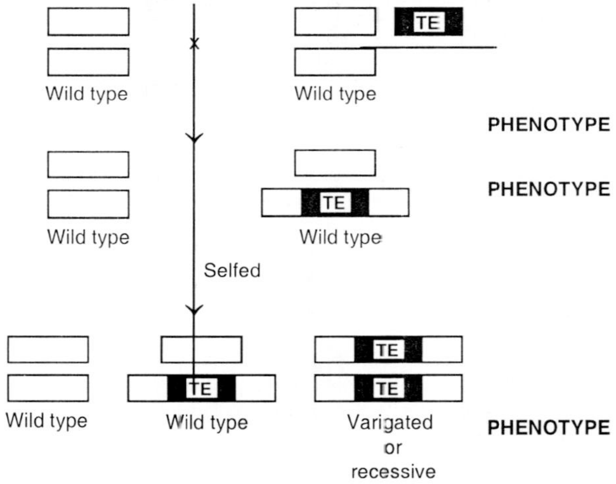


**FIGURE 3.12:** Non–targeted approach

Once the homozygous mutant is recovered, a genomic library can be created and screened with a radioactive labeled probe of the transposable element sequences. By isolating clones with homology to the transposable element probe, adjacent sequences from the gene of interest will also been cloned since the mutation is generated by the integration of the element within the gene sequence. The region that flank the element can then be sub–cloned and used as probes to screen a genomic library from normal, non–mutant plants to isolate wild type gene.

**Importance of Transposon Tagging:** The technique is now being adopted in wide range of plant species to isolate genes involved in developmental and metabolic processes, which have been very difficult to approach, by traditional biochemical methods. Genes controlling resistance to several plant diseases, floral initiation and development, as well as synthesis and detection of hormonal signals are among those being targeted with transposons. The technique is also used to isolate nuclear photosynthetic genes in corn. Successful tagging of a gene involved in thylakoid membrane organization has already been reported

(Martiennsen et al. 1989).

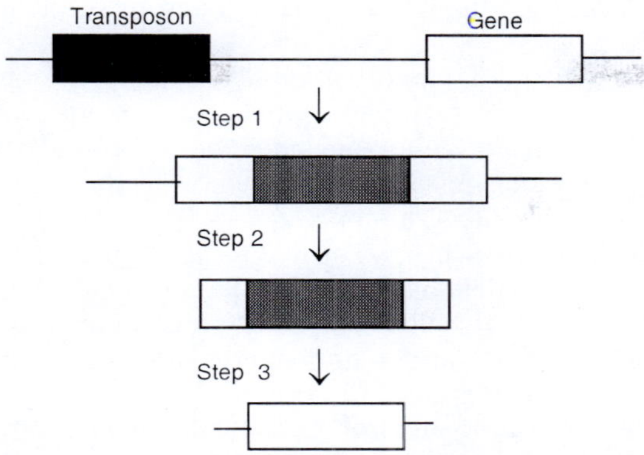


**FIGURE 3.13:** The major steps in isolating gene by transposon tagging.

## AGROINFECTION

Agroinfection a term coined by Grimsley et al. (1986) and subsequently also referred to as agroinoculation (Elmer et al. 1988), can be broadly defined as the introduction of plant infectious agents to plants via *Agrobacterium* (Grimsley and Bisaro 1987).

Using this technique, gene transfer to the cereal, *Zea mays* was demonstrated. The germinivirus maize streak virus (MSV) was chosen to test this approach. Neither isolated viral particles nor isolated native or cloned DNA were ever shown to yield symptoms. The only possible way of viral transmission was, via the insect vector *Cicadulinamblia*, a leafhopper indigenous to Asian and African countries. Strains of *Agrobacterium tumefaciens* were constructed carrying dimers of MSV DNA in their T–DNA and when these were inoculated in the stem of young maize plants, symptoms developed. Inoculation of the leaves with the same bacterial strains did not incite a viral infection.

This was an important discovery because it showed for the first time that *Agrobacterium* could transfer DNA to cereal cells. It was even demonstrated that the efficiency of T–DNA transfer probably does not differ from the frequency of gene transfer in a dicotyledon showing wound response. But this method is not different from normal Agrobacterium mediated transformation in cereals. In cereals, agroinfection leads to the transfer of the virus–carrying T–DNA with the wound–adjacent cell layers. The virus is subsequently released, replicates and spreads systematically.

### Biolistic Plant Transformation

Macro projectiles can be defined as any macroscopic object that can be accelerated to high velocity, which can thereby accelerates the desired micro projectiles placed on its surface. A biolistic target can be defined as any organelle, cell or tissue into which a substance is to be delivered via high–velocity micro projectile. Indeed the biolistic approach has advantages and potential for general applicability. Micro projectile bombardment employs high velocity metal particles to deliver biologically active DNA into plant cells. Devices have been described and tested with promising results. Typically macro projectiles take the shape of a bullet–like plastic cylinder or a disk–like metallised film. The method is based on the principle that high velocity bombardment of cells with micro projectile, made of tungsten or gold (in the range of 1 to a few microns in diameter) onto which DNA is precipitated.

Merits of Biolistic Plant Transformation: The ability to deliver foreign DNA into regenerable cells, tissues or organs appears to provide the best method, at present for achieving truly genotype–independent transformation in many agronomic crops, by passing Agrobacterium host–specificity and tissue culture–related regeneration difficulties. The technique is (i) it is easy to handle; (ii) one shot can lead to multiple hits; (iii) cell survives the intrusion of one particle; (iv) the genes coated on the particle have biological activity; (v) target cells can be as different as pollen, cell culture cells, cells in differentiated tissues and meristems; (vi) they can be located at the surface or deeper layers of organs; (vii) the method depends on physical parameters.

**Biolistic Plant Transformation** has been successfully been done in onionskin cells, corn, rice, wheat and soybean. There is no apparent difference in the efficiency of biolistic transformation of monocotyledons vs. dicotyledons. The most important target tissues have proven to be biolistically transformable, including embryogenic cell suspension, meristems, embryos and pollen. Organelles have also been biolistically transformed as shown in mitochondria of yeast and in chloroplast of Chlamydomonas. The recent evidence indicates that tobacco chloroplasts are biolistically transformable.

The basic principle of particle bombardment for gene transfer is to use high velocity micro projectile to incorporate the introduced genes. The outer cell layers are penetrated in order to introduce genetic material into living cells. The micro projectiles are commonly DNA coated, gold or tungsten particles and are accelerated to a very high initial velocity (1,400 ft. per sec). These micro projectiles, normally 1–3 $\mu$m in diameter are carried by a 'macro projectile' or the 'bullet' and accelerated into the living target cells. This technique is variously named as particle acceleration method, the bioblaster method, and micro projectile bombardment the particle gun method, the micro projectile method, and the gene gun method. As the DNA is being 'shot' into cells, it represents a type of biological ballistics; hence the term 'biolistics' is given.

The purposes of gene transfer via micro projectiles includes: efficient inoculation with infectious nucleic acids; studies of gene regulation based on transient expression of introduced DNA in target cells, cell lineage analysis using chimeric transformants expressing visual marker genes, reproducible transformation of cellular organelles and, regeneration of transgenic organisms expressing useful genes following selection for stably transformed target cells.

The biolistic process currently in widest use employs a gunpowder charge to drive a plastic cylindrical macro projectile. Numerous laboratories have built variations of this system like using pressurizes nitrogen to drive the macro projectile. Low–mass cylindrical macro projectile has also been used, driven by compressed air. A macro projectile accelerator has been built with a thin disk that is accelerated by the hot gases (produced by a high voltage discharge).

Different biolistic acceleration mechanisms are available today like gas entrainment, transferred impulse, macro projectile acceleration, centripetal acceleration and electrostatic acceleration. The acceleration by macro projectile has proved wide following utility.

- Acceleration by compressed air.
- Acceleration by transferred impulse.
- Acceleration by macro projectile.
- Acceleration by centripetal force.
- Acceleration by an electric field.

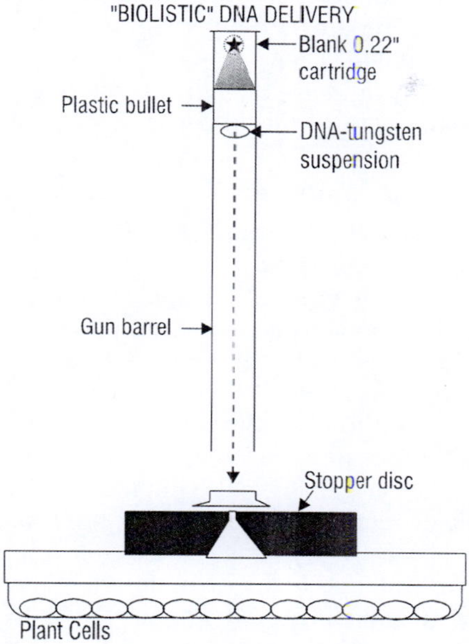


**FIGURE 3.14:** The gene gun.

Following points are important while planning for transgenic plant production using micro projectile techniques.

- Penetrable target tissues should be capable of plant regeneration.
- Bombardment and culture conditions should maximize the recovery of stably transformed cells.
- Tissue culture duration for target preparation, selection and regeneration should be minimized to reduce undesirable somaclonal variation.
- Effective selectable markers, reporter genes and transient expression frequency should be stressed for efficient recovery of stably transformed cells.

Several other factor might affect the efficiency of foreign gene delivery by this technique, namely, the physiological state and morphological character of the cells (cell wall thickness, size of cell aggregate, etc.); the DNA–carrying ability of the tungsten particles; the effective damage and; the degree of expression of the foreign gene in the bombarded cells.

DNA has been delivered and expressed by particle bombardment method in a number of plants including onion, corn, immature zygotic embryo of rice, wheat and suspension cultures of cotton and soybean and even in root section of Arabidopsis. No apparent difference appears in the efficiency of biolistic transformation of monocotyledons versus dicotyledons.

The high transformation efficiency could be related to the following:

- Only some of the cells are in a **competent state** with a high capacity to express the gene transferred into them. Competence here is viewed as the level of metabolic activity or the cell cycle stage and not as a function of some specialized cells. This competence state could, for instance, be reached in cells with high DNA repair activity. The tungsten particles used for bombardment may induce DNA damage, thus, possibly induce repair mechanisms, i.e. possibly increase the number of competent cells.
- This may be a consequence of introducing a large number of DNA molecules into each cell.
- Particle bombardment may introduce DNA directly into the nucleus.

There are several fundamental advantages of this process over other plant transformation techniques. The biolistic process appears to be effective regardless of species or tissue type; it is rapid and very simple procedure; it should facilitate the direct transformation of totipotent tissues such as pollen, embryo, meristems and morphogenic cell culture, and it appears to be uniquely suitable for organelle transformation. Since intact cells can be used, many difficulties encountered with the use of protoplasts are circumvented. Moreover, thousands of particles are accelerated at the same time, causing multiple hits resulting in transfer of genes into many cells simultaneously.

There are some disadvantages to the biolistic process as it requires special equipment and is still in the early stages of its development. Consequently, delivery efficiencies are still not high. In addition, potential users should be prepared to spend some time adapting existing protocols to their specific species or tissue of interest.

Moreover, the exact transformation frequency by this technique (transformed/survivor) is extremely difficult to calculate due to: (1) uneven distribution of the particle over the plated cells. (2) Inability to obtain monolayers of plated cells (tissue composed of cell aggregates). (3) The difficulties encountered when counting the actual number of cells in the tissue before bombardment.

### *Limitations and Problems Biolistic Plant Transformation*

**Physical Factors:** There are presently undefined biolistic variables, which can cause significant fluctuations in efficiency from day to day and from bombardment to bombardment. These fluctuations are frustrating and require replications of all treatments within an experiment, and often experiments need to be repeated. The particles presently used, as micro projectiles are irregular in shape and size therefore, only a fraction of these particles are likely to be effective. The majority of particles are either too small (failing to penetrate) or too large (causing cell death).

**Biological Factors:** These factors influencing micro projectile penetration into cells and subsequent cellular survival are not well understood and better understanding of these biological variables affecting penetration and survival should lead to major improvements in transformation efficiency. Efficiencies might also be enhanced by co–delivery of protein with the DNA, which might enhance transport to the nucleus or active integration into chromosomes. More attention needs to be paid to the biology of explants prior to, and following bombardments.

## Microinjection–Mediated Gene Delivery

Microinjection uses micro capillaries and microscopic devices to deliver DNA into defined cells in such a way that the injected cells survives and can proliferate. Early use of this approach was made in studies of plant viruses but the widespread work is in progress in plant science today owes more to the technology and success of the approach in animal engineering with recombinant genes. Microinjection helps in stable gene transfer to cells in culture for integration, expression and inheritance of marker gene introduced into a plant e.g., *Brassica napus*. In this plant target was a small somatic embryo derived from microspore culture.

The **advantages** of this technique are (i) the quantity of DNA delivered can be optimized; (ii) the experimenter can decide into which cell to deliver DNA; (iii) delivery is precise and predictable even into the cell nucleus and is under visual control; (iv) cells of small structures can be precisely targeted; (v) defined micro injected cell can be micro cultured; (vi) it offers an approach to transformation open for every species and variety with sexual propagation. The **disadvantage** is that only one receives DNA per injection and handling requires more skill and instrumentation.

## Micro Laser

A micro laser beam focused into the light path of microscope can be used to burn holes into cell walls and membranes. This helps in incubating the perforated cells in DNA solutions and can serve as a basis for vector–independent gene transfer into walled cells (Weber et al. 1988). It also helps to overcome the barrier imposed by plant cell walls. The laser can be focused to dimensions of less than 1 μm. By focusing the beams inside of a cell, manipulation on the nucleus or organelles can be performed without opening the cell membrane. In order to introduce DNA into a cell through a hole in the wall and plasma membrane, a differential pressure is generated between the inside and outside of cells by placing them in hypertonic buffer. This gradient facilitates the uptake of material into a cell. Incorporation of DNA through a laser hole into a cell has been verified by suspending cells in hypertonic buffer containing DNA labeled with bisbenzimide. After laser irradiation of individual cells, fluorescence became associated with the inside of these cells, indicating uptake of DNA (Weber et al. 1988).

## Electroporation

Using these technique naked DNA molecules can be introduced into protoplasts or cell of both dicots and monocots by electrically induced changes in membrane permeability (Jones et al. 1987). The electric field provides a driving force for the introduction of DNA into protoplast. After the removal of pectin from the plant cell wall increases the amount of DNA, which can be introduced by electroporation. The pectin inhibits DNA transfer physically or chemically. Calcium ions within the pectin fraction may precipitate DNA and so prevent its further movement through the wall. Pectinase treatments reduce the size of cell aggregates and cells accumulate exogenous DNA. Results from other laboratories have demonstrated that tobacco mosaic virus can be introduced into pectinase–treated cells. Intact bacteria and yeast cells can be transferred at high frequencies using electroporation.

# 4. Techniques of Recombinant DNA Technology for Production of Biologicals

Biotechnology is used to develop new medicines and treatments for farm and companion animals. Spending on animal health research and development totaled more than $400 million in 1999, more than 80 per cent of which, was in the development of pharmaceutical and biotechnology products. The Center for Veterinary Biologics at the U.S. Department of Agriculture has licensed 93 veterinary biotech products. In 1999, more than $4 billion was spent on animal health products in the United States, about two–thirds for livestock and farm animal care and the remainder on pets. The world's top 20 animal health companies recorded approximately $550 million in international sales of animal health biologicals in 1999. The following are the important aspects of animal health biotechnology:

- Different strains of rabies vaccines are available today that protect companion animals as well as wildlife from contracting and spreading this deadly disease.
- Genetic mapping and the development of DNA makers are being used to identify genes in chickens that have developed a resistance to Marek's disease, a virus–induced disease similar to cancer.
- Recombinant protein vaccines are under development to control coccidiosis in chickens, which seriously impairs their growth.
- DNA cloning technology is being used to research treatments for cryptosporidiosis, a parasitic disease found mostly in calves, but also in children and immuno–suppressed individuals, such as those inflicted with AIDS.
- Sodium hyaluronate products, used in the treatment of joint disease in horses and other animals, decrease the effect of inflammatory enzymes in cartilage. In addition, a patented biofermentation process is now available that allows these therapies to be given intravenously.
- A biological antitoxin prepared from the blood of healthy horses helps prevent and treat tetanus in domestic animals.
- Many animal feed products are enhanced with proteins to boost nutrition and control disease.
- Injectable products are being used to protect cattle from 36 stages of internal and external parasites, including *Ostertagia ostertagi*, nematodes and trematodes.
- A vaccine for brucellosis, or Bang's disease, is available for use in female cattle. This contagious cattle disease causes abortion and infertility in cattle and undulant fever in humans.

- Monoclonal antibody technology is being used to develop treatments for canine lymphoma.
- Other antibody technology is being used to develop diagnostics for bluetongue, avian leucosis, feline infectious peritonitis and feline immunodeficiency virus.
- Using gene therapy, scientists have restored the vision of dogs with a version of Leber congenital amaurosis, an untreatable condition that causes almost complete blindness.
- Drugs already on the market treat anemia and diabetes in companion animals.
- A vaccine is available for feline leukemia, a deadly and debilitating disease.
- Gene–based drugs and vaccines are being developed to treat arthritis, infectious disease and cancer in companion animals.
- Diagnostics are available that better detect disease in livestock and pets.

In general, the level of foreign gene expression in transgenic animals is low as compared with that in genuine transgenic animals. However, in transfectgenic and somatotransgenic animals there is the prompt gene expression, whereas, in genuine transgenic animals the expressed protein cannot be recovered from the mammary gland until adulthood. In cattle, for instance, it takes approximately two years. By using a retroviral vector, Archer *et. al.*, (1994) demonstrated the introduction of human growth hormone (GH) gene in the goat mammary gland and its successful secretion in the milk from the following day. However, the concentration of human GH secreted into the goat milk was so low that their practical use must await further refinement: a 100 to 1000 fold induction would have to be achieved for efficient protein production. Most successful application to gene therapy comes from the studies in skin. Originally, in vivo transdermal drug and antibody delivery was attempted for various type of cancer in this tissue.

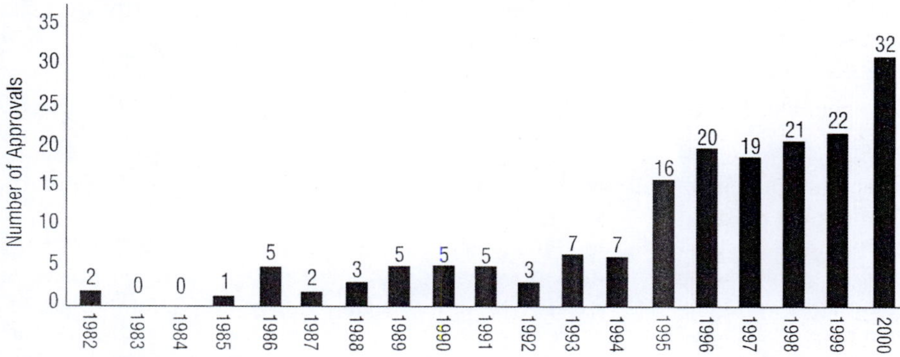


**FIGURE 4.1:** Presentation of new biotech drugs and vaccine approvals (year–wise).

**TABLE 4.1:** World Federal Agencies (WFA) regulating animal health

| Government Regulatory Agency | Scope of Regulatory Authority |
|---|---|
| Department of Agriculture, Center for Veterinary Biologics | Regulates veterinary biologicals, veterinary vaccines and veterinary diagnostic test kits. |
| Food and Drug Administration, Center for Veterinary Medicine | Reviews and approves new pharmaceuticals and feed additives. |
| Environmental Protection Agency, Office of Prevention, Pesticides and Toxic Substances | Regulates pesticides and topical products that kill fleas and other parasites. |

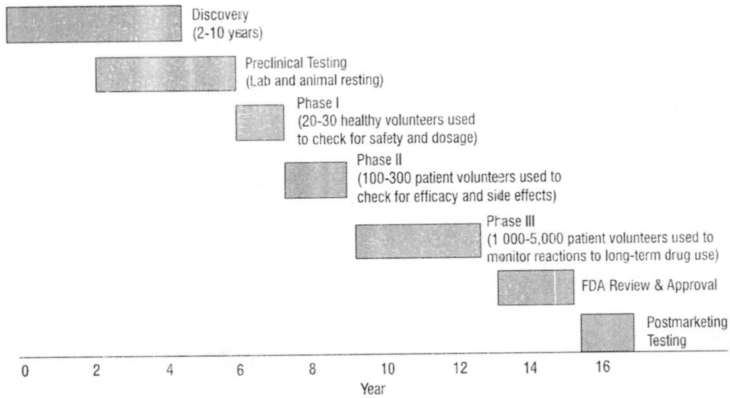


**FIGURE 4.2:** Presentation of biotech drug discovery process (Source Ernst and Young, LLP, Biotechnology Industry Report, 2000).

## RECOMBINANT PROTEIN PRODUCTION

Biotechnology played important role in development of successful fermentation technology, producing recombinant proteins and peptides. These products have justified their uses as absolute supply, relative supply, safely specificity and economics. Naturally derived biological products, such as factor VIII, hGH, etc., of human origin are prone to contamination from their source. The rDNA products ensue safely as for as their use is concerned. In case of diagnostic products i.e., antibodies, the most significant aspect is specificity. The recombinant–derived proteins provide a higher level of sensitivity and specificity than had been available previously. Proteins, which are produced in minute quantity by living organisms, the genes for these, have been identified, cloned and expressed at high level in host cells.

### Bio–activity of Expressed Protein

The activity of expressed protein depends on the type of the host cell used and is dependant on choice of the host, post–translational changes, scale–up of the recombinant culture, biology of the host cell substrate, relative or absolute degree to which the post–translational changes are needed. For example erythropoietins requires some key post–translational modifications and for this purpose and mammalian cells are used. For production of insulin, interferons, human growth hormones etc there is no such requirement. During synthesis and before they under to final stage, the bioactive molecules undergo several changes including folding, secretion and modification within the host cells. Host cells like mammalian yeast; bacterial cells have significant differences in the degree to which proteins undergo such post–translational changes. Mammalian cells are most suitable because they provide post–translational changes, which resemble those found in the natural protein. The proteins obtained from mammalian cells are of biomedical importance. The concern is for cellular properties, associated with tumorigenicity or transformation that inherit characters to several mammalian continuous cell lines. These properties are found in DNA of the cell. Mammalian viruses, as well as oncogenes are absent in cell strain such as $W_1$–38 and MRC–5, bacterial and yeast cells.

## Synthesis and Secretion of Recombinant Proteins

Synthesizing and secreting recombinant proteins by transgenic animals are thought to be attractive systems for the production of valuable pharmaceutical proteins. Most studies aim to express genes in mammary gland, although the protein production in some other organs has been reported. As an example of extra–mammary gland expression, functional human haemoglobin was produced in transgenic swine blood. Using a bladder specific uroplakin gene promoter, human GH was produced into the urine of transgenic mice.

Genes encoding pharmaceutical proteins with milk gene promoters can be introduced into germ lines via microinjection, and expressed specifically in the mammary gland. Although so far many transgenic animals that produce recombinant pharmaceutical proteins in the milk have been reported, a limited number of studies have attained high level expression. A large scale production of pharmaceutical proteins by this transgenic system may be brought to realization in the near future.

## Xenotransplantation

Progress of transplantation therapy in terminal organ failure has resulted shortage of human donor organs. This has led one to conduct extensive studies in search of alternative organ sources including xenotransplantation in which donor organs from different animal species are utilized. As far as xenograft transplantation is concerned, a suitable donor animal species would be pigs because their organs are almost of the same size as those of humans with anatomical and physiological similarities.

The currently available immuno–suppressive agents that are being used for chronic treatments do not suppress the hyper acute rejection. One of the approaches to resolve this problem is the use of inhibitory proteins. Complement system is known to have several such species–specific regulatory proteins to suppress complement activation (RCA). They include decay accelerating factor (DAF) membrane co–factor protein (MCP) and homologous restriction factor (HRF–20 or CD59). If pig organs express human RCAS, the transplanted pig organs may not substantially activate human complement system. Producing the transgenic animals harbouring and expressing human RCA genes in mice and pigs tested this hypothesis. Transplanted organs from the transgenic pig expressing human DAF functioned seemingly normally for long periods in primate recipients for up to two months.

Alternatively, one could possibly diminish the major histocompatibility antigen by diverting the reaction catalysed by α 1,3–galactosyltransferase towards that by α1,2–fucosyltransferase. It is the α 1,3–galactosyltransferase in pigs that catalyses the production of the αGal antigen, a cell surface major histocompatibility antigen considered to be responsible for the hyper acute rejection in humans. Over expression of the α 1,2–fucosyitrinsferase, which competes with the α 1,3–galactosyltransferase for the same substrate, should result in reduced substrate supply for the latter, and thereby lead to lowered production of the aGal antigen. Indeed, transplanted organs from transgenic mice carrying and expressing the human α1,2–fucosyltransferase showed prolonged survival time in ex vivo studies on xenograft rejection.

## PRODUCTION OF BIOLOGICALS

Some of the important biologicals, that are the products of recombinant DNA technology, include hormones, growth factors, erythropoietins, interferons, interleukines, monoclonal antibodies, vaccines, and immunomodulators, etc. Some of these that find key areas of market potential have been discussed here.

### Production of Human Insulin

Diabetic patients suffer because they are unable to produce sufficient quantity of insulin from their pancreas. For treatment of such patients exogenous source of insulin are required for replacement therapy. Insulin was first derived (1982) using recombinant technology for human use from *E. coli*.

Theoretically, there are two ways in which one could go about producing the insulin molecule. Insulin consists of two different protein chains, the so–called **A chain** and the **B chain**. One could produce the normal precursor of insulin, proinsulin that is found in the pancreas. *Proinsulin* is a molecule that contains insulin but also contains an extra connecting peptide linking the two chains together. In the pancreas gland, this so–called connecting peptide is clipped out, leading to the production of insulin that is then released into the blood circulation system. One can mimic this process today in the laboratory and actually even in production, but up to the present it is not being used to produce human insulin on a large scale. Presently the A chain and the B chain are made individually, and then are coupled in the plant to produce the bioactive insulin.

As the process involves chemical synthesis of genes for A and B chains, therefore, the two chains cloned separately were linked to a plasmid near end of β–galactosidase gene of *E. coli*. During fermentation both the chains of A and B chains of insulin are expressed. The further procedure is to purify and combined these two chains in vitro. Earlier, advent of rDNA insulin was studied, but later on replacement therapy of insulin was employed from animal (porcine) origin. In both the cases, the difference was in amino acid sequences, that existed in both these forms, is in the carboxyl terminus of β–chain. Threonine is present in the human form and alanine in the porcine form. It has been observed that certain patients are allergic to animal hormones. The porcine derived insulin is highly immunogenic. Patients receiving porcine derived insulin develop circulating anti–insulin antibodies, which neutralize insulin in the circulation, and are responsible for fat atrophy. Compared to this recombinant derived human insulin is less immunogenic, eliciting lower frequency of allergin antibodies. The properties of recombinant–derived insulin are functionally similar to that of porcine–derived insulin.

In separate plasmids, the genes have been introduced individually for the A chain and B chain of insulin, and then these plasmids have been transformed into bacterial cells. The *Escherichia coli* are then grown in large fermentation facilities. The product that is initially made is a large chimeric **protein consisting of the A chain or B chain** attached to the end of a naturally occurring *E. coli* protein. This protein is subjected to a cleavage reaction in which the A chain and the B chain are chemically cleaved away from the rest of the chimeric molecule. Then, following several further purification steps, these two chains are combined, and the biosynthetic human insulin is recovered and purified.

Large amounts of the gene product of this plasmid accumulate in the *E. coli*. A thin–section electron micrograph of *E. coli* producing human insulin polypeptide shows dense areas, which are deposits of that protein within the cell. The protein is produced in very substantial amounts and can occupy a major portion of the cell.

This is one of the advantages of biotechnology today–by using appropriate control systems and regulatory systems on the plasmid being dealt with, one can make the protein of interest a major portion of the total protein of the microorganism. It can become a very efficient process. It is a **crystalline protein**, and it has all of the characteristics of the insulin that is circulating in all of our bodies.

There are at least two advantages to being able to produce human insulin as opposed to continuing to use the pork and beef insulin that is currently used in many diabetic patients.

1. **First**, the chemical structure of pork and beef insulin differs slightly from that of human insulin.
2. The **second** advantage relates to the fact that currently produced pork and beef insulin are really by–products of the meat industry.

Some of the plasma proteins are albumin, globulins–$\alpha$, $\beta$, $\gamma$, lipoproteins–$\alpha$, $\beta$, plasminogen, fibrinogen, prothrombin and transferrin. **Albumin** is a protein and now manufactured using recombinant–DNA technology and many of the genes for other proteins in the plasma protein series have also been cloned.

## Production of Human Growth Hormone

The human growth hormones are synthesized easily and cheaply by using genetic engineering techniques. Somatostatin (brain hormone originating from hypothalamus and is a peptide chain consisting of 14 amino acid units) is the first human hormone synthesized by genetic engineering. It acts to inhibit the release of the human growth hormone.

Current available methods of biotechnology have not made possible to synthesize genes for peptides longer than 30 amino acid units, because it is not economic. It requires isolation of mRNA from its natural source, encoding a sequence for the desired product, and a cDNA prepared by means of reverse transcriptase method. After replication of DNA it is transformed into appropriate bacterial cell like *Escherichia coli*, which is used for production of Growth Hormone (hGH). Such hormones are used for treatment of growth hormone deficient children. Usually, human growth hormone is made of 191 amino acids. The translation product of mRNA carries a signal peptide. Production in *E. coli* involves the assembly of a gene that is partly synthetic and partly natural. Signal sequence is not cloned.

Complementary DNA prepared from hGH mRNA has single cleavage site in the sequence and codes for amino acid 23 and 24 of hGH. Restriction enzymes are used to cleave the large DNA fragment (amino acids 24–191). It is combined with chemically synthesized DNA fragments (amino acids 1–23) containing ATG (initiation codon). The complete gene is inserted into the modified version of pBR322–incorporatin the lac operon. Fermentation is carried out in *E. coli* host. It is followed by isolation of the product from intact cell. The first form of recombinant–derived hGH, possessed additional methionine residue at the amino terminus (met–hGH product). At functional level met–hGH expressed in *E.coli* is equivalent to natural hGh, however, it differs in its immunogenicity. This difference is due to changed amino terminal structure. When it is obtained in second generation, this difference is lacked in its residue.

## Production of Growth Hormone Releasing Factor (GRF)

GRF synthesized in the hypothalamus and is transported to pituitary glands. GRF is destroyed while transportation in the blood. However, its concentration in the pituitary remains sufficient to regulate growth and nitrogen balance in body which is due to the production of growth hormone. GRF usually works in concert with negative control factors i.e., **somatostatin** to maintain growth hormone release at correct concentration. This level is not achieved when recombinant hGH is given exogenously, but is achieved using GRF therapy. GRF therapy modifies recombinant proteins. It helps in less degradation and a development of new system for slow release of materials.

## Production of Erythropoietins (EPO)

Human kidney produces certain hormones, which are responsible for activating the growth and differentiation of RBC precursors in the bone marrow. Reduced level of hemoglobin in anemic patient is reversed by the giving EPO to the patient. EPO at commercial level is produced using rDNA technology. It has been observed that if decreased level of hemoglobin is increased by blood transfusion, there are risk possibilities of hepatitis B/HIV infection. To overcome this problem, EPO is administered. EPO may provide safety and efficacy.

Urine of the severally anemic individuals contains nucleotide sequence. It is the good source of EPO. A biologically active hormone is obtained by cloning the gene for EPO and expressed it in CHO (Chinese hamster ovary) cells. The culture supernatants are subsequently subjected to purification for its use.

EPO derived from recombinant CHO (glycosylated) is biologically active. It is administered to patient, which shows significant increase in hemoglobin concentration. This remains devoid of toxic effects or organ dysfunction and do not induce antibodies in recipients. It is produced in mammalian continuous cell lines hence, has no risk of residual contaminants such as nucleic acid or viruses. Recombinant CHO cells have been produced at commercial level.

## Interferons (IFN)

Nasal secretion and serum of normal human beings have the ability to neutralize certain viruses e.g., influenza A and B, vaccinia, mumps and Newcastle viruses. This is due to the presence of certain enzymes like alkaline and acid phosphatases, and properdin in serum and secretions. They have the viricidal activities. The cells in response to viral infection synthesize **interferon** a non–specific antiviral agent. It inhibits intracellular replication of viruses.

### *Interferon or Antiviral Substances*

Interferon provides **first line of defense** against virus infection; in animals treated with antiserum to interferon, virus infections progress much more rapidly and cause much more severe disease than in untreated animals. Interferons are cellular protein. Healthy cell do not contain detectable amounts of interferon; the formation of interferons must be included. These cellular proteins are host cell species–specific. Sometimes this specificity is very narrow. The recombinant DNA technology has made it possible to identify the genes that produce different type of interferons, to the clone the genes, and to mass production of the various subclasses of these particular proteins.

**Nature of Interferon**

Human interferons are proteins that are about 145 amino acid long (molecular weight about 17,000). Three types of interferon are known: IFN–α, IFN–β (produced predominantly by leukocytes and fibroblasts respectively) and IFN–γ or earlier known as *immune* or type–2 interferon, which is produced by unsynthesized lymphoid cells in response to mutagens and by sensitized lymphocytes when stimulated with specific antigen.

**Assay of Interferon**

The antiviral activity of interferon is usually assayed by exposing cells to serially diluted interferon preparations for about 12 hours and then infecting them with standard amount of virus known to produce a certain number of plaques. The titer of an interferon preparation is the reciprocal of that dilution that reduces this number by 50 per cent.

**Nature of Interferon Genes**

There are 14 different human IFN–α–genes, one IFN–β– gene and one IFN–γ– genes. All these genes encode proteins that are exactly the same length, namely 166 amino acids long. The nucleotide sequence are altered to avoid the formation of secondary structures, which effect expression of the genes Having a cloned gene coding for a particular IFN, it is possible to modify the gene structure and allow its expression in a recombinant vector. The properties of new IFNs can be analyzed. It may be possible to create IFN with varied enhance activity relative to the authentic material.

### *Biological Activity of Interferon*

Interferons are powerful cellular regulators that interact with cell through specific glycoprotein receptors. There are separate receptors from IFN–α, IFN–β–and IFN–γ–. The binding of interferon to their receptors activates second messengers that induce formation of a series of proteins. Some of these proteins induce an antiviral state; others are surface glycoprotein that cause the cells to be recognized by the cells of the immune system, other have functions that have yet to be identified. Interferons interfere with virus multiplication; this is the property for which they are named. They interfere with the ability of parental or early viral messenger RNA molecules to be translated which results in the synthesis of specific type of proteins, no progeny viral genomes are formed, and infection is aborted.

Interferon affects variety of cell functions, which are mediated by the proteins, the formation of which is induced by exposure to interferon. Most remarkable of these effects are those mediated by IFN–γ–. This interferon is a lymphokine secreted by activated T lymphocytes. It activates macrophages, stimulates B cells, increases fibroblast and endothelial cell resistance to many non–viral parasites, and modulates cell surface proteins central immune regulation.

### *Production of Interferon*

Biotechnological production of interferon is based on recombinant DNA technology. In this human interferon gene are inserted in *E. coli*. One litre of culture can yield a quantity equivalent to that obtained from leukocyte of 100 donors.

Post–translational modifications are required to convert the primary translational product into functional proteins. This includes removal of signal sequence and glycosylation in case of IFN–α and IFN–γ. Although IFN–α and IFN–γ are believed to be glycosylated, when produced in bacteria without their carbohydrate chains, they seem to posses full biological activity. Natural IFN are synthesized with a hydrophobic N–terminal signal sequence, which is removed during processing for secretion. It is possible that the natural signal sequence functions to allow secretion to the periplasmic space in *E. coli*, but for IFN–β, the presence of signal can lead to increased instability and degradation of properties. Intermolecular disulphide bridges exist in IFN–α and IFN–β, and one of this important for biological activity. In contrast, IFN–γ does not appear to form such a disulphide bond.

## TUMOUR NECROSIS FACTOR (TNF) AND LYMPHOTOXIN (LT)

Tumour necrosis factor and lymphotoxin are the anticancer drugs of polypeptide in nature. They are produced by human cells and are responsible for recognition of tumour cells. They destroy them selectively. They have some extensive sequences in common. The genes for both TNF and LT have been cloned and bacterially synthesized.

### Interleukines (IL)

Interleukines are natural immunomodulators. They mediate communication between leucocytes and other cells. For normal functioning of immune system a complete set of cell–to–cell interaction, mediated by lymphokines or cytokines, or by cell surface proteins i.e., T cell receptors, are responsible. On interaction with macrophage and T–helper cells, the macrophage secretes a cytokine called interleukin–1. It in turn activates T–helper cell proliferation to the higher level of activity. Interleukin–2, along with IFN–γ and IL–1 (a product of macrophage) are important for differentiation of cytotoxic lymphocytes. Both the interleukines types are also important for antibody production by β–lymphocytes. Virus infection or any pharmacological agents may suppress IL–2 production by their effect on lymphocyte population. It may be possible to eliminate immunosuppressive effect by adding exogenous IL–2.

Use of genetic engineering technique (rDNA technique), made it possible to clone IL2 genes from a various species. An analog of the human IL–2 gene (continuous chain of 133 amino acid residue) has been synthesized by using genetically engineered *E.coli*.

### Other Products

Biotechnology (rDNA technology and gene cloning techniques) is used to develop new medicines and treatments for humans. The important products are Factor VIIIc, epidermal growth factor, $rCD_4$, Fibronectin, β–endorphin etc. These are being produced at commercial level by isolating specific DNA sequence.

## VACCINES

Killed or attenuated microorganisms used for artificial active immunization are called vaccines. The objective of vaccination is to provide effective immunity by establishing adequate level of antibodies and a primed population of cells, which can rapidly expand on renewed contact with antigen.

Vaccination with living organisms has the advantage that if administered by the route employed in natural infection, the immunity stimulated will include all the characteristics of immune response induced by that infection itself. These include cell–mediated immunity, the production of circulating antibody and perhaps more importantly, for many infections the production of circulating (secretary) antibody at the site of infection. Vaccines in which purified immunogens capable of eliciting protective response are identified and selectively used in isolation are referred to as subunit vaccines. However, such purification of antigens is potentially dangerous and inefficient. Therefore, there is considerable impetus to isolate the relevant genes, encoding the desired antigens and introduce them into the microbial host like *E. coli B. subtilis* or yeast.

The contribution of biotechnological research so far has been to produce the following types of vaccine:

- Vaccine using genetically engineered organisms.
- Synthetic peptide vaccines
- Minicells as vaccines
- Disabling mutations resulting in suitable pathogen (organism) for vaccine.

Whatever the method of biotechnology used, vaccines produced by these new techniques have definitive advantages over the current methods of vaccine production. These are:

- more effective vaccines with increased response and high specificity.
- long–lasting immunity;
- no side effects, less toxicity;
- production methods easier and cheap; and
- novel vaccines.

## Sub–unit Vaccines

Biotechnology has provided new class of vaccines called subunit vaccines, which are created by using purified antigen sub–units, which are produced through r–DNA technology. Genes that code for the desired antigen in the immunization process are introduced into the non–pathogenic organism. A sub–unit vaccine has no potential for reverting to a pathogenic organism because only limited number of components from original pathogen is present in the vaccine. A sub–unit vaccine should also have fewer side effects, as it does not have any infectious agent.

Various types of vaccines have been developed for diphtheria, pertussis, tetanus, herpes, HIV, Lyme disease and melanoma. Vaccines, which provide poor financial returns and therefore, are not attractive for commercialization are referred to as **orphan vaccines**. Vaccines for the Third World fall in this category. There are two types of vaccines: killed vaccines and live vaccines.

## Killed Vaccines

Suspensions of organisms in their virulent phase are killed by heat or antiseptic, care being taken not to denature important antigens by excessive heat or strong detergents. Two or more injections of these are given preferably at 4 weeks or longer intervals. Examples of killed vaccine are TAB vaccine for enteric fever, whooping cough vaccine, salk vaccine of poliomyelitis, antirabies vaccine etc.

### TAB Vaccine

This is prepared from smooth culture of *S. typhi*, *S. paratyphi* A and *S. paratyphi* B. It is the suspension of these organisms containing 1000 million bacilli of *S. typhi*, 750 million bacilli each of *S. paratyphi* A and *S. paratyphi* B per millilitre sterilized by heating at 60°C for thirty minutes and preserved with 0.5% phenol. It is used for vaccination against enteric fever. The vaccine provides immunity lasting for about 2 years. Booster dose of 0.5 ml is given after every two years to provide a lasting immunity.

### Whooping Cough Vaccine

This should be given to every child to prevent whooping cough. Formalin or formalin methiolate killed *Bordetella pertussis* organisms in a suspension of 200,000 million organism/ml are given in three doses of 1ml each intramuscularly at interval of 4 weeks. Vaccination of infants has been found to reduce the infection rate from 90% to 100% in endemic area.

### Salk Vaccine of Poliomyelitis

This is a formalin–inactivated vaccine containing all the three polioviruses, type 1, type 2 and type 3. Two doses of 1 ml of it are given intramuscularly at an interval of 3 weeks. Thereafter, two booster doses are required.

### Antirabies Vaccine (Sample Vaccine)

It is prepared from sheep brain infected with fixed rabies virus. A 4% suspension of brain in which the virus is inactivated by 0.5% phenol at 37° C is used. A new rabies vaccine has been introduced from virus grown in human diploid cells (HDC). This vaccine is near to perfect as regards potency, safety, tolerability but very expensive to produce.

### Purified Chick Embryo Cell (PCEC) vaccine

It is similar in potency to HDC vaccine. This virus was adopted to chick embryo cells by 25 passages, inactivated by 0.025% β–propiolactone and purified by continuous zonal centrifugation. A complete course of vaccinations for adults and children consists of a total of 6 injections given intramuscularly one on each of the days 0, 3, 7, 14, 30 and 90.

### Meningococcal Polysaccharide Vaccine

It is composed of purified group A and group C meningococcal polysaccharide 50 micrograms each lyophilized to be reconstituted in physiological buffer 0.5 ml. It is given subcutaneously.

## Live Vaccine

Attenuation results in loss of pathogenicity without loss of antigenecity of the microorganisms, which is usually achieved by growing the microorganisms repeatedly on artificial media, sometimes in adverse conditions or in unnatural hosts. Such organisms multiply within the body, like the pathogens, without producing the diseases yet **inducing immune responses** like their pathogenic counterparts. It is essential to assess the safety and efficacy of recombinant live vaccines. For this purpose extensive clinical testing of the parental virus, biology of the cellular receptor for the virus and stable attenuation of the parental virus should be tested. The immune response occurs largely at the site of natural infection. Some of the commonly used live vaccines are as follows:

### Sabin Poliomyelitis Vaccine

Attenuation of poliovirus is obtained by repeatedly growing relatively avirulent strains on monkey kidney cell lines. All the three types 1, 2 and 3 strains are used. It is given orally in 3 doses each of 0.15 ml (3 drops) spaced an interval of 4–8 weeks.

### B.C.G. vaccine

*Bacile Calmette et Guerin* (BCG) is a bovine strain of *Myco. tuberculosis* that had been cultured for many years on a bile potato medium by two French scientists – Calmette and Guerin. It is given intracutaneously in dose of 0.1 ml containing 0.05–0.1 mg. of moist weight of bacilli (1–2 million viable bcilli).

### Live Attenuated Measles Vaccine (Schsarz strains)

It is prepared from chick embryo tissue culture. Each dose of the reconstituted vaccine contains not less than 1000 TCID of measles virus. Reconstituted 0.5 ml of vaccine is injected subcutaneously. It confers long lasting immunity in most subjects.

### Small Pox vaccine

It is prepared from vaccinia virus obtained from vesicles of artificially infected calves for this purpose. The crude pulp obtained by curettage is homogenized and treated with 2 parts of 1% phenol and then diluted 4 times with glycerol saline or phosphate buffer and stored at 10° C or freeze dried. The latter is recommended for fieldwork, it can be reconstituted immediately before use without loss of potency. Vaccine can also be prepared from cultures on chick embryo.

## PROCESSES FOR RECOVERY AND PURIFICATION OF BIOLOGICALS

Several chromatographic purification methods are available like chromato–focussing, reversed–phase chromatography, hydrophobic interaction chromatography, charge transfer chromatography, size–exclusion chromatography (molecular sizing), ion–exchange chromatography and affinity chromatography. These processes are used to get cell culture and yeast derived products because these products are secreted directly into the medium. The recovery process involves cell separation followed by purification. The recombinant protein products, after fermentation or cell culture, can be recovered by the bacterial (*E.coli*) derived products. It involves the lysis of bacterial cell. This further involves in the release of protease. Protease may affect the stability of the product. To overcome with this problem, the cell lysis is followed by rapid purification. Protease may be present in traces, which develops a major problem during purification. The major impurities present in rDNA products are of varied nature. The important are tabulated below:

**TABLE 4.2:** Assay methods for impurities of rDNA products

| Impurities | Assay methods |
|---|---|
| Aggregated proteins | SDS–PAGE |
| Amino acid substitution | Amino acid analysis, Peptide mapping, Mass spectrometry |
| DNA | DNA hybridization, Spectrophotometry, protein binding |

| Impurities | Assay methods |
|---|---|
| Endotoxins | Bacterial endotoxin tests, pyrogen test |
| Formyl methionines | Peptide mapping, HPLC, Mass spectrometry |
| Host–cell proteins | SDS–PAGE, Immuno assays |
| Media proteins | SDS–PAGE, Immuno assays and HPLC |
| Monoclonal antibodies | SDS–PAGE, and Immuno assays |
| Oxidized methionines | Peptide mapping, HPLC, Mass spectrometry |
| Protein mutants | Peptide mapping, HPLC, Mass spectrometry |
| Proteolytic cleavage | Iso–electric focusing, SDS–PAGE, Mass spectroscopy, and Amino acid analysis |

The contaminants are also present in rDNA products, which may be of microbial or of virus origin. These are evaluated by microbial limit test, sterility test or CPE, hemadsorption, reverse transcriptase activity method respectively.

# 5. Animal Cell and Tissue Culture

The animal tissue culture is a technique in which the tissue/cells are taken out from an animal's body and grown, and maintained in laboratory in a suitable medium. It is only in the past few decades that man with the immense knowledge and techniques of science at his command has been able to venture into the deep bionomics of cell and learned more and more about its applications to the well being of human beings. Tissue culture is not a science but an art made up of a wide range of activities, which made a rough distinction between fundamental and applied research. At present, tissue culture is an indispensable tool, particularly in the fields of medical and agricultural sciences.

## ADVANTAGES AND DISADVANTAGES OF TISSUE CULTURE METHODS

There are several advantages and disadvantages of animal tissue culture; therefore, this technique should not be used indiscriminately.

### Advantages

1. The behaviour of cells in culture can be interpreted and regulated properly.
2. There remains homogeneity of cell types (achieved through serial passages)
3. It is economical because smaller quantities of reagents are needed than *in vitro*.
4. The legal, moral and ethical questions of animal experimentation are avoided.

### Disadvantages

1. It needs controlled physiochemical environment (pH, temperature, osmotic pressure, $O_2$, $CO_2$, etc.)
2. It is ten times more expensive for same quantity of animal tissue; therefore reasons for its use should be compelling.
   a) Unstable aneuploid chromosome constitution
1. Controlled and defined physiological conditions (constitution of medium, etc.) are needed

## PRIMARY AND ESTABLISHED CELL LINES

A piece of tissue from the organism is usually quite complex and may contain, besides the specialized cells of the tissue, connective tissue cells, a variety of blood cells and reticuloendothelial cells. When a culture is first set up, these all survive for at least a day or two. In usual culture conditions many cells begin to migrate almost immediately.

The first cells to do this are often macrophages, followed by fibroblasts, which migrate in a radial manner from the explant. Many of the specialized cells, however, remain immobile in the connective tissue stroma. Nerve cells, for example, usually remain in the explant although axons may migrate out from it.

The subsequent fates of these cells vary enormously. Many of them are quite short–lived, e.g., most of the blood cells die and disappear from the culture within two or three days. Other cells, such as neurons and muscle cells, frequently persist in culture, sometimes for months without dividing at all and then eventually die too. Still other cells begin to divide rapidly and continue to do so for sometime. However, many of these also die after a period, which varies from weeks to months. If the cells multiply repeatedly for a long time, they can often be `passaged'. First obtaining the cells in suspension and then inoculating this into a new culture vessel along with fresh medium usually achieve this. As soon as cells have been `passaged' in this way the culture is designated as **primary cell line.**

Primary cultures is used advantageously because of:

1. The expense and inconvenience of maintaining a established cell stocks is obviated.
2. They are particularly suitable for vaccine production since the probability of *in vitro* transformation of cells to malignancy is minimized.
3. Massive quantities of tissue can be obtained conveniently for short–term studies.
4. They are hardy and can be sustained well in media of relatively simple composition.
5. Their degree and range of sensitivity to viral infection may exceed that of the common established cell lines.
6. The use of primary cell cultures has following inherent disadvantages:
7. The may be contaminated with latent viruses (e.g., foamy virus in monkey kidney tissues).
8. Long–term experiments concerning the biological properties of cells cannot be carried out.

Mixed populations of cells may confuse interpretation of experimental findings (Merchant et al., 1964).

Frequently, primary cell lines go on dividing at quite a high rate for a long time and can be passaged repeatedly. Even after quite a large number of passages, some primary cell lines cease to proliferate and die out. However, sometimes cell lines can be cultured for such a long time that they apparently have developed the potential to be sub cultured indefinitely *in vitro.* Such cell lines are called **established cell lines.** As a general rule, a line is not designated an established cell line unless it has been subcultured at least 70 times at intervals of 3 days between subcultures.

Established cell lines behave in a remarkably similar manner irrespective of their origin.

1. They have short doubling timings, of the order of 12 to 20 hours.
2. They are invariably aneuploid.
3. They have similar nutritional requirements, whatever their origin.
4. They grow too much higher densities than primary cell lines.
5. They do not usually show much evidence of spatial orientation.
6. It is rare for them to show obvious specialized functions.
7. They will often grow from dilute inocula.

Most of them can be established in suspension cultures, whereas, it is exceptional for primary cell lines to grow in suspension.

## KINETICS OF CELL GROWTH

### Established Cell Lines

Cells in culture display very much the same type of growth pattern as microorganisms. In particular, they show the classical growth demonstrated by cultures of bacteria, yeast and protozoa. The culture exhibits following phases:

1. **Lag phase:** This is the phase when practically no growth occurs and may last for some hours to a few days before growth commences.
2. **Logarithmic (log) phase:** During this phase growth proceeds steadily with the population doubling every 15 to 20 hours in the case of fast growing cells
3. **Stationary phase:** At the end of log phase maximum population is reached and the growth becomes stationary.

The growth rate of cultured cells is not necessarily always the same. It may vary to some extent from one cell strain to another but in particular it varies considerably with environmental conditions such as pH, temperature and osmotic pressure.

### Primary Cell Lines

Primary cell lines do not behave in quite the same ways as established cell lines. When established cell lines have become stationary the medium in which they have been grown is usually inadequate for maintenance of a fresh inoculum of cells. However, it can often be observed that cultures of primary cell lines stop growing before the medium is exhausted. The reason for this is a phenomenon called **contact inhibition,** which was first described by Abercrombie and Heaysman. They observed that when certain kinds of normal cells, mainly fibroblasts, came into contact they were immediately immobilized. This is the very common property of primary cell lines. Not only is cell movement affected by contact inhibition; cell division, DNA synthesis, RNA and protein synthesis are much reduced or eliminated when a cell sheet becomes confluent. When primary cell lines become transformed, one of the most outstanding features of the transformed cells is the absence of great diminution of contact inhibition. This is the reason why transformed cells grow to higher densities and pile up on each other.

### Interaction among Cells

Besides contact inhibition, the most interesting interaction is the ability of like cells to recognize and adhere to each other. Holtfreter showed that when amphibian embryonic cells were disaggregated in calcium and magnesium free salt solution and the separated cells were then allowed to mix together, they first adhered in a random manner and then sorted themselves out into groups of like cells. Eventually the entire embryo was reconstructed. Moscona demonstrated the same phenomenon with mammalian and avian cells.

## GENETICS OF CULTURED CELLS

Primary cell lines usually retain their diploid karyotype. Transformed cell lines show a great variation in karyotype. Shortly after transformation the incidence of tetraploid cells increases; aneuploid cells then make their appearance and stable established cell lines commonly have an aneuploid karyotype with a wide spread of chromosome numbers. The emergence of aneuploidy is apparently due to non–

disjunctive cell division. Later on, in the established cell lines the morphology of the individual chromosome also changes; it is because of chromosome breaks, fusion and translocations, e.g., many mouse cell lines contain metacentric chromosomes, whereas, it is typical of the mouse that all the chromosomes are normally acrocentric. Mutant cells also arise in the tissue culture.

### Metabolism

In cultured cells, the metabolism of glucose proceeds by way of glycolytic and Krebs pathways as in the tissues of the intact organism. However, many cells show a marked tendency to accumulate lactic acid and keto–acids in the medium. Cultured cells have very variable needs for oxygen and considerable part of the energy requirements can be met by the breakdown of 6–carbon carbohydrate such as glucose to 3–carbon compounds such as lactic acid and pyruvic acid, a process which requires no oxygen. Cells can be grown in the complete absence of oxygen for rather short periods of time. Tissue culture cells can synthesize lipids from simple chemical substances such as acetate. It is not known whether fat themselves can be taken up from the medium and stored. Some cells are capable of synthesizing cholesterol.

Cultured cells can build nucleic acids readily from simple compounds in a medium (such as formate, glycine and bicarbonate). They possess all the enzymes necessary for this purpose.

## ANIMAL CELL AND TISSUE CULTURE

Animal cell cultures have many applications including their use as model systems for biochemical, physiological and pharmacological studies and the production of growth factors, blood factors, monoclonal antibodies, interferons, enzymes, vaccines, and hormones.

### Mechanical Technique

This includes the forcing of the tissue through cheese or silk cloth or shaking the tissue with glass beads in an appropriate buffer. This technique causes considerable cell damage and results in a low cell yield.

### Biochemical Technique

This technique has largely overcome the problems caused by mechanical dissociation. Collagenase and hyaluronidase in a calcium free medium have been used for hepatocyte isolation. The liver is thinly sliced and then incubated with the enzymes.

Animal cells may be broadly subdivided into:

1. Those which remain viable only when attached to a solid substrate (e.g., primary cultures, normal diploid fibroblast cell strains), and
2. Those that will proliferate in fine suspension (e.g., human tumour cells, Hela cell lines and hybridomas).

The material to, which cell stick, must be nontoxic, sterilizable and preferably transparent so that the cells may be observed microscopically. The surfaces, which have been used successfully, include plastic, glass, teflon tubing and DEAE–sephadex. Cells may be detached from such surfaces by mild trypsin digestion.

**TABLE 5.1:** Some examples of the applications of animal cell and tissue culture

| Cell type | Process investigated |
|---|---|
| Monocytes and macrophages | Pinocytosis and phagocytosis |
| Blood lymphocytes | Karyotype analysis for detection of genetic defects in humans |
| Normal and transformed fibroblasts | Surface adherence properties of normal and malignant cell membranes |
| Kidney tubule epithelial cells | Differentiation of monolayers; electrical and vectorial transport of solutes; monoclonal antibody production |
| Myceloma cells and ß–lympohcytes | Purification and characterization of specific membrane proteins e.g. α–and ß–adrenergic receptors, dopamine receptors |
| Kidney epithelial cells | To investigate relationship between membrane polarity and budding properties of envelope RNA viruses |
| Transformed leucocytes, fibroblasts and either lymphocytes or lymphoblastoid cells | Cells are infected with Sendai virus to produce Â, ß, and interferon respectively |
| Transformed Hela cells, mammalian cells | Radiation therapy and the design of radiosensitisers and radio protectors |
| Mouse fibroblasts | Acute and chronic toxicity testing and metabolism of xenobiotics; vaccine production |
| Primary monkey kidney cells | Production of polio vaccines; hormone secretion |
| Fibroblasts, mammalian brain cells | To identify chemicals capable of including chromosome aneuploidy |

## ANIMAL TISSUE CULTURE MEDIA

### Culture Media Containing Naturally Occurring Ingredients

In spite of vast use of chemically defined media in tissue culture, it is still necessary in most undertakings to depend on naturally occurring substances derived from the organism. The various kinds of such media, which are in use, may be:

(i) blood plasma, (ii) blood serum, (iii) tissue extract. and (iv) complex natural media.

#### *Blood Plasma*

The first tissue cultures were made by Harrison (1907) in clotted frog lymph. Burrows (1910) substituted a coagulum prepared from chicken plasma. After many years it was found that plasma provided a complete nutrient in which cells could survive and multiply slowly for extended periods under conditions that resembled in many respect those found in body. Plasma is still being used to advantage for the following purposes:

1. To provide a nutritive substrate and a supporting structure for many types of cultures, just as it also provides a matrix for new cells during the repair of injury in the body.
2. To provide a means of conditioning the surface of glass for better attachment of cells.
3. To provide a means of protecting cells and tissues from excessive traumatic damage during subculture.

4. To provide some degree of protection from sudden changes in the environment at times of fluid change.
5. To provide localized pockets of conditioned medium around cells.

For culture work, plasma from the adult chicken is preferred to mammalian plasma because it forms a clear, solid coagulum even when diluted several times. Mammalian plasma is either too opaque for good optical work or else it fails to produce solid clots.

The plasma is obtained by centrifugation of whole blood before coagulation takes place. The tissue is then placed in a small quantity of the plasma and coagulation encouraged by addition of a small amount of tissue extract or thrombin. This is done because the cells in culture require a solid support for continued growth and activity. In case of fowl the blood is obtained from the wing, heart or carotid artery. In case of mammals it is obtained from carotid artery and heart.

### Blood Serum

Blood serum (*plasma minus fibrinogen*) with or without other nutritive substances may be used either as the entire culture medium or as the fluid phase of a medium consisting partly of a plasma coagulum.

For many years it was assumed that whole serum was toxic, that plasma was useful only as a supportive structure and that the nutritive requirements of the cells were supplied by the embryo extract that was usually added to the medium. Eventually, however, it was found possible to cultivate tissues in serum alone without plasma or tissue extract. Serum does provide some of the growth factors or some of the physical conditions, or both, that are presently lacking in synthetic media.

**Preparation of Chicken Serum:** The fluid plasma from which the serum is prepared should be completely coagulated. The plasma is coagulated deliberately by adding to each tube a drop or two of embryo tissue extract or an equivalent amount of thrombin and leaving the tubes to incubate for several hours at 37° C. The coagulated plasma is broken up into fragments and it is ground in a mortar with sterile quartz sand. After grinding the serum is separated by centrifugation.

**Preparation of mammalian serum:** The mammalian blood is left at room temperature for an hour. The clot is removed by a glass rod and then centrifuged for 30 minutes at 3000 rpm and the serum is separated.

### Serum Free Media

The use of serum in culture media is not so common because it has following disadvantages:

1. The quality of serum varies from batch to batch and deteriorates within one year. Therefore, every batch of serum needs fresh testing.
2. If more than one–cell types are used, each may require different serum batch, therefore, many batches are to be maintained and co–culturing may be difficult.
3. The demand of serum usually exceeds the supply for a variety of reasons.
4. When cell culture is used for downstream processing to recover cell products, the presence of serum is an obstacle to purification;
5. Serum increases the cost of the medium.
6. Serum may stimulate undesirable growth and may even inhibit growth in

some cases.

**Advantage of Serum Free media**

1. It has the ability to make a medium selective for a particular cell type, since each cell type appears to require a different recipe.
2. It has high degree of purity of reagents and water.
3. It needs high degree of cleanliness of all apparatus.

**How to Develop Serum Free Media?**

1. A known recipe for a related cell type may be used.
2. Individual constituents of the serum may be altered, until the medium is optimized.
3. The development and assessment constituents are time consuming process. Sometimes it takes about three years for development of a new medium.
4. In the second approach, an existing medium like **RPMI 1640** or **Ham's F12** or **DMEM (Dulbecco's modified minimal essential medium)** is taken and a shorter list of constituents like selenium, transferrin, albumin, insulin, androgen, hydrocortisone, estrogen, etc. is used for manipulation.

## *Tissue Extracts*

Carrel (1912) discovered that embryo tissue extract had remarkable powers of promoting cell growth and multiplication in cultures of connective tissue cells from chick embryo heart. Since Carrel's experiments, there have been many attempts to determine the chemical nature of the substances responsible for the stimulating effect of embryo extract.

The growth promoting activity appeared to be associated particularly with fractions containing predominantly nucleoproteins of the ribonucleic acid type. Fractions relatively high in deoxyribonucleic acid appeared much less active. Active nucleoprotein fractions from adult chicken heart, brain, liver and spleen have been used but no indications of organ specificity were observed. The activity of the fraction also did not depend on their total nucleic acid content or on the age of the individuals from which they were prepared.

**Preparation of Embryo Extract:** Chick embryo extract is made from 10 to 11 days old embryos (before the calcifying mechanisms have become too active). The embryos are removed from the egg, and then homogenized in a motor driven homogenizer. Six to eight embryos and a measured quantity of balanced salt solutions (e.g., 2.0 ml per embryo) may be processed at one time. After homogenization, it is centrifuged and further diluted 10 to 20 times. Embryo extract may be stored indefinitely after it has been dried from the frozen state.

## Complex Natural Media

Some of the complex natural media are as follows:

**Supplemented Hanks–Simms medium:** Weller and co–workers (1952) in their earlier work with polioviruses made excellent use of a combination of 3 parts Hanks's balanced salt and 1 part Simm's ox serum ultra filtrate. For roller tube cultures of various human and animal tissues (embryonic, infant and adult), the complete medium consisted of Hanks–Simms solution (85%), beef embryo extract (10%), horse serum inactivated at 56° C for 30 minutes (5 to 20%), penicillin (50 μg/ml) and streptomycin (50 μg/ml).

**TABLE 5.2:** Different commonly used media (with serum), for animal cell and tissue culture, and their constituents (all quantities in mg/l)

| Component | Eagle's MEM | Dulbecco's modification | Ham's F12 | CMRL 1066 | RPMI 1640 | 199 | L15 | Fischer's | Way–mouth MB 752 |
|---|---|---|---|---|---|---|---|---|---|
| **AMINO ACIDS** | | | | | | | | | |
| L–alanine | – | – | 8.90 | 25.0 | – | 25.0 | 225 | – | – |
| L–arginine (free base) | – | – | – | – | 200 | – | 550 | – | – |
| L–arginine–HCl | 126 | 84.0 | 211 | 70.0 | – | 70.0 | – | 15.0 | 75.0 |
| L–asparagine | – | – | – | – | 50.0 | – | – | – | – |
| L–asparagine–$H_2O$ | – | –15.0 | – | – | – | 260 | 11.4 | – | |
| L–aspartic acid | – | –13.3 | 30.0 | 20.0 | 30.0 | – | – | 60.0 | |
| L–cysteine (free base) | – | – | – | – | – | – | 120 | – | 61.0 |
| L–cystine | 24 | 48.0 | – | 20.0 | 50.0 | – | – | – | 15.0 |
| L–cystine. 2Na | – | – | – | – | – | 23.7 | – | 23.7 | – |
| L–cysteine. HCL.$H_2O$ | – | – | 35.1 | 260 | – | 0.0987 | – | – | – |
| L–glutamic acid | – | –14.7 | 75.0 | 20.0 | 66.8 | – | – | 150 | |
| L–glutamine 292 | 584 | 146 | 100 | 300 | 100 | 300 | 200 | 350 | |
| Glycine | – | 30.0 | 7.50 | 50.0 | 10.0 | 50.0 | 200 | – | 50.0 |
| L–histidine (free base) | – | – | | | 15.0 | – | 250 | – | 128 |
| L–histidine HCl.$H_2O$ | 42.0 | 42.0 | 21.0 | 20.0 | – | 21.9 | – | 81.1 | – |
| L–hydroxyproline | | – | – | 10.0 | 20.0 | 10.0 | – | – | – |
| L–isoleucine | 52.0 | 105 | 3.94 | 20.0 | 50.0 | 20.0 | 125 | 75.0 | 25.0 |
| L–leucine | 52.0 | 105 | 13.1 | 60.0 | 50.0 | 60.0 | 125 | 30.0 | 50.0 |
| L–lysine. HCl | 73.1 | 146 | 36.5 | 70.0 | 40.0 | 70.0 | 93 | 50.0 | 240 |
| L–methionine | 15.0 | 30.0 | 4.48 | 15.0 | 15.0 | 15.0 | 75.0 | 100 | 50.0 |
| L–phenylalanine | 33.0 | 66.0 | 4.96 | 25.0 | 15.0 | 25.0 | 125 | 67.0 | 50.0 |
| L–proline | – | – | 34.5 | 40.0 | 20.0 | 40.0 | – | – | 50.0 |
| L–serine | – | 42.0 | 10.5 | 25.0 | 30.0 | 25.0 | 200 | 15.0 | – |
| L–threonine | 48.0 | 95.0 | 11.9 | 30.0 | 20.0 | 30.0 | 300 | 40.0 | 75.0 |
| L–tryptophan | 10.0 | 16.0 | 2.04 | 10.0 | 5.0 | 10.0 | 20.0 | 10.0 | 40.0 |
| L–tyrosine | 36.0 | 72.0 | 5.40 | 40.0 | 20.0 | – | – | – | 40.0 |

| Component | Eagle's MEM | Dulbecco's modification | Ham's F12 | CMRL 1066 | RPMI 1640 | 199 | L15 | Fischer's | Way–mouth MB 752 |
|---|---|---|---|---|---|---|---|---|---|
| L–tyrosine. 2Na | – | – | – | – | – | 49.7 | 373 | 74.6 | – |
| L–valine | 47.0 | 94.0 | 11.7 | 25.0 | 20.0 | 25.0 | 100 | 70.0 | 65.0 |
| **VITAMINS** | | | | | | | | | |
| L–ascorbic acid | – | – | – | 50.0 | – | 0.050 | – | – | 17.5 |
| Biotin | – | – | 0.0073 | 0.010 | 0.200 | 0.010 | – | 0.010 | 0.02 |
| D–Ca–pantothenate | 1.00 | 4.00 | 0.480 | 0.010 | 0.250 | 0.010 | 1.00 | 0.500 | 1.00 |
| Calciferol | – | – | – | – | – | 0.100 | – | – | – |
| Choline chloride | 1.00 | 4.00 | 14.0 | 0.500 | 3.00 | 0.500 | 1.00 | 1.50 | 250 |
| Folic acid | 1.00 | 4.00 | 1.30 | 0.010 | 1.00 | 0.010 | 1.00 | 10.0 | 0.40 |
| i–inositol | 2.00 | 7.20 | 18.0 | 0.050 | 35.0 | 0.050 | 2.00 | 1.50 | 1.00 |
| Nicotinamide | 1.00 | 4.00 | 0.04 | 0.025 | 1.00 | 0.025 | 1.00 | 0.50 | 1.00 |
| Pyridoxal. HCl | 1.00 | 4.00 | 0.062 | 0.025 | – | 0.025 | – | 0.50 | – |
| Riboflavin | 0.10 | 0.40 | 0.038 | 0.010 | 0.20 | 0.010 | – | 0.50 | 1.00 |
| Thiamin. HCl | 1.00 | 4.00 | 0.34 | 0.010 | 1.00 | 0.010 | – | 1.00 | 10.0 |
| Vitamin $B_{12}$ | – | – | 1.36 | – | 0.005 | – | – | – | 0.20 |
| Pyridoxine HCl | – | – | 0.062 | 0.025 | 1.00 | 0.025 | – | – | 1.00 |
| Cholesterol | – | – | – | 0.200 | – | – | – | – | – |
| Para–amino benzoic acid | – | – | – | 0.050 | 1.00 | 0.050 | – | – | – |
| Nicotinic acid | – | – | – | – | – | 0.025 | – | – | – |
| Menaphthone sodium bisulphite $3H_2O$ | – | – | – | – | – | 0.019 | – | – | – |
| Dl–Â tocopherol $PO_4$.2Na | – | – | – | – | – | 0.01 | – | – | – |
| Vitamin A acetate | – | – | – | – | – | 0.115 | – | – | – |
| Riboflavin | – | – | – | – | – | – | 0.10 | – | – |
| $PO_4$.2Na Thiamin mono $PO_4$,$2H_2O$ | – | – | – | – | – | – | 1.00 | – | – |
| **INORGANIC SALTS** | | | | | | | | | |
| $CaCl_2$ (anhyd.) | 200 | 200 | – | 200 | – | – | – | – | – |

| Component | Eagle's MEM | Dulbecco's modification | Ham's F12 | CMRL 1066 | RPMI 1640 | 199 | L15 | Fischer's | Way–mouth MB 752 |
|---|---|---|---|---|---|---|---|---|---|
| $CaCl_2.2H_2O$ | – | – | 44.0 | – | – | 186 | 186 | 91.0 | 120 |
| $Fe(NO_3)_3.9H_2O$ | – | 0.10 | – | – | – | 0.10 | – | – | – |
| KCl | 400 | 400 | 224 | 400 | 400 | 400 | 400 | 400 | 150 |
| $KH_2PO_4$ | – | – | – | – | – | 60.0 | 60.0 | – | 80.0 |
| $MgCl_2.6H_2O$ | – | – | 122 | – | – | – | – | – | 240 |
| $MgSO_4.7H_2O$ | 200 | 200 | – | 200 | 100 | 200 | 400 | 121 | 200 |
| NaCl | 6,800 | 6,400 | 7,599 | 6,799 | 6,000 | 8,000 | 8,000 | 8,000 | 6,000 |
| $NaHCO_3$ | 2,200 | 3,700 | 1,176 | 2,200 | 2,200 | 350 | – | 1,125 | 2,240 |
| $Na_2H_2PO_4.H_2O$ | 140 | 125 | – | 140 | – | – | – | 78.0 | – |
| $Na_2HPO_4$ (anhyd) | – | – | – | – | – | 47.5 | 190 | 60.0 | – |
| $Na_2HPO_4.7H_2O$ | – | – | 268 | – | 1,512 | – | – | – | 566 |
| $CuSO_4.5H_2O$ | – | – | 0.00249 | – | – | – | – | – | – |
| $FeSO_4.7H_2O$ | – | – | 0.834 | – | – | – | – | – | – |
| $ZnSO_4.7H_2O$ | – | – | 0.863 | – | – | – | – | – | – |
| $CaNO_3.4H_2O$ | – | – | – | – | 100 | – | – | – | – |
| **OTHER COMPONENTS** | | | | | | | | | |
| D–glucose | 1,000 | 4,500 | 1,802 | 1,000 | 2,000 | 1,000 | – | 1,000 | 5,000 |
| D–galactose | – | – | – | | – | – | 9,000 | – | – |
| Lipoic acid | – | – | 0.21 | – | – | – | – | – | – |
| Phenol red | 10.0 | 15.0 | 12.0 | 20.0 | 5.00 | 17.0 | 10.0 | 5.00 | 10.0 |
| Sodium pyruvate | – | 110 | 110 | – | – | – | 550 | – | – |
| Hypoxanthine | – | – | 4.10 | – | – | 0.30 | – | – | – |
| Linoleic acid | – | 0.084 | – | – | – | – | – | – | 25.0 |
| Putrescine 2HCl | – | – | 0.161 | – | – | – | – | – | – |
| Thymidine | – | – | 0.73 | 10.0 | – | – | – | – | – |
| Cocarboxylase | – | – | – | 1.00 | – | – | – | – | – |
| Coenzyme A | – | – | – | 2.50 | – | – | – | – | – |
| Deoxyadenosine | – | – | – | 10.0 | – | – | – | – | – |

| Component | Eagle's MEM | Dulbecco's modification | Ham's F12 | CMRL 1066 | RPMI 1640 | 199 | L15 | Fischer's | Way–mouth MB 752 |
|---|---|---|---|---|---|---|---|---|---|
| Deoxycytidine HCl | – | – | – | 10.0 | – | – | – | – | – |
| Deoxyguanosine | – | – | – | 10.0 | – | – | – | – | – |
| Diphosphopyridine nucleot | | | | | | | | | |

**Supplemented bovine amniotic fluid medium:** Milovanic and co–workers (1957, 1958) used the following medium:

1. Bovine amniotic fluid (37.5%), horse serum inactivated at 56° C for 30 minutes (20%), bovine embryo extract (5%), Hanks balanced salt solution (37.5%), streptomycin (100μ /ml), penicillin (100μ/ml) and mycostatin (100μ/ml).
2. Serum–supplemented yeast extract medium
3. Various human cell lines and strains from other species have been successfully grown in this medium. The medium consists of:
4. Yeast extract medium (76 parts) [composition: 10 parts of 1% solution of Difco's yeastolate, 2.5 parts of 10% glucose solution and 87.5 parts of Hanks balanced salt solution], human serum (20 parts), and 1.4 per cent sodium bicarbonate solution (4 parts).

**Serum–supplemented lactalbumin hydrolysate and yeast extract medium:** This medium consisted of: Earle's saline containing lactalbumin hydrolysate (0.5%), yeast extract (0.1%), and human or ox serum (10–20%).

### Chemically Defined Media

Earlier, the nutritive media for the cultivation of animal cells *in vitro* were derived from the organisms and consisted of blood plasma, blood serum, tissue extracts, etc. The complexity and variability of these naturally occurring materials made it difficult to use them in experiments designed to determine the nutritive substances required by the cells and the effect of particular substance upon them. Morgan et al. (1950) published the composition of a more adequate medium No. 199. The medium included almost complete complement of amino acids and vitamins as well as several nucleic acid constituents and certain intermediary metabolites and accessory growth factors. Glutamine was included because of the importance placed upon it by Fischer's group and Tween–80 was used as a water soluble source of fatty acid and as means of dissolving the fat soluble vitamins and cholesterol in a minimal concentration of ethyl alcohol. The medium also contained the salts of Erle's balanced salt solution, glucose, ferric nitrate and phenol red.

**TABLE 5.3:** Examples of selective media (modified from Freshney, 1987)

| Cells of cell line | Medium |
|---|---|
| Bronchial epithelium | LHC |
| Endothelium | MCDB 130 |
| Fibroblasts | MCDB 202 |
| Fibroblasts | MCDB 110 |
| Keratinocytes | MCDB 153 |
| Mammary epithelium | MCDB 170 |
| Prostate epithelium (human) | WAJC 404 |
| Prostate epithelium (rat) | WAJC 404 |
| Small cell lung cancer | HITES |

### Other Tissue Culture Media

**Medium No. 612:** It was observed that medium 199 had a very high oxidation–reduction potential. Healy and co–workers attempted to bring it near the physiological range by increasing very considerably the levels of three reducing agents present in the medium namely cysteine, glutathione, and ascorbic acid.

Cysteine and glutathione increase the rate of cell multiplication and improve the appearance of the cells; the ascorbic acid has no apparent effect. Healy increased the levels of cysteine and glutathione 2,600 and 200 times respectively. He designated this medium as medium No. 612.

**Medium No. 635:** This medium was devised by omitting from medium No. 612 all the purines, pyrimidines as well as adenosinetriphosphate, adenylic acid, ribose and deoxyribose. This medium gives a better growth response than that of medium No. 612, but the cultures rarely survive more than 40 days.

**Medium No. 858:** This medium includes all the deoxyribonucleosides, coenzymes, and sodium glucoronate in addition to medium No. 635. This medium contains only L–form amino acids, and yields ten–fold increase in the population of culture cells in 7 days. Addition of 10 to 20 per cent of horse serum yields 20 to 30–fold increase in 7 days.

**Medium No. 866:** Medium No. 866 is similar to that of medium No. 858 supplemented with three fatty acids namely linoleic acid, linolenic acid and arachidonic acid. This medium does not improve the growth of culture cells.

**Medium CMRL–1066:** This medium is identical to medium No. 858 except that the fat soluble vitamins (A, D, E and K), ferric nitrate and sodium bicarbonate included in medium No. 858 have been omitted. Five B vitamins (thiamine, riboflavin, niacin, niacinamide and sodium pantothenate) were added. It also contains 0.02 mg percent n–butyl parahydroxybenzoate, which offers protection from certain moulds. Antibiotics are added according to the need.

Eagle found that 12 amino acids are essential for the growth of both L–strain and HeLa–strain cultures even in the presence of small quantities of whole or diazlyzed serum. These amino acids are arginine, cystine, histidine, isoleucine, leucine, lysine, methionine, phenylalanine, threonine, tryptophan, tyrosine, and valine. Recent attempts have been made to determine the vitamin requirements of cells cultivated in chemically defined media containing dialzed serum.

Glucose is generally added to tissue culture media as an energy source, other carbohydrates have also been tested. Harris and Kutsky have shown that chick heart fibroblasts can utilize D–fructose or D–mannose as well as D–glucose.

Interesting studies have also been made of the inorganic requirements of cells *in vitro.* Harris found that carbon dioxide is beneficial for the outgrowth of cells from explanted chick embryo tissues. Eagle has shown that sodium, potassium, magnesium, calcium, chloride and $H_2PO_4$ ions are essential for the survival of L– and HeLa cells.

## LABORATORY FACILITIES FOR TISSUE CULTURE

For animal tissue culture laboratory facilities/methods are needed, they are: sterile handling, incubation, preparation of glassware, media and tissue, wash–up, sterilization and storage. The clean area for handling should be available at one end of the room and the wash up and sterilization at the other end, with preparation, incubation and storage in between.

**TABLE 5.4:** Tissue culture facilities (modified from Freshney, 1987)*

| Minimum requirements (essential) | Important features required (beneficial)<br>Other useful additions required |
|---|---|
| **Sterile area:** clean and quiet, no thorough traffic, separate from animal house and microbiological labs | Filtered air (air conditioning)<br>Piped $CO_2$ and compressed air |

| Minimum requirements (essential) | Important features required (beneficial) Other useful additions required |
|---|---|
| **Preparation area** | Hot room with temperature recorder<br>Store room for bulk plastic |
| **Storage areas** | Microscope room<br>Containment room for biohazard work. |
| *liquids*–ambient, 4°–20°C | Dark–room<br>Liquid $N_2$ storage tank ( 500 1) |
| *glassware* (shelving) | Service bench adjacent to culture area |
| *plastics* (shelving) | Separate preparation room. |
| *small items* (drawers) | Separate sterilizing room |
| *specialized equipment* (slow turnover), cupborard (s) | Cylinder store |
| *chemicals*–ambient, 4°–20°C (share with liquids but keep chemicals in sealed container over desiccant) | |

*(For details also refer to the book *Culture of Animal Cells: A Manual of Basic Techniques*, R. lan Freshney, Alan R. Liss Inc., New York).

### Other Facilities

1. The **laminar flow cabinets** or **hoods** are required for cleaning bench. It facilitates the maintenance of aseptic and sterile conditions, so that no separate room for sterile handling is needed and a laminar flow cabinet can be kept in an unspecialized laboratory space. The aseptic rooms may be designed in such a way so that filtered air may be supplied from the ceiling. It helps to sterile the complete working area (room).
2. **Incubation** is generally done in incubators or thermostatically controlled hot rooms. In hot rooms, racks are designed for the placement of cultures (tubes, Petri plates, flasks, etc.).
3. The ready–made media are available in market and may be purchased or may be prepared in laboratory.
4. The **washing and sterilization** are being done outside the tissue culture laboratory. These procedures generate moisture (increased humidity) and heat.
5. Space for **storage** is also required for the following: (filters, large receiver flasks).

## THE SUBSTRATE ON WHICH CELLS GROW IN LABORATORY

Most of the vertebrate cells cultured *in vitro* are grown as monolayers on an artificial substrate. Spontaneous growth in suspension is restricted to **hemopoietic cell lines, rodent ascites tumour** cell lines and a few other selected cell lines. With these exceptions, most cells need to spread out on a substrate in order to proliferate. Inadequate spreading will inhibit cell proliferation. Cells shown to require attachment for growth are said to be **anchorage dependent.**

However, cells, which have undergone **transformation,** frequently become anchorage independent and can grow in suspension. (Note: In microbiology and plant cell culture, transformation means uptake of new genetic material but in animal cell culture, particularly mammalian cell culture, transformation means phenotypic modification of cells in culture, without involving the uptake of new genetic material, which is described as **transfection** to distinguish it from transformation). Whenever, a substrate is needed, it may be **adhesive** (glass, plastic, palladium, etc. or a metallic surface; these surfaces may sometimes be variously modified) or **non–adhesive** (agar, agarose, methocel) and will be briefly discussed in this section.

**Glass as a Substrate:** Glass is the most commonly used substrate, in the form of slides, coverslips, flasks and test tubes, which may or may not be variously modified. It is cheap, easily washed without losing its growth supporting properties.

**Disposable plastic as a substrate:** Polystyrene is often used as the most common and cheapest substrate, to be discarded after single use. This plasticware is supplied sterile and is not autoclavable. However, it is hydrophobic and needs to be treated by gamma irradiation or some chemical or with electric are to produce a charged surface, which thus becomes wettable. Other plastics used for tissue culture include the following: polyvinylchloride (PVC), polycarbonate, polytetrafluoroethylene or teflon (PTFE), thermanox (TPX), etc.

**Palladium and metallic surface as substrates:** For 90% of animal cell and tissue cultures, glass or plastic is employed as a substrate. For instance, palladium deposited on agarose (using electron microscopy shadowing equipment) was used as a substrate for growth of fibroblasts and gila. Stainless steel discs and other metallic surfaces were also used.

## Treatment of Substrate Surfaces

The substrate surfaces should be treated to improve cell attachment. The treatments with culture medium meant for another culture, or with purified fibronectin (1ng/ml) added to the medium or with collagen are commonly used. Coating with collagen can be achieved, using rat tail collagen or other commercially available forms, and pouring the collagen solution over the surface of the dish, draining the excess, and permitting the residue to dry. After coating with collagen, the substrate may be sterilized under UV and used either immediately or may also be stored. Sometimes on the inert coating of collagen (or gelatin or poly–D–lysine), a monolayer of an appropriate cell type may be used to provide a matrix for the maintenance of some specialized cells.

## Feeder Layers on Substrate

Sometimes, on the substrate surface a monolayer of special kind of cells is used, partly to supplement the culture medium for growth and partly for conditioning of the substrate by cell products. These monolayers are called feeder layers, since they also feed the growing cultures. The feeder layer may consist of mouse embryo fibroblasts, normal fetal intestine, glial cells, etc. and has been effectively used for selective growth of breast and coelomic epithelium, central and peripheral neurons, and also for growing cells used for the production of transgenic animals through transfection.

## The Gas Phase for Tissue Culture

The animal cell and tissue cultures also require for their growth a gas phase consisting of $O_2$ and $CO_2$ even though cultures (cell culture *vs* organ culture) may vary in their oxygen requirement. These gases also influence the pH and the $HCO_3^-$ ion concentration of the culture.

## CULTURE PROCEDURES

The major steps involved for culture are preparation and sterilization of glassware, equipment, reagents and media. The apparatus and liquids meant for tissue culture should be reserved for that purpose alone. The washing of glassware should also be done separate.

### Preparation and Sterilization of Glassware and Apparatus

The following procedure and precautions should be followed:

1. Do not allow soiled glassware to dry, and always collect soiled glassware in water with sodium hypochlorite. It will prevent microbial contamination.
2. A non–toxic and effective detergent meant for your area should be used.
3. After washing, glassware should be thoroughly rinsed in tap water and then in deionized or distilled water and then should be dry inverted. The dried glassware should be sterilized by dry heat (in oven at 160°c for 1 hr).
4. Automatic washing machines are also used.

### Preparation and Sterilization of Reagents and Media

Reagents and media should be produced in pure form and should be sterilized either by moist heat i.e. by autoclaving (if heat stable) or by membrane filtration (if heat labile). During autoclaving (moist heating by steam), the container may be kept sealed (if borosilicate or polycarbonate) or the caps may be left slack (if soda glass) to avoid breakage. The liquid layer in container may go down due to evaporation and should be restored with sterile distilled water after autoclaving. You may also use ready made pre sterilized culture media available in market.

**TABLE 5.5:** Sterilization of equipment and apparatus (modified from Freshney, 1987)

| Item | Sterilization |
|---|---|
| Apparatus containing glass and silicone tubing | Autoclave* |
| Dispenser tubing for compupet | Autoclave |
| Disposable tips for micropipettes | Autoclave |
| Filter–Millipore, Sartorius | Autoclave–do not prevac. or postvac. |
| Glass bottle with screw caps | Autoclave |
| Glass coverslips | Dry heat |
| Glass slides | Dry heat |
| Glassware | Dry heat+ |
| Instruments | Dry heat |
| Magnetic stirrer bars | Autoclave |
| Pasteur pipettes–glass | Dry heat |
| Pipettes–glass | Dry heat |
| Screw caps | Autoclave |
| Silicone tubing | Autoclave |
| Stoppers–rubber, silicone | Autoclave |
| Test tubes | Dry heat |

* Autoclave = 100 kPa (15lb/in$^2$); 121° C for min., + Dry heat = 160°C/l hr.

**TABLE 5.6:** Sterilization of liquids (modified from Freshney, 1987)

| Solution | Sterilization | Storage |
|---|---|---|
| Agar | Autoclave* | Room temperature |
| Amino acids | Filter + | 4°C |
| Antibodies | Filter | –20°C |
| Bacto–peptone | Autoclave | Room temperature |
| Bovine serum albumin | Filter–use stacked filters | 4°C |
| Carboxylmethyl cellulose | Steam–30 min** | 4°C |
| Collagenase | Filter | –20°C |
| DMSO | Self–sterilizing, aliquot into sterile tubes | Room temperature–keep dark; avoid contact with rubber or plastic. |
| EDTA | Autoclave | Room temperature |
| Glucose–20% | Autoclave | Room temperature |
| Glucose–1.2% | Filter | Room temperature |
| Glutamin | Filter | –20°C |
| Glycerol | Autoclave | Room temperature |
| HEPES | Autoclave | Room temperature |
| HCI IN | Filter | Room temperature |
| Lactalbumin hydrolysate | Autoclave | Room temperature |
| Methocel | Autoclave | 4°C |
| $NaHCO_3$ | Filter | Room temperature |
| NaOH IN | Filter | Room temperature |
| Phenol red | Autoclave | Room temperature |
| Salt solutions (without glucose) | Autoclave | Room temperature |
| Serum | Filter–use stacked filters | –20°C |
| Sodium pyruvate 100mM | Filter | –20°C |
| Transferrin | Filter | –20°C |
| Tryptose | Autoclave | Room temperature |
| Trypsin | Filter | –20°C |
| Vitamins | Filter | –20°C |
| Water | Autoclave | Room temperature |

* Autoclave=100 kPa (15 lb/in$^2$) 121° C for 20 min., + Filter =0.2 µm pore size, ** Steam = 100°C for 30 min.

## Preparation of Animal Material

It is essential that all sort of contamination should also be avoided during experimental manipulations by collecting tissue aseptically or sterilizing it after removal. Embryo or other contents of uterus are free from contamination while other adult tissues are contaminated with bacteria. Note that embryo should be removed carefully under aseptic conditions eliminating the possibility of contamination. The adult tissue may be washed with a solution having high concentration mixture of antibiotic. The mixture is a *balanced salt solution* containing 100 units of penicillin and 0.5 mg of streptomycin or neomycin per ml. Further handling of tissue should be carefully done under fully aseptic conditions, which can be created either through preparing an aseptic room or through the use of *laminar flow cabinet* or *hood*. The use of hoods has greatly facilitated the handling of material under aseptic conditions.

## PRIMARY CULTURE, CELL LINES AND CLONING

### Disaggregation of Tissue and Primary Culture

A Primary cell culture may be obtained by either by allowing cells to migrate out from the tissue or by disaggregating the tissue mechanically or enzymatically for the production of suspension of cells. The normal untransformed cells can survive and proliferate easily and they produce *primary culture* when attached to a substrate. These cells are to be obtained by **disaggregation**. It is achieved by following methods:

1. **Physical disruption:** It involves cutting the tissue into pieces.
2. **Enzymatic digestion:** Several enzymes are being used for disaggregation like, trypsin and pronase. Collagenase, elastase, mucase, pancreatin. pappain may also be used for this purpose. Enzymatic disaggregation is labour intensive and involves damage of cells.
3. **Treatment with Chelating Agents:** Chelating agents are used for production of cell suspensions from established cultures. Epithelium tissues which needs $Ca^{++}$ and $Mg^{++}$ ions for its integrity, can be treated with chelating agents, e.g., citrate and ethylene–diamine–tetra–acetic acid (EDTA).

### Isolation of Tissue

For tissue isolation sterilize the site with 70% alcohol, remove the tissue aseptically. The isolated tissue may be stored in a refrigerator before transferring it to BSS or to a culture medium, because viable cells may be recovered from chilled tissue several days after explantation. Freshney (1987) described various protocols for isolation of tissues like mouse embryo, hen's egg, human biopsy material. For details refer to his book.

### Enzymatic Disaggregation

The mechanical or enzymatic disaggregation gives a much higher yield of representative cells in a short duration. The embryonic tissue disperses more readily and gives higher yield than the newly born or adult. In several tumours, the tissue (being fragile) is mostly destroyed during enzyme treatment and viable cells are obtained with difficulty.

#### *Disaggregation by Trypsin (for Embryonic Tissue) or Trypsinization*

Trypsinization may be of two types: (1) warm trypsinization and (2) cold trypsinization. Crude trypsin is the most common enzyme used for disaggregation. It can be tolerated by a variety of cells and is effective for many tissues. Its residual activity is neutralized by serum of the medium or by a trypsin inhibitor (e.g. **soybean trypsin inhibitor**) in case of a serum free medium. The cells are exposed to the warm enzyme (36.5° C) for a minimum period. The dissociated cells are collected every half an hour. The trypsin is removed by centrifugation after 3–4 hours. This time is required for complete disaggregation. Alternatively cold trypsinization may be used which involves soaking of tissue in trypsin at 40° C (to allow penetration of enzyme), followed by incubation at 36.5°C for a shorter period.

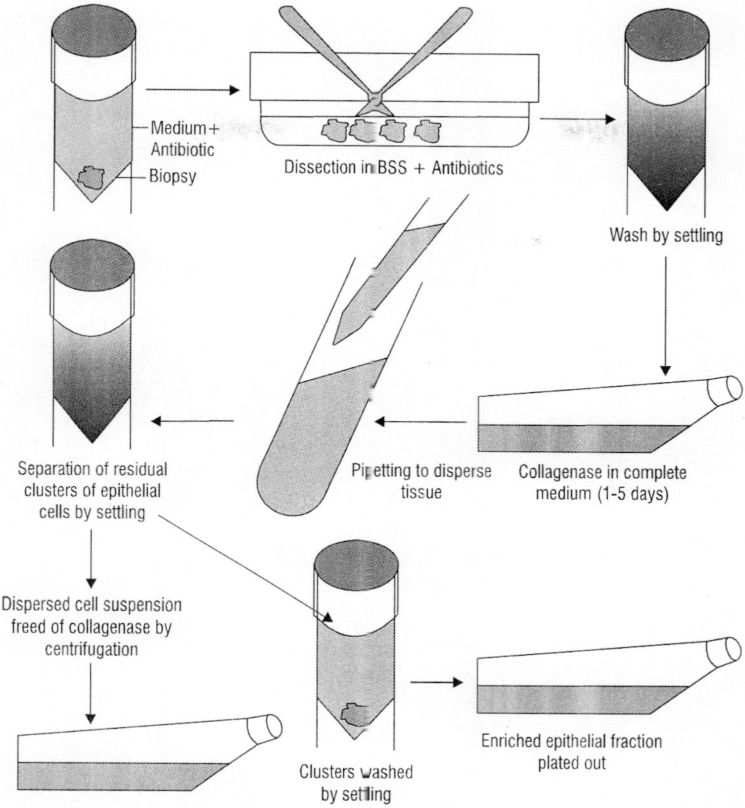


**FIGURE 5.1:** Stages in disaggregation of tissue for primary culture by collagenase.

### *Disaggregation by Collagenase (for Embryonic, Normal and Malignant Tissues)*

Sometimes the disaggregation by trypsin may be damaging for epithelial cells or it becomes ineffective for fibrous tissues. The intracellular matrix contains collagen, therefore, collagenase has proved effective for disaggregation of several normal and malignant tissues. These tissues may be sensitive to trypsin. Crude collagenase (with contamination of nonspecific proteases) is often used with a finely chopped tissue in complete medium. When tissue is disaggregated, collagenase is removed by centrifugation, and the cells may be seeded and cultured at a high concentration.

### Mechanical Disaggregation

The cell suspension is more quickly obtained than in the enzymatic disaggregation. The mechanical disaggregation is **force based technique** in which the tissue is carefully sliced and the cells which spill out are collected. In this technique the cells are either pressed through the sieves of gradually reduced mesh or forced through a syringe and needle, or repeatedly pipetted. Sometime there may be a mechanical damage of the cells. The mechanical disaggregation provides good yield of cells in a shorter time. Steps involved in one of the methods of mechanical disaggregation.

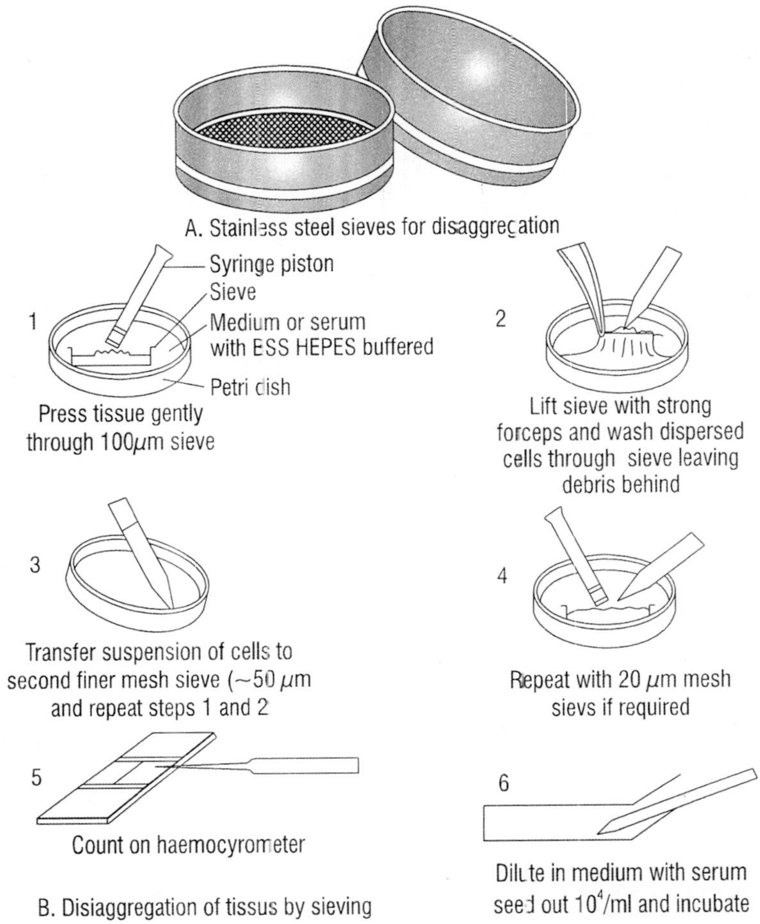


**FIGURE 5.2:** (a) Stainless steel sieves suitable for disaggregation of tissue, (b) Various steps in disaggregation of tissue by sieving.

### *Separation of Viable and Nonviable Cells*

The dissociated cells obtained as above, usually described as primary cells. They grow well when seeded on culture plates at high density. These are known as **adherent primary cultures** in which the nonviable cells are removed at the first change of medium.

The primary cultures can also be maintained in suspension in which the non–viable cells are gradually diluted out, when cell proliferation starts. The non–viable cells can be removed from primary disaggregate using centrifugation technique using a mixture of Ficoll and sodium metrizoate.

### *Somatic Cell Fusion*

Fusion of different types of cells has been successfully achieved in animals to produce hybrid cells. The technique is useful because it helps to study of the control of gene expression, differentiation, to know the problem of *malignancy*, viral replication, gene mapping, and production of hybridomas for antibody production.

The cell fusion in animals has been studied within the body e.g., myoblasts fusing to form multinucleate muscle fibres. They can also be fused in vitro to produce binucleate heterokaryons (cells fuse, but two types of nuclei do not fuse) or uninucleate hybrid cells.

In 1960s, in France, hybrid cells were obtained from mixed cultures of two different cell lines derived from mouse. Later many successful attempts of cell fusion using widely different taxa were made. Sendai virus is often used to induce such a cell fusion. The cell fusion can be induced by polyethylene glycol (PEG). Removal of surface carbohydrates is also, sometimes, required for successful cell fusion.

## TISSUE CULTURES

### Primary Explantation Techniques

Harrison (1907) and Carrel (1912) developed this technique for cultivation of pieces of fresh tissue derived from the organism. This technique is also used for embryo and organ culture, but are variously modified to become specialized techniques. The different explantation techniques are:

(i) Slide cultures, (ii) Carrel flask cultures and (iii) Roller test tube cultures.

### Slide or Coverslip Cultures

This is the oldest method of tissue culture and is still quite widely used. The technique is being used for limited purposes, specially for morphological studies and for cinemicrographic investigations. The slides or cover slips are prepared by placing a fragment of tissue (*explantation*) onto a coverslip, which is subsequently inverted over the cavity of a depression slide. The advantages and disadvantages are as follows:

#### *Advantages of Slide or Coverslip Cultures*

1. It is simple and relatively inexpensive.
2. Cells in living state are spread out in a manner suitable for microscopy and photography.
3. Cells grow directly on coverslip and can be fixed and stained to make permanent slides.

#### *Disadvantages of Slide or Coverslip Cultures*

1. Supply of oxygen and nutrients is rapidly exhausted, so that the medium quickly becomes acidic and requires transfer of rapidly growing tissue.
2. Sterility can not be maintained for a long period.
3. One can culture very small amount of tissues.

**Single coverslip with plasma clot** (Harrison, 1907)

*This technique has following steps:*

1. Prepare medium in two parts, one containing 50% plasma in BSS (balanced salt solution) and the other containing 50% embryo extract in serum;
2. Under sterile and aseptic conditions, using a capillary pipette, place one drop of plasma containing solution in the centre of each of one, or more coverslips (22mm);

3. Transfer a fragment (one or two pieces) of tissue (called explant) to this drop without crushing the tissue;
4. Add the embryo extract containing solution and mix thoroughly before clotting starts and then locate the explant;
5. Place two small spots of petroleum jelly (using a glass rod) near the concavity of a depression slide and invert this slide over the coverslip; apply gentle pressure, so that jelly sticks to coverslip;
6. Allow culture medium to clot; (vii) turn over the slide and seal the margins of coverslip with paraffin;
7. Label and incubate at 37°C.

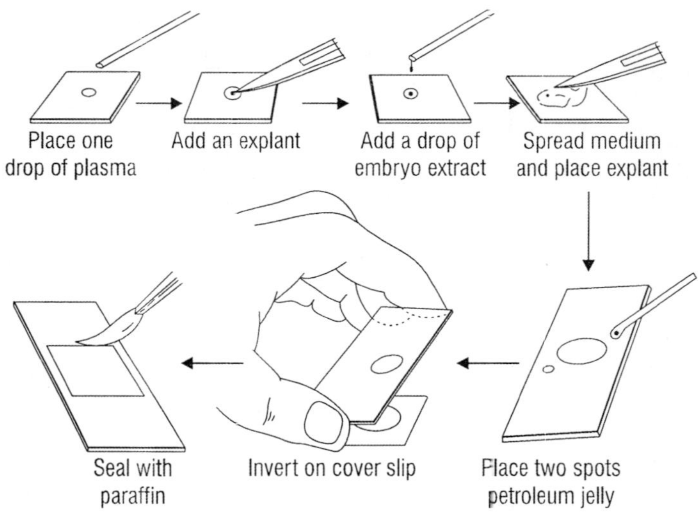


**FIGURE 5.3:** Preparation of single cover slip cultures.

## Double coverslip with Plasma Clot (by Maximov)

This technique resembles the single coverslip method and is completed in following steps. A large depression slide is used and the entire preparation is attached to it by petroleum jelly and wax in such a way that the small coverslip is not in contact with the slide at any point. The technique includes the following steps:

1. A small drop of BSS is placed on a large coverslip (40mm).
2. A square or round coverslip (22mm) is placed over BSS in the centre of large coverslip.
3. The above two steps are then followed by the steps listed above for single coverslip method.

## Single Coverslip with Liquid Medium (Lying and Hanging Drop Cultures)

Following steps are involved in this method:

1. Prepared explants are placed in culture medium in a watch glass.
2. The explants are then drawn into the tip of a capillary pipette
3. One explant is deposited in the centre of each coverslip.
4. The liquid medium can be spread out in a very thin circular film with the explant protruding above the surface.

5. A depression slide with petroleum jelly is applied immediately and preparation turned over with a quick flip to prevent the fluid from running into the crevice between the slide and the coverslip.
6. The coverslip is sealed and the slide incubated at 37°C, upright or inverted; the tissue grows on the coverglass.

**Post –care of slide cultures:** Single coverslip cultures are very useful for short term studies but they are difficult to handle subsequently except by a process of transfer. Therefore, **double coverslip method** is recommended. It is desirable to leave the explant in its original location to obtain long term tissue cultures. This needs washing, feeding, patching and transfer of cultures.

### *Washing and Feeding Double Coverslip Cultures*

Washing and feeding involves the following steps:

1. Remove seal using razor blade and remove large coverslip with small one still attached and flip it over, so that culture is uppermost in orientation.
2. Detach small coverslip from the large coverslip using needle and forceps
3. Transfer it to a Columbia staining dish (watch glass or Petri dish may also be used) containing balanced salt solution (BSS).
4. Remove small cover slip treated as above (one at a time) and place it on a large coverslip with culture up.
5. While removing small coverslip from dish with BSS, the amount of BSS brought with it may be controlled by the rate of removal.
6. Too much liquid will interfere in feeding operation and too little will allow air bubbles.
7. Feed the culture by adding a drop of feeding solution (e.g., BSS: serum: 50% embryo extract = 1:1:1) to the small coverslip.
8. Attach a clean depression slide, using petroleum jelly, as done earlier.

## Flask Cultures

The Carrel flask technique is meant for the establishment of a strain from fresh explants of tissue because it has excellent optical properties for microscopic examination. The polystyrene culture flasks can also be used for this purpose. There are two types of flask techniques:

1. **Thick clot cultures:** It allows rapid growth suitable for short term cultures.
2. **Thin clot cultures:** It can be maintained for a considerable period of time.

### *Advantages of Flask Culture*

1. The tissue can be maintained in the same flask for months or even years.
2. A large number of cultures can be easily prepared.
3. A large amount of tissue can be grown with large amount of medium.

### *Preparation of Flask Cultures*

1. Place upto six D3.5 Carrel Flasks in a rack with their necks flamed and pointing to the right.
2. Place a drop of plasma on the floor or flask and spread this plasma out in a circle.
3. Use spatula and transfer the desired number of explants to the plasma and allow clotting to occur.
4. After clotting of plasma and fixing of explants in position, add extra medium.

5. For thick clots 1.2ml of dilute plasma and for thin clots 1.2ml of dilute serum is added instead of plasma; the whole thing is left for clotting.
6. The flasks are gassed with gas phase (5% $CO_2$ in air).

### *Renewal of Medium*

It is essential to remove the medium periodically and is to be replaced in flask cultures in the following manner:

1. Old fluid is drawn off with the help of a pipette.
2. 1.2ml of fluid medium is added as replacement.
3. The flask is gassed as above.

### *Transfer of Culture*

For transfer, the culture grown in a flask culture is to be removed and cut into pieces. These pieces are then used for replantation as usual.

### Test Tube Cultures

The feeding, patching and transfer of culture is done as in other primary explantation techniques. It is the cheap technique and can be used for preparing a large number of cultures, which can be placed in stationary racks or roller drums. This technique provides poor optical property for microscopy. Difficulty also arises in quantitation due to the curvature and high risk of contamination. Cultures on plasma clots in test tube are prepared just like those in flasks, but tissues may be grown on the wall of test tube without a plasma clot. Test tubes may also be used for developing suspension cultures.

## ORGAN CULTURE

Pieces of an organ or whole organ can be cultured in vitro. The main objective is to maintain the architecture of the tissue and direct it towards normal development. In this technique, it is essential that the tissue should never be disrupted or damaged. It requires careful handling. Media used for growing organ cultures are generally the same as those used for tissue culture. The techniques of organ culture can be divided into (i) those employing a **solid medium** and (ii) those employing **liquid medium.**

### Culture of Embryonic Organs

Embryonic organs culture is easier than to normal organs from adult animals. Following are the three techniques used for embryo culture.

### Organ Cultures on Plasma Clots

It involves following steps:

1. Prepare a plasma clot by mixing 15 drops of plasma with five drops of embryo extract in a watch glass.
2. Place the watch glass on a pad of cotton wool in a Petri dish; cotton wool is kept moist to prevent excessive evaporation from the dish.
3. Place a small, carefully dissected piece of tissue on top of the plasma clot in the watch glass.

The technique has now been modified, and a raft of lens paper or rayon net is used on which the tissue is placed. Transfer of the tissue can then be achieved by moving the raft easily. Excess fluid is removed and the net with the tissue placed again on fresh pool of medium.

### Organ Cultures on Agar

Media solidified with agar are also used for organ culture and these media consist of 7 parts 1% agar in BSS, 3 parts chick embryo extract and 3 parts horse serum. Defined media with or without serum are also used with agar. The medium with agar provided the mechanical support for organ culture. It does not liquefy. Embryonic organs generally grow well on this medium, but from adults will not survive.

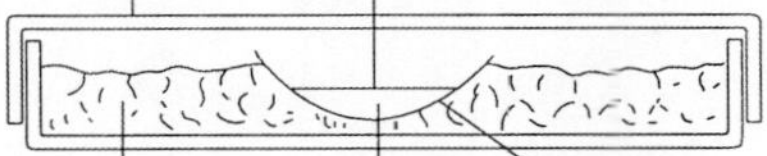


**FIGURE 5.4:** The classical organ culture technique.

The culture of adult organs or parts from adult animal is more difficult due to their greater requirement of $O_2$. A variety of adult organs (e.g. liver) have been cultured using special media (e.g. T8; sec Paul, 1975) with a special apparatus **(Towell's Type II culture chamber).** Since serum was found to be toxic, serum free media were used, and the special apparatus permitted the use of 95% $O_2$.

## WHOLE EMBRYO CULTURE

### Culture of Chick Embryo (Spratt, 1956)

In this technique, 40–hours old embryos were used and the embryo development could be followed for another 24–48 hours *in vitro* before the embryo dies.

Steps of Chick Embryo Culture

1. Prepare a suitable defined medium and added to sterile watch glasses, placed on moist absorbent cotton wool pads in Petri dishes (as for organ culture).
2. Incubate hen's eggs at 38°C for 40–42 hours to provide about a dozen embryos.
3. The shell is wiped with alcohol and broken into a sterile evaporating dish containing 50 ml chick saline or BSS.
4. A circular cut is made (using scissors) into the vitelline membrane around the blastoderm and the latter is transferred to a Petri dish containing BSS
5. The adherent vitelline membrane is removed with the aid of forceps and the embryo is examined under the microscope to determine the stage of development.
6. The blastoderm is transferred to the top of the medium in the watch glass prepared in (1) above.
7. The blastoderm is spread on agar gel (ventral side down) and the excess BSS is removed.
8. Culture is incubated at 37.5°C.

## ANIMAL TISSUE CULTURES IN BIOMDICAL RESEARCH–GENETICS

Cell culture methods have come to be used extensively in both applied and theoretical genetic studies. The various genetic applications are as follows:

## Karyological Studies

The karyotype of a population of cells or an organism is the catalogue of the chromosomes of a typical' or an average cell. While karyotyping is a useful technique in experimental biology, for example for identifying the species of origin of cell–lines, its most important application in medicine is in the early diagnosis of some congenital abnormalities. Mostly these are abnormalities resulting from the inheritance of an unbalanced complement of chromosomes, such as Klinefelter's syndrome (XXY), Turner's syndrome (XO) or Down's syndrome (mongolism) in which chromosome number 21 is represented by 3 instead of 2. Some other anomalies such as translocations or deletions can also be recognized.

## Identification and Study of Hereditary Metabolic Disorders

The precise nature of many of the genetic diseases can be determined only by demonstrating the metabolic abnormality, commonly the absence of an enzyme in tissue. Disorders such as absence of catalase in acatalasia, of alpha–D–galactose–1–phosphate, UDP glucose uridyltransferase in galactosaemia or the failure to decarboxylate branched chain keto–acid in maple syrup urine disease can be demonstrated in primary fibroblast cultures. In some instances, the heterozygous locus, a deficiency of the transferase enzyme per cell can be measured. Some human genetic disorders which can be studied in this way are: glycogen storage diseases, hemocystinuria, Hunter's syndrome, Hurler's syndrome, I–cell disease, Niemann–pick disease, orotic aciduria, Refsum's disease, xeroderma pigmentosa, etc.

## Somatic Cell Genetics

Somatic cell genetic studies have three main aims: (i) study of the behaviour of heterokaryons following fusion of parent cells with different properties, (ii) study of the behaviour of hybrid cells derived from heterokaryons with the view to eliciting information about interaction between different genes, and (iii) studies with a view to mapping loci on different chromosomes. These studies can be made in tissue cultures.

## Tissue Cultures in Biomedical Research–Virology and Host–Parasite Relationships

The ability of tissue cultures to support the growth of viruses and to reveal their presence by lesions which are in some cases specific has been applied in virology for the following purposes:

1. The study of host–parasite relationships.
2. The detection and identification of viruses.
3. The production of viruses for vaccine manufacture.

The study of the relationships between viruses and cells have been greatly aided by the use of the tissue culture methods. These techniques are also useful for detection and identification of viruses from the suspected cases. Of very great importance in recent years have been the development of vaccines against poliomyelitis which are made from the viruses grown in tissue culture. In addition to viruses other intracellular parasites which have been grown in tissue culture and studied are: Rickettsiae, mycobacteria (especially tubercle and leprosy bacilli), lepraemurium, parasitic protozoans, etc.

## Tissue Cultures in Biomedical Research–Cancer

Tissue culture techniques have been very useful at understanding the problem of cancer. Since culture of canine lymphosarcoma by Beebe and Ewing (1906), and unbroken trail of pioneers such as Carrel, Warburg, Earle, Rous, Gey, Murray, and others worked on this problem. The breakthrough was the development of quantitative assay procedures for tumour viruses stimulated by the exploitation of cell culture methods by virologists. Since Rous demonstrated the infective nature of the fowl sarcoma, many other infective cancer producing agents have been recognized.

# 6. Hybridomas Technology and Monoclonal Antibodies

## HYBRIDOMA

The decade of the 70's saw major biological innovations-recombinant DNA (rDNA) and hybridoma techniques-and these innovations have been the driving force behind the biotechnology revolution. In addition to the emergence of several hundred firms to develop and commercialize the new age pharmaceuticals and diagnostics, fundamental advances have been made in the understanding of the structure, function, differentiation, and disease in biological systems. New antigens and biomolecules hitherto unknown have been discovered, and previously known molecules have been made available in quantities large enough for systematic research and development applications in biology and medicine.

**Hybridoma** - a biotechnological tool- is the new path for achieving the goal of complete immunization of human body from infections. In 1975, a new era in the immunology was launched with the discovery of the hybridoma technique, a method of creating pure and uniform antibodies (immunoglobulins) against a specific target (antigen). To understand the basic concepts of hybridoma technology and its role in production of monoclonal antibodies.

Hybridomas are the hybrid cells of myceloma (cancer) cells with antibody producing cells (lymphocytes) from an immunized donor (animal). The hybrid cell or hybridoma resulting from this fusion has the ability to multiply rapidly and indefinitely in vitro and to produce an antibody of predetermined specificity, known as **monoclonal antibody (Mabs)**.

### Monoclonal B-cells

B-cell *hybridomas* are aneuploid/polyploid cells derived from fusing B cells and myeloma cells. T-cell lymphoma. Hybridoma technology precipitated a wide range of applications in fundamental and applied immunology. Prior to this development a number of naturally occurring transformed B cells or *plasmacytomas* were available secreting antibodies of single species. The binding specificities of most of these antibodies were poorly understood. One B cell normally secretes one species of immunoglobulin. Kohler and Milstein combined the two desirable properties, i.e., specific antibody secretion and continuous cell division into one fused hybridoma cell derived from the union of a myceloma cell and a spleen cell obtained from a mouse immunized with the desired antigen. The hybrid cells were selected and grown exploiting a drug selection procedure to eliminate myceloma cells. B-cells die naturally after a short period of time in tissue culture.

*Hybrid-Hybridomas* are cells obtained by second derivative fusion secreting bispecific monoclonal antibodies and parental antibody species. Two types of hybrid-hybridomas are recognized. *Triomas* are fusion products of a hybridoma cell with a spleen cell while *quadromas* are derived by fusing two hybridomas. Thus, resolution and isolation of individual monoclonal antibodies from such monoclonal cultures is necessary for practical applications, *Transfectomas* are lymphoid cells transfected or injected with desired immunoglobulin genes obviating the cell fusion process of hybridoma technology.

**TABLE 6.1:** A chromology of progress in monoclonal antibody technique

| Year | Key Development |
|---|---|
| 1975 | Monoclonal antibodies with predetermined specificity |
| 1978 | In Vitro Immunization |
| 1980 | Human monoclonal antibodies |
| 1980 | FDA approves first monoclonal based in vitro test. |
| 1981 | Bispecific monoclonal antibodies by cell fusion |
| 1984 | Affinity fusions |
| 1984 | Chimeric human (C)-mouse (V) monoclonals |
| 1985 | Recombinant antibody-enzyme molecules |
| 1986 | FDA approves OKT3 for in vivo therapy Humanized mosaic or CDR grafting |
| 1987 | Catalytic monoclonal antibodies |
| 1989 | Transgenic production of antibodies in plants<br>Transgenic production of IgM with human μ chain<br>Single domain antibodies (dAbs) bs MAbs by rDNA methodology |

## Types of MAbs

Monoclonal antibodies (MAbs) are immunoglobins of a single species derived from coned B lymphocytes, its fused derivatives or transformants. MAbs are secretory products of monoclonal cultures. Some monoclonal cultures are known secreting mixtures of Immunoglobulin species. Myeloma fusion partners used in the early days of hybridoma technology secreted endogenous immunoglobulin chains and thus a resulting hybridoma secreted and assembled more than one species of immunoglobulins.

The most common type of monoclonal antibody is the *monospecific bivalent* form belonging to the IgG class of immunoglobulins. Monoclonals of all the four sub classes of IgG are known. The structure of a MAb deserves some description to understand the modifications to be enumerated subsequently.

Typically, a MAb of the IgG class has two heavy (H) chains and two light (L) chains linked covalently by disulfide bonds and further stabilized by non-covalent interactions to form a "Y" shaped molecule. The amino terminal half of the heavy and light chains are called *variable* (V) regions due to the amino acid sequence variability between different antibodies. Hypervariable sites are hot spots in the V region exhibiting high sequence variability. Two identical heavy and light chains are synthesized by the B-cell due to allelic exclusion resulting in two identical antigen binding domains in antibodies. The specificity and affinity of antigen binding are determined by the variable region. The remaining portions of the chains are the *constant* (C) regions. Amino acid sequences in these portions are constant between antibodies of a given class or sub-class. The heavy has three constant region domains-$C_H1$, $C_H2$ and $C_H3$-as well as a hinge region. The stem of the IgG molecule is referred to as the Fc region, which bears the two Fab arms. Unique modifications of the monospecific bivalent monoclonal antibodies of the IgG type are diagrammatically represented and some are discussed in greater detail in other sections. Monoclonal IgM antibodies have been of particular use in agglutination assays and blood group typing.

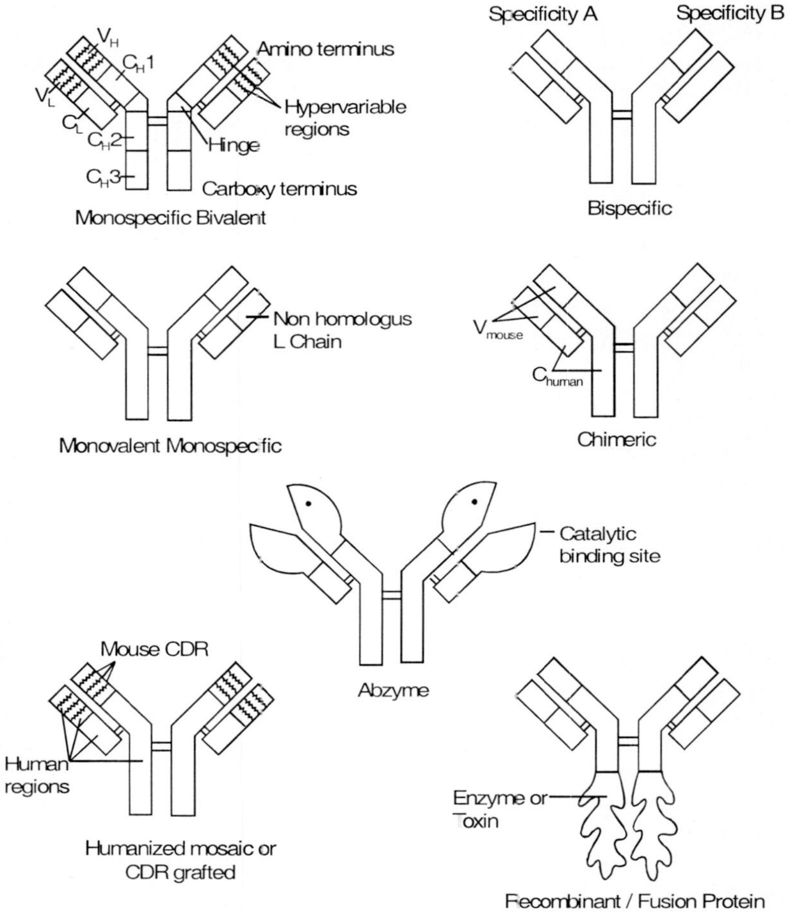


**FIGURE 6.1:** Diagrammatic presentation of modifications of monoclonal antibodies.

### Bispecific MAbs

Bispecific MAbs are secreted by hybrid hybridomas and are generated by a heterologous assembly of the two different heavy chain species resulting in an antibody with two separate antigen binding specificities in one molecule. *Monovalent Monospecific* antibodies are formed when one homologous and one heterologous H and L chain association occurs resulting in one normal Fab arm and one inactive Fab arm respectively. A direct method to generate such antibodies with augmented therapeutic effects is described utilizing myelomas secreting L chains as parental fusion partners of immunized spleen cells.

### Chimeric MAbs

Chimeric MAbs incorporate antibody domains from at least two different animal species engineered by recombinant DNA methodology. One chimeric antibody of interest results from the fusion of the mouse variable regions with human constant domains to produce less immunogenic human-mouse antibodies for *in vivo* applications. A further extension of this chimeric approach is to graft only the hyper variable regions of the mouse variable domain on to a human antibody. This technique is called complementary determining region (CDR) grafting to generate humanized mosaic antibodies. While chimeric and humanized antibodies are distinguished based on the extent of mouse domains retained, the two terms are often used interchangeably in the literature.

New term **humanized mosaics** has been introduced to describe the latter category of engineered MAbs. Another category of MAb modification is the *recombinant* or fusion protein Mabs. These are novel fused proteins produced by genetically replacing the constant region or a portion thereof by non-Fc functionalities such as enzymes.

## SOMATIC CELL FUSION

Barski et al. (1960) in laboratory of Virology and Tissue Culture of the Institute of Gaustave Roussy, at Villejuif (France); observed during their culture experiments of two stocks of tumour mouse cells that a new cell type was formed. This type had morphological characteristics and growth pattern that are different from those of parent cells. The nuclei of these new cells contained number of chromosomes equal to the sum total of chromosome numbers of the two parent cells. Unfortunately, the frequency of cell fusion was very low, between $10^{-4}$ to $10^{-6}$.

Latter, the frequency of fusion of animal cells was increased by various workers using biotic and abiotic fusing agents. Okada (1962), succeeded in increasing the frequency of fusion between animal tumour cells by using the Japanese Hemagglutination Virus (JHV). He has also reported that JHV inactivated by ultraviolet radiation retained the ability to induce somatic cell fusion, so any complication resulting from using an infective virus could be avoided.

Harris and Watkins (1965), succeeded in somatic cell fusion between human HeLa cell and tumour mouse cells using inactivated Sendai virus as fusing agent. Weiss and Green (1967) noticed a peculiar feature in some somatic hybrid in which chromosomes of one of the parent cells gradually eliminated or lost. In somatic hybrid of human and mouse cells, human chromosomes generally lost after fusion while in somatic hybrid of monkey and mouse cells the chromosomes of monkey are eliminated. Davis (1981) used polyethylene glycol as fusing agent and could get somatic hybrids of animal and plant cells separately belonging to two different species and even hybrid of animal and plant cells.

## HYBRIDOMA TECHNOLOGY

The hybridoma technology for the production of standardized antibodies of a given class, specificity, and affinity has provided scientists with a tool that permits the analysis of virtually any antigenic molecule. Such reagents (antibodies) can be made in unlimited amounts whenever needed, thus, making them readily available to all investigators.

Kohler (1974) successfully produced a hybridoma by fusing a P3 myceloma cell (resistant to azaguanine) and a lymphocyte (from the spleen of a mouse) immunized against sheep-red blood cells. The experiment consisted of immunizing mice against red blood cells (the antibodies produced against this antigen are easily detected in the serum by means of an assay developed in 1963 by Jerne and Nordin), then mixing mouse myceloma P3 cells with spleen cells from immunized mice in the presence of polyethylene glycol. These chimeric cells, called hybridomas retained the property of immortality of the myceloma cells as well as that of secreting an antibody specific for an unknown antigen; resulting from the fusion of an antibody-secreting cell and of a tumour cell. They are capable of growing indefinitely in culture and of producing at the same time a particular species of antibody (a monoclonal antibody).

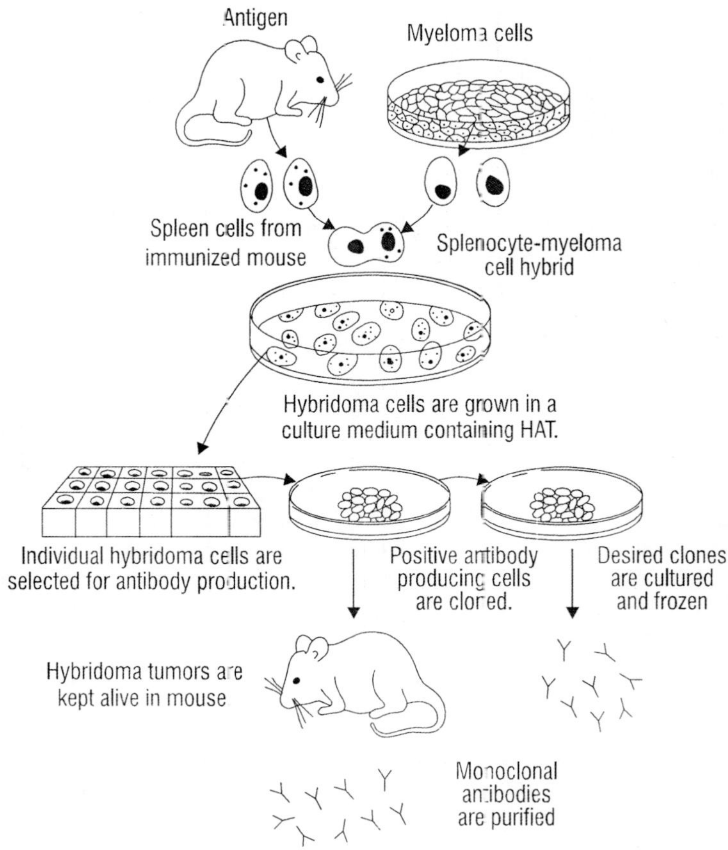


**FIGURE 6.2:** Technique for the Production Monoclonal Antibodies. Antigen-stimulated spleen cells are fused with special mutant myeloma cells, yielding hybridomas. Each of them secretes a single, *Monoclonal* antibody. Once the hybridoma secreting the desired antigen is identified, it is cloned to generate many antibody-secreting cells that yield the huge quantity of a single antibody needed in medicine or science. Some hybridoma cells may be stored frozen and later cloned for antibody production or kept alive in laboratory animals.

## PRODUCTION OF MONOCLONAL ANTIBODIES

The hybridoma technique seemed to be useful and applicable in producing monoclonal antibodies which is for affinity purification, tumor imaging, immunodiagnostics, and cancer treatment. They can also be useful in describing as well as predicting and optimizing secretion in antibody production system. The monoclonal antibody can be defined as a chemical reagent of known structure that can be reproduced at will, whereas conventional antiserum is a variable mixture of reagents that can never be reproduced, once the original supply is exhausted. The monoclonal antibody was purified by passage through chromatography columns containing the known antigen and, within 6 to 12 months, unlimited quantities of a single antibody could be obtained with a degree of purity and homogeneity. The clones selected can be stored by freezing technique.

It is now possible to obtain consistently large number of hybrids and one can select the lines producing the antibody of choice. The critical component in the production of hybridomas is the preparation of cells for fusion. Where rodents are used as a source of cell for fusion, immunization protocols must be devised to optimize the proliferative response to antigen in the spleen. Plasmablasts appear to be more suitable as fusion partners than mature plasma cells.

The fusion and culture of hybridoma is quite straight forward. The scheme is equally useful when cultures of in vitro immunized cells are used as a source of plasmablasts. The production technique of monoclonal antibodies can be divided into three steps:

## Step I: Fusion and Culture of Hybridomas

- Boost mice with an intravenous injection of antigen 72 hours before use. On the day of fusion, collect the spleen and make a free cell suspension by injecting and flushing the spleen with sterile serum-free medium. Centrifuge the cells and re-suspend in lysine solution. When sterile water is used, suspend the cell in 1 ml of water and immediately dilute in 20 ml of medium. Any delay will result in the loss of plasmablasts and a consequent reduction in antibody-producing hybrids.
- Maintain plasma cells in cell culture and feed into fresh flasks and medium for 16-24 hours before fusion to ensure that they are in early phase of growth at the time of fusion. At the time of use, collect the cells, centrifuge, and re-suspend in serum-free medium.
- Count both myceloma and spleen cells and then mix in the appropriate ratio. Depending on the properties of the tumour cells, the ratio of spleen to tumour cells may vary from 5:1 to 2:1.
- Following mixing, the cells are centrifuged into a loose pellet by spinning at 1000 rpm for 10 to 15 minutes. Remove the supernatant and overlay the pellet with 1 ml of PEG. For 3 minutes, mix the PEG into the pellet. In doing so breakup the pellet into uniform small clumps.
- Following fusion, dilute the cells in 30 ml of serum-free medium with the first 10ml medium being added and mixed at 1 ml per minute. Slow dilution reduces the risk of osmotic disruption of the fused cells. Centrifuge the cells and resuspended in complete medium containing HAT and then dispense into 96-well tissue culture plates, 10 plates for each 108 splenocytes used in the fusion. Added 106 thymocytes to each well to serve as feeder-cells. The latter step has proven critical for optimizing that outgrowth of newly formed hybrids. Feeder-cells and 2-ME in the medium exert a synergistic effect.
- In carbon dioxide incubator with high humidity, incubate the cultures for 3-4 days with rapidly growing cultures between changes of culture medium, replace half the medium with fresh HAT medium every 4th day.
- Identify and mark the wells containing hybrid colonies (hybridomas) on day 9 or 10 and then allow the colonies to grow to 500 or more cells. In rapidly

growing cultures, supernatants can be collected and assayed for antibody activity by day 12-14. Where appropriate, collect and test the supernatants from the largest colonies first; test the remainder 2-4 days later.

- After each test for antibody, transfer positive cell lines to 24 well plates. Add 3-5 ×106 feeder cells to each well to promote rapid cell growth. Maintain cells in static culture for a minimum of 2 weeks by removing 50-75% of the hybrid cells after 2-3 days interval. The maneuver select for stable, rapidly growing, antibody-producing hybrids. Slow growing hybrids and hybrids that cease to synthesize antibody are eliminated.
- After 2 weeks, make duplicate cultures and allow the cells to overgrow and die. Collect the supernatants and assay for the presence of antibody. Take cultures producing antibodies of the desired specificity from the master plate and expand into 6-well plates (2-3 wells for each cells lines). Harvest the cells twice for preservation in liquid nitrogen. Replenish the cultures with fresh medium and allow the cells to overgrow and die. Collect final supernatant for further analysis. The 6-well plates are essential for minimizing labour and time during this phase of hybridoma production.

### Step II: Cloning and Preservation of Hybridomas

- Following preliminary selection of hybrids, screen the final supernatants in detail to identify antibodies of immediate interest. Take the parent cell lines from the freezer and clone by limiting dilution. When viability is good, clone the cell lines immediately. Take the remainder of the excess cells and culture them in a T75 flask for 1-2 days. As soon as enough cells are present, harvest the cells and freeze one ampoule to replace the ampoule used for cloning. The culture should not be allowed to proliferate more than necessary to avoid change in composition of the cell line at this early stage of processing. As in the initial step, use feeder cells to promote growth.
- At 6-8 days mark wells containing a single colony. At 12-14 days assay supernatants from the marked wells. Transfer 24-48 positive cultures to 24 well plates and maintain in static culture as described previously.
- After noting which cloned lines are stable, expand 4-6 clones of each cell line into 6-well plates for cell preservation and production of antibody. Record which lines are stable and which are unstable.
- Preserve 4-6 clones from each cell line in liquid nitrogen.

### Step III: Production of Antibody

To produce antibody, culture the cloned cell lines in vitro or grow in ascites from the mice. After these steps, proper method is selected for assaying monoclonal antibodies in supernatants of hybridomas. Following methods are in use for assaying monoclonal antibodies:

(a) Enzyme-linked Immunosorbent assay (ELISA) (b) Radioimmunoassay (RIA). (c) Immunofluorescence (d) Cytotoxicity (e) Flow Cytometry, etc.

## APPLICATION OF MONOCLONAL ANTIBODIES

Monoclonal antibodies, in addition to being highly effective detectors of cancer cells, are under consideration as potential means of tumor therapy. The possible roles for monoclonal antibodies include the delivery of high concentration of cell-killing radioactive chemicals, drugs, or toxins to cancerous tissues, as well as the use of these antibodies to locate and attack tumor cells.

For effective delivery of anticancer agents linked to monoclonal antibodies, the combination (conjugates) should penetrate and destroy the parts of a tumor that contribute to its growth and then bind directly to most, if not all, individual tumor cells. Additionally, anticancer drugs may be released around the tumor cells (extracellularly) following monoclonal antibody attachment to tumor cell surfaces.

Hybridomas can be stored by freezing technique. Hybridoma banks have been established in some institutes and laboratories in order to meet research needs. Many pharmaceutical firms are interested to the highest degree in the large-scale production of monoclonal antibodies from these hybridomas. Keeping in view their applications in various fields of medical science:

- Dose determination of a medicine can be carried out by using the monoclonal antibodies of an animal, immunized against this particular medicine?
- They are used to detect allergies, to carry out hormone tests, to diagnose viral diseases, to detect certain types of cancers, to monitor the presence or appearance of malignant cells after surgical or radio-therapeutic treatments.
- The purification of complex mixtures or substances, the biological role which is important (proteins, hormones, toxins, etc.) could also be carried out with monoclonal antibodies.
- The use of these antibodies was also envisaged for the labelling and precise identification of specialized cells such as neurons in order to gain better knowledge of the way in which these cells associate and operate.
- Monoclonal antibody technique is also of great value in the area of the structure of cell membrane as membrane proteins are hard to purify.
- In the field of direct therapy, serotherapy can be made more effective with the administration of a monoclonal antibody.
- Monoclonal antibodies could also be used in the preparation of very specific vaccines, particularly against certain viral strains and against other parasites.
- Monoclonal antibodies could also neutralize the action of lymphocytes responsible for the rejection of grafts and destroy the auto-antibodies produced in auto-immune diseases.
- In association with medicinal substances, they could considerably increase the effectiveness of the latter on the target cells, while avoiding the serious side-effects of cancer therapies.

Many European and North American firms are interested in the applications of monoclonal antibodies. In California, for-instance, certain companies are preparing diagnostic kits designed for the screening of certain lethal diseases. It is anticipated that the future support of some aspects of this hybridomas based monoclonal antibody technology will be tailored to the needs of each developing country in the world.

# 7. Gene Therapy

Genes discovered for their roles in development are also critical in human disease. Excitement has come from finding genes that guide cell differentiation and pattern formation and that are linked to human disease. For example, mutation in β-catenin contributes to human colon and other cancers. Mutations in human *Sonic hedgehog* result in holoprosencephaly; mutations in hedgehog receptor patched lead to polydactyly and *Spina bifida* in Gorlins syndrome and to basal cell carcinoma, medulloblastoma and rhabdomyosarcoma. Mutations in the tinman-related human gene NKx 2.5 cause heart disease. Cell cycle regulators such as APC and Rb are critical in human cancer and are now viewed in the context of their normal developmental roles, APC in the Wnt signaling pathway and Rb in eye development.

Gene therapy is the treatment or prevention of disease by gene transfer, and is regarded by many as a potential evolution in medicine. This is because gene therapies are aimed at treating or eliminating the cause of disease, whereas most current drugs treat the symptoms. This radical improvement is possible because the gene-based approach can provide superior targeting and prolonged duration of action. Hence, in comparison with other forms of therapy, it will permit biological effects that are more subtle and better localized to the most appropriate cells. Ultimately, this will translate into substantial improvements in therapeutic ratio and cure-rate of diseases that are presently untreatable or poorly managed.

The first clinical studies involving gene transfer began in 1990 and since then gene therapy has become the focus of a whole new industry. **TABLE 1.1** indicates the wide range of diseases and target cells for which clinical studies are ongoing. All clinical studies involve gene addition rather than correction or replacement of defective genes, which is technically more difficult. All clinical protocols approved to date involve gene transfer only to somatic cells rather than germ-line cells. Gene transfer to these somatic cells can take place either *ex vivo* or *in vivo*. In the *ex vivo* approach, cells are removed from the patient for transfaction and the therapeutic entity comprises engineered cells. This offers the advantage of more-efficient gene transfer and the possibility of cell propagation to generate higher cell doses. It is patient-specific and more costly. The *in vivo* approach involves direct administration of the gene-transfer vector to patients. It is therefore not patient-specific thus conferring advantages of reduced cost, logistics and infrastructure requirements.

## GENE-TRANSFER SYSTEM

There are main types of gene-delivery vectors: viral, non-viral and physical.

### Viral Vector

The most advanced viral vectors are retrovirus (Rv), adcnovirus (Ad) and adeno-associated virus (AAV). Substantial efforts are made in developing poxviruses for genetic vaccines and herpes simplex virus.

## Non-Viral

Non-viral approaches fall into three main categories involving: naked DNA, DNA complexed with cationic lipids and particles comprising DNA condensed with cationic polymers (in some cases contained in liposomes).

**TABLE 7.1:** Gene therapies in Phase II clinical studies

| Disease | Vector | Gene transfer | Gene | Therapeutic strategy |
|---|---|---|---|---|
| Cancer | Cationic lipid | Tumour cells *in vivo* | HLA B7 | Immunogenicity enhancement |
| Cancer | Cationic lipid | Tumour cells *in vivo* | IL-2 | Immunogenicity enhancement |
| Cancer | Cationic lipid | Tumour cells *in vivo* | Ad E1A | Apoptosis induction |
| Cancer | Ad | Tumour cells *in vivo* | P53 | Apoptosis induction |
| Cancer | Ad | Tumour cells *in vivo* | Ad death genes | Killing by lytic virus |
| Cancer | Rv | Tumour cells *in vivo* | TK | Killing by enzyme and/or prodigy |
| Cancer | Rv | Fibroblasts ex vivo | IL-12 | Immunogenicity enhancement |
| Limb ischaemia | Naked DNA | Muscle cells *in vivo* | VEGF | Angiogenesis stimulation |
| Cystic fibrosis | AAV | Airway cells *in vivo* | CFTR | Provision of functional protein |
| HIV | Rv | T cells ex vivo | Chimeric TCR | Retargeting killer T cell |

## PHYSICAL

The physical methods involve needle-free infectors and electroporation. Currently, because of their highly evolved and specialized components, viral system are by far the most effective means of DNA delivery, achieving high efficiencies (usually <90%) for both delivery and expression.

In fact, around 75% of recent clinical protocols involving gene therapy use recombinant virus-based vectors for DNA delivery (http:/www.-Wiley.com/genetherapy/clinical/vectors.html.). As yet, however, no definitive evidence has been presented for the clinical effectiveness of any gene therapy protocol (Anderson, 1998). The impotence of current methodology is attributable to the limitations of viral-mediated delivery, including toxicity, restricted targeting of specific cell types, limited DNA carrying capacity, production and packaging problems, recombination and high cost. Furthermore, the toxicity and immunogenicity of viral systems also hamper their routine use in basic research laboratories. For these reasons, non-viral systems, especially synthetic DNA delivery systems, have become increasingly desirable in both basic research laboratories and clinical settings.

## VIRAL VECTORS

### Adenovirus

Adenovirus (Ad) gives more efficient gene transfer *in vivo*. It can transfer genes to both proliferating and quiescent cells. Following delivery, transgene expression is at high level but is transient, being low or undetectable in most tissues after two weeks. This is because Ad vectors do not integrate and for safety reasons are disabled for replication Ad vectors is not suitable for the long-term correction of chronic diseases but is adequate for direct cell killing, most immunotherapy strategies and some acute diseases. The first generation vectors have an insert-size limit of ~7.5 kb.

**TABLE 7.2:** Advantages and disadvantages of gene-transfer vectors

| Vector | Advantages | Disadvantages |
|---|---|---|
| Adenovirus | Very high transfaction efficiency ex vivo and *in vivo*<br>Transfaction proliferating and non-proliferating cells<br>Substantial clinical experience acquired<br>Efficient retargeted transfaction demonstrated | Repeat dosing ineffective owing to strong immune responses<br>Insert-size limit of 7.5 kb Manufacture, storage, QC are moderately difficult<br>Short duration of expression |
| Retrovirus | Fairly prolonged expression<br>High transfaction efficiency ex vivo<br>Substantial clinical experience ex vivo<br>Low immunogenicity | Low transfaction efficiency *in vivo*<br>Insert-size limit of 8 kb<br>Transfects only proliferating cells<br>Safety concern of insertional mutagenesis<br>Manufacture, storage, QC are extremely difficult |
| Lentivirus | Transfects proliferating and non-proliferating cells<br>Transfects hematopoietic stem cell | Safety concerns from immunodeficiency virus origins<br>Manufacturing, Storage, QC are extremely difficult<br>Insert-size limit of 8 kb<br>No clinical experience |
| AAV | Efficiently transfects a wide variety of cells *in vivo* very prolonged expression *in vivo*<br>Low immunogenicity | Insert-size limit of 4.5 kb<br>Manufacture, QC are very difficult<br>Little clinical experience<br>Safety concern of insertional mutagenesis Repeat dosing affected by neutralizing antibody responses |
| Naked DNA | Manufacturing, storage, QC are simple and cheap very low immunogenicity<br>Clinical efficacy demonstrated in critical limb ischaemia<br>Very good safety | Very short duration for expression in most tissues<br>Very inefficient transfaction ex vivo and *in vivo* Retargeting transfaction very difficult |
| Cationic lipids | Relatively simple manufacturing, storage, QC Efficient transfaction ex vivo<br>Low immunogenicity<br>Good safety profile | Inefficient transfection *in vivo*<br>Very short duration of expression<br>Little clinical experience<br>Retargeting transfection difficult |
| Condensed DNA particles | Relatively simple manufacturing, storage, QC Efficient transfaction ex vivo<br>Low immunogenicity<br>Good safety profile<br>Retargeted transfection | Inefficient transfaction *in vivo*<br>Very short duration of expression<br>No clinical experience |

| Vector | Advantages | Disadvantages |
|---|---|---|
| | demonstrated | |

**Note:** All viral vectors share the disadvantage of safety concerns arising from the generation of replication-competent virus during manufacturing or after administration to patients.
*Abbreviation*: QC, quality control.

The promoter used most frequently with Ad is derived from cytomegalus virus (CMV), which gives strong expression in many cell types. Ad gives particularly efficient gene transfer to the liver such that dissemination from the site of local infection (such as tumours) and consequent liver transfaction, is the most serious safety concern. This is being addressed by attempts to target both viral infection and transgene expression following delivery. Recent work has established that Ad infection can be efficiently retargeted *in vitro* through several cell-surface proteins and many promoter systems preferential transcriptional in certain cell types or tumours retain reasonable specificity in the context of Ad vectors.

**TABLE 7.3:** Target cells for gene transfer in ongoing clinical studies

| Disease | Target cells |
|---|---|
| Cancer | Tumour cells, antigen-presenting cells, blood progenitor cells, T cells, fibroblasts, muscle cells |
| Inherited monogenic disorders | Lung epithelial cells, macrophages, T cells, blood progenitor cells, hepatocyte, muscle cells |
| Infectious diseases | T cells, blood progenitor cells, antigen-presenting cells, muscle cell |
| Cardiovascular diseases | Endothelial cells muscle cells |
| Rheumatoid arthritis | Sinovial lining cells |
| Cubital tunnel syndrome | Nerve cells |

The most serious limitation of Ad vectors stems from their tendency to elicit strong immune and inflammatory responses. Single, large doses of Ad provoke neutralizing antibody responses directed in proteins of the viral particle, which prevent binding to target cells and abrogate gene transfer upon repeat dosing by systemic administration routes in animals.

Most Ad vectors are derived from the Ad5 serotype. Recent studies indicate that 55% of adult humans have pre-existing anti-Ad antibodies capable of neutralizing *in vitro* infection by Ad5. It is not yet clear whether these pre-existing (low titre) antibodies will interfere with gene transfer upon a first systemic Ad administration but it appears that even high titer neutralizing antibodies in the blood do not reduce gene transfer by repeated intra-tumoral infections of Ad vectors. Attempts to circumvent the neutralizing antibody response to Ad by oral tolerization, shielding key epitopes with hydrophilic polymers or blockade of the co-stimulatory molecule CD40 ligand at an early stage. First-generation Ad vectors deliver the transgene along with many residual viral genes, expression of these leads to cytotoxic-T-lymphocytes (CTL) responses directed to transfected cells. In certain tissues, CTL-mediated destruction of the expressing cells contributes transgene silencing. Ad vectors have now been developed, with most or virtually all residual Ad genes deleted, which can accommodate inserts up to 30 kb. Although these "gutted" vectors appear to show greatly reduced immunogenicity and much more prolonged expression of homologous transgene in mice, they present additional problems in recombinant virus construction and manufacture.

## Retrovirus

Retrovirus (Rv) vectors, unlike Ad, transfect by integrating the transgene into the target-cell chromosome. This usually leads to more prolonged transgene expression but at reduced levels. Transgenic expression ceases within days to weeks and tends to be shorter *in vivo*. This silencing phenomenon is probably due to methylation in the vicinity of the promoter and incorporation of the insertion site into condensed chromatin, in which the transgene is inaccessible to the transcription machinery. The Rv provirus penetrates the nucleus only at mitosis, so transfaction is restricted to profile-rating cells. This has been major limitation to Rv vectors.

Rv vectors have insert size limit of 8kb. Several tissue-specific promoters work well in Rv but retargeting transfaction is proving to be much more difficult to achieve than with Ad. It is clear that Rv can be retargeted to bind cell-surface proteins other than its natural receptor but efficient retargeted transfaction has not yet been reported. Most first generation Rvs do not express residual viral genes, which is partly why Rv is much less immunogenic than Ad. Nevertheless, it is clear that the expression of heterologous transgene leads to CTL responses that eliminate the expressing cells and reduce the effectiveness of repeat dosing. This occurs even in symptomatic AIDS patients, who are seriously immunocompromised indicates that this is a serious and general problem for gene therapy. Rv gives inefficient gene transfer for most cell types *in vivo*, partly because it is rapidly inactivated by the human complement system.

**TABLE 7.4:** Distribution of gene-transfer clinical protocols approved by or submitted to regulatory authorities in North America and Europe[a]

| Disease | Number of protocols | Vector | Number of protocols |
|---|---|---|---|
| Cancer | 216 | Rv | 159 |
| Infectious | 25 | Ad | 58 |
| Monogenic diseases | 49 | AAV | 4 |
| Cardiovascular diseases | 8 | Poxviruses | 19 |
| Rheumatoid arthritis | 2 | HSV | 1 |
| Cubital tunnel syndrome | 1 | Naked DNA | 16 |
| Total | 300 | Gene gun | 1 |
| *In vivo* transfer | 144 | Electroporation | 1 |
| Ex vivo transfer | 156[b] | Naked RNA | 1 |
| | | **Total** | **300** |

*Abbre.:* Rv, retrovirus; Ad, adenovirus; AAV, adeno-associated virus; HSV, herpes simplex virus.

The most extensive use of Rv vectors has been with ex vivo applications, for which cells can be cultured to allow efficient transfaction and for which sustained expression is generally required. Ex vivo applications require smaller quantities of virus. This is an important consideration because Rv has presented serious manufacturing challenges, with difficulties in achieving high titres and retaining infectivity through concentration and storage. Despite these problems more clinical experience has been obtained so far with Rv than any other vector, largely due to its capability for longer expression. Recent improvement in manufacturing processes and the use of envelope proteins from other viruses, notably vesicular stomatitis virus, to package the recombinant Rv. Genomes promise to allow higher viral titres, greater complement resistance, transfaction of a wider range of cell types and more efficient transfaction *in vivo*. Good progress has been made recently in developing vectors from lentivirus (Lv) such as HIV and SIV, which are capable of integrative transfaction of several quiescent and post-mitotic cell types as well as proliferating cells.

### Adeno-associated Virus

Adeno-associated virus (AAV) is also capable of integrating into the target cell chromosome. The integration process is not well known but wild-type AAV integrates exclusively into a single site on human chromosome 19. It appears that AAV recombinants integrate much less efficiently and more randomly and many show long-term persistence in unintegrated forms. AAV gives sustained transgene expression upon *in vivo* administration compared with other vectors. Expression of homologous genes has been detected two years after infection in mice and several months after infection in dogs primates and man. Long-term correction of hemophilia has been achieved in a dog model of disease with a single injection of an AAV vector expressing a gene for clotting factor IX.

Immunogenicity of AAV vectors suggest 32% of adult human possess pre-existing antibodies that neutralize transfaction *in vitro* by the AAV2 serotype from which most AAV vectors are derived. Studies in mice suggest injection with doses giving therapeutic levels of transgene expression leads to the generation of neutralizing antibodies that will reduce the effectiveness of repeat dosing. CTL responses to AAV-transfected cells tend to be weak or undetectable, partly because in addition to the transgene the vectors contain only short viral sequences required for packaging. This contributes to prolonged expression, which has been observed in lung, muscle, liver and the central nervous system in mice. AAV transfers genes efficiently to both quiescent and proliferating cells. Some promoters with cell-type restriction retain their specificity in AAV and preliminary data suggest efficient retargeted transfaction will prove to be feasible for this vector. This resilience of AAV particles is a major advantage for *in vivo* applications. Indeed, substantial transfaction and long-lasting transgene expression in the epithelial and sub-epithelial cells of the upper intestinal tract have recently been reported following oral administration of an AAV vectors.

The main disadvantages of AAV concern insert size and manufacturing, the virus can accommodate insert up to 4.5 k and manufacturing processes have required the use of helper viruses (usually Ad), which present problem of low titre, contamination and costly purification procedures. Clinical protocols involving AAV have been initiated so far, but the result to date suggest transgene expression with this vector is sustained in man, as it is in animals.

## NON-VIRAL APPROACH

### Naked DNA

The direct injection of DNA (*e.g.* uncomplexed DNA) into cell nucleus is the most simple and therefore appealing approach to gene delivery. One drawback of the approach, however, is that microinjection can be achieved only cell at a time which limits its use to applications in which individual cell manipulation is desired and possible, such as producing transgenic organisms. Though relatively efficient, the method is also rather slow and laborious and therefore neither appropriate for research with large numbers of cells nor practical for DNA delivery *in vivo*.

Mann et al. (1999) have developed a pressure mediated naked oligonucleotide transfaction method using controlled, nondistending pressure. When they tested the approach on cardiovascular tissues, >50% efficiency of delivery was achieved. Whether or not this approach can be adapted to noncardio-vascular tissues with plasmid remains to be seen. Similarly, hydrodynamic force (rapid injection via tail vein) has been applied to deliver naked DNA to hepatocyte with delivery efficiencies of 40%. Another method, ultrasonic nebulization has been used to deliver DNA-lipid complexes to cells.

## Chemical Methods

The use of uptake-enhancing chemicals, which is the easiest, most versatile, most effective and most desirable of the DNA delivery methods, was demonstrated more than 30 years ago. The general principle is based on complex formation between positively-charged chemicals (usually polymers) and negatively charged DNA molecules. These techniques can be broadly classified on the basis of chemical involved in to those based on 2-(dimethylamino) ether (DEAE)-dextran, calcium phosphate, artificial lipids, protein, dendrimers or others.

The earliest chemical methods for DNA delivery were introduced in the late 1950s. These techniques used high salt concentration and polycationic proteins to enhance nucleic acid entry into the cell. Over a 10-year-period starting in 1965, a variety of other chemicals were introduced including DEAE-dextran and calcium phosphate, which interact with DNA to form DEAE-dextran-DNA and calcium phosphate-DNA complexes respectively. After the complexes are deposited onto cells, they are internalized by endocytosis.

DEAE-dextran and calcium phosphate methods are simple, effective and still widely used in the laboratory for *in vitro* transfaction. Even so, both methods are hampered by cytotoxicity and the difficulty of applying them to *in vivo* studies. In addition, DEAE-dextran can be used neither with serum in culture medium nor for stable transfaction. The calcium-phosphate method also suffers from variations in calcium phosphate-DNA sizes, which causes variation among experiments.

Since the early 1970s, many other polymers have been demonstrated to increase DNA uptake by cells. The most noticeable improvement is the development of artificial lipid-based DNA delivery systems. Felgner and colleagues developed the cationic lipid Lipofectin in 1987. Lepofectin-DNA complexes can be handled easily and therefore became one of the first chemical systems that could be used in animals.

In addition, DNA has been successfully complexed with cationic, anionic and neutral liposomes as well as various mixtures thereof. New lipid formulations and new recipes of blending have yielded continuous improvements and have been used to form complexes with PLL and DNA. 6Al4 is an *E. coli* DNA binding protein and invasin is the cell binding protein from *Yersinia pseudotuber-culosis*. Together, Gal4-invasin, PLL and DNA complexes have been demonstrated *in vitro* to deliver DNA in an invasin receptor-specific fashion. One drawback of this approach is that bacteria fusion protein many present immunological problems when used *in vivo*.

Polyamidoamine (PAMAM) dendrimers are capable of condensing DNA and delivering it to a variety of cell lines with minimum cytotoxicity. Szoka and

Colleagues (1996) have improved the expression level 50-fold by partially degrading dendrimers. Since their introduction, PAMAM and other dendritic polymers have gained popularity because of their versality and simplicity in transfaction.

Choi and Colleagues (1999) have conjugated linear polyethylene glycol (PEG) with a dense globular PLL dendrimer forming PEG-block-PLL-dendrimer, a water-soluble- polyconic spherical complex. Other chemicals (mostly polymers) have also been explored for enhancing gDNA delivery, polyethylenimine (PEI) and its conjugates such as mannose-PEI, transferring-PEI and linear PEI (IPEI), gelatin and chitosan, methscrylate/methacrylamide polymers and pyridinium surfactants.

Recently, biocompatible controlled-release polymers have also been introduced into the synthetic DNA delivery field. Several groups have successfully encapsulated naked DNA into biodegradable poly (D, L-lactide-co-glycolide) (PLGA) microparticles and PLGA microspheres for long-term and controlled DNA release. DNA has also been encapsulated into highly biocompatible poly (ethylene-co-vinyl acetate) (EVAc) matrices resulting in the controlled and predictable release of bioactive DNA for several months. Mooney and Colleagues (1999) have recently demonstrated a significant tissue response using PLGA to deliver functional DNA locally.

## PHYSICAL METHODS

### Needle-Free Injection

Two devices have been developed that allow gene delivery by injection without needles. The first uses a high-pressure helium stream to deliver DNA coated onto gold particles, directly into the cytoplasm and is often referred to as the **gene gun**. The other called the intrajet or **jet gun** uses liquid under high pressure for delivery into interstitial spaces. The gene gun gives a moderate transfection efficiency for a variety of cell types *ex vivo* and *in vivo* a moderate transfection efficiency for cells in the skin. These skin cells include Langerhans cells, which are antigen penetrating cells that normally capture foreign antigen from invading organisms and present them to naïve T cells to elicit primary immune responses. Gene gun requires much smaller DNA doses. The liquid-based infectors are at earlier stage of development but give stronger immune responses.

### Electroporation

An alternative approach is to use high-voltage electrical pulses to transiently permeabilize cell membranes, thus permitting cellular uptake of macromolecules. This process, called electroporation, was first used to deliver DNA to mammalian cells in 1982. Since that time, electroporation has been used to deliver DNA to myriad cell types *in vitro*. It is one of the most efficient gene transfer method but is limited because of the high motality of cells after high-voltage exposure and difficulty in optimization.

Although electroporation is difficult to apply *in vivo*, some progress has been made in skin corneal endothelium and muscles. An electrical permeabilization method using low-voltage, high frequency electrical pulses reported by Rizzuto et al. (1999), produced similar DNA uptake in muscle with only transient tissue damage, represents a significant improvement.

### Protection and Intracellular Release of DNA

It is essentially to protect DNA from both extracellular and intracellular degradation during the long journey to the cell nucleus. Extracellularly, DNA can be protected by complex formation with various polymers and lipids (described above). If delivered systemically, however, DNA is susceptible blood clearance. A process is known as opsonization it removes 80-90% of hydrophobic particles in blood and is thus a major limiting factor for DNA delivery using artificial lipids.

Intracellularly, DNA must escape from normal endosomal pathway, which lead to degradation. Chloroquin, which raises endosomal pH, has been used to decrease DNA degradation by inhibiting lysis. Branched cationic polymers such as PEI or starburst dendrimers, also promote early release of DNA. Some attempts have been made to bypass endosomal pathways, subunits of toxins such as, Diphtheria toxin and Pseudomonas exotoxin have been incorporated into DNA complexes to enhance transfection but this approach can be limited by the inherent toxicity and immunogenicity of the subunits. Viruses have evolved to escape and/or bypass endocytosis with high efficiency. For example, the viral capsid of the adenovirus is capable of association with the endosomal membrane which results in pore formation and lysis of the endosome. Similarly, hemagglutinin HA2 peptide from influenza virus can insert itself into the endosomal membrane which also leads to lysis. Incorporation of these viral components can enhance transfection efficiency.

The cytosol is an unfriendly environment for DNA. Movement of DNA through the cytosol toward the nucleus occurs by diffusion. It is a slow process during which DNA must be protected. Coating with PEG can stabilize plasmid DNA encapsulated in lipid particles and protect them from degradation by nucleases. It has been shown that block copolymer, PEG-PLL, are highly resistant to DNase I attack.

## RECENT DEVELOPMENT AND FUTURE PROSPECTS

At the present time the gene therapy concepts concern the following:

- Direct killing of tumour cells with genes delivered by Ad vectors for local management of cancer.
- Delivery of naked DNA by injection or by gene gun for preventative vaccination against infectious diseases.
- Naked DNA delivery of genes promoting angiogenesis for cardiovascular disorders.
- AAV delivery for chronic disorders such as hemophilia and anemia.

There is no single vector at present, with generic utility for all types of diseases but Ad remains the most suitable vector for *in vivo* therapeutic concepts that require only short-term expression especially where a single dose will suffice for clinical benefit. The incorporation of elements for targeted cell uptake and expression, and reduced immunogenicity, will no doubt extend the utility of AD, but probably not to the long-term treatment of chronic diseases. Over the next few years, AAV is likely to be the most commonly used vector for long-term replacement of defective proteins at least in cases requiring only a small transgenic.

For all integrating vectors, and perhaps for most monintegrating vectors, transgene expression appears to be limited and usually silenced by incorporation into condensed and transcriptionally inactive chromatin both *in vivo* and ex vivo. Chromatin opening elements, which resist transcriptional silencing by modifying the structure of chromatin, offer solution to this problem. These elements are of two types (a) locus control regions (LCRs) and (b) ubiquitously acting chromatin opening elements (UCOEs). LCRs open chromatin and confer sustained expression in a tissue-specific manner. UCOEs confer the same benefits in a wide range of tissues. The existence of LCRs was established some years ago and LCRs giving specific expression in several different tissues have since been identified. They represent a promising approach to achieving sustained expression in the cell lineage most appropriate for clinical benefit following gene delivery to stem cells. UCOEs were discovered recently and are likely to improve transgene expression for a wide range of vectors.

Combining chromatin opening elements with the inducible promoter systems developed recently shouted allow full exploitation of the latter. These promoters allow induction of transgene expression in response to exogenous substances. An orally active small molecule can induce the most advanced system. It has been shown to enable low base like expression and respected cycles of induction to high levels over several months for the erythropoietin gene delivered using AAV to mice or monkeys. These results represent proof of principle for one of the long-standing technical objectives of gene therapy; long-term persistence of a silent transgene with expression induced at will.

Non-viral vectors giving efficient gene transfer *in vivo* are not yet available. However, there is every prospect that non-viral vectors with broad utility will be developed in the next few years, particularly with flexible approaches such as condensed DNA particles. In the long term, non-viral approaches are more attractive for commercial products because they offer crucial advantages over viruses of superior targeting, low immunogenicity and reliable, large-scale manufacture at an acceptable cost. For the wide utility of non-viral vectors, however, it will be necessary to prolong transgene expression substantially because this lasts only a few days. This is partly because present non-viral systems lead to neither integration nor extrachromosomal replication and maintenance. Progress is being made in the development of replicating episomal vectors (REVs) based on oncogenic viruses such as EBV and SV40 and on mammalian artificial chromosomes (MACs). At present, REVs give inadequate stability and safety. MACs are too large for efficient manipulation, manufacture and delivery.

Over the next few years, no doubt, most new clinical protocols will involve gene-addition therapy but substantial progress has been reported recently towards gene replacement and gene-correction therapy. AAV has been shown to give significantly greater efficiency of DNA replacement by chromologous recombination between flanking sequences than any other vector, perhaps as a result of its ability to transfer large numbers of single-stranded DNA molecules to the nucleus. Heritable and accurate replacement in 1% of human fibroblasts *in vitro* has been observed. An alternative approach uses delivery of double-stranded DNA-RNA chimeric oligonucleotides, with RNA sections to provide targeting homology for a DNA region with a single-base mismatch to the target. This chimeroplastry approach leads to gene correction through mismatched repaid, which might give greater efficiencies than homologous recombination. Approximately 10% correction in primary blood cells has been reported and 40% correction in liver cells following *in vivo* administration in the rat. This technology is not yet robust but, when fully developed, gene replacement or gene correction therapy will prove to be preferable to gene-addition therapy for many diseases.

Substantial progress has been made in the past eight years in overcoming the major inadequacies of first generation gene transfer vectors. No doubt the genomics revolution will provide an over-increasing range of genes and information on their causal relationship with diseases, rendering more and more of these amenable to gene therapy.

# 8. Microbes in Pharmaceutical Industry

Microorganisms are one of the most important partners in our life. Their presence is felt by human beings in one way or the other through the major part of the day. They may be called worst of our enemies but at the same time they are among the best of our friends. Hence, one can not forget the gifts, which have been given by these organisms to this world except the harmful effects.

## SINGLE CELL PROTEIN

The cultivation of unicellular microorganisms as a direct source of food for man was suggested in the early 1900s. The expression **single cell protein** (SCP) was coined at the Massachusetts Institute of Technology around 1966 to depict the idea of microorganisms as food sources. Although SCP is a misnomer in that proteins are not the only food constituent represented by microbial cells, it obviates the need to refer to each product generically as in **algal protein**, **yeast cell protein**, and so on. Although SCP is a potential and real source of food for man differs from the other products.

In addition to meeting these criteria, SCP can be produced on a variety of waste materials. Among the overall advantages of SCP, over plants and animal sources of proteins are the following:

- Microorganisms have a very short generation time, and can thus provide a rapid mass increase.
- Microorganisms can be easily modified genetically to produce cells that bring about desirable results.
- The protein content is high.
- The production of SCP can be based upon raw materials readily available in large quantities.
- SCP production can be carried out in continuous culture and thus be independent of climatic changes.

The greater speed and efficiency of microbial protein production compared to plant and animal sources may be illustrated as follows: a 1,000-lb steer produces about 1 lb of new protein/day; soybeans (prorated over a growing season) produce about 80 lb; while yeasts produce about 50 tons.

### Microorganisms and their Fermentation Substrates

Single cell protein (SCP) can be obtained from certain members of algae, cyanobacteria, yeasts, molds and higher fungi. These plant species are harvested and dried for use of protein source. The SCP is characterized by high percentage of protein by weight (sometimes upto 85%). For the production of SCP from plant sources, a variety of raw materials are required which serve as carbon source including solar radiation (for photoautotrophic plant species). The important are

$CO_2$, $NaHCO_3$, sunlight (*Spirulina maxima*, protein = 60%), $CO_2$ and sunlight (*Scenedesmus acutus*, protein 55%), Bagasse (*Cellulomonas*, protein 85%), molasses (*Saccharomyces cerevisiae*, protein 50%). The SCP is being manufactured under different commercial names like *Brovile*, *Marmite*, *Pruteen*, etc. Pruteen is manufactured by Imperial Chemical Industries, U.K. from the bacterium *Mythylophilus methulotropous* which is grown on methyl derived from natural gas. The company produces about 50,000 tons of Pruteen every year as animal feed. Similarly Rank Hovis McDougall Company, U.K. produces SCP under the trade name *Mycoprotein* using potato, wheat, cassava as carbohydrate source for protein production. A large number of algae yeast, molds, and bacteria have been studied as SCP sources. Among the most promising genera and species are the following:

- **Algae:** Spirulina maxima, Chlorella spp., and Scenedesmus spp.
- **Yeasts**: Candida guilliermondii, C. utilis, C. lipolytica, and C. tropicalis; Debaryomyces kloeckeri; Torulopsis candida, T. methanosorbosa; Pichia spp.; Kluyveromyces fragilis; Hansenula polymorpha; Rhodotorula spp.; and Saccharomyces spp.
- **Filamentous fungi**: Agaricus spp.; Aspergillus spp.; Fusarium spp.; Penicillium spp.; Endomycopsis fibuligera; and Trichosporon cutaneum.
- **Bacteria:** Bacillus spp.; Cellulomonas spp.; Acinetobacter calcoaceticus; Nocardia spp.; Methylomonas spp.; Aeromonas hydrophilia; Alcaligenes eutrophus (Hydrogenomonas eutropha), Mycobacterium sp., and Rhodopseudomonas sp.

## Environmental Conditions for Growth

Of these groups, yeasts have received by far the most attention. The choice of a given organism is dictated in large part by the type of substrate or waste material in question. The cyanobacterium *Spirulina maxima* grows in shallow waters high in bicarbonate at a temperature of 30° C and pH 8.5-11.0. It can be harvested from pond waters and dried for food use. This cell has been eaten by the people of the Chad Republic for many years. Other cyanobacteria require sunlight, $CO_2$, minerals, water, and proper growth temperatures. However, the large-scale use of such cells as SCP sources is said to be practical only in areas below 35 °C latitude, where sunlight is available most of the year.

## Choice of Fermentable Materials

Bacteria, yeasts, and molds can be grown on a wide variety of fermentable materials including food processing wastes (such as cheese whey and brewery, potato processing, cannery, and coffee wastes); industrial wastes (such as sulfite liquor in the paper industry and combustion gases); and cellulosic wastes (including bagasse, newsprint mill, and barley straw). In the case of cellulosic wastes, it is necessary to use organisms that can utilize cellulose, such as a *Cellulomonas* sp. or *Trichoderma viridae*. A mixed culture of a combination of *Endomycopsis fibuligera* and a *Candida* sp. such as *C. utilis* has been employed, in which the former effects hydrolysis of starches and the latter subsists upon the hydrolyzed products to produce biomass. Some other representative substrates and fermenting organisms are listed below:

**TABLE 8.1:** Some examples of substrate materials that support the growth of microorganisms in the production of SCP

| **Substrates** | **Microorganisms** |
|---|---|
| $CO_2$ and sunlight | *Chlorella pyrenoidosa, Scenedesmus quadricauda, Spirulina maxima* |
| n-Alkanes, kerosene | *Candida intermedia, C. lipolytica, C. tropicalis, Nocardia* sp. |
| Methane | *Methylomonas* sp. (*Methanomonas*), *Methylococcus capsulatus sp.Trichoderma* |
| $H_2$ and $CO_2$ | *Alcaligenes eutrophus* (*Hydrogenomonas eutropha*) |
| Gas oil | *Acinetobacter calcoaceticus* (*Micrococcus cerificans*), *Candida lipolytica* |
| Methanol | *Methylomonas methanica* (*Methanomonas methanica*) |
| Ethanol | *Candida utilis, Acinetobacter calcoaceticus* |
| Sulfite liquor wastes | *Candida utilis* |
| Cellulose | *Cellulomonas* sp., *Trichoderma viride* |
| Starches | *Endomycopsis fibuligera* |
| Sugars | *Saccharomyces cerevisiae, Candida utilis, Kluyuveromyces fragilis* |

*Spirulina* (a cyanobacterium) offers a rich source of SCP. *S. platensis* contains about 60% crude protein and is rich in tryptophan, lysine and vitamin $B_{12}$. In India, NBRI, Lucknow and CFTRI, Mysore have started mass culture and utilization of this cyanobacterium as a source of SCP. *S. platensis* is cultured in sewage water containing 9 g/l of sodium bicarbonate and 1 g/l of sodium nitrate. The average yield of the microorganism varies between 6 $g/m^2$/day and 9 $g/m^2$/day on dry weight basis. Similarly A. M. M. Marugappa Chettiar Research Centre, Madras has isolated *Spirulina fusiformis* and produced SCP after filtration from cloth filters (30 mesh/inch). SCP is also produced from paper industry waste, because the industry faces a serious problem in disposing one of the waste products - sulphide liquor. If this pollutant is discharged in the river, it quickly depletes the water's oxygen reserves and causes death of fishes. The sulphide liquor is fed to *Paecilomyces* molds. This process is known as **Pelilo**. This process purifies the waste liquids from paper factories but also yields a rich harvest of microbes which are sold as animal feed.

## SCP Products

While the products of fermentation of other foods in this section consist of an altered or modified substrate, the product of SCP fermentation consists solely of the microbial cells that grow upon the substrate. The cells may be used directly as a protein source in animal feed formulations, hereby freeing animal feed, such as corn, for human consumption; or they may be used as a protein source or food ingredient for human food. In the case of animal feed or feed supplements, the dried cells may be used without further processing. As noted above, whole cells of *Spirulina maxima* are consumed by man in at least one part of Africa.

### *Chemical Composition of SCP*

Chemical analyses of the microorganisms evaluated for SCP reveal that they are comparable in amino acid content and type to plant and animal sources with the possible exception of methionine, which is lower in some SCP sources. All are relatively high in nitrogen. For example, the approximate percentage composition of N on a dry weight basis is as follows; bacteria 12-13, yeast 8-9, algae 8-10, and filamentous fungi 5-8. In addition to proteins, microorganisms contain adequate levels of carbohydrates, lipids, and minerals, and are excellent sources of B vitamins. The fat content varies among these sources, with algal cells containing the highest levels and bacteria the lowest. On a dry weight basis, nucleic acids average 3-8% for algae, 6-12% for yeasts, and 8-16% for bacteria. B vitamins are high in all SCP sources. The digestibility of SCP in experimental animals has been found to be lower than for animal proteins such as casein.

### *Forms of SCPs for Human Use*

For human use, the most likely products are SCP concentrates or isolates that may be further processed into textured or functional SCP products. To produce functional protein fibres, cells are mechanically disrupted, cell walls removed by centrifugation, proteins precipitated from disrupted cells, and the resulting protein extruded from syringe like orifices into suitable menstra such as acetate buffer, $HClO_4$, acetic acid, and the like. The SCP fibres may now be used to form textured protein products. Baker's yeast protein is the only product of this type presently approved for human food ingredient use in the United States, but as Litchfield has noted, it is reasonable to assume that similar products from other food-grade yeasts such as *Candida utilis* and *Kluyveromyces fragilis* will ultimately be approved.

Success has been achieved in rat-feeding studies with a variety of SCP products, but human-feeding studies have been less successful, except in the case of certain yeast cell products. Gastrointestinal disturbances are common complaints in man following the consumption of algal and bacteria SCP, and these and other problems associated with the consumption of SCP have been reviewed elsewhere. When gram-negative bacteria are used as SCP sources for human use, the endotoxins must be removed or detoxified.

### Harmful Effects of SCP Use

The high nucleic acid content of SCP leads to kidney stone formation and/or gout in man. As noted above, the nucleic acid content of bacterial SCP may be as high as 16%, while the recommended daily intake for man is about 2 g. The problems are caused by an accumulation of uric acid, which is sparingly soluble in plasma. Upon the breakdown of nucleic acids, purine and pyrimidine bases are released. Adenine and guanine (purines) are metabolized to uric acid. Lower animals can degrade uric acid to the soluble compound allantoin (they possess the enzyme uricae) and consequently the consumption of high levels of nucleic acids does not present metabolic problems to these animals as it does to man. While high nucleic acid contents presented problems in the early development and use of SCP as acid precipitation, acid or alkaline hydrolysis, or by use of endogenous and bovine pancreatic RNases.

**TABLE 8.2:** Examples of recombinant DNA systems used to modify gene expression

| Microorganism | Product | Change |
|---|---|---|
| *Clostridium* gene in *Bacillus* | Cellulase | Amplification of secretion through chromosomal DNA amplification |
| *Saccharomyces cerevisiae* | Recombinant protein albumin | Fusion to a high-production protein |
| *Saccharomyces* | Heterologous protein | Use of the inducible strong hybrid |

| Microorganism | Product | Change |
|---|---|---|
| *cerevisiae* | | promoter $UAS_{gal}$/CYCl |
| *Streptomyces coelicolor* | Actinorhodin | Modification of gene transcription |

## ALCOHOLIC FERMENTATION

Ethyl alcohol is considered to be the oldest product of fungi. Many varieties of fungi bring about this alcohol through fermentation of sugar. Yeasts (*Saccharomyces* spp.) are most commonly used in this connection. There are two species of *Saccharomyces*: one is the top fermenter (*S. cerevisiae*) and the second is bottom fermenters (*S. carlsbergonsis*). Alcohol is usually produced by fermentation of molasses, which is an aerobic process. Microorganisms utilize the carbohydrate present in molasses to obtain energy for growth and metabolic activities lead to the formation of alcohol. The reaction takes place via the Embden Mayerhof Paranas (EMP) pathway.

There are three main by-products of alcohol industry- carbon dioxide, higher alcohol (isoamyl alcohol, butyl alcohol, collectively named as fusel oil) and alcohol. Alcohol is often used in manufacturing of ether, acetaldehyde, acetic acid butanol, etc. Ethyl alcohol is one of the best known solvent of hundreds of different materials from which various compounds are synthesized. As a solvent, it is used in manufacture of such widely diverse materials as dye, soaps, plastics, polishes synthetic rubber etc.

Butyl alcohol and acetone are also manufactured by the fermentation, which is brought about by the bacterial activity (*Clostridium acetobutylicum*) on molasses. Both chemicals are of great industrial use.

**TABLE 8.3:** Immobilized microbial cells, their products and reactions

| Microorganisms Used | Product / Activity | Process Type |
|---|---|---|
| *Bacillus* spp. | Bacitracin | |
| *Brevibacterium ammoniagenes* | Coenzyme A | Enzymes |
| *Candida tropicalis* | Degradation of phenol | |
| *Corynebacterium dismutans* | L-alanine | Amino acids |
| *Corynebacterium glutamicum* | L-glutamic acid | |
| *Curvularia lunata* and *Corynebacterium simplex* | Prednisolone | Steroids |
| *Escherichia coli* | L-tryptophan | |
| *Escherichia coli* | Pantothenic acid | Vitamins |
| *Lactobacillus delbrueckii* | Lactic acid | |
| *Microbacterium ammoniaphila* | L-lysine | |
| *Micrococcus* spp. | Denitrification | |
| *Penicillium chrysogenum* | Penicillin | Antibiotics |
| *Pseudomonas aeruginosa* | Heavy-metal sorption (uranium, plutonium) | |
| *Pseudomonas* spp. | Degradation of paranitrophenol | Environmental management |
| *Saccharomyces cerevisiae* | Ethanol | Fermentation products |
| *Streptomyces clavuligerus* | Cephalosporins | |
| *Streptomyces fradiae* | Protease | |

Coal, oil and natural gas are commonly used in industry and in our homes as fuels. All three fuels were made millions of years ago when the dead remains of plants and animals were acted upon by microbes and by pressure from the rock layers deposited above.

It has been predicted, however, that our reserves of coal, oil and gas will be exhausted early next century if we continue to use them at the present rate. It is essential, therefore, that alternative sources of fuel be found. Several alternatives are available, one being to produce fuels on a large scale using microbes. Two such fuels are alcohol and methane. Microbially-produced fuels are a 'renewable resource' because once they have been used up, more fuel can be made whereas coal, oil and natural gas cannot be replaced once they have been used.

## MICROBES IN ORGANIC ACID PRODUCTION

A number of organic acids are produced by microorganisms, specially moulds and bacteria through the process of fermentation. These are as follows:

- **Citric acid:** It is the most important product and is now produced commercially after fermentation by moulds like *Aspergillus niger*, *Citromyces* spp. and certain species of *Mucor*. Apart from these fungi, *Penicillium* can also produce citric acid from sugar. The acid is used in manufacturing citrates, flavouring citrates, soft drinks, etc.
- **Gluconic acid:** *Aspergillus niger* synthesizes gluconic acid from sugar. Calcium gluconate cures cow's suffering from milk fever. It is also used in increasing calcium content of egg and in the preparation of toothpaste.
- **Gallic acid:** *Aspergillus gallomyces* and *Penicillium glaucum* synthesize gallic acid.
- **Kojic acid:** It is extracted from the mycelium of *Aspergillus oryzae*.
- **Acetic acid:** Bacteria like *Mycoderma aceti* synthesize acetic acid from fermentation of sugary solution and is used in vinegar production. *Acetobacter aceti*, *A. xylinum* and *A. rances*.

## MICROBES IN ENZYME PRODUCTION

Various microorganisms like bacteria, fungi and protozoa produce different kinds of enzymes. They produce extracellular enzyme to break those organic and inorganic compounds which have high molecular weight. Simpler substances can easily be assimilated through cell membranes of microorganisms.

*Bacillus mesentericus* is used for commercial production of amylase and protease. *Bacillus macerans* and *B. polymyxa* are also used for amylase production. Bacterial protease is employed to desize cotton and degum silk.

Fungi (*Aspergillus oryzae*) also synthesize amylase. Amylase is a starch digesting enzyme and is generally used in conversion of starch to fermentable sugars in alcohol industries. The commercial scale production of amylase is carried out by yeasts. Yeast (*Saccharomyces cerevisiae*) also produces an important enzyme called invertase. The enzyme is used in production of confections. It brings about a partial hydrolysis of sugar into glucose and fructose. Apart from these important enzymes, fungi, such as *Penicillium glaucum* and *Aspergillus niger* also produce enzyme like pectinase and cellulose oxidase respectively.

## MICROBES IN MEDICINE

Microorganisms and other products are of an enormous significance in medicine. Their curative nature has long been known. Some important drugs synthesized by microorganisms are listed below.

### Antibiotics

An antibiotic is a substance which is produced by living organisms and has the capacity to inhibit the growth or metabolic activities of other living organisms. Green alga Chlorella produces chlorellin- an antibiotic- which is known to inhibit the growth of several species of bacteria and is effective against other algae. Bacteria and fungi are the well known groups which produce various kinds of antibiotics (see chapter Antibiotics in this book). Apart from these microorganisms, certain protozoans, e.g., Paramecium aurelia also produce antibiotic.

**TABLE 8.4:** Major Antibiotics Produced by Actinomycetes, ther Bacteria, and Fungi and Microorganisms Affected

| Microorganism | Microbial/Antibiotic Group |
|---|---|
| *Streptomyces nodosus* | Amphotericin-B |
| *Streptomyces halstedii* | Carbomycin |
| *Streptomyces aureofaciens* | Chlorotetracycline |
| *Streptomyces venezuelae* | Chloramphenicol |
| *Streptomyces griseus* | Cycloheximide |
| *Streptomyces erythraeus* | Erythromycin |
| *Streptomyces kanamyceticus* | Kanamycin |
| *Streptomyces antibioticus* | Oleandomycin |
| *Streptomyces rimosus* | Oxytetracycline |
| *Streptomyces fradiae* | Neomycin-B |
| *Streptomyces niveus* | Novobiocin |
| *Streptomyces noursei* | Nystatin |
| *Streptomyces griseus* | Streptomycin |
| *Bacillus polymyxa* | Polymyxin-B |
| *Bacillus licheniformis* | Bacitracin |
| *Cephalosporium acremonium* | Cephalosporin |
| *Aspergillus fumigates* | Fumigillin |
| *Penicillium griseofulvum* | Griseofulvin |
| *Penicillium chrysogenum* | Penicillin |

### Ergotin

Ergot of rye (the sclerotia of *Claviceps purpurea*) has been used as a drug to induce uterine contraction in case of delayed child birth. This, like almost any other drug, can even fatal if taken frequently or in excess quantity.

### Vitamins

They are synthesized by microorganisms like bacteria and fungi. In 1958, H. A. Barker and H. Weissbarch discovered that cell-free-extract of *Clostridium tetanomorphum* produce vitamin B. It has been discovered recently that *Streptomyces griseus* and *S. aurofacius* produce vitamin B12. This vitamin is used as a preventive measure against nervous disorder. Fungi like *Ashbya gossypi* and *Eremothecium ashpyi* synthesize the vitamin riboflavin which is an important component of vitamin B complex group.

### Glycerin

It has been noted that glycerin can be produced commercially by the action of certain yeasts, e.g., *Saccharomyces ellipsoideus*, on sugar.

## MICROBES AS TOOLS FOR BIOLOGICAL RESEARCH

Some of the microorganisms are widely used for academic studies and research work.

### Genetical, Cytological and Bio-chemical Studies

Microorganisms such as *Chlorella, Anacystis nidulans, Chlamydomonas, Nostoc, Anabaena, Neurospora, Saccharomyces, Ascobolus, Sordaria*, etc are the important tool in aforementioned researches. Protozoa are essential for the studies of geology and palaentology.

*Gibbrella fujikuroi* produces an important growth hormone called gibbrellic acid, which has its great importance in physiological studies. *Penicillium notatum* produces antibiotic penicillin, which also shows activity like a plant hormone. It has been shown to affect plant metabolism by promoting several hormone-regulated phenomena. It shows Gibberellin-like activity, Cytokinin-like activity and Auxin-like activity. The work was carried out in India by Dr. S. Mukherji and Dr. S.S. Purohit (see Plant Physiological Research in India, Edited by S. P. Sen, 1988).

**TABLE 8.5:** Examples of Modified Proteins Created through Recombinant DNA Technology

| **Bacterium Used** | **Characteristic Modified** | **Result** | **Mechanism** |
|---|---|---|---|
| *Actinoplanes* | Glucose isomerase thermostability | Enhanced stability of soluble and immobilized forms | Amino acid substitution of arginine for lysine at position 253 |
| *Escherichia coli* | Protein separation characteristics | 96-fold difference in modified partition coefficient | Fusion with partitioning tetrapeptide |
| *Escherichia coli* | Protein structure and crystallinity | Unique macromolecular polyamide copolymers | Modification of leader region in message |
| *Escherichia coli* | Biopesticide activity | Modified molting of fall armyworm | Insertion of *LacZ* gene and modification of hormone |
| *Escherichia coli* | Enzyme oxidation resistance | Alpha-antitrypsin more resistant to oxidation | Replacement of a critical oxygen-sensitive methionine with valine |
| *Escherichia coli* | Anticoagulant polypeptide alteration | Improve the 65 amino acid polypeptide hirudin | Substitution of asparagines residue 47 with lysine or arginine |

**TABLE 8.6:** Recombinant Proteins Produced by *Saccharomyces cerevisiae*

| | |
|---|---|
| Vaccines | Hepatitis B virus surface antigen, Malaria circumsporozoite protein, HIV-1 envelope protein |
| Diagnostics | Hepatitis C virus protein, HIV-1 antigens |
| Human Therapeutic Agents | Epidermal growth factor, Insulin, Insulinlike growth factor, Plated-derived growth factor, Proinsulin, Fibroblast growth factor, Granulocyte-macrophage colony stimulating factor, $\alpha_1$-antitrypsin, Blood coagulation factor XIIIa |

# 9. Development of Industrial Microorganisms

Natural selection has adapted various 'wild-type' bacterium or yeast cell to its environment and to compete with other species of organisms. Various industrial products have been produced by the industrial microorganisms in form of desired commercial metabolites in huge quantity relative to its other end products. For making a normal metabolic product in an amount required at commercial level the wild type organism isolated from natural environment, requires genetic programming. Elimination of undesirable properties or introduction of entirely new ones can be achieved by these changing its genetic information. Such an attempt may result into shunting of more carbon and possibly other metabolites through metabolic sequences to a desirable end product formation. So to achieve the genetic improvement of industrial strains mutation and/or recombination proved more helpful.

## MUTATION

Mutations [Latin *mutare*, to change] were initially characterized as *altered phenotypes or phenotypic expressions*. Long before the existence of defect proof that a mutation is the result of a stable, heritable change in the nucleotide sequence of DNA, geneticists predicted that several basic types of transmitted mutations could exist. They believed that mutations could arise from the alteration of single pairs of nucleotides and form the addition or deletion of one or two nucleotide pairs in the coding regions of a gene.

The expression of a mutation can be observed if it produces a detectable altered phenotype. A mutation from the most prevalent gene form, the **wild type**, to a mutant form is called as **forward mutation**. Later, a second mutation may make the mutant appear to be a wild-type organism again. Such a mutation is called a **reversion mutation**. In this the organism reverted back to its original phenotype.

A true **back mutation** converts the mutant nucleotide sequence back to the wild-type arrangement.

The wild-type phenotype also can be regained by a second mutation in a different gene, a **suppressor mutation**, which overcomes the effect of the first mutation. Sometimes, however, mutations may occur and not alter the phenotype for a variety of reasons. When a large deletion and insertion mutations exist, and affect only one base pair in a given location are called **point mutations**.

### Mutant Detection

When collecting mutants of a particular organism, one must know the normal or wild type characteristics so as to recognize an altered phenotype. Usually, *mutations are rare*, about one per $10^7$ to $10^{11}$ cells, therefore, it is necessary to have a very sensitive detection system and one has to look at perhaps thousands to millions of colonies or clones. Mutation detection systems in bacteria and other haploid organisms are straightforward, because any new allele should be seen immediately, even if it is a recessive mutation.

Sometimes detection of mutants is direct. If albino mutants of a normally pigmented bacterium are being studied, detection simply requires visual observation of colony color.

Other direct detection systems are more complex. For example, the **replica plating** technique is used to detect auxotrophic mutants. It distinguishes between mutants and the wild-type strain based on their ability to grow in the absence of a particular biosynthetic end product.

Once some sort of detection method is established, mutants are collected.

### Mutant Selection

**Selection Technique:** An effective selection technique uses incubation conditions under which the mutant will grow, because of properties given by the mutation, whereas the wild type will not. It involves reversion mutations or the development of resistance to an environmental stress.

**Resistance Selection Technique:** It follows a similar approach. Often wild type cells are not resistant to phage attack or antibiotic treatment, so it is possible to grow the bacterium in the presence of the agent and look for surviving organisms. Consider the example of a phage-sensitive wild type bacterium. When the organism is cultured in medium lacking the virus and then plated out on selective medium containing phages, any colonies that form will be resistant to phage attack and very likely will be mutants in this regard. Resistance selection can be used together with virtually any environmental parameter; resistance to bacteriophages, antibiotics, or temperature is most commonly employed.

**Substrate Utilization Technique:** This technique is also employed in bacterial selection. Many bacteria catabolize only a few primary carbon sources, with these bacteria, it is possible to select mutants by a method similar to that used in resistance selection. The culture is plated on medium containing an alternate carbon source. Any colonies that appear can use the substrate and are probably mutants.

### *Induction of Mutation*

Mutations can be induced in the laboratory by exposing the culture to mutagenic agents i.e., ionizing radiations, ultra violet light, nitrous acid, nitrosoguanidine, purine and Pyrimidine analogues and other mutagenic agents.

## THE SELECTION OF MUTANTS PRODUCING HIGH YIELD OF PRIMARY METABOLITES

Corynebacterium glutamicum and Brevibacterium flavum produces amino acids as primary metabolites. Corynebacterium glutamicum and Brevibacterium flavum produces one of the important amino acid i.e. lysine. In these wild type organisms, the production of lysine is regulated in such a way that, it is produced to a level to meet the needs. The fermentation is based on bacterial biosynthesis of amino acid and genetic regulation. In these bacteria lysine is one end product. Beside this these are also responsible for the production of some other amino acids like methionine and theonine.

Such bacteria works on the mechanisms, which is known as feedback, control. Feedback control is the regulatory mechanisms involved in the inhibition of the activity of key enzymes in biosynthetic route and the repression of their synthesis by the end products produced which are present in the cell at required concentration to meet the needs of the organisms. That is, when these two amino acids accumulates in excess in the organisms then their synthesis is inhibited by the enzymes activity; conversely, a shortage of threonine or lysine increases their rate of synthesis.

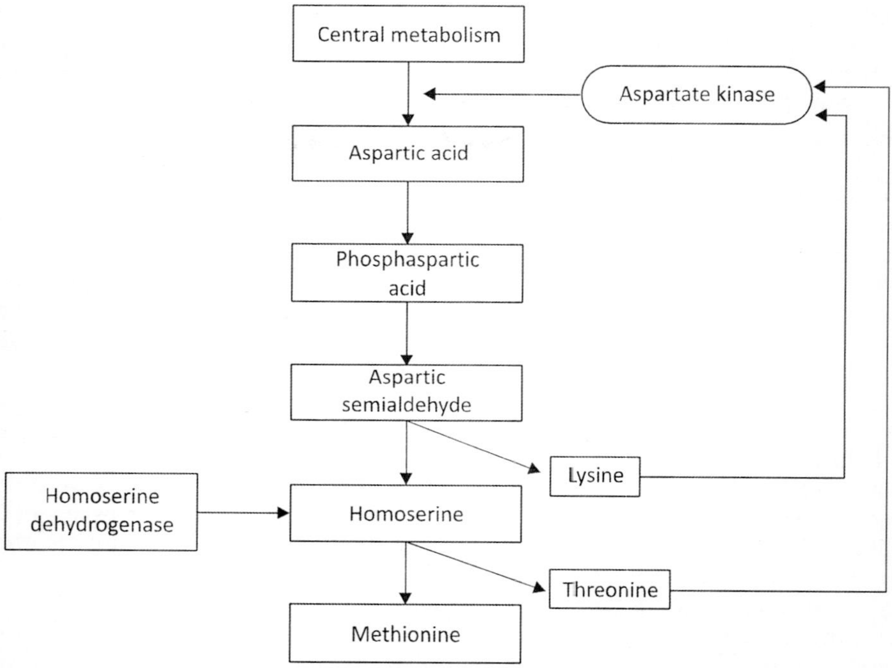


**FIGURE 9.1:** The control of lysine production in *Corynebacterium glutamicum* (Feedback inhibition).

Isolation of two kinds of mutants can help to overproduce lysine beyond the own needs of the bacterium. In one kind a mutation in the gene coding for the enzyme homogenize dehydrogenase abolishes the enzyme activity and thereby prevents the bacterium from making threonine (one of the inhibitory products) and methionine. During this pathway lysine does not interrupt the biosynthetic route from aspartic semialdehyde to the formation of lysine. In homoserine supplemented medium mutant, who could not catalyse the conversion of aspartic semialdehyde to homoserine is capable of growth. Such organisms are known as homoserine auxotrophs. When this auxotrophs, or nutritionally deficient mutant, is cultured in a medium with just enough threonine and methionine to support growth but not enough threonine to cooperate with lysine in shutting of asparate kinase's activity, the pathway to lysine continues to operate at full speed. These mutants have been selected by testing thousands of those colonies, which are being treated with a suitable mutagen. These auxotrophs grow only when particular growth factors like methionine and threonine are supplied in the medium by some external source. Beside this these auxotrophs can be isolated more conveniently by the inoculation of culture into a medium that has been mutated and which lacks the appropriate growth factors as well as which contain penicillin. This kills only proliferating bacteria and the auxotrophs cannot grow and survive, whereas the unmutated bacteria are killed by this activity.

Mutant has an altered form of aspartate kinase itself, which does not react with lysine and has therefore lost its sensitivity to feedback inhibition. This causes a high level accumulation of lysine in fermenter. Therefore these mutants must be selected so that they could resist lysine analogue S-(2-aminoethyl) cysteine (AEC), which resembles lysine very closely and which minimizes the regulatory effect of it, inhibiting aspartate kinase even if no lysine is being synthesized. For getting a higher yield it must be noted that several richer media have to be tested and detected as well as there must be an evaluation of high yielding mutant.

## THE SELECTION OF SECONDARY METABOLITE PRODUCING MUTANTS

It is very difficult to devise a procedure, which can isolate mutants, which over produce secondary metabolite because till today the information available on control of production is not known or if known is very less as well as such products are not required to the producing organisms for their growth. For example, at specific stages in life cycle of various molds, actinomycetes, and spore forming bacteria various important antibiotics are being synthesized. Still the genetic regulation is poorly understandable. Besides this the production of antibiotics depend on various other factors also like, the availability of carbon, nitrogen and free phosphate, synthesis of chemical building blocks which are being needed for the production of antibiotics etc. The yield of an antibiotic also depends on various kinds of genes, so it is impossible to find the individual mutations that can increase the yield from wild strains. Nowadays various genetically improved strains have been developed through mutations and selections. In this process the culture is treated with a mutagen and thousands of colonies obtained by these are examined properly. And from these colonies mutant, who display significantly increased productivity, is being served as the starting point for mutagenesis and screening. So by this way high yielding improved antibiotic producing strain is produced.

But the production of mutant variety by this method is very slow and labour intensive, and its results are unpredictable. This is so because not only the genes of the producer but also the culture conditions influence the antibiotic levels strongly.

### *Recombination*

Plasmid DNA which is introduced into plant cells via particle bombardment undergoes recombination with other plasmid molecules as well as with the host chromosomal DNA. Recombination occurs whenever new gene arrangements are formed through exchange, elimination or insertion of DNA. Plasmid DNA can either insert cleanly into plant chromosomal DNA as a single entity or large linkages of plasmids can form and subsequently integrate (the full range of intermediate combinations is possible). Integration of a large chain of plasmids can result in a single insertion event with a high copy number of introduced plasmid DNA. With transformed event tissue obtained via particle bombardment, a single integration event rarely results in the introduction of a single plasmid. When large numbers of plasmids integrate into the same site, the plasmids may recombine with each other prior to integration into genomic DNA.

Although the exact fate of DNA introduced via particle bombardment is not clear, it appears that plasmid DNA molecules rearrange or recombine with each other to yield linkages of both whole plasmids and plasmid fragments. Recombination has been applied for the production of various industrial strains.

Besides these various other methods have also been applied for the production of various important industrial strains. Some of these methods include, Protoplast fusion, *in vitro* recombinant DNA technology etc. These have been discussed in detail in the following chapters of the book.

# 10. Culturing of Industrial Microorganisms

For the production of the desired product it is necessary that the growth of the producing organism must be significant. For the production of these fermentation products there is necessity for the proper growth of the fermentation organism as well as the involvement of the large-scale production tanks (fermenters) in which the organisms can be grown with required temperature, pH, dissolved oxygen and substrate concentration. For the production of fermentation products, industrial fermenters are very well designed to attain the best possible growth and biosynthesis conditions for important microbial cultures being used in the fermentation industry. For a successful fermentation programme inoculum development, cultural conditions employed and growth rate of the producing organisms are regarded as vital aspects.

## INOCULUM DEVELOPMENT

Standard culture preservation techniques have been employed to preserve the initial stock of the important cultures, which is metabolically active, free from contamination. This culture has to be sub-cultured from initial stock cultures stored in well aseptic conditions. Basically these cultures are the spores and these may be mixed with inert support, which is sterile and microbe free such as soil or sands etc. These spore suspensions can also be stored under liquid nitrogen or can be lyophilized in appropriate media.

To obtain a pure inoculum in sufficient volume for the increasing production capacity in the fast growing logarithmic phase and to achieve high population density inoculum development programme for culturing of the producer organisms has been employed in succession. The volume of culture increases nearly ten times with each successive stage in the progression of the inoculum.

For reproducing high yields of spores the quality and reproducing ability of inoculum are critical factors. Mutation may also be done among these cells to achieve better yield. Obviously, stock culture maintenance of mutant fermentation strains should be rigorously controlled so that least possible number of back mutant forms occurs in the inoculum production.

The problems, which a fermentation industry face is the inoculum, build up which is a critical task in the inoculum development. Many industrially important fungi used in the fermentation industry are spore forming and filamentous. These may give rise to various growth forms in submerged culture, which may range from filamentous homogenous cultures to discrete, compact pallets of mycelium in the broth. Adequate aeration must also be there for the growth of these filamentous types with viscous broths. In pallet forms, mycelium at the centre may not be accessible to nutrients and oxygen, due to limitations of diffusion. As small biomass may be used to a very high density of inoculum build up so spore production is more significant for the development of inoculum.

## BATCH CULTURE

The growth of the fermentation organism occurs in a closed system in a batch culture. Important nutrients given to these organisms are being utilized for the reproduction and the accumulation of these metabolic products. These results in continually changing conditions in the culture in which the composition of the cell wall and the concentrations of a number of cytoplasmic components vary as growth of the microorganisms proceed.

When organisms are inoculated into a medium, a typical growth curve obtained as the microbial culture will pass through *logarithm exponential, stationary* and *decline* phases.

Lag phase is the period of adaptation to new conditions or prior to active growth. Inoculating exponentially growing cells reduce during this period, and the growth rate gradually increases at a constant rate. The growth rate of these microbes is most rapid under optimal environmental conditions. The optimal growth rate is the genetic property of the strain. The period of maximum growth rate is known as log or exponential phase.

The time taken for the population to double in size is referred as "doubling time", or mean generation time, which is the average time taken by a single cell to divide into two daughter cells. The two parameters that is the doubling time and mean generation time are identical under favourable conditions, but if some daughter cells are non-viable the mean generation time is less than the observed doubling time.

If at zero time is the mean generation time is No, then after one doubling time the population is 2 No, after two doubling time it is 4 No. = $2^2$ No, and after n doubling times is $2^n$ No i.e., the population at time t-$N_t$=$2_t$/td $N_o$, where td is the doubling time, therefore

$$\text{Log } N_t/n_o = (t/td) \log^2$$

The growth rate constant (K) in doublings/hour is the slope of the growth curve in the exponential phase and can be expresses as

$$K = t/td = (\log N_t - \log N_o)/t.$$

The growth rate of the microbes also decreases due to the nutrient limitation because of the depletion in the substrate coupled with accumulation of toxic metabolites. The specific rate of the cells eventually ceases from maximum to the stationery phase. Until the substrate concentration declined to a very low level, the specific growth rate in batch culture is not affected. The amount of biomass in the stationery phase is also an important parameter and it depends on the composition of the medium and the efficiency of the organisms to convert the substrates into daughter cells.

Where the growth is limited by the exhaustion of that limiting component in a medium, the concentration of the cells at the stationery phase, can be expressed as

$$x = Y.SR$$

*where,*
x = Cell concentration
Y = Yield factor for the limiting substrate
SR = Initial substrate concentration in the medium

Y represents measure of efficiency of the cell in converting substrate into biomass. The biomass in a batch can be given by

$$x = Y(SR - S)$$

*where,* $S$ = Residual concentration at that point in time.

*Thus,*

$$Y = \frac{X}{(S_R - s)}$$

## CONTINUOUS CULTURE

There are three transitional periods between growth phases have been included in the typical bacterial growth curve in a batch culture. In this condition all the cells are not in identical physiological state. Time is required for some cells to catch up with others. In terms of physiological conditions, the growth curve includes young, actively metabolizing cells as well as cells which are at the stage of dying, with the cells in between these phases. The effect of physical conditions as well as chemical substances on the organism is also not similar or identical in all phases of growth. In a commercial process it is obviously desirable to maintain a microbial population in the logarithmic phase of growth in a constant environment. The growth rate of a culture, and its physiological conditions can be maintained absolutely constant by the help of process of continuous culture, which is applied usually in a Chemostat. Nutrient medium, which is just, made flows into the culture vessel from a reservoir of a sterile medium at a constant rate. The volume in the culture vessel is kept constant by a device that allows the culture medium, together with the accumulated waste products and, older and dead cells to leave the culture vessel at the same rate. By maintaining a fixed limiting concentration of a particular nutrient (i.e. growth limiting chemicals component) in the medium, the level of the growth can be controlled. The remaining constituents essential for growth of the producer organism are added in the medium in excess of requirement. The growth-limiting nutrient is also added into the medium at a required concentration for attaining the maximum growth.

## THE KINETICS OF CONTINUOUS CULTURE

At different dilution rates of the medium the chemostat is usually applied and operated. The dilution rate is the ratio of inflowing amount of nutrient medium per hour to the volume of the culture.

Therefore, the dilution rate (D) can be put into

$$D = \frac{F}{V}$$

*Where*
$D$ = Dilution rate ($h^{-1}$)
$F$ = Flow rate ($dm^3/h$)
$V$ = Volume ($dm^3$)

The change in the cell concentration in the fermenter, in a given period becomes.

$$\frac{dx}{dt} = \text{Growth} - \text{Output}$$

$$\frac{dx}{dt} = \mu x - Dx$$

During this process the medium is added continuously at a suitable rate and by the production of new biomass or the output of biomass is balanced by the growth of the cells, the displacement of culture is balanced. The steady state can be expressed as

$$\frac{dx}{dt} = 0$$
$$\mu x = Dx$$
$$\mu = D$$

The steady state condition in a system is controlled by the flow of medium through the system (i.e. Dilution rate), and the mechanism underlying this balance, is essentially, the relationship between μ and residual limiting substrate concentration (Monad, 1942)

$$\mu = \frac{\mu_{max} S}{K_S + s}$$

From the equation, we know that: μ = D

*Therefore,* $$D = \frac{\mu_{max} S}{K_S + s}$$

Where s is the steady state residual concentration in the medium, and this can be represented as:

$$\bar{S} = \frac{K_2 D}{\mu_{max}\ S}$$

By the help of dilution rate the substrate concentration in the chemostat can be determined. Virtually, depletion of the growth of a substrate to a fixed concentration is equal to the rate of the dilution.

The concentration of cells in the system at steady state can be described as

$$X = Y(S_R - s)$$

Where X represents the steady state cell concentration

From the equations above,

$$X = \frac{Y(S_R - K_S D)}{\mu_{max} - D}$$

Thus it becomes apparent that, the operational variables (i.e. SR and D) control the steady state of biomass concentrations.

Within the range of dilution rates in the chemostat, maximum stability could be attained, over which cell density changes only slightly with changes in dilution rates. As D approaches μ max the concentration of residual limit in substrate required to support the increased growth rises significantly, and therefore, the residual substrate rises and biomass depletes. Ultimately, a dilution rate is reached at which x equals to zero. This dilution rate is referred to as Dcrit. And can be expressed as

$$Dcit = \frac{\mu_{max} \cdot S_R}{K_S + S_R}$$

In a chemostat system, as the substrate limiting factor controls the cell growth concentration, Dcrit., will be slightly lower than the $\mu$ max.

## FED BATCH CULTURE

Nutrient substrates are fed continuously with the fresh medium and cultural fluid in the closed batch culture. So, the closed batch culture where these are fed is being represented by fed batch culture. In fed-batch culture, the volume of culture in the system increases with the time.

The biomass concentration at stationary phase (X max) in a batch culture, under growth limiting conditions can be described as:

$$X\ max - Y.S_R$$

The initial inoculum is supposed to be insignificant in comparison with the final biomass.

If the fresh medium where allowed to enter the system at a dilution rate less than X-max, all the substrate gets exhausted.

$$F.Sr \ - \ \mu\frac{X}{Y}$$

Where, X is the total biomass present in the vessel.

The concentration of the cells 'X' remains constant, even if the total biomass 'X' increases with the time.

$$\frac{dx}{dt} = 0$$

and, therefore $\mu = D$

such a system then represents a quasi-steady state. As time advances, the volume of culture increases with decrease in the dilution rate. Therefore, D can be represented as

$$D = \frac{F}{V_O - F.t}$$

*where* F = flow rate
Vo = Initial volume of culture
T = time that fed-batch conditions have been operating.

By increasing exponentially the flow rate, dilution rate can be kept constant. Most industrial processes operate 'Fed-batch' culture systems with significant extension of productive period of a traditional fed batch process without inherent disadvantage of a continuous process.

## CULTURE MEDIA

Culture medium and the fermenter and mode of operation of the fermenter greatly influence the performance of the process organisms employed in the fermentation process. The chemical environment of the producer organism should be such that, it not only supports the growth of the organisms, but also optimum product formation. The culture medium used for the culture of the microbes must also meet the requirements of the fermentation process.

The essential nutrients required for growth of the organism, energy, building of cell components, and biosynthesis of fermentation products etc. are being provided by the culture medium. It provides water, sources of carbon, nitrogen, inorganic nutrient, vitamins, amino acids, precursors, buffers, antifoams etc. If large quantities of carbon and nitrogen are channeled through the metabolic pathways of microorganisms, good yield of fermentation products occurs. Autotrophs require simple inorganic/salts, water and nitrogen and obtain their requirements for carbon from $CO_2$ of the air or from carbonates and bicarbonates. Fastidious microbes like lactic acid bacteria lacks the synthetic ability and hence require preformed nutrients. The vast majority of commercially significant microorganisms are chemo-organotrophically. Carbon and nitrogen are supplied as complex mixtures of cheap natural products or by products of some process. The carbon sources include products or by products of some process like starch, glucose, whey, sulphite waste liquor, malts, beet molasses, cane molasses, vegetables oils and hydrocarbons. The nitrogen sources include, ammonia, ammonium salts, nitrates, corn steep liquor, peanut granules, dried blood, soya meal, distiller's solubles and yeast extract.

Antifoam agents or we can say the sources of alcohols and fatty acids are required for proper maintenance of the nutritive balance of the medium. Any specific requirements, such as amino acids, or vitamins may be supplied in pure forms i.e. complex plant or animal extracts, which are more economical. Phosphates were used as buffering agents although external pH control is normally employed. Potassium phosphates, magnesium sulphate, and ammonium sulphate or phosphate can meet inorganic nutrients. Calcium carbonate is some times included either to neutralize the acid produced during fermentation or to prevent rapid decline in pH that occur during growth.

Crude constituents are also employed because these crude constituents are providing the growth requirement of the process organisms and the additional media sources are also sometimes included with cell hydrolysates or autolytes as growth factors.

## PRECURSORS AND INDUCERS

There are certain substances, which generally serve to improve the yield or quality of the product. These substances are known as the precursors. These precursors are incorporated without any major change into the molecule of the fermentation product. For example phenyl acetic acid and cobalt are being added for penicillin G and Vitamin $B_{12}$. It also acts as inducer that determines the particular product formation among several possibilities i.e. corn steep liquor may yield various penicillin, but the addition of phenyl acetic acid determine penicillin G production. Proteases for various proteins are another such examples, a-amylases for starch, cellulase for cellulose, pectinase for pectin and penicillin acylase for phenyl acetic acid.

## REPRESSORS

The growth of the microorganisms is restricted by the formation of catabolic enzymes and secondary metabolite production, are being subjected to repressive effects by certain limiting factors. Media allowing restricted growth provides high product yield as major portion of carbon and other components of the medium are shunted to product formation rather than to growth. Repression of the formation of enzymes and secondary metabolites has been shown by rapidly utilized carbon and nitrogen sources. Depending on the control system involved media may be carbon, phosphate or nitrogen limited or some times, limited by other compounds. For example, the nitrogen sources such as Soyabean meal and proline for streptomycin; production is probably due to their slow utilization, thus avoiding nitrogen metabolite repression. Replacement culture with fungal mycelium is also one of the examples for restricted growth medium. *Aspergillus niger* for gluconic acid production is first grown on a medium that supports a rich growth as well as product formation, then the mycelium is separated and placed in a fresh medium high in carbon substrate (sugar) but lacking combined nitrogen so that additional growth cannot occur. So, those substances, which are being employed for the repression of the industrial cultures, are known as the repressors.

# 11. Industrial Fermenters

According to need of the industry varying environment requirement, and types of product various kinds of industrial fermenters have been developed and designed. They provide optimum growth and metabolism of industrial processed strains for production of quantities of commercially desired product. Fermenter maintains: aerobic/anaerobic conditions, temperature at the desired level, culture in pure state throughout the process to support the growth for optimum product formation. While designing a fermenter, an operation/designing engineer considers various criteria in his mind. An ideal fermenter should have the provisions and control over various operations like temperature, pH adequate aeration and agitation (to meet the metabolic requirements of the microorganisms), less evaporation loss, low power consumption, proper sampling facilities, use of cheapest material, operated aseptically for a long period throughout the operation, vessel of similar geometry (smaller and larger vessels), vessel with smooth internal surfaces, vessel with minimal use of labour during operation, harvesting, cleaning, suitable for variety of reactions, etc.

Important types of industrial fermenters are stirred, baffled, and aerated tanks types. These are provided with systems of temperature, pH, and foam formation control. Most of the commercial fermenters have mechanical agitation system which is design to get proper mixing and oxygen transfer, while a few are design in such a way which depend on very high gas inputs without mechanical agitation. The design of a fermenter (bioreactor) also depends on the type of environment required for an organism, its structure and the product. Therefore, following are the major types of industrial fermenters: (1) mechanically agitated fermenter, (2) Non-mechanically agitated fermenter and (3) Non-agitated fermenter.

## CONTINUOUS CULTURE

For continuous culture of organisms required for process should have a reservoir of sterile medium. Culture is kept in vessel with constant volume. It should also be provided with a collecting vessel in which culture leaving the fermenter accumulates.

## THE COMPONENTS OF A FERMENTOR

The main subdivisions of a bench-scale fermentor are as follows:

- Base components including drive motor, heaters, pumps, gas control, etc.;
- Vessel and accessories;
- Peripheral equipment such as reagent bottles;
- Instrumentation and sensors.
- These components perform the following functions combinedly:
- Provide operation free from contamination;
- Maintain a specific temperature;
- Provide adequate mixing and aeration;

- Control the pH of the culture;
- Allow monitoring and/or control of dissolved oxygen;
- Allow feeding of nutrient solutions and reagents;
- Provide access points for inoculation and sampling;
- Use fittings and geometry relevant to scale-up;
- Minimize liquid loss from the vessel;
- Facilitate the growth of a wide range of organisms.

## PARTS OF FERMENTORS

### Vessel

The vessel is a single-walled cylinder of borosilicate glass with a flat bottom or as a glass-jacketed system having a typically round bottom. The top plate is made from stainless steel and is compressed onto the vessel flange by an easily released clamping system. The vessel glass and the top plate is separated by a seal. Port fittings of various sizes are provided for insertion of probes, inlet pipes, exit gas cooler, cold fingers, sample pipes, etc. These work by compressing the sides of the probe/pipe against the seal. A special inoculation port will have a membrane seal held in place with a collar. Culture can be withdrawn into a sampling device or a reservoir bottle via a sample pipe situated in the bulk of the fermentor fluid. A gas sparger is also fixed into the top plate and this terminates in a special assembly, which ensures that incoming air is dispersed efficiently within the culture by the flat-bladed 'Rushton-type' impellors fixed to the drive shaft. A drive motor is attached for the purpose of stirring to the drive shaft and is usually fitted directly to the drive hub on the vessel top plate. An exit gas cooler is also provided to the vessel, which works like a simple Liebig condenser to remove as much moisture as possible from the gas leaving the fermentor to prevent excessive liquid losses during the fermentation. It also prevents the wetting of the exit air filter.

A narrow platinum resistance temperature sensor is also attached with the vessel, which control the temperature either by direct heating with a heater pad or by circulating warm water around the vessel jacket. A cold finger is also used to cool the vessel contents, if direct heating is used.

The sensors which fit into the vessel do so either by direct coupling with a thread on the body of the electrode, as is the case in the gel-filled type of pH electrode, or by a special fitting on the vessel top plate, as in the case of the dissolved oxygen electrode (polarographic type).

## PERIPHERAL PARTS AND ACCESSORIES

### Reagent Pumps

Pumps are used for the essential parts of the instrumentation system for measuring pH and antifoam and control. Peristaltic pumps are basically used and the flow rate is normally fixed with a timed system of control. Flow rates will depend on the bore of the tubing used and it is a good idea to move the tubing every few days so that one piece does not become worn out by the pump rollers. A dip tube goes to the bottom of the bottle. A filter in the top of the reservoir bottle is kept open during autoclaving to prevent a build-up of pressure. For extremely high accuracy of addition, the reservoir bottles are placed on analytical balances, which can be used to determine how much reagent has been pumped in a given time.

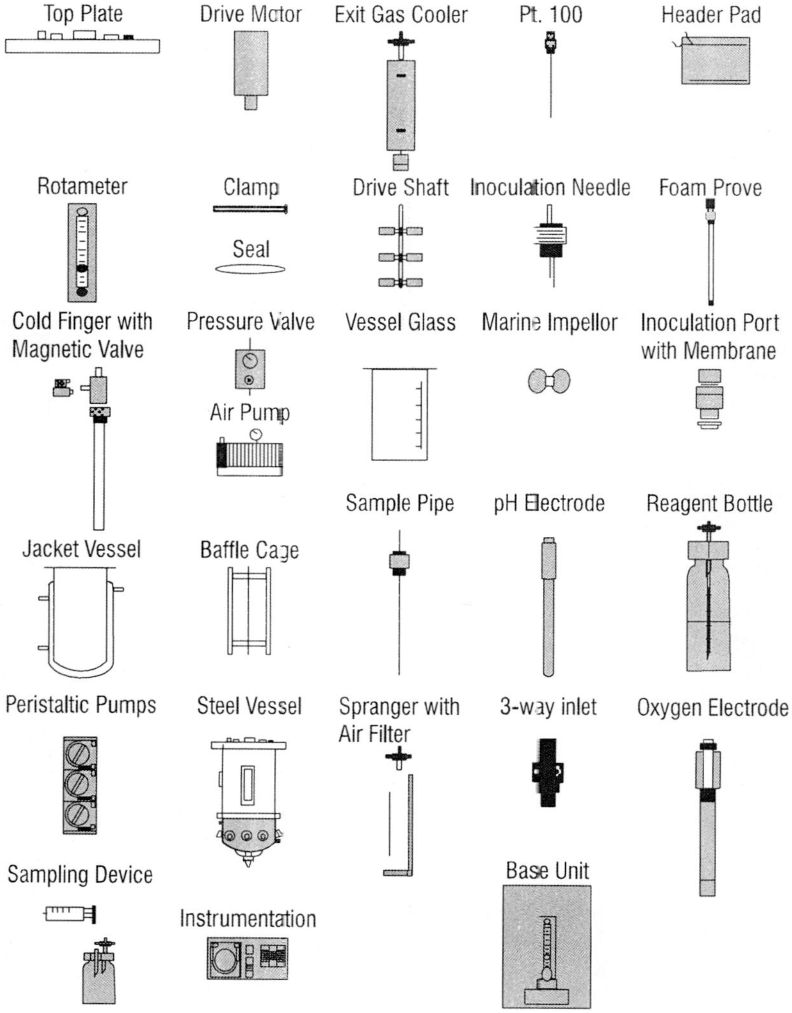


**FIGURE 11.1:** Illustration of major components of fermentor.

## Medium Feed Pumps and Reservoir Bottles

Medium feed pumps are often used with variable speeds so as to give the maximum possible range of feed rates. The reservoir bottles must be changed several times during a long continuous culture experiment, so it is usual to fit an aseptic coupling in the tubing. An effluent pump is often used to remove culture fluid from the fermentor vessel into a storage reservoir bottle.

## Rotameter/Gas Supply

A laboratory air supply is provided from a separate pump, which must be oil free. A variable area flow meter or rotameter is used to control the airflow rate into the fermentor vessel. A pressure regulator valve is also attached before the rotameter, which ensures safer operation. A sterile filter (usually 0.22 μm) is used as a bridge between the tubing from the rotameter and that connected to the air sparger of the fermentor.

### Sampling Device

This allows culture fluid to be removed aseptically during the fermentation at intervals decided by the user. The frequency of sampling and the size of each sample are determined empirically according to the needs of the experiment.

## ALTERNATIVE VESSEL DESIGNS

There are various types of alternative vessel designs commonly found for the fermentation processes. Usually, alternative vessel designs are tried when the standard vessel configuration does not allow adequate growth of the organism. Shear forces in fermentors with turbine impellors often disrupt for example animal cells. These are also used when the scale-up criteria require a different design of bioreactor. For example production of large quantities of single-cell protein is cheaper on a large scale if airlift fermentors are used to eliminate energy costs associated with a drive system. A number of these special designs are available at the bench/pilot scale of operation to allow small-scale research into the suitability of a particular method.

### Air Lift

By the use of airlift the need for a stirrer system can be minimized or can be neglected. A tall, thin vessel with an aspect ratio of around 10:1 (height to base diameter) is the best shape of the airlift. Sensors can be mounted in a steel base section, from a collar at the base of the conical section or from the vessel top plate. A common use of airlift fermentors is the growth of shear-sensitive cells such as plant and animal cultures. Also, the design has been used for the production of large amounts of biomass as single-cell protein.

During the fermentation process the culture fluid is mixed and aerated by a jet of air, which enters into the vessel from the base of the vessel. A hollow pipe or draft tube in the centre of the vessel, which rises the air, which is full of bubbles to move upwards to the top of the vessel.

A large headspace is provided above the top of the draft tube, which allows easy gas transfer from the liquid to the gas phase and causes the density/specific gravity of the liquid to increase. It then descends to the bottom of the vessel and the descending liquid returns to the base of the vessel, which is then regassed.

### Fluidized Bed

Inm fermentation process the microbes/cells are trapped in a physical medium and then put into the vessel by a mesh. The medium is recirculated with a pump, which is easily adapted to give a continuous/semi-continuous flow to allow the trapped cells to effect chemical changes on constituents of the medium without being washed out along with spent medium.

### Hollow Fibre

In this cells are embedded in fibres contained in a cartridge, which are bathed in circulating culture medium. This is often used for mammalian cell culture, where anchorage-dependent cell lines can be perfused with oxygenated medium. An extension of this method is to use cartridges containing two different bundles of fibres as a separation 'membrane' between a pair of fermentor vessels and allow dissolved gases and metabolites to be exchanged without cells crossing the barrier to allow, for example, It is used for the study of toxicity or interactive studies.

### In Situ Sterilizable Fermentors

For safety and insurance purposes, vessels above the recommended size are usually made of stainless steel and are sterilized *in situ* using house steam or an electrical steam generator, which may be built into the base unit. The heating for the vessel is normally done through a double jacket, which can either be the full length of the vessel or cover just the bottom. The bottom section contains large port fittings for electrodes and usually has some kind of steam-sterilizable sampling device/harvest valve. The mechanical seal and drive shaft usually enter from the bottom on this size of vessel. In some fermentors, the air inlet filter and connecting pipe to the air sparger are also provided.

### Containment

If the organism to be cultured is pathogenic or genetically modified then an *in situ* sterilizable fermentor may have to be altered so that any microorganism cannot release in the environment. It is provided with special features such as a double mechanical seal, magnetic coupling, additional air filters, extra foam control systems and special sampling devices. There are several different categories of containment (1-3 for pathogens and 1-4 for genetically modified organisms). Requirements differ at each stage and relate to building and working protocols as much as the design of the fermentor.

## ADDITIONAL ACCESSORIES AND PERIPHERALS

### Feed Pumps

For fed-batch or continuous applications a separate feed pump is used. The type of pump depends on the size of the vessel, the required flow rates, the nature of the additive (e.g. viscous, vorrosive, etc.) and the speed at which the addition should be made.

#### *Peristaltic Pump*

The pumps used for medium feed are typically variable speed units, which can accept signals from an external source such as a computer or process controller to allow adjustment of feed rates based on, for example, respiratory quotient (RQ) or optical density (OD). This is the most common type of addition pump used in fermentor applications. The key advantages of peristaltic pumps are as follows:

- The liquid being pumped is not contaminated by any part of the pump.
- The liquid being pumped does not contaminate the pump.
- A wide range of flow rates is possible using tubing of different bores.
- The accuracy of delivery is good over a wide range of flow rates.
- The pump can run safely even when the tubing is empty.

The disadvantages are:

- The flow is pulsed and cannot 'push' against high backpressure.
- The tubing wears over time and flattens, reducing accuracy and increasing the risk of bursting.

- Silicone tubing is not suitable for all liquids and alternative tubing material may be too rigid to allow correct peristaltic action.

The problem with tubing wear can be partly overcome by leaving extra length in the tubing when connecting it to the fermentor vessel and moving a new length into the pump head every few days to stop wear in just one small section.

### Syringe Pump

This type of pump is typically used in medical applications for the precise delivery of drugs. Most now allow for computer control. In this case, a syringe of various sizes from 1μl to 100 ml, is filled with the liquid to be added and secured onto the syringe pump. The advantages of syringe pumps are:

- Very precise and accurate delivery; non-pulsile flow.
- Very low flow rates possible.
- Can work against backpressure.
- Can pump very viscous fluids, e.g. oils.
- The disadvantages are:
- Must be replaced or refilled regularly.
- Range of flow rates not as wide as with peristaltic pumps.
- Usually more expensive.

### Diaphragm/Positive Displacement Pump

It uses a flexible membrane and soft tubing to displace the liquid and control its flow but also employs a piston and pump heads of fixed volume. A stainless steel pump head is connected to soft tubing and autoclaved. It is then located onto the pump and held securely. Advantages of this type of pump include:

- Accurate delivery.
- More choice in type of tubing for corrosive liquids, for example.

**Disadvantages are:**

- Less flexibility on flow rate ranges for each pump head.
- Stainless steel chamber and diaphragm may not be suitable for some liquids.
- Pulsed flow even more pronounced than with a peristaltic pump.

## Exit Gas Analysis

The exit gas may need to be conditioned (e.g. moisture removed) before going into the analyzer, depending on the type of instrument used.

### Infra-Red Carbon Dioxide Analyzer

An infrared detector measures the amount of radiation reaching it after the gas absorbs some. This type of detector allows continuous measurement with good sensitivity, accuracy and selectivity.

### Paramagnetic Oxygen Analyzer

A dumb-bell of gas-filled spheres is linked to a support mechanism suspended in a magnetic field created by permanent magnets. If oxygen is present in the test gas, its attraction to the magnetic field will cause the dumb-bell to be displaced. The movement is detected using a mirror in the centre of the balance system, which displaces a beam of light shone via the mirror onto a photocell. The signal form the photocell will be proportional to the amount of oxygen present in the sample.

A low flow rate is needed for this system to work properly, so inlet gas is typically pumped into the detector cell with most being discarded through bypass pipe work.

### *Mass Spectrometer*

With the help of mass spectrometer the analysis is rapid, so it allows a multi-inlet system to be used, which 'shares' the analyzer between separate vessels within a bank of fermentors. This is metal plate, which generates a tiny electric current whenever a gas ion strikes it. This signal can be amplified and displayed. By re-tuning the magnetic field very quickly (milliseconds) different ions can be detected and measured in a sequential way, which is so fast that it appears to give an almost simultaneous measurement.

## TYPES OF REACTORS

### Mechanically Agitated Stirred Tank Reactors

These are the most widely used bioreactors for industries. There are several problems in this kind of fermenter. The agitator shaft rotates and there is a micro clearance between the shaft and stuffing box. This is an important factor in its design. It cannot be neglected and overtime it increases due frictions. To ensure aseptic condition in the fermenter special arrangements are needed to avoid contamination and which needs additional expenses. It has high shearing and tearing effect of impeller blades. Mechanically agitated stirred tank reactor is not suitable for mycelial fungal strains or for tissue cultures. Since it is made of a very large capacity vessels, therefore, it is difficult to achieve uniform mixing. To overcome this problem, various other types of bioreactors have been developed. They are: (1) air-lift bioreactors, (2) air-lift fermenter with external loop, (3) continuous fermentation stirred tank, (4) cyclone type bioreactor, (5) deep jet fermenter, (6) fluidized bed bioreactor, (7) hollow fibre bioreactor, (8) ICI Air-lift fermenter with internal loop, (9) membrane recycle bioreactor, (10) modified air lift bioreactor, with both internal and external loop type (IIT Kharagpur), (11) packed bed bioreactor. The important one are discussed below:

### Air-lift Bioreactor and Modified Air Lift Bioreactors

This kind of fermenter works on principle of an air lift pump. It is of two kinds: (1) internal loop type (2) external loop type. The reactor's volume is determined by its capacity, kinetic data, and specific growth rate of the organism used. The rate of airflow of the reactor depends on the volumetric mass transfer coefficient in the reactor system. Most of the industries are using bioreactor developed by IIT-Kharagpur. It is a uniform cylindrical cross type and has an internal loop or external loop riser configuration, diverging converging.

The external loop riser configuration is adjustable and the change in the configuration improves the $O_2$ transfer rate vis-à-vis mass transfer coefficient for a particular rate of airflow. This helps provide required particular dissolved $O_2$ concentration for specific microbial system. This reactor reduces the operating cost for pumping air through the bioreactor.

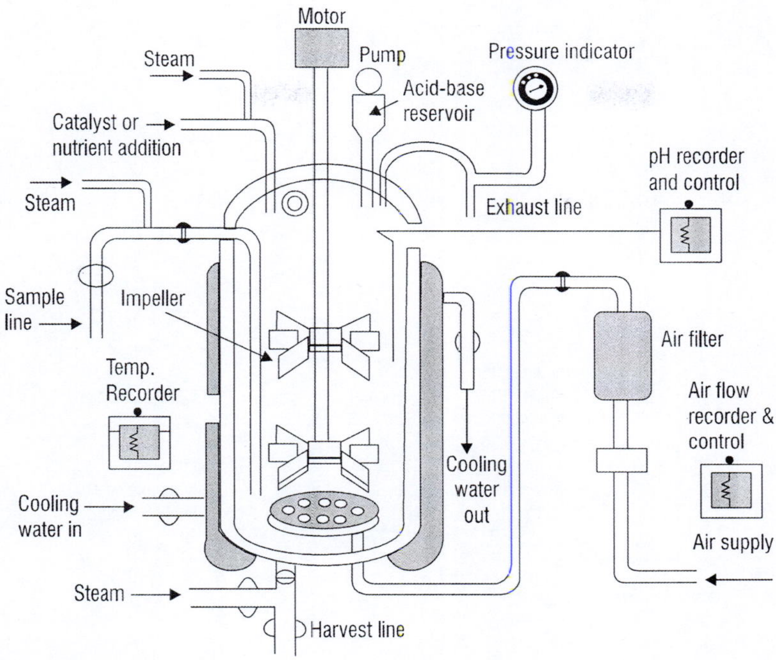


**FIGURE 11.2:** A typical bioreactor.

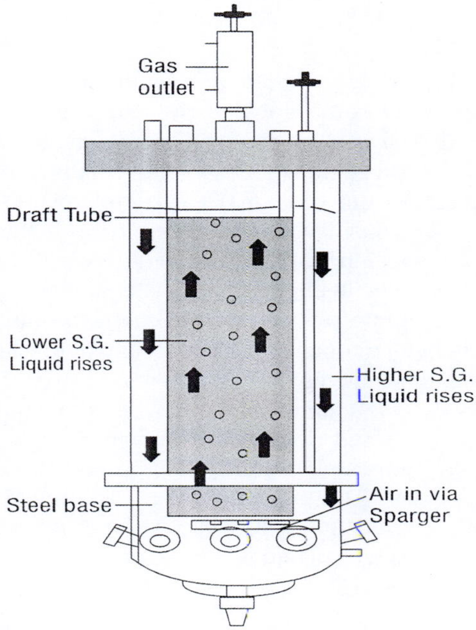


**FIGURE 11.3:** Air lift fermenter.

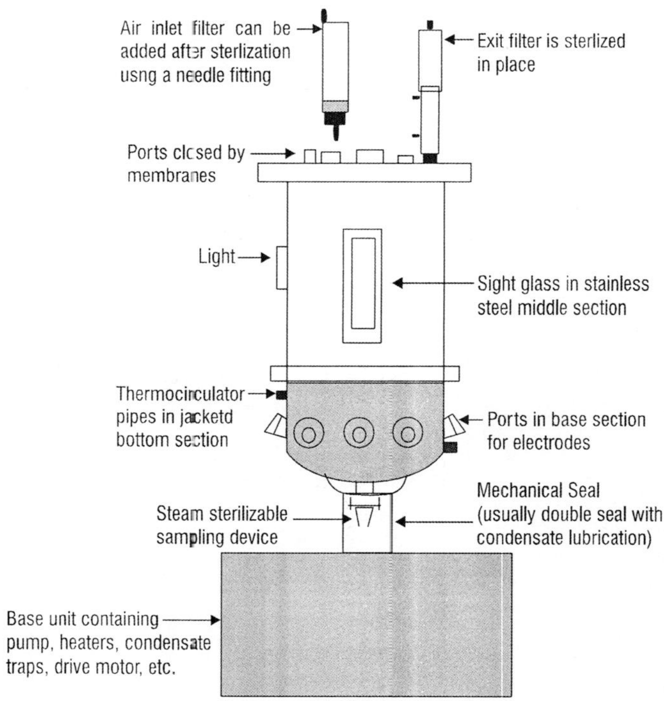


**FIGURE 11.4:** Illustratration of major parts of an in situ sterilizable fermentor vessel.

## Novel Seesaw Bioreactor

Foam formation occurs while aeration and agitation in the bioreactor, which disturbs the physical system and blocking the flow passages and the cell mass attaches to the foam, thereby reduces the cell density in the broth. Cells in contact with bubbles may be destroyed. In such cases aeration is to be done from outside the reactor. This increases the cost of the system. Novel Seesaw bioreactor is design in such a way to consider certain inherent problems. In this bioreactor aeration of broth is done by surface diffusion which avoids bubbling and reduces the cause of foam formation. Shear effect is also negligible.

The reactor has minimum two bioreactor chambers and two gas chambers. Each bioreactor chamber is connected to gas chamber, which controls the gas pressure inside the chamber by means of valve and blower. The gas is alternately forced into each bioreactor chamber. Gas forces the liquid levels inside the bioreactor for alternate up and down ward movement. The system is a closed circuit. Inflow of gas is from outside, and supply is controlled. It has the provision to remove excess of gas. The bioreactor has additional solid surfaces for improving the aeration efficiency. The reactor has also controlling device to control pH, dissolved $O_2$, $CO_2$ level and temperature.

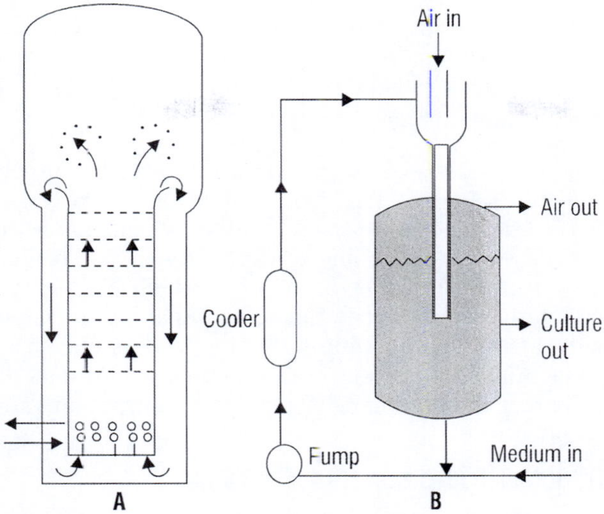


**FIGURE 11.5:** Different types of fermenter (A) Air-lift fermenter with internal loop. (B) Deep jet fermenter.

## DIFFERENT TYPES OF FERMENTOR INSTRUMENTATION

Various kinds of fermentor instrumentation are being used in fermentation technology. These can range from simple analogue control modules arranged as a 'stack', to powerful, embedded microprocessors which operate as on-board computers to directly operate the complete system.

### Analogue Controllers – Rack System

For compact, flexible instrumentation the individual modules for the measurement and control of each parameter are made to plug into a rack system. A bus system in the racking transmits signals from one module to another to allow measured values to be passed to controllers and for one controller to influence another, e.g. oxygen levels altering speed is one of the such examples.

### Analogue Controllers – Separate Modules in Housings

The modules may be stacked one on top of another and can be supplied with 'pass through' power connections so that only one power socket is needed. In this each measurement/control module has a separate housing and is operated in isolation. This system allows gradual building up or swapping of instrument modules between several fermentors.

### Digital Controllers – Embedded Microprocessor

The measurement sensors link directly to the single control module and several parameters are displayed immediately on a single screen. Control is by direct action on heaters, valves, etc. (Direct Digital Control or DDC). The microprocessor is permanently embedded in the instrumentation and may even be a single chip. Operation is usually via a simple menu system.

### Digital Controllers – Process Controllers

The controller is usually 'programmed' using simple commands to input set-point values, etc. Usually from a production control environment, a complete process controller is added to a housing containing all the signal processing and control actuators for the fermentors. It controls in the same way as an embedded controller but is essentially a 'plug-in' component.

### Digital Controllers – Direct Computer Control

The processing power and speed of modern PCs made the process more than capable of replacing separate PLCs or embedded controllers. . A special Input/Output (I/O) card is needed for the computer which allows the input values to be accessed by the measurement and control software and sends signals out to operate relays, e.g. to turn a heater on or off. In this case, there is no external instrumentation or processor between the actuators and the computer. A printed circuit board with operational amplifiers for the probe input signals is the only electronic part which may be present in the fermentor base unit. The advantage with this system is that the computer display and control software is integrated totally with the fermentor.

## COMMON MEASUREMENT AND CONTROL SYSTEMS

### Speed Control

Speed control basically relies on the feedback from a tachometer, which is located within the drive motor and determines the power delivered by the speed controller to maintain the speed set-point value, which is set by the user. A digital display shows the actual speed in rpm, as determined by the tachometer signal. Where speed is used to control the level of dissolved oxygen, an external signal from the oxygen controller can have an effect on stirrer speed. An absolute maximum and minimum value for speed can be set on the speed control module to limit the effects of the oxygen controller.

### Temperature Control

A thermocirculation system around a vessel jacket has been used for the temperature control because it is the most complex of all the methods. For simple direct systems such as a heater pad, it is simply a matter of fixing the heater, setting the desired temperature and switching on. Cooling is normally via a cold finger and flow of cooling water is controlled via the action of a solenoid valve.

The Pt-100 sensor provides the feedback signal, which causes the controller to take one of the following actions:

- Heat at full power as the actual temperature is some way below set point;
- Pulse the heater power as the actual temperature is close to set point;
- Turn on the cooling valve as the actual temperature is above set point.

The heating and cooling is controlled in exactly the same way as a directly heated system. The jacket provides a large surface area in contact with the vessel wall for heat exchange. Good temperature control can be achieved from approximately 5-8°C above the ambient temperature or 5°C above the temperature of the cooling water. Counter-cooling with water ensures stable temperature control when operating near ambient temperatures.

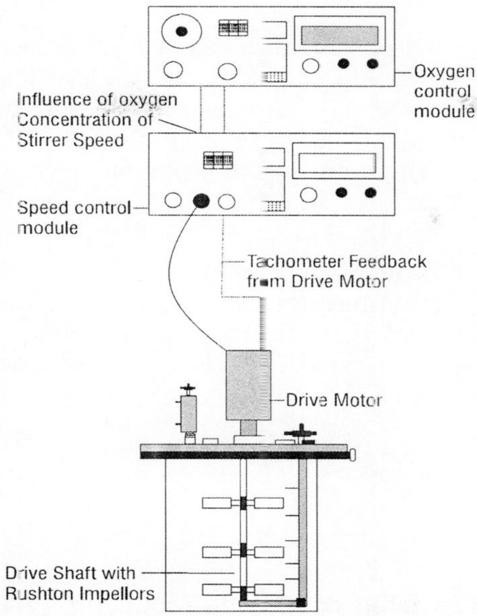


**FIGURE 11.6:** Illustration of speed control system.

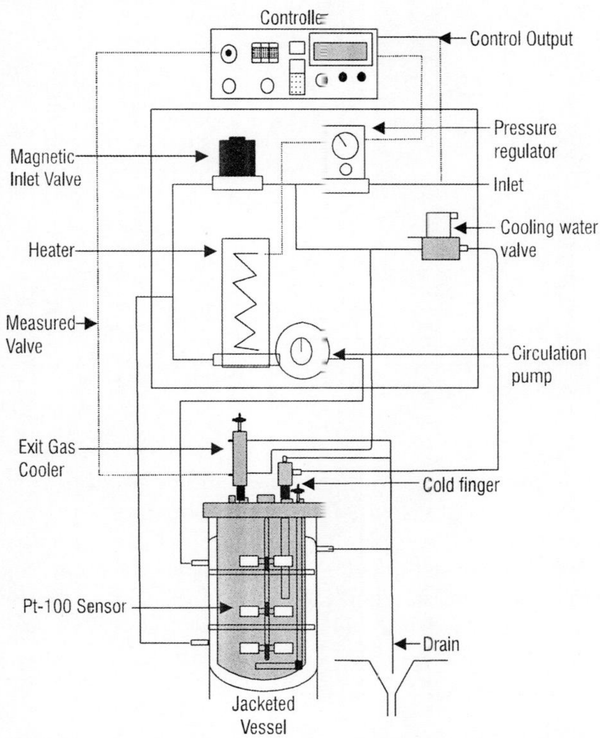


**FIGURE 11.7:** Illustration of temperature control system using water circulation.

## Control of Gas Supply

A pressure regulation valve ensures that air reaches the vessel at a maximum of 1.5 bar and rotameter is used to control the actual flow rate of air through the fermentor. A valve at the bottom of the rotameter is turned and the indicator 'ball' in the rotameter tube rises or falls in proportion to the valve position. The end of the sparger is typically a ring with small holes through which the air is forced. A 'head space' of around 20% is normally left between the culture level and the vessel top plate. For certain types of fermentation, e.g. mammalian cell culture, a gas mixing station can be used to pre-mix several gases before they are introduced into the fermentor.

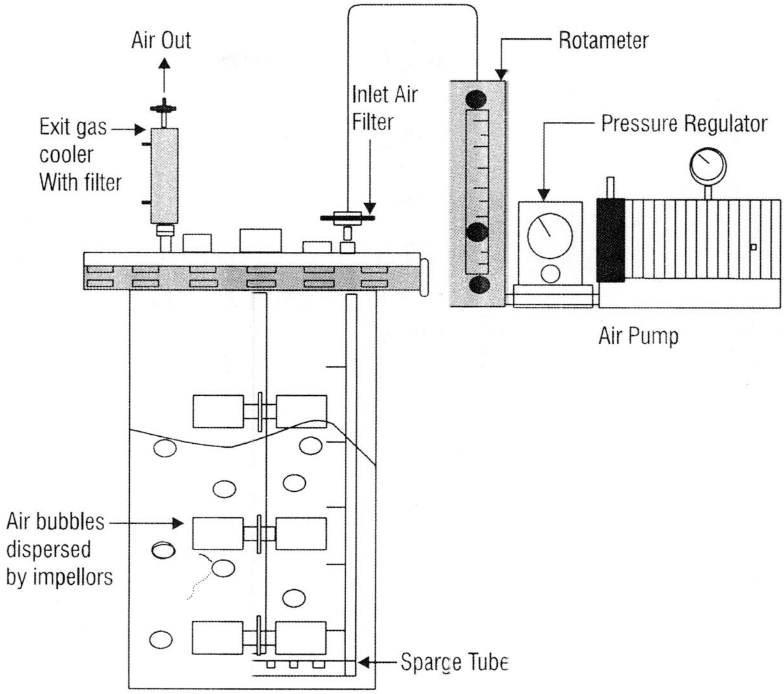


**FIGURE 11.8:** Illustration of gas supply system.

## Control of pH

pH control is achieved by the addition of either acid or alkali to correct changes in the pH of the culture during growth. The controller uses a pH electrode (either glass or gel type) to sense these pH changes and provide a feedback signal. The reagent bottles are connected to the fermentor via transfer lines of silicone tubingNormally, solutions of between 0.2 to 0.5 M of acid and base are used as a starting point.

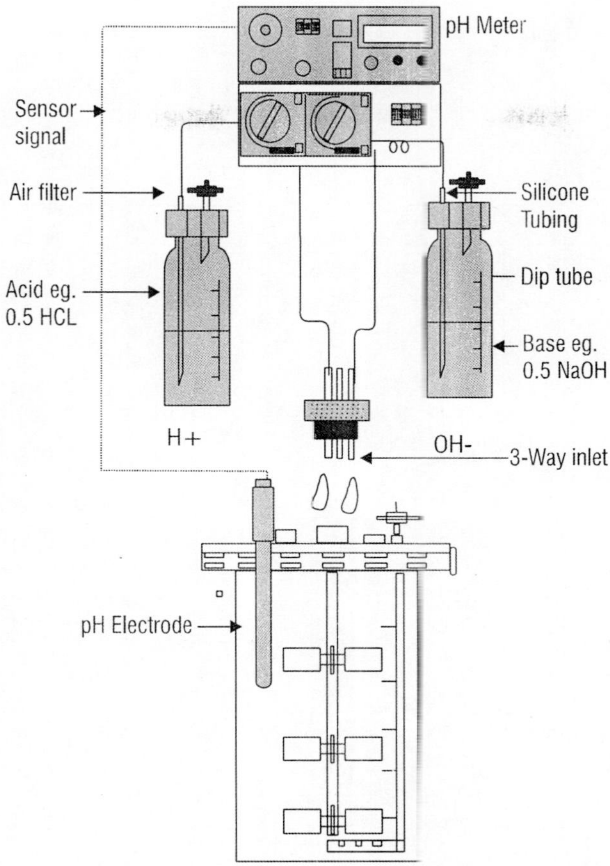


**FIGURE 11.9:** Illustration of pH measurement and control.

## Control of Dissolved Oxygen

Control of dissolved oxygen is the most important and the most difficult parameters. The electrodes used to measure dissolved oxygen are of two distinct types. These are, Galvanic electrodes, which are generally simple, cheap, slow and have a limited life and Polarographic electrodes, which are generally more complex, accurate, rapid and robust but expensive.

The polarographic electrode is used more commonly as it requires a voltage to polarize the anode and cathode of the detecting cell. This electrode must be connected to its relevant module, which in turn must be switched on.

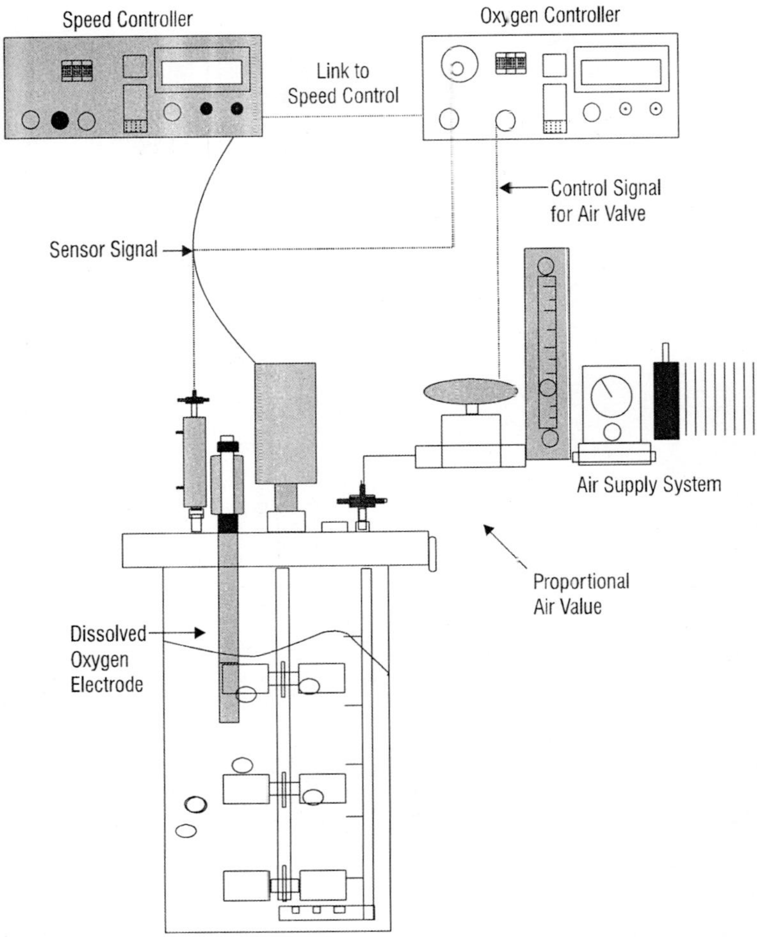


**FIGURE 11.10:** Illustration of dissolved oxygen measurement and control.

## Antifoam Control

If foam is allowed to build up unchecked it can eventually reach the exit gas filter so blocking it and providing a path for contamination, so antifoam reagents are used. Antifoam reagents like mineral oils, vegetable oils or certain alcohols are commonly used in the fermentation process. Commercial preparations are available for use with cultures, which ultimately provide materials for pharmaceutical manufacture. The key thing with using oils is that they can form a 'skin' on the surface of the culture and interfere with gas transfer at the liquid/for interface.

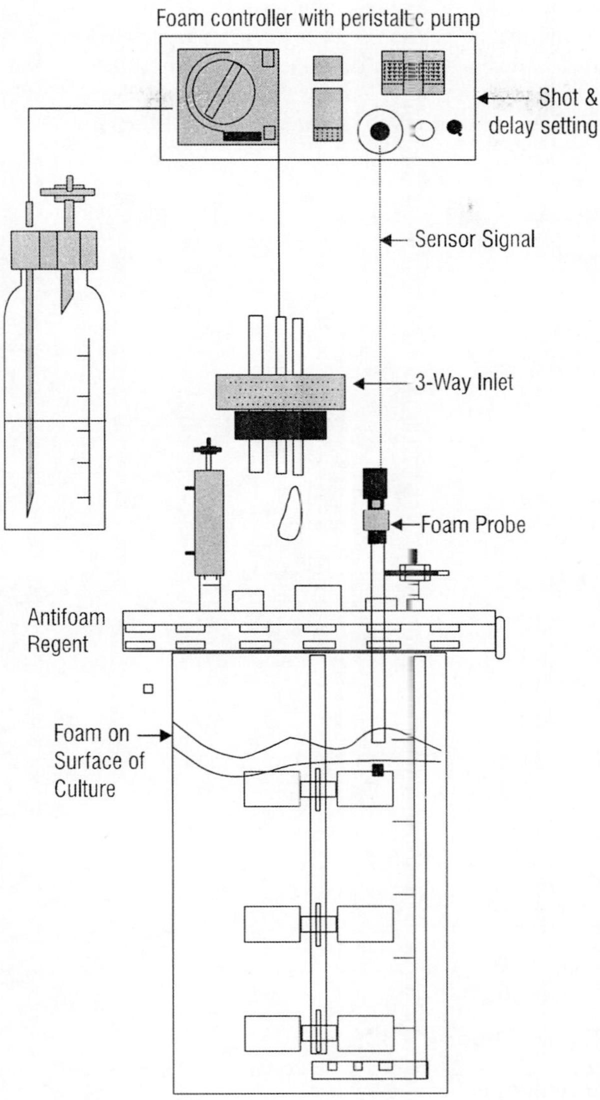


**FIGURE 11.11:** Illustration of foam control system.

## SIMPLE CONTINUOUS CULTURE

There are a number of different ways of setting up a continuous culture. The method of choice has the virtue of being able to be added to any fermentor vessel without modification or the addition of extra controllers beyond the acquisition of a feed pump and a harvest pump. The feed pump is variable speed but the harvest pump can be fixed speed as long as its speed is greater than that of the feed pump. The harvest pipe, being sited at the surface of the culture, has the disadvantage of taking off only the top of the culture (which may not be representative of the bulk culture) but the advantage of helping to prevent build-up of foam. If foam is a problem then antifoam reagent can be added to the medium feed at a background level, e.g. 1 part in 20,000. Samples can be taken by putting a 'Y' piece in the take-off tubing and placing a sampling device in one arm of the 'Y'.

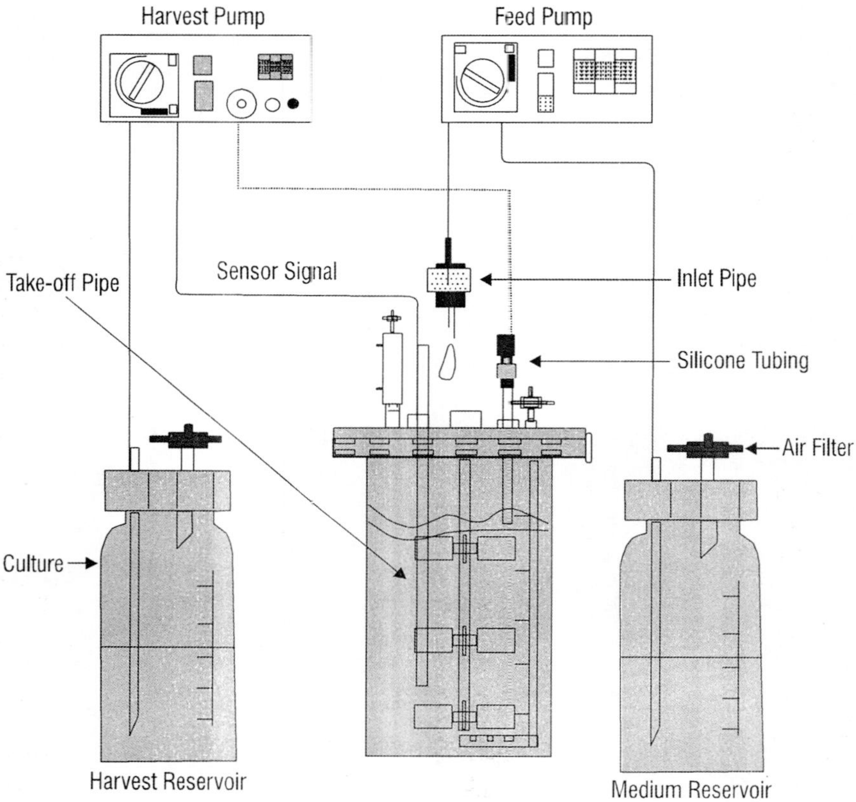


**FIGURE 11.12:** Illustration of simple continuous culture.

## FERMENTOR PREPARATION AND USE

### Scaling-up of Fermentation

Before designing and planning any fermentation process, it is necessary to scale-up experimentation required for fermentation process. Small to big differences exist in design and efficiency of bioreactor. There is a vast difference between the laboratory fermentation process/conditions and pilot plant. Therefore, detailed study is required to know the conditions, which virtually exists during production. Small-scale experiments are the starting point for scale-up procedure. Few changes are required in small scale–plan before employing it for large-scale production. For development of bioreactor for bioprocess development many factors like geometric of reactors, kinematics and dynamic similarities, thermodynamic, micro kinetic, transport, flow and diffusion (i.e., shear, mixing, mass transfer, and heat transfer etc.) should be studied and planned properly.

Other criteria are for required production at large scale are efficient mixing in large vessel, aeration, improved strains, type of nutrient, proper sterilization conditions, separation of growth inhibitory substances, regulation of secondary metabolite biosynthesis, carbon and nitrogen source control, etc.

## Disassembly of the Vessel

For carrying out the fermentation process, it is necessary to disassemble the vessel. The fermentation unit is shut down from the control unit and transfer lines and cable connections are removed. The vessel should be re-autoclaved and the culture should be disposed of as laid down by departmental safety procedures. The complete vessel including its each and every part is disassembled.

## Cleaning

The vessel should be rinsed several times in distilled water to remove any loose culture residues. During cleaning, an examination of any chips or cracks in the vessel glass can be carried out and a replacement can also be made, if necessary. Vessels must be stored clean and dry. In use, any spillages of reagents or medium should be wiped up immediately with a damp cloth and not be allowed to dry out. The pump heads and covers must be thoroughly cleaned if a tube breaks and reagent leaks out.

## Preparations for Autoclaving

Before proceeding for the autoclaving, the vessel seal should be removed and lightly greased with suitable silicone grease. On replacing the seal, it must be correctly located so that there is no chance of any part lifting or kinking. All fittings such as pipes etc. are fitted. Ports not in use have 'stoppers' fitted, and their O-ring seals should have a light coating of silicone grease applied. The pH and dissolved oxygen electrodes should be fitted, taking care not to damage them by careless insertion into the port. Both must be tightly capped to prevent moisture getting into the electrical contacts. Reagent bottles are prepared in a similar way to the fermentor vessel.

The exit gas cooler should be fitted to one of the wider ports. The air outlet line must be kept open during autoclaving. A short length of silicone tubing must be fitted to the air sparger inlet pipe with a 0.2 μm disposable filter mounted on top. The tubing between the sparger pipe and the filter must be clamped shut during autoclaving.

## Autoclaving

On a steel tray or in an autoclave basket the vessel and reagent/sampling bottles already connected by silicone tubing are assembled together. It should check that all the outlets and inlets are properly closed. A quantity of liquid is certain to be lost during autoclaving (approximately 10%) so either the medium is over-diluted to compensate for this or sterile, distilled water is added afterwards to restore the volume. Some form of indicator such as autoclave tape should be included to provide a warning if the correct sterilization procedure has not been carried out. Autoclaving at 121°C for 30 min is normally considered adequate for vessel sterilization to minimize damage to the constituent chemicals of the medium. The autoclave used must have good pressure equalization during the cooling down phase of operation to prevent medium being boiled off. The vessel and any accessory bottles must be allowed to cool completely before handling.

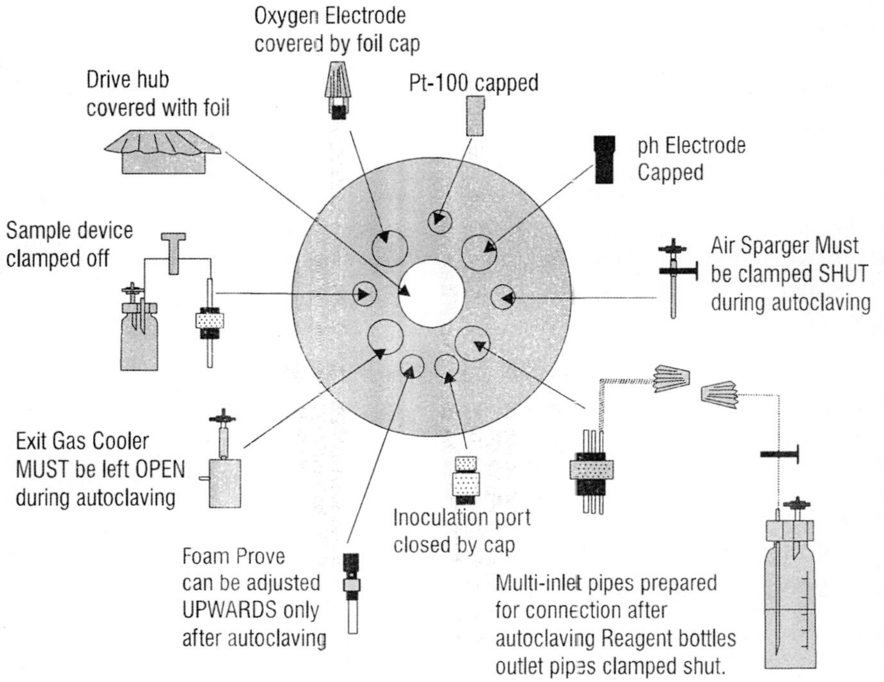


**FIGURE 11.13:** Illustration of vessels and accessories which are prepared for autoclaving process.

## Set-up following Autoclaving

The air sparger is connected to the rotameter by a piece of silicone tubing from the top of the filter to the air outlet of the rotameter. The air sparger line is unclipped between the metal pipe and the air filter. The exit gas cooler is connected to the water supply, either directly or via the fermentor base unit. The tubing for water in, out and drain is connected to the vessel jacket for a water system, and the water turned on so that the vessel jacket is filled. Alternatively, any pads or heater cartridges are connected to the base unit or temperature control modules.

Aseptic connections must be made first if the reagent bottles were autoclaved separately from the vessel. The clamps are removed so liquid can flow freely. The Pt-100 temperature sensor is connected to the control module and the pH electrode by removing the shorting cap and screwing in the cable. The dissolved oxygen electrode is connected to the appropriate cable. Connections to the foam probe are made, usually one wire into the electrode and one on the vessel top plate to make a circuit. Setting the temperature control at this stage will ensure the fermentor is ready to inoculate. The maximum stirrer speed to be used and the maximum airflow required on the rotameter are set. The fermentor is now ready to inoculate.

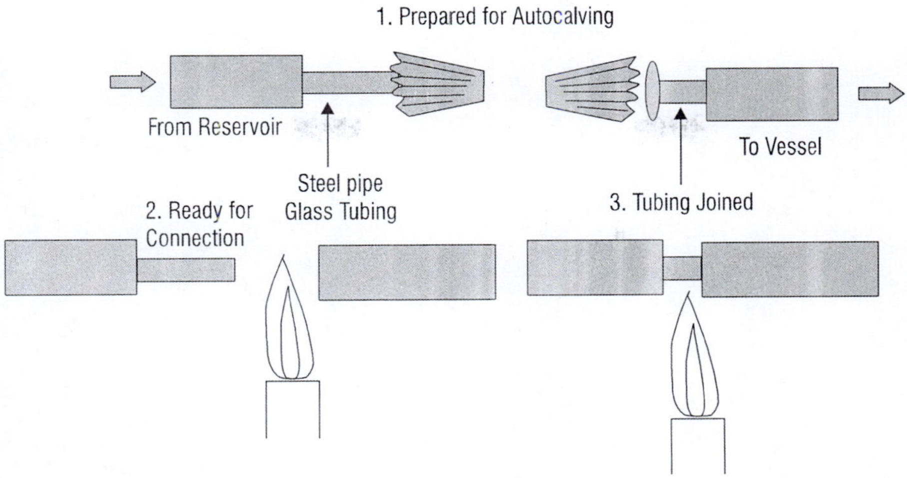


**FIGURE 11.14:** Illustration of how tubes are connected aseptically.

## Sampling from a Fermentor Vessel

Aeration is stopped during this process to prevent 'surging' of culture into the syringe. This method can be prone to contamination and is not suitable for organisms with any risk to operators. This sampling device is a bottle, which has two metal needles/pipes permanently, fixed through both the metal and rubber seals of its cap. A syringe is fitted to the air filter after autoclaving. The sample device is usually attached to the vessel top plate so that the glass bottle hangs down vertically beneath the cap and a supply of bottles of the same size are autoclaved ready for use. When a sample is to be taken, the clamp on the sample pipe tubing is released, and the syringe pulled back to create a partial vacuum in the sample bottle. Culture will flow into the sample bottle. When the required volume of sample has been taken, the syringe is pushed in to clear the line and the sample tubing is clamped shut again. The sample bottle with the culture is quickly removed and capped with a sterile cap from a new glass bottle, which replaces it under the sampling device.

## Inoculation of a Fermentor Vessel

The inoculum, which should normally be not more than 5-10% of the total culture volume, is aseptically transferred to a sterile, disposable syringe of a suitable size. The port fitting is removed and held vertically to prevent contamination of the bottom end. The syringe needle is quickly pushed through the membrane and the inoculum transferred into the vessel. The vessel may be actively aerated during this procedure to minimize the risk of a contaminant getting into the vessel. The syringe needle is quickly withdrawn and the silicone membrane reseals. The port fitting is now replaced. Aseptic connection can be made to an inlet pipe.

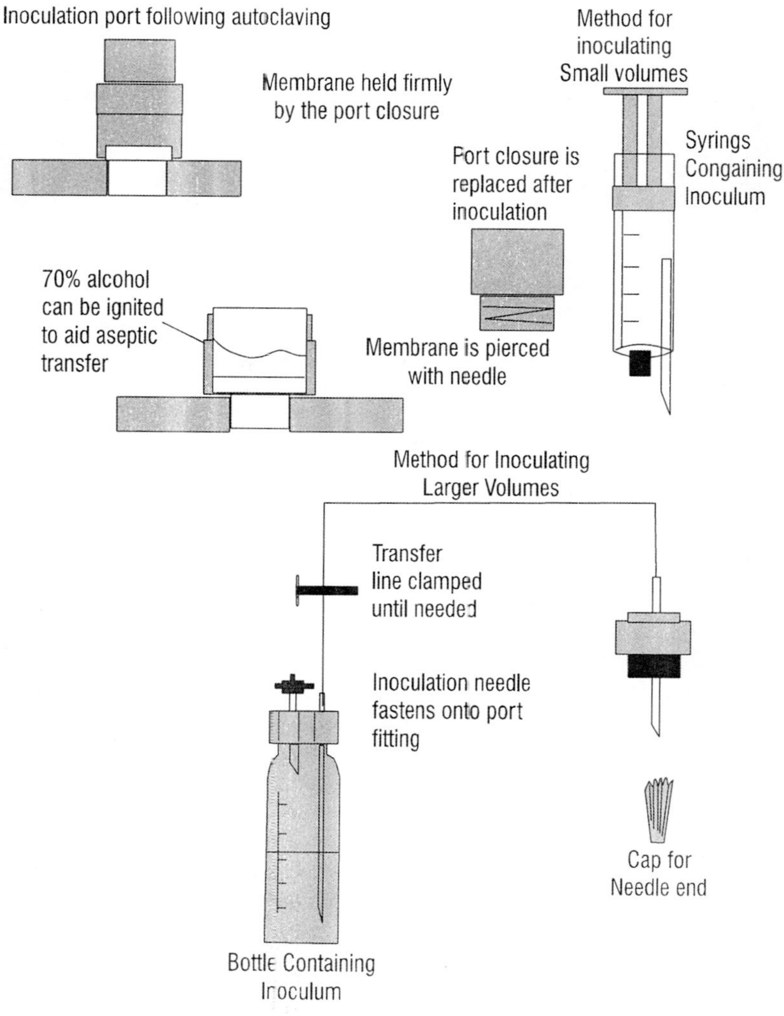


**FIGURE 11.15:** Illustration of inoculation process of a fermentor vessel.

## MAINTENANCE OF ASEPTIC CONDITIONS

Bioreactor should remain contamination free. The contamination is caused by microorganisms and results in a variety of consequences e.g., depletes nutrients, nature of the final product, reduction in yields, degradation of the desired metabolite, interference with downstream recovery etc. Therefore, the bioreactor should be designed in such a way so that there should be minimum or no contamination occurs. Integrated efforts should be made while designing bioreactor, use of materials and layout, effective operating procedure, scheduled checks and trained personnel. Contamination may be minimized by using sterile inoculum; adequately sterilized nutrient medium, the design, material and fabrication of the bioreactor (aeration and agitation systems, sampling and monitoring ports, and foaming) air liquid transfers. The operation of sterilization and aseptic additions of industrial plants involve complex procedures and perhaps, manipulations of hundreds of valves. The following are some of the important factors to be considered for an aseptic fermentation:

- For proper aseptic condition in the bioreactor, regular preventive maintenance of vessels is essential.
- The fermenter must be designed properly in such a way that the fermenter and its contents are steam sterilizable.
- Entry and exit of steam should be proper.
- The agitated drive shaft should have double mechanical aseptic seal. Seals should be of best quality to withstand with host water lubrication.
- Filtered and sterilized air should be supplied to bioreactor.
- The vessel should be operated at positive pressure.
- Addition or transfer of liquid systems in a vessel should be designed in aseptic manner.

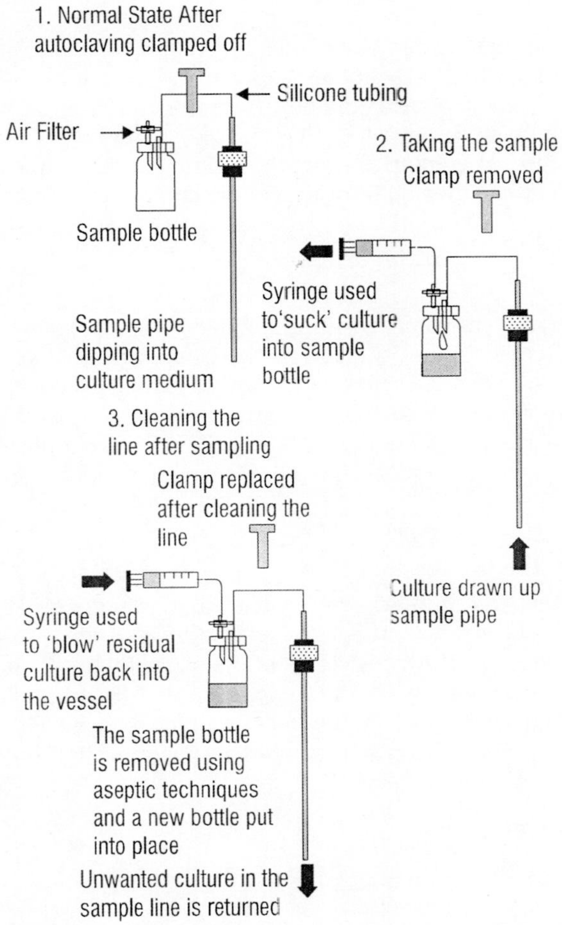


**FIGURE 11.16:** Illustration of sampling.

## TRANSFER OF OXYGEN

Oxygen is required for microbial culture activity throughout fermentation because aerobic microbes need oxygen at higher rate. The oxygen present in the culture is even utilized by them within 15 seconds. Oxygen supply to microbial processes needs simultaneous aeration and mixing. Mixing is done by agitation. Yeasts and bacteria do not require mixing of the cultures by separate agitation. Well-placed air spargers are used for such processes. Actinomycetes and molds are filamentous microbes. They require mechanical mixing. Aeration and mixing in submerged cultures are very much related to optimum rate of air flow and optimum impeller frequency. Saturation of dissolved oxygen concentration is achieved by sparging air into the vessel throughout the course of fermentation process. The transfer of oxygen from the gaseous phase proceeds by molecular diffusion. The rate of oxygen transfer into a solution can be expressed as follows:

$$\frac{dC_L}{dt} = K_La(c^*-C_L) \qquad (1)$$

Where $C_L$ is the dissolved oxygen concentration (m mol $dm^3$) t is the time, (h), C* is the saturated dissolved oxygen concentration, KL is the mass transfer coefficient (cm $h^{-1}$), a is the gas/liquid interface area in mass transfer and $dC_L/dt$ is the oxygen transfer rate (m $mol^{-3}h^{-1}$). The term $K_L$ a represents volume coefficient of oxygen transfers characterizing the aeration capacity of the system. The oxygen requirement of a given biological system can be represented as:

$$\frac{dC_L}{dt} = K_La(c^*-C_L)-rX \qquad (2)$$

where 'r' is the rate of respiration of microorganisms and X is the concentration of cells. If the rate of oxygen consumption exceeds the rate of absorption, the whole system will become oxygen limited. Above two equations hold for an ideally mixed system i.e., homogenous gaseous and liquid phases; in fermenters usually the oxygen concentration differs different places. If the oxygen concentration in the liquid is zero, equation (1) can be simply represented as:

$$\frac{dC_L}{dt} = K_La.C^* \qquad (3)$$

which, gives the maximum rate of oxygen transfer under given conditions. $K_La$ value depends on different factors like vessel design, degree of agitation in a stirred vessel, rate of airflow, viscosity of the culture broth and the available antifoam agent in the medium. The bioreactor should be designed in such a way that during operation microbes should get sufficient oxygen. The above equation (3) reveals that the rate of oxygen transfer is maximal if the concentration of dissolved oxygen is zero.

### *Critical Concentration (C.crit)*

It is the lowest dissolved oxygen concentration at which the specific oxygen uptake is maximum. If dissolved oxygen concentration is increased, the specific oxygen uptake rate also increases up to a maximum rate. Therefore, low oxygen concentration below C.crit will bring about a decline in the rate of respiration. Hence, it is necessary to keep a fermenter and culture in a fully aerobic state. The available dissolved oxygen concentration should remain above the C.crit.

Cultures of filamentous organisms have non homogenous environment. In such environment local oxygen shortage arise, which ultimately increases critical concentration value (apparent critical concentration value, AC.crit). The value of AC.crit in a few microbial processes may be quite high. IT may be varied. Thus, the optimum aeration conditions for microbial product formation may be different from those giving biomass production. Hence, it may be necessary to operate a fermenter at a dissolved oxygen concentration either below critical concentration or at apparent critical concentration (i.e., below or above C.crit).

The value of $K_{La}$ in stirred vessel increases with rising superficial air until it reaches a certain critical value. The critical value corresponds to the point of impeller chocking i.e., where the impeller rotates virtually only in the air.

The value of $K_{La}$ does not depend on the bubble size. Increase in $K_{La}$ occurs with very small bubbles. A relative increase or decrease in the bubble surface area can be attributed to surface active agents (i.e., sodium lauryl sulphate, Tween 80, silicon oil etc.) that can alter the $K_{La}$ value.

The viscosity of the medium depends on the nature of the microorganisms. Filamentous microbes increase medium viscosity. They also affect the degree of agitation. The oxygen transfer rate considerably reduces with increase in viscosity of the culture broth. The viscosity of the broth decreases with increasing agitation. Therefore, it is essential that, the agitation regime should govern good oxygen transfer in a region of high shear but also circulates the broth through the high shear zone. At higher viscosity the transfer rate does not rise even on increasing the impeller frequency.

# 12. Production of Antibiotics

In the original sense; any microbial product which, in low concentrations (of the order of micrograms/millilitre), is capable of inhibiting or killing (susceptible) microorganisms. The term was first used by Waksman in the 1940s; it distinguished the (then newly-discovered drug penicillin from the synthetic) sulphonamides which had been developed in the 1930s. Difficulties in terminology become apparent when the penicillins (and other natural antimicrobial products) were subjected to chemical modification or synthesized de novo in the laboratory.

Currently, the term antibiotic is often extended to include drugs such as the Sulphonamides, Nalidixic Acid, and the semi-synthetic Penicillins. Many antibiotics are Chemotherapeutic agents but others are too highly toxic to be of any clinical value; certain antibiotics (e.g. colicins) are used for non-therapeutic purpose. Most of the antibiotics used as chemotherapeutic agents are active against bacterial pathogens; some (e.g., griseofulvin) are antifungal, others (e.g., Fumigillin, suramin) are antiprotozoal, while a few (e.g. the rifamycins) have limited antiviral activity.

The idea of using one microorganism to combat the evil effects of another is by no means a new one. Pasteur was probably the first to describe clearly the production of an antibiotic, and, during the intervening years, many claims have been made as to the therapeutic value of bacterial products.

Production of antibiotics (antibacterial or antifungal compounds) is one of the largest and most important microbiological industries. Interest in utilization of antibiotics for therapy began in 1929 when Alexander Fleming found that a mold has contaminated his cultures of a pathogenic bacterium, *Staphylococcus aureus*, and had killed bacteria in its immediate vicinity leaving a clear ring of lysed bacteria. The mold was a *Penicillium* (later found to be *P. notatum*). Fleming named the unidentified **lytic principle penicillin.**

**TABLE 12.1:** Some antibiotics produced by microorganisms

| Antibiotics | Microorganisms |
|---|---|
| Amphotericin B | *Streptomyces nodosus* |
| Bacitracin | *Bacillus licheniformis* |
| Carbomycin | *Streptomyces halstedii* |
| Chlorotetracycline | *Streptomyces aureofaciens* |
| Chloramphenicol | *Streptomyces venezuelae* |
| Erythromycin | *Streptomyces erythreus* |
| Fumagillin | *Aspergillus fumigatus* |
| Griseofulvin | *Penicillium griseofulvin, P. nigricans, P. urticae* |
| Kanamycin | *Streptomyces kanamyceticus* |
| Neomycin | *S. fradiae* |
| Novobiocin | *S. niveus, S. spheroides* |
| Nystatin | *S. noursei* |
| Oleandomycin | *S. antibiotics* |

| Antibiotics | Microorganisms |
|---|---|
| Oxytetracycline | *S. rimosus* |
| Penicillin | *Penicillium chrysogenum* |
| Polymyxin B | *Bacillus polymyxa* |
| Streptomycin | *S. griseus* |
| Tetracycline | Dechlorination and hydrogenation of chlorotetracycline; direct fermentation in dechlorinated medium |
| Vira A (Adenine arabinoside) | *S. antibiotics* |

## SCREENING OF ANTIBIOTIC PRODUCERS

The search for antibiotics in the pharmaceutical industry presents a good example of how screening procedures are employed to select microorganisms for industrial applications. The discovery of new antibiotics results from laborious searches. Sample from many sources, including soil from around the world, are examined as potential sources of antibiotic-producing microorganisms; countless strains of microbial isolates are tested by pharmaceutical laboratories. One of the best penicillin-producing strains of *Penicillium* was isolated from an orange purchased at a road side fruit stand. Several antibiotic-actinomycetes were isolated from a manure-enriched pasture.

A useful antibiotic-producing strain must produce metabolites that inhibit the growth or reproduction of pathogens. This essential property can be assayed by using test strains and examining whether the isolate being screened produces substances that inhibit the growth of these test organisms. A positive result in such a primary screening procedure in no sense ensures the discovery of an industrially useful antibiotic-producing strain. It simply identifies those strains of microorganisms that have the potential for further development.

The search for new antibiotics still goes on, and new ones are frequently announced. Genetic engineering has opened up several new possibilities for employing microorganisms to produce economically important substances. The mutation and selection approach is hit or miss, whereas the use of recombinant DNA technology permits the purposeful manipulation of genetic information to engineer a microorganism, which can produce high yields of a variety of products. The use of biotechnology has great social consequences.

## β-LACTAM ANTIBIOTICS

The naturally occurring penicillins have been included in β-lactam family. Antimicrobially active β-lactam compounds are represented by penicillins and cephalosporins. The penicillins fused to a thiazolidine ring, whereas the cephalosporins have the β-lactam fused to dihydrothiazine ring. The discovery of the 7α-methoxy cephalosporins provided the first new, naturally occurring β-lactam nucleus following the isolation of Cephalosporin C.

### Penicillins

Not only did penicillin developed interest in antibiotics and launch the antibiotic industry, but penicillin is still the most widely used antibiotic. Most penicillins are produced by species of *Penicillium* while others are semi synthetic. It is produced from *Penicillium notatum* (penicillin F) and *Penicillium chrysogenum*. Numerous species of Streptomyces also produce penicillin N together with cephalosporins, cephamycins, clavams, or cephapanems. It is the drug of choice, when infection is caused by organism susceptible to it. Penicillin is effective against Gram-positive bacteria and also against some of the larger viruses and rickettsia. Penicillin is a generic term applied to an entire group of antibiotics, which are closely related in structure.

### *Chemical Nature and Biosynthesis*

Penicillins are N-acyl derivatives of 6-aminopenicillanic acid (6-APA). They are synthesized in two types of fermentation. Type-1 penicillins synthesized with non polar side chains as the only β-lactam, and type 2 penicillins N (2) is synthesized along with other β-lactams. The type I i.e. penicillins with non polar side chain is found in fungi. The nature of the side chain ring is determined by addition of a suitable side chain precursors to the fermentations i.e., addition of phenyl acetic acid as precursor to the fermentation produces benzylpenicillins.

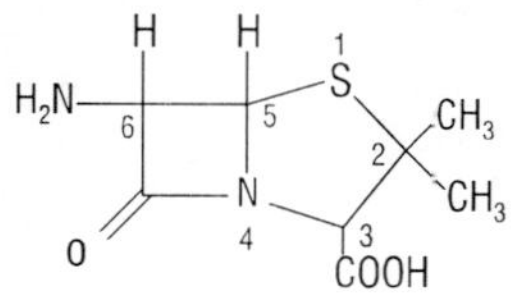

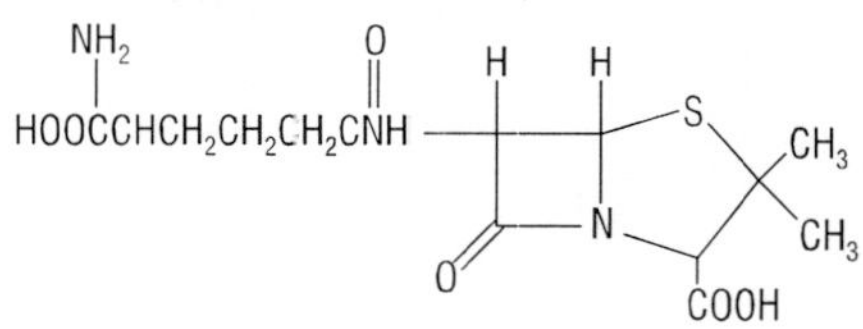

The naturally occurring penicillins have the following basic structure and differ in the R groups. The R group could be provided by precursor with the structure $RCH_2COOH$ or its salt, amide or ester, although $RCH_2CH_2NH_2$ also could serve as precursors. The various R-group side chains may effect relative potencies, to a certain extent, the relative activities against sensitive microorganisms. Several species of penicillin molds produce penicillins with non-polar side chains.

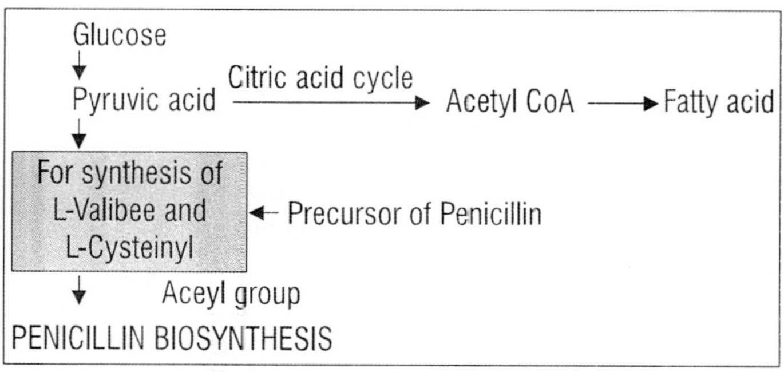


**FIGURE 12.1:** Biosynthesis of Penicillin.

Penicillin is produced along synthetic pathways not required for growth but superimposed upon those pathway required for maintenance. Key intermediates in the pathway are pyruvic acid, valine and cystine. Penicillin production is at its maximum level when the citric acid cycle is not actively taking place or if a great deal of pyruvic acid is not being converted to acetyl CoA and then to fatty acid. The accumulating pyruvic acid can then be available for synthesis of L-valine, to form a dipeptide precursor of penicillin, L-cysteinly-L-valine. This last intermediate is then combined with the acyl group from phenyl acetic acid in the case of penicillin G or acyl groups from other specific precursors in the synthesis of the other penicillins. Many of the details concerning penicillin biosynthesis are still unknown.

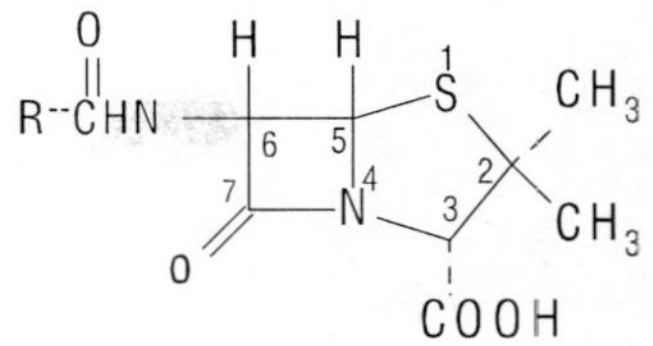

| Penicillins | R |
|---|---|
| Benzylpenicillins/Penicillins G | 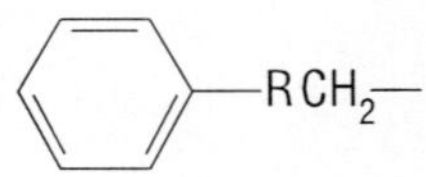 |
| Phenoxymethylpenicillins/Penicillin V | 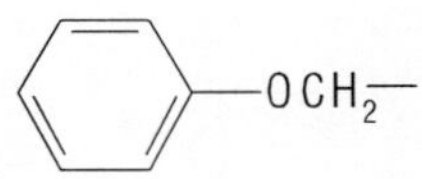 |
| 2-Pentenylpenicillin/Penicillin F | $CH_3CH_2CH{=}CHCH_2$— |
| n-Amylpenicillin/Dihydropenicillin F | $CH_3(CH_2)_4$— |
| n-Heptylpenicillin/Penicillin K | $CH_3(CH_2)_6$— |
| p-Hydroxylbenzylpenicillin/Penicillin X |  |

**FIGURE 12.2:** Different types of penicillins produced by fermentation of *Penicillium chrysogenum*

## *Commercial Production of Benzylpenicillins (Penicillin G)*

The original *P. notatum* Straub isolated by Fleming gave low yield of penicillin. Finally a strain of *P. chrysogenum* was isolated and was found to give superior yields of penicillin. Strains of *Penicillium* may produce more than 180 times as much penicillin as the original isolate with which the breeding program was initiated. Penicillin is produced in submerged vat cultures using a selected strain of *P. chrysogenum*. High yielding strains have been developed using mutation or genetic engineering techniques. Yield has been increased from less than 10 μg/ml to about 30 μg/ml. These strains have high titre, improved growth, tolerance to side chain precursors, ability to accumulate intracellular sulphate, acetyltransferase activity, and other properties as per the need.

## *The Inoculum*

Master stock (spores) is the major source of inoculum. Spores above or below a critical level show differences in morphology and productivity. Spores of *P. chrysogenum* are developed from vegetative mycelium which is prepared in a series of inoculum development steps. The high yielding strains are genetically unstable. Spores from working stock cultures are suspended in water or non-toxic lauryl sulphonate.

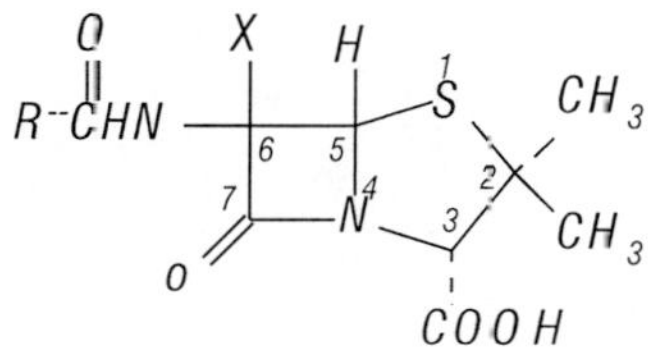

| Penicillins | X | R |
|---|---|---|
| Benzylpenicillin | --H | 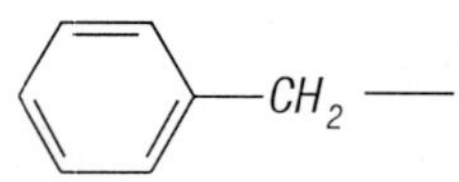 |
| Methicillin | --H | 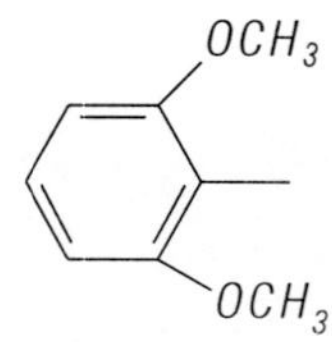 |
| Ampicillin | --H | 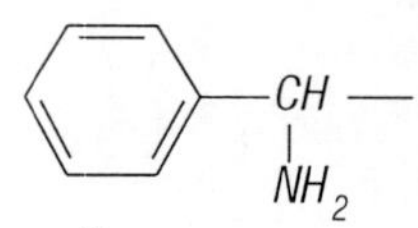 |
| Ticarcillin | --H | 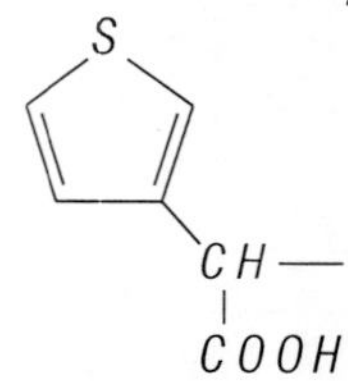 |
| Piperaicillin | --H | 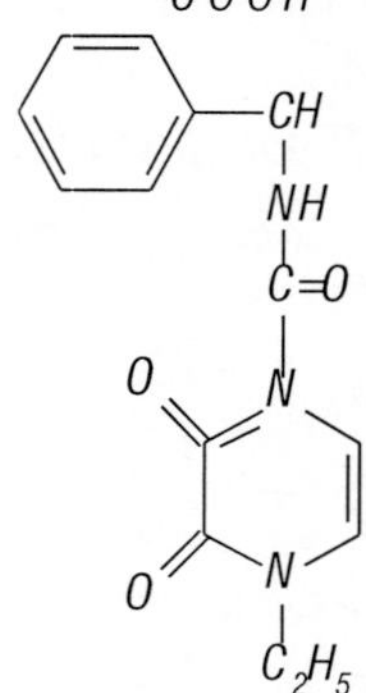 |
| Temocillin | --$OCH_3$ | 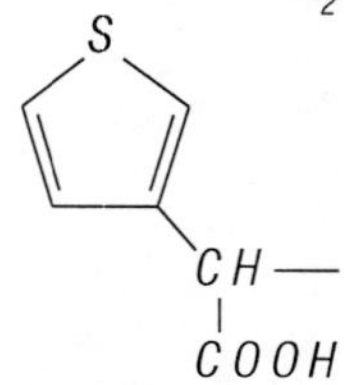 |

**FIGURE 12.3:** Various types of semi synthetic penicillin.

These are added to flasks containing wheat bran and nutrient solutions. A shake flask culture of 4 days old is inoculated into a seed tank for 3 days. It acts as inoculum. Heavy mycelial growth is obtained with agitation and aeration. Care should be taken to avoid all sort of contamination. The main purpose of inoculum development process is to increase the concentration of fungal mycelium.

### The Medium

A typical production medium includes glucose, sodium phenyl acetate, corn step liquor (CSL), ammonium sulphate, and potassium hydrogen phosphate. Lipid nutrients are used by the fungus during fermentation. They also utilize oil as an antifoam agent. These nutrients contribute to both the quantities of mycelium and product yields, but high levels are deleterious in both the early and late stages of fermentation. The on set of production phase can be marked by the addition of phenyl acetic acid (PAA) by continuous feed.

**Fermentable carbohydrate:** The principal carbohydrate in precursor molecules (phenethylamine and phenyl amine). It is needed for penicillin production and are present in the corn steep liquor. It is the most favourable carbon source for penicillin production because it is assimilated slowly, giving the fungus a steady carbon source over a period of time and does not lead to the accumulation of organic acids. In general 65% of the consumed carbon is used for maintenance, 25% for mycelial growth, and only 10% for penicillin production (the final yield is in the range of 3.5% based on carbohydrate consumed). If glucose is available in less quantity from lactose it leads to starvation conditions required for the penicillin formation.

**TABLE 12.2:** The major constituents of typical medium (Jackson) includes:

| | |
|---|---|
| Fermentable carbohydrate | Corn steep (3.5%), Lactose (3.5%), glucose (1%) |
| Potassium dihydrogen phosphate | 0.4% |
| Organic nitrogen source | |
| Phenyl acetic acid precursor | |
| Edible oil | 0.25% |
| Calcium carbonate (acts as buffer) | 1% |
| Organic salts | |
| pH after sterilization | 5.5 to 6.0 |

**Aeration (Oxygen supply):** Supply of oxygen in a bioreactor is the limiting factor penicillin biosynthesis. If oxygen availability falls below critical level, it will reduced the growth. Aeration conditions should be adjusted in such a way that fast growth can be achieved only until the cell production reaches maximum density.

**Biomass Production:** Production of penicillin depends upon biomass production therefore, it is desirable to have a high biomass concentration in the vessel. The medium contains carbon compounds and other nutrients and they allow organisms to grow initially near the maximum specific growth rate. This rapid growth rate causes a considerable increase in the initial oxygen uptake rate and carbon dioxide evolution rate. Penicillin production is increased by increasing and improving mycelial biomass. It is achieved by increasing the agitation rate and power.

**pH:** It is maintained near neutrality by calcium and magnesium carbonate in the medium by phosphate buffer. It is also controlled by adding sodium hydroxide or sulphuric acid in the medium.

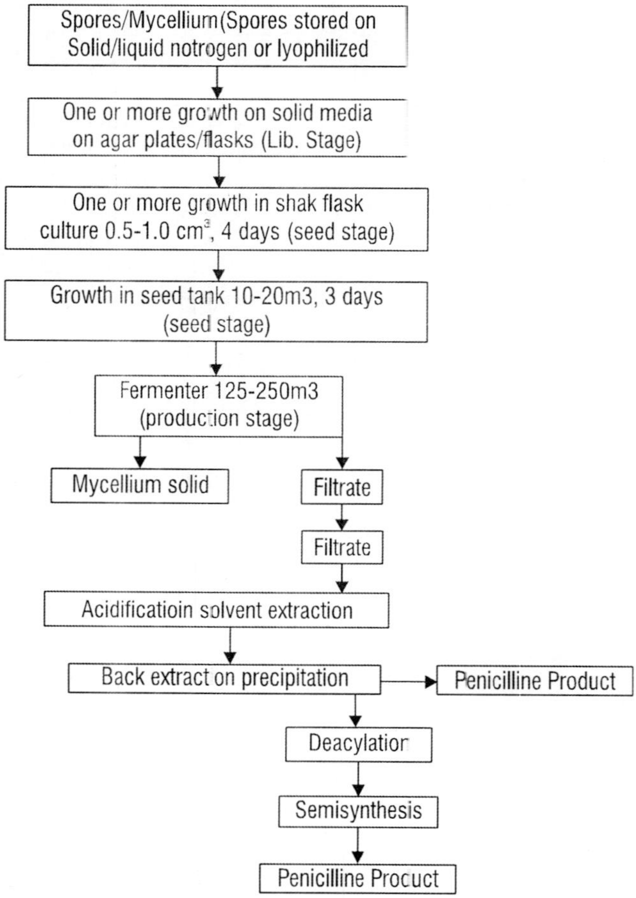


**FIGURE 12.4:** Presentation of flow chart of penicillin production.

### *Procedure*

About 30,000 liters of the medium are placed in a tank, sterilized and inoculated with an aqueous suspension of *P. chrysogenum* conidia. During the incubation period, the medium is aerated with sterile air and agitated. The tank is equipped with devices which allow the continuous addition of glucose syrups, sodium hydroxide, and sulphuric acid to maintain the pH between 6.8 and 7.4, cooling coils to maintain the temperature between 23°C and 25°C, devices for the introduction of antifoam agents, and pumps for the addition of the acyl or precursor. The most commonly used precursor is phenyl acetic acid. The other precursors in the form of a salt, amide, or ester of the corresponding acid or amine may be added to yield penicillins bearing the desired acyl group. At first, the mycelium grows actively and utilizes lactic acid and organic nitrogen compounds as sources of carbon. Ammonia is formed during the breakdown of these nitrogenous compounds. Ammonia is now actively assimilated and the pH is lowered. Active penicillin production is associated with this phase of lactose and ammonia utilization. Penicillin production ceases when the lactose is exhausted.

The duration of the incubation period is about 5 to 6 days. After incubation, the mycelium is filtered from the liquid medium which contains the penicillin. The penicillin is mixed with a solvent (amyl or butyl acetate) and the resulting emulsion is centrifuged to extract the acetate solvent which now contains the penicillin. A phosphate buffer is added to the acetate, and the penicillin is extracted with the phosphate buffer by centrifugation. This last step may be repeated, using successively smaller quantities of liquid. Butanol is added to the aqueous mixture, and the potassium salt of penicillin is crystallized from the solution. Potassium penicillin may be further purified and used in this form or it may be converted to procaine penicillin.

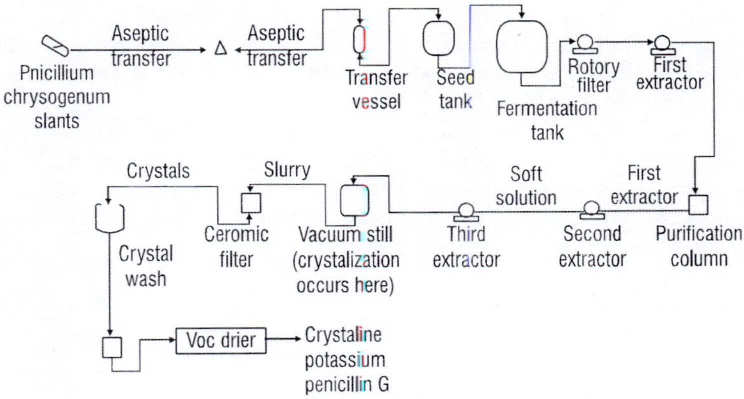


**FIGURE 12.5:** Commercial production of Penicillin.

## *Recovery and Purification*

Harvested culture broth includes penicillin G along with a variety of other metabolites. Vacuum filter is used for separation of mycelium from the broth on a rotary. The procedure is conducted under aseptic conditions. Penicillin (acid form) is extracted under controlled condition of temperature pH, sterility of the penicillin containing filtrate. Care should be taken to minimize chemical and enzymatic degradation of the penicillin in the pumps, collecting vessel, and the extractor. Conversion of penicillin to the anionic form occurs at low pH (2.0 to 2.5). The lowering of pH is done by adding phosphoric acid or sulphuric acid. The mixture is fed to Podbieliak counter current solvent extractor, with an organic solvent such as amyl acetate, methyl isobutyl ketone, or butyl acetate. For removal of pigments and other impurities from solvent containing penicillin, it is treated with active charcoal. The charcoal is separated from the extract on a precoated rotary vacuum filter and washed with solvent. The product is back extracted into water from solvent by adding potassium or sodium hydroxide to form its salt. So obtained aqueous solution is further subjected to repetition of the process to acid purification. The end product of penicillin is then crystallized into sodium or potassium penicillin.

## *Semisynthetic Penicillins*

The semisynthetic penicillins contain 6-aminopenicillanic acid (6-APA) nucleus. 6-APA is obtained by fermentation in the absence of precursor. This procedure is not suitable because it reduces the yield and also contaminated with different lipid substances. 6-APA is prepared on large scale by removal of the phenyl acetate side chain from penicillin G. Semisynthetic β-lactam antibiotics now, have been synthesized from penicillin G, by converting into other β-lactam antibiotics. Enzyme acylases of microbial origin is used for this purpose.

**Enzyme penicillin amidases (acylases):** This enzyme catalyses the removal of the side chain from benzylpenicillins. It is a membrane bound enzyme found in *E.coli* and *Alcaligenes*. It has high specificity for penicillin G. Several types of enzymatic deacylation processes are used. Penicillin is treated with intact bacterial cell or with isolated enzyme. Immobilized cells or enzymes are also used. They are advantageous because they can be reused (batch process) or used continuously (column process).

6-APA is also used for production of new penicillins like Methicillin, nafcillin, and isoxazolyl penicillins, the broad spectrum penicillins, amoxicillin, epicillin, and cyclacillin; antipseudomonal penicillins such as carbenicillin and piperacillin, azocillin, and mezlocillin. These are obtained through chemical acylation with appropriate acid chlorides.

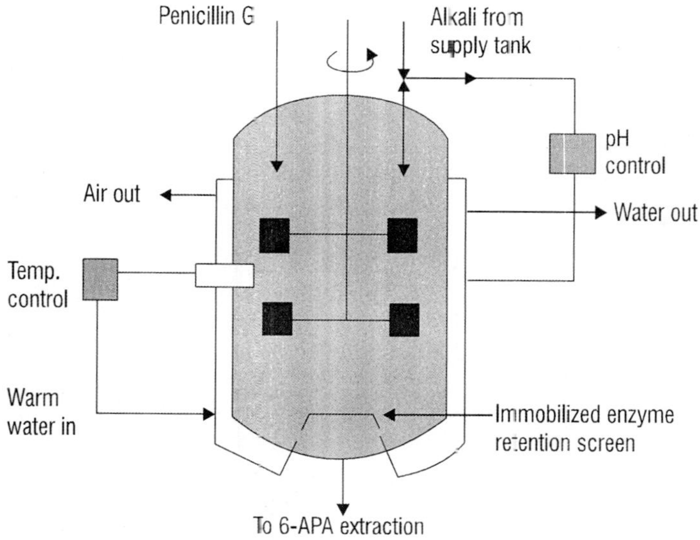


**FIGURE 12.6:** Modified batch process stirred tank-reactor.

**Commercial Production of 6-APA using Penicillin G. Acylase:** The production is based on deacylation reaction, for this purpose different types of reactors are used. The conventional stirred tank reactors used for the original cell bound enzymation process with certain modification. The 6-APA is extracted from the cooled enzymation liquors. pH of the liquor is adjusted to 2.5 by adding hydrochloric acid. Methyl isobutyl ketone (0.5 vol.) added to extract phenyl acetic acid and residual penicillin G. After separation, the pH of the aqueous phase is adjusted to pH 7.8 using sodium hydroxide. It is than evaporated under vacuum to obtain 6-APA concentration of 12-15%. 6-APA is precipitated by adjusting the pH to isoelectric point 4.3. Crystals washed successively with water and acetone, and dried under vacuum.

## Cephalosporins

A Cephalosporin C is made as the fermentation product of *Cephalosporium acremonium*, but this form of the antibiotic is not potent enough for clinical use. The cephalosporin C molecule, however, can be transformed by removal of an α-aminoadipic acid side chain to form 7-α-amino-cephalosporanic acid, which can be further modified by adding side chains to form clinically useful products with relatively broad spectra of antibacterial action. We are at present into the so-called third-generation cephalosporins, such as moxalactam, which have been developed to combat bacteria that produce enzymes capable of degrading penicillins and cephalosporins.

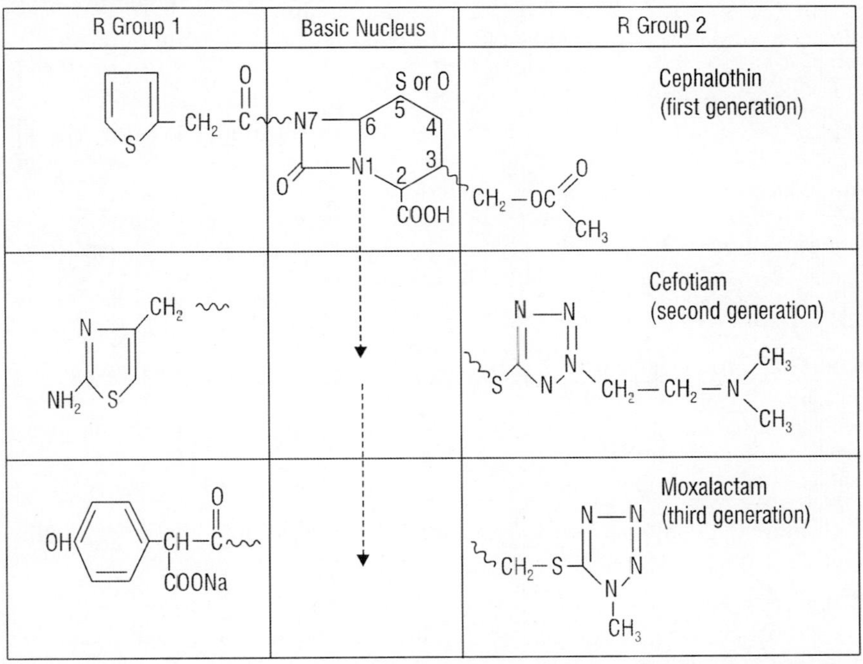


**FIGURE 12.7:** The structure of cephalosporins. Like penicillin, they have a beta-lactam ring (red), but a different main ring (yellow). However, unlike penicillins, they have two sites for placement of R groups (at positions 3 and 7). This makes possible several generations of molecules with greater versatility in function and complexity in structure.

## AMINOGLYCOSIDE ANTIBIOTIC

Amino glycoside antibiotics are a representative group of streptomycin. They were discovered by Waksman (1944) for treatment of bacterial infections. About 152 naturally occurring aminoglycoside antibiotics have been isolated from culture filtrated of various bacteria and actinomycetes. These are mixtures of all closely related water-soluble basic pseudosaccharides. They have a broad antibacterial spectrum.

Streptomycin is used for treatment of tuberculosis. Kannamycin is a second generation of aminoglycosides and is used for treatment of serious infections caused by different gram negative pathogenic bacteria. Gentamicin is another member of this group. It has expanded antibacterial spectrum of aminoglycoside antibiotics to the member of Pseudomonadaceae family and is hardly affected by other antibiotics.

Single bacterial strain may produce many analogous compounds of aminoglycoside antibiotics. For example kannamycin is produced by *Streptomyces kanamyceticus* and gentamicin by *Micromonospora purpurea* etc. Therefore, it is difficult to separate a single aminoglycoside antibiotic from the culture filtrate in the pure state. The techniques like column chromatographic separation with alumina, silica gel, cellulose, ion-exchange cellulose (e.g., carboxymethyl-cellulose, phosphocellulose and sulphoethyl cellulose) or iron exchange gels (e.g., C M Sephadex, SP Sephadex and CM Sepharose) HPLC, etc. have been used to separate and isolate these closely related compounds.

## Streptomycin

Streptomycin and various other antibiotics are produced using strains of *Streptomyces griseus*. It is an important chemotherapeutic agent active against gram positive and gram negative including mycobacteria. As a modification of streptomycin, dihydrostreptomycin was prepared by hydrogenation of streptomycin with platinum as catalyst. Dihydrostreptomycin is commercially more successful because it is thought to be less toxic.

**TABLE 12.3:** Types and producing organism of streptomycine

| Type of Streptomycin | Produced by organism |
|---|---|
| Streptomycin (by fermentation) | *Streptomyces griseus*, *S. humidus*. |
| Mannosidostreptomycin (Streptomycin B), hydroxystreptomycin (reticulin) | *S. reticuli* |
| Mannosidohydroxystreptomycin | *Streptomyces* species. |

**Chemical Composition:** It contains three sugars each derived from glucose, with C.H.O. and N elements.

### *Medium*

**Carbon Source:** Starch, dextrin, glucose, glycerol and other economically available materials.

**Nitrogen source:** Natural agricultural by products, soybean meal, corn steep liquor, cotton seed flour, casein hydrolysate, or yeast and its extract. Inorganic nitrogen salts like ammonium sulphate, and ammonium nitrates are also used.

Apart from these mineral salts and vegetable or animal oil are often fed to the fermenter.

**TABLE 12.4:** The Hockenhul medium

| | |
|---|---|
| Glucose | 2.5% |
| Extracted soybean meal | 4% |
| distillers dried soluble | 0.5% |
| Sodium chloride | 0.25% |
| pH | 7.3-7.5 (before sterilization) |

**Inoculum:** High yielding mutated *Streptomyces griseus* spores are maintained as soil stocks or lyophilized in a carrier such as sterile skim milk. The spores from these stock culture are inoculated into a Sporulation medium. It provides enough sporulated growth to initiate liquid culture build up of mycelial inoculum in flasks or inoculum tanks.

### Fermentation Process

As in penicillin fermentation, spores of *S. griseus* are inoculated into a medium to establish a culture with a high mycelial biomass for introduction into an inoculum tank, with subsequent use for mycelial inoculum to initiate the fermentation process in a tank. The basic medium for the production of streptomycin contains soybean meal as the nitrogen source, glucose as the carbon source, and sodium chloride. Yield in production vessel, strongly responds to high aeration and agitation condition. Temperature range of 25-30°C (approximately 28°C) and pH range between 7 and 8 (peak production 7.6-8) are required. Fermentation lasts 5 to 7 days and the streptomycin fermentation passes through 3 phases.

The classic fermentation process for the production of streptomycin involves three phases.

- **Phase I:** Rapid growth of *S. griseus*, with production of mycelial biomass. Proteolytic enzymatic activity of *S. griseus* releases ammonia to the medium from the soybean meal, causing a rise in pH. During this initial fermentation phase there is little production of streptomycin. This phase is characterized by the proteolytic activity from soybean meal resulting into release of ammonia into the medium (pH raises to 7.5). Carbon nutrients of the soybean meal are utilized for growth. Glucose of the medium is slowly utilized with slight production of streptomycin.
- **Phase II:** A little additional production of mycelia. The glucose added in the medium and the ammonia released from the soybean meal are consumed during this phase. The pH remains fairly constant (7.6 to 8).
- **Phase III:** It is the final phase of the fermentation, after depletion of carbohydrates from the medium, streptomycin production ceases and the bacterial cells begin to lyse. There is a rapid increase in pH because of the release of ammonia from the lysed cells.

### Recovery and Purification

In the end of the fermentation, the mycelium is separated from the broth by filtration and the streptomycin is recovered. For recovery, the broth is acidified, filtered and neutralized. It is then subjected to column containing a cation exchange resin to adsorb the streptomycin from the broth and subsequently the column is washed with the water and streptomycin eluted with hydrochloric acid before concentration *in vacuo* almost to dryness. The streptomycin is dissolved in methanol and filtered. Acetone is added to the filtrate to precipitate the antibiotic. In the end the precipitate is washed with acetone and dried *in vacuo* before being dissolved in methanol for preparation as pure streptomycin chloride complex. The purification consists of adsorbing the streptomycin onto activated charcoal and eluting with acid alcohol.

## TETRACYCLINES

Tetracyclines are significant chemotherapeutic agent. Tetracyclines include two types: chlortetracycline and oxytetracycline. It was first prepared by catalytic dehydrogenation of chlortetracycline. It was biosynthetically produced by *Streptomyces aureofaciens*. Demethyltetracyclines and decarboxyamide are also included in tetracycline group. These are tetracycline derivatives.

## Production of Tetracycline

It is commercially produced using fermentation processes. Two strains of *Streptomyces aureofaciens*: ATCC 13908-13911 and Ncl B-9114 are used for industrial production. The stock cultures is maintained for long periods in the form of spores. Spores are either lyophilized or kept under liquid nitrogen. The bioreactor and other equipments are made of stainless steel. To obtain high production, bioreactor should be well aerated and should have sufficient agitation. Bioreactor requires special device for continuous monitoring. Automatic control of physiological parameters of the culture regulates the fermentation by sterile feeding of nutrients and pH adjustment. The content of nutrients, growth of the culture, production of the antibiotic, morphology of the culture and sterility are usually tested. Liquid nutrient media are usually sterilized for 40 minutes at 120 °C.

### *Medium*

The tetracycline production is affected by the ratio C:N source in the medium. The medium contains glycerol, starch and sucrose as carbon sources, ammonium salts as nitrogen source, soybean meal in combination with mineral salts serve as nitrogen source. The medium also includes peanut meal, cotton seed meal, corn steep liquor, amino acids, casein, meat extract etc. For high production, it is essential that the content of chloride ions in the medium should remain low. Synthetic media is also used for tetracycline production used for the production. It is composed of deionised corn steep liquor/other raw materials freed from chloride ions. Optimal temperature (28 °C) and pH ranges 5.5-6.5 (5.8-6.0) varies.

### *Inoculum*

Type of inoculum (quality of spores or vegetative inoculum i.e., its age, genetic homogeneity, metabolic activity) used for production determines biosynthesis of tetracycline. In spite all favourable conditions required for fermentation type of inoculum plays important role in yield. Optimum yield is obtained from medium containing inoculum for 24 hrs. Optimum pH value of inoculum is 5.6-6.2. Amount of inoculum be in the range of 2-10%.

**Aeration:** Submerged cultures of *S. aureofaciens* require intensive aeration from the beginning of the cultivation. Interruption in aeration during the first hours result in a substantial decrease of production.

### *Production*

Production of tetracycline is obtained in submerged culture. It includes three phases.

- **Phase I:** It is a growth phase and is characterized by rapid consumption of nutrients. Cell mass increases in this phase. It is influenced by the phosphate concentration in the medium. During production phase of fermentation secondary mycelium (thin hyphae) regulates contents of phosphate ions in the medium. The production type of the mycelium is not formed in the presence of excessive phosphate ions.

- **Phase II:** It is a production phase. The growth rate is decreased and sometimes ceases. Maximum amount of antibiotic is produced in this phase.
- **Phase III:** In this phase the production becomes low. Mycelium is fragmented and lysis starts.

### *Isolation*

Tetracycline is obtained from the filtrate of the acidic medium by adsorption on active material and subsequent elution. Tetracycline is precipitated in the form of salts with metals of alkaline earth. Extraction is carried from the broth in the presence of a carrier i.e., quaternary ammonium compounds. Tetracycline forms complex salt with divalent $Ca^{2+}$ and $Mg^{2+}$ (at pH higher than 7). This salt is poorly soluble in water, however, readily soluble in organic solvents. After the separation of the organic phase, the tetracycline is transferred back to the aqueous phase. In the end it is salted out or crystallized.

## CHLORAMPHENICOL

Chloramphenicol was isolated from *Streptomyces venezuelae* in 1947. It is one of the oldest known antibiotics and has marked inhibitory activity Gram positive and Gram negative cocci, bacilli, some mycobacteria, actinomycetes, etc. It is also effective against anaerobic bacteria (*Lymphogranuloma psittacosis* group and *Vibrio cholerae*). It is especially active against *Haemophilus influenza*, *Salmonella typhi* and *Bacterioides fragilis*. Though it is quite effective antibiotic, however, it has severe side effects. It affects bone marrow, therefore, it is strongly restricted for human therapy.

### Chloramphenicol Production

### *Medium*

For submerged aerobic fermentation complex media are used. Chloramphenicol is obtained by *Streptomyces venezuelae*. For 30 litre fermenter, about 18 liters of fermentation medium is required. Medium is sterilized at 120°C for 1 hr.

**Carbon source:** Glycerol (1%) is used, because it gives higher titres and is superior to glucose, maltose, or lactose. Sometimes molasses and dried distillers grain slops (0.5%) are also used.

**Nitrogen sources:** Proteinaceous materials like yeast fermentation solubles, hog stomach residue preparations etc. Tryptone is also a best source of nitrogen.

**Mineral nutrient:** Sodium chloride (0.5%), zinc and iron in traces

**pH:** It is adjusted to 7.5 with NaOH.

### *Production*

After inoculation in the fermenter the peak yield and the maximum cell mass is obtained within 3-4 days at 25°C. The highest obtained antibiotic level varies from was 200 to 300 mg/L. In defined media based on glycerol, DL serine, and lactate, *S. venezuelae* grows slowly, however, it gives relatively high yield after 10-15 days fermentation.

### *Isolation and Purification*

It is extracted from the clarified broths. It is filtered at a slightly acidic pH (about 4.0), at a neutral or slightly alkaline pH (8.5-9.0) with ethyl – or amyl acetate. The extracts are concentrated, diluted with kerosene, and then washed with diluted acids (acetic acid), alkalis (sodium bicarbonate) and water. Lipids with petroleum ether are removed from evaporated residue. The crude product is decolorized by passing the organic solution through a short column of charcoal or alumina. The purified product is dissolved in hot water, decolorized, clarified, and cooled to induce crystallization. The product is recrystallized from ethylene or an ether petroleum ether mixture.

## GRISEOFULVIN

It is a less toxic, systemic, antifungal antibiotic and used orally for the treatment and control of fungal diseases of skin, hair, and nails caused by various dermatophytes. It is produced from genus *Penicillium: Asymmetrica fasciculate* and *Asymmetrica divaricata*. The strains are the mutants of *P. patulum, P. raistrickii*, or *P. urticae*. They are maintained on cracked corn in sand. Agar plates can also be used. The strains with conidia yields maximum product.

### Inoculum

Both spore suspension and vegetative cells have been reported as suitable inocula for laboratory-scale fermentations.

### Medium

There are three types of media, (1) sporulation inoculum, (2) germination medium for vegetative inoculum and (3) vegetative inoculum.

- **Sporulation medium:** It provides inoculum suitable for direct inoculation of shake-flask culture. The medium contains (in g/L) whey powder lactose (30.0), whey powder nitrogen (50.0), $KH_2PO_4$ (4.0), KCl (0.5), CSL solids (3.8).
- **Germination Medium for Vegetative Inoculum:** It contains (in g/L) protopeptone (20.0), malted cereal extract (10.0), glucose (40.0), soluble starch (20.0), $NaNO_3$ (3.0), $KH_2PO_4$ (1.0), $MgSO_4.7H_2O$ (0.5) and $FeSO_4.7H_2O$ (0.02).
- **The Seed Stage (Vegetative) Medium:** It contains (in g/L) C.S.L. nitrogen (3.0), brown sugar (20.0), chalk (10.0), maize oil (10.0), Hodag M.F. (0.3).
- **Typical Production Medium:** It contains (in %) corn steep liquor – N source (0.175), $CaCO_3$ (0.4), $KH_2PO_4$ (0.4) and KCl (0.15).

### Production

For production of griseofulvin nutrient level in the fermenter is the limitation factor. For cell production it is necessary to maintain aerating potential (good aeration) of the fermenter and control of both pH and available nitrogen. Good yield could be obtained when carbohydrate level (glucose, lactose, sucrose, up to 12%) is sufficient. Phosphate (0.4-0.5%) and chlorine are also required during fermentation in sufficient amount for good yield otherwise the obtained product will be dechlorogriseofulvin (devoid of antifungal property). The carbohydrate glucose syrup (50%) is added stepwise after the first 16-20 hrs (10-12%). pH maintained in the range of 6.8-7.2 with continuous glucose addition. The fermentation is carried at 25°C with an air flow of 1 vvm. The yield at the end will be 6-8 g/L after 10 days. Further improvements (12-15.5 g/L) can be had from the addition of various methyl donors (chlorine salts, methyl xanthate, and folic acid) to the medium.

### Extraction and Purification

For extraction of griseofulvin from the cells they are treated with organic solvent and removed by filtration, then the wet mycelium is treated with butyl acetate followed by subsequent evaporation. The crude griseofulvin is then washed with chloroform. It is recrystallized from aqueous methanol. This procedure gives yield up to 58%.

If the extraction of mycelium is done with methylene chloride (after evaporation and crystallization) about 88% pure product can be obtained with 95% yield.

After recrystallization with acetone improved product with 99% purity of 99% is obtained.

## MACROLIDE ANTIBIOTICS

Macrolides are group of lipophillic antibiotics with a macrocyclic lactone ring in their structure. Macrolides are active against β-lactam or aminoglycoside. Various types of macrolide antibiotics are available, the important are erythromycin, josamycin, leucomycin, midecamycin, oleandomycin, spiramycin and tylosin. These have been successfully produced at commercial scale.

### Erythromycin

Erythromycin is produced from Streptomyces (*S. erythraeus*, *S. griseoplanus*), and Micromonospora. During fermentation process different types of erythromycin are produced like erythromycin-A (major product), B, C and D (as minor products).

#### *Inoculum*

The stable inoculum is prepared from suspension of at least $10^8$ viable cells/mL. The Sporulation of *S. erythraeus* is carried out on tryptone agar slants. The cells are harvested from agar plates and the suspensions is taken in sterile water. It is stored at 4°C.

#### *Medium*

Production media for erythromycin production include ingredients (in%) sucrose (5), corn steep liquor (0.5), soybean oil meal (1.5), yeast (1.0), NaCl (0.5), $CaCO_3$ precipitate (0.3), pH (7.0-7.2). For obtaining high yield the fermentation medium should be of proper composition and concentration, with optimum pH and viscosity along with proper concentration of inorganic phosphate, nitrogen and metal ions, as well as organic nitrogen and carbon, and substrate precursor.

#### *Fermentation*

Fermentation is carried out in a stirred tank fermenter. It should have proper aeration, agitation, pH, temperature, viscosity, and dissolved oxygen concentration. Production of erythromycin occurs when growth reaches the stationary phase and consumption capacity of inorganic phosphate, ammonium nitrogen and metal ions as well as nitrogen sources, and a high concentration of dissolved oxygen by *Streptomyces*. The addition of n-propanol as precursor increases erythromycin titre.

The quality of inoculum enhances erythromycin yield. It also depends on the aeration and agitation conditions in the fermenter. For agitation of 1000 litre media in a fermenter 1.85 kW power input is required, and the same volume is aerated at 0.4 vol/vol/min. from 0-12 hrs, and at 0.9 vol/vol/min from 12 hrs till harvest. Carbon dioxide is added at 0.1 vol/vol/min or 11% of the inlet air. The supplement of $CO_2$ do not affect pH and growth, but increases the erythromycin yield to 40% of the control.

### *Isolation and Extraction*

**Isolation:** It is extracellularly produced in the fermentation broth. Mycelium is separated from the broth by filter press, centrifuge, or drum filter with a filter aid. The acidic condition helps to separate mycelium from the broth.

**Extraction:** The erythromycin is extracted from the filtrate using a water immiscible organic solvent (methyl-iso-butyl-ketone or ethyl acetate). It is than transferred to acidic water. pH is adjusted with HCl, $H_3PO_4$; acetic acid or citric acid. The purification and concentration is carried with ion exchange resin i.e., amberlite 50. Further purification step antibiotic absorbed on resin is eluted by a mixture of organic solvent and water at pH 3 to 8.

## RIFAMYCINS

This new group of antibiotic obtained from *Nocardia mediterranea*, *Streptomyces* sp. *Micromonospora locustis*, *M. halophytica* etc. The isolated product has five active substances i.e., rifamycins A, B, C, D and E. Rifamycin B is most active. It is spontaneously transforms to more active products like rifamycins O, S and SV. These are microbiologically active. *In vivo* rifamycin SV shows good activity, tolerability and solubility properties (treats infections from gram positive bacteria).

### Inoculum

Stock cultures of *Nocardia mediterranea* is maintained in sealed ampoules in lyophilized state. Cultures of a frozen vegetative mycelium of selected high producing colonies in small vials at a low temperature are highly stale over a long period of storage.

### Medium

The vegetative medium for rifamycin includes (in g/L) peanut meal (25), soybean meal (10), $(NH_4)SO_4$ (10), propylene glycol (5), glucose (110), glycerol (added when glucose level falls below 5%) (60). Fish meal (20-40 g/L ) may also be used as nitrogen source. The composition of the vegetative media is an important factor influencing the final yield. It depends on types of strain used. The best yields can be obtained when 5% inoculum of a preculture grown for 24 hrs or with a 1.25% inoculum of 40 hrs old cultures is used in aerated fermenters at 28°C. It gives an yield of 6-7 g/L. During fermentation *N. mediterranea* has high oxygen requirement in a definite period of growth-production cycle. Therefore, an airflow ranging from 0.8-1.5 l/min is essential.

### Fermentation

It is carried out in conventional bioreactors of 20-100 $m^3$ volume. It is equipped with turbine stirrers, baffles, air ring spargers, and heating-cooling systems. In a fermenter, consumption of glucose, growth of mycelium and production of antibiotic occur at the same time. At initial stages of growth, consumption of glucose and nitrogen is slow, however, it becomes very fast when mycelial growth starts. The demand of oxygen is also very high at this stage. Metabolic activity of mycelium changes the initial pH 6.2-6.4 to 6.6-6.7, later it decreases to 5.5-5.7. pH again rises slowly until the end of fermentation (optimal pH 6.2-6.4). The fermentation lasts for 300 hrs.

## Recovery and Purification

Purification of broth after harvest is done at pH 7.5-8.0. The filtrate is acidified (pH 2.0). Extraction of the product be carried with chloroform. Re-extraction from chloroform be done in phosphate buffer of pH 8 followed by acidification of the aqueous solution and extraction with ethyl acetate from which it crystallizes.

# 13. Production of Vitamins

Vitamins are organic substances that are essential in small amounts for the health, growth, reproduction, and maintenance of one or more animal species, which must be included in the diet since they either cannot be synthesized at all or in sufficient quantities in the body. Each vitamin performs specific function, hence, one cannot replace or, act for, another. In general, the human body cannot synthesize them, at least in large enough amounts to meet its needs. However, vitamin D is exception; when a person is exposed to ultraviolet rays, the vitamin is synthesized from its precursor, which is found in the skin.

## VITAMINS: GENERAL ASPECTS

**Nomenclature and Classification:** There is no universal agreement on the nomenclature of the vitamins. But the modern tendency is to use the chemical name, particularly in describing members of the B complex. In this book, the most common designations are used. Today, vitamins are generally classed as (1) fat-soluble *vs* water-soluble vitamins, (2) vitamin B complex, and (3) vitamin like substances.

**Fat-Soluble *vs* Water-Soluble Vitamins:** Several phenomena of vitamin nutrition are related to solubility. Vitamins are soluble in either fat or water. Consequently, it is important that both nutritionists and consumers be well informed about solubility differences in vitamins and make use of such differences in programs and practices. Based on solubility, vitamins may be grouped as follows:

**TABLE 13.1:** Groups of vitimins on the basis of solubility

| The Fat-Soluble Vitamins | The Water-Soluble Vitamins |
|---|---|
| Vitamin A | Biotin |
| Vitamin D | Choline |
| Vitamin E | Folacin (folic acid) |
| Vitamin K | Niacin (nicotinic acid; nicotinamide) |
| | Pantothenic acid (vitamin B-3) |
| | Riboflavin (B-2) |
| | Thiamin (vitamin B-1) |
| | Vitamin B-6 (pyridoxine; pyridoxal; pyridoxamine) |
| | Vitamin B-12 (cobalamins) |
| | Vitamin C (ascorbic acid; dehydroascorbic acid) |

It is necessary that vitamin C is the only member of the water-soluble group that is not a member of the B family.

**Chemical composition:** The fat-soluble vitamins contain only carbon, hydrogen, and oxygen, whereas the water-soluble B vitamins contain these three elements plus nitrogen.

**Occurrence:** Vitamins originate primarily in plant tissues which includes vegetable products (cabbage, carrot, orange, lemon, corn or maize oil, and corn or maize bran etc.) and microorganisms (brewers yeast, bakers yeast etc.). Vitamins are also obtained from animal products like fish ox or calf liver oil, eggs etc. Vitamins C and D are present in animal tissues only. If an animal consumes foods containing them or harbors microorganisms that synthesize them. Fat-soluble vitamins can occur in plant tissue in the form of a pro-vitamin (or precursor of a vitamin), which can be converted into a vitamin in the animal body. But no pro-vitamins are known for any water-soluble vitamin. Also, the B vitamins are universally distributed in all living tissues, whereas the fat-soluble vitamins are completely absent from some.

Storage: The fat-soluble vitamins are stored in appreciable quantities in the body, whereas the water-soluble vitamins are not. Any of the fat-soluble vitamins can be stored wherever fat is deposited; and the greater the intake, the greater the storage.

**Excretion:** The fat-soluble vitamins are excreted exclusively in the faces via the bile. The water-soluble vitamins are excreted primarily in the urine, although limited amounts may be present in the faeces. This difference in pathway of excretion reflects the difference in solubility.

## VITAMIN B COMPLEX

With the exception of vitamin C, all of the water-soluble vitamins can be grouped together under the vitamin B complex, of which there are the following nine: biotin, choline, folacin (folic acid), niacin (nicotinic acid; nicotinamide), pantothenic acid (vitamin $B_3$), riboflavin (vitamin B-2), thiamin (vitamin B-1), vitamin $B_6$ (pyridoxine; pyridoxal; pyridoxamine), vitamin $B_{12}$ (cobalamins)

Vitamin $B_{12}$ belongs to a group of compounds known as cobalamins. Chemically, vitamin $B_{12}$ is composed of a central cobalt atom coordinated by a porphyrin-like group called corrin. The axial coordination sites are occupied by a base, dimethylbenzimidazole, and a cyano group. Other cobalamins are also known. The compounds containing bases different from dimethylbenzimidazole are known as $B_{12}$ analogs.

### Vitamin $B_{12}$ (Cobamide) Production

Vitamin $B_{12}$, like so many other members of the B complex, is not a single substance; rather, it consists of several closely related compounds with similar activity. The term *cobalamins* is applied to this group of substances because all of them contain cobalt. Vitamin B-12, which is the most active member, is cyanocobalamin, named after the cyanide ion in the molecule. Other chemically related compounds known to have vitamin $B_{12}$ activity include hydroxocobalamin, nitritocobalamin, and thiocyanate cobalamin.

**Chemistry:** Vitamin $B_{12}$ is the largest and the most complex of vitamin molecules. The main part of the molecule consists of a porphyrin ring containing cobalt as the central element. A cyanide (CN) group may be attached to the cobalt, in which case the compound is called cyanocobalamin (or vitamin $B_{12}$); the commercially available form of the vitamin, little of which occurs naturally. The cyanide group attachment to the cobalt can be replaced by a hydroxy group (-OH), giving hydroxocobalamin, the common naturally occurring form of the vitamin; or it can be replaced by a nitrite group ($-NO_2$), giving nitritocobalamin, a form found in certain bacteria.

Vitamin $B_{12}$ is the only vitamin produced by fermentation. For the commercial production of riboflavin, chemical or semisynthetic procedures are feasible.

Various types of microorganisms like of *Pseudomonas* (i.e. *P. denitrificans*) and *Propionibacterium* (*P. shermanii* and *P. freudenreichii*) are used for commercial production of vitamin $B_{12}$. These were preferred to Streptomyces and other species because of their rapid growth and high productivity of selected mutants.

Vitamin $B_{12}$ production includes a laboratory-scale fermentation process from *Pseudomonas denitrificans* and pilot plant-scale fermentation process from *Propionibacterium shermanii.* The process based on carbohydrates of the culture media.

## Vitamin $B_{12}$ from Pseudomonas denitrificans

Mutant strains have been developed from the original wild-type of *P. denitrificans* for commercial production of vitamin $B_{12}$.

**Medium:** Medium for laboratory-scale fermentation process for vitamin B12 from *Pseudomonas denitrificans* includes (in g/L) beet molasses (60.0), brewers yeast (1.0), Nz amine (1.0), $(NH_4)_2HPO_4$ (2.0), $MgSO_4.7H_2O$ (1.0), $MnSO_4.7H_2O$ (0.2), $ZnSO_4.7H_2O$ (0.02), $Na_2MoO_4.2H_2O$ (510-3), agar (25.0), and pH (to 7.4). The inoculated medium is incubated at 28°C for 4 days.

Seed culture medium includes the same composition as mentioned above but it is devoid of agar. It is incubated for 3 days at 28°C on rotary shaker.

Production culture medium includes (in g/L) beet molasses (100.0), yeast (2.0), $(NH_4)_2HPO_4$ (5.0), $MgSO_4.7H_2O$ (3.0), $MnSO_4.H_2O$ (0.2), $CO(NO_3)_2.6H_2O$ (0.188), $ZnSO_4.7H_2O$ (0.02), $Na_2M_0O_4.2H_2O$ (5 × 10-3), 5,6-DBI (0.025), and pH (to 7.4). The media inoculated with the seed culture is incubated for 90 hrs at 29°C with agitation (420 rpm) and aeration (1v/v/m).

**Carbohydrate source:** Sugar beet molasses containing 5-10% betaine (trimethyl glycine). It stimulates the production of vitamin $B_{12}$ by promoting synthesis of Ala-synthetsae. It helps in production of δ-aminotvulinic acid (the first intermediate of the pathway of cobalamin biosynthesis).

The cobalamin biosynthesis by *Pseudomonas denitrificans* require external supply of 5,6-dimethylbenzimidozole (5,6-DB1) and cobalt salt.

**Fermentation:** Aerated condition with agitation bioreactor is used for growth of *Pseudomonas* and biosynthesis of vitamin $B_{12}$.

## Vitamin $B_{12}$ from Propionibacterium shermanii

**Medium: Maintenance medium** for *P. shermanii* includes (in g/L) tryptone (10.0), yeast extract (10.0), filtered tomato juice (200.0), agar (10.0) and pH (to 7.2). The inoculated media is incubated for 4 days at 30°C.

**Seed Culture Medium:** This is different types and prepared according to stages.

- **First stage medium:** It is similar to maintenance medium but is devoid of agar. It is incubated for 2 days at 30°C without agitation.
- **Second stage medium:** It includes corn steep liquor, 20 g/L, glucose, 90 g/L and pH maintained at 6.5. Stainless steel bioreactor is used. The medium is incubated at 30°C for 24 hrs without aeration, pH is adjusted to 6.5.

**Main culture medium:** The production media includes (in g/L) corn steep liquor (40.0), glucose (100.0), $COCl_2.6H_2O$ (0.02), and pH (to 7.0). It is incubated at

30°C. The first 80 hrs is allowed to proceed without aeration, with slight nitrogen with agitation and slight aeration (0.1 v/v/m). pH is adjusted to 7.0.

Propionic organisms are grown on carbohydrate based media in an *unaerated* conditions. These organisms produce cobaltocorrinoids. Cobalt supplement is necessary for cobalamin production. It also depends on the internal formation or external supply of 5,6-dimethylbenzimidazole (5,6-DBI). The mutant strains of *P. shermanii* synthesis their own 5,6-DBI. It increases the yield up to 65 mg/L on pilot-scale.

Aeration promotes formation of 5,6-DBI, but it decreases vitamin $B_{12}$ biosynthetic pathway at one of the stages. Therefore, at first stage (80 hrs) fermentation is conducted in absence of oxygen (anaerobic condition), but with little agitation until all the sugar in the media is utilized for the growth and the formation of cobinamide. This kind of operation will have no repressive effect. The subsequent stage (next 88 hrs) is followed with agitation and slight aeration. Aeration at this stage induces biosynthesis of 5,6-DBI and cobinamide is converted in to cobalamin.

### *Isolation and Purification*

The extraction process depends on the release of unstable cobalamin from the cells lysed and subsequent treatment of this with cyanide to convert into more stable cyanocobalamin. Cells are separated from the culture broth. Cells are lysed by heat treatment at 80 to 120°C for 10-30 minutes at pH 6.5-8.5. The cells on lysis release various cobalamins. These are then solubilized with potassium-cyanide in the presence of sodium nitrite. The obtained cobalamins gets converted into cyanocobalamin.

The purification of the product is done using adsorption method for substances like amberlite IRC 50, Dowex 1x2, alumina, silanized silica gel, and Amberlite XAD2. It is then followed by elution with water-alcohol or water-phenol mixtures.

## VITAMIN $B_2$ (RIBOFLAVIN) PRODUCTION

Riboflavin is present in virtually all living cells; and, like niacin, it has an essential role in the oxidative mechanism in the cells. Chemically, riboflavin is composed of an alloxine ring linked to an alcohol derived from the pentose sugar ribose. In living cells, riboflavin generally occurs combined either with phosphoric acid, or with phosphoric acid and adenylic acid, both of which may be combined with specific proteins (i.e., protein bound FMN and FAD). It results from the combination of flavin, iso-allozine, and a polyol, D-ribitol, derived from D-ribose and are found in the form of two coenzymes i.e., flavin-mononucleotide (FAM-riboflavin 5'-phosphate) and flavin adenine dinucleotide (FAD-riboflavin 5-adinosyldiphosphate).

**Properties:** In pure form, riboflavin exists as fine orange-yellow crystals, which are bitter testing and practically odorless. In water solutions, it imparts a greenish-yellow fluorescence. It is sparingly soluble in water. (It is much less soluble in water than thiamine). It is heat-stable in neutral or acid solutions, but it may be destroyed by heating in alkaline solutions. It is easily destroyed by light, especially ultraviolet light; for example, it may be destroyed by sunlight striking milk kept in glass bottles. (Milk in cartons is protected against such losses). Synthetic riboflavin should always be kept in dark bottles. Because of its heat stability and limited water solubility, very little riboflavin is lost in processing and cooking foods.

It is produced from bacteria: *Clostridium acetobutylicum* and *Candida flareri*. They produce 100 mg/L in 4 days and 600 mg/L in 7 days respectively. Two mutant strains of *Ascomycetes* spp. (*Eremothecium ashbyii* and *Ashbya gossypii*), and one recombinant bacterium i.e., *Bacillus subtilis* can also produce vitamin B2. These organisms are not affected by the presence of $Fe^{+}$ in the production medium. *Ashbya gossypii* is most preferable strain than *Eremothecium ashbyii*.

## Production by Fermentation of *Ashbya gossypii* (Nrrl Y 1056)

The hyphae of *Ashbya gossypii* can accumulate large amounts of riboflavin. Under optimal condition and in suitable culture medium *Ashbya gossypii* can produce riboflavin up to 10-15 g/L after 10 days. The crystals of the product sometimes appear outside the mycelium.

### *Medium*

The production medium includes (g/L) corn steep liquor (45.0), peptone (35.0), soybean (45.0), and glycine (3).

- **Carbon source:** Glucose, sucrose, and maltose.
- **Vegetable oil:** Preferred for growth.
- **Nitrogen source:** Peptones are produced by enzymatic hydrolysis of collagen like proteins (e.g., gelatin, pancreatic hydrolysate) and corn steep, yeast extract, or distillers solubles. **Vitamins:** Biotin, inositol, thiamine.) are better source for nitrogen. Glycine (1-3 g/L) is also used as precursor of guanine. It increases the yield to 10-30%.

### *Fermentation*

The fermentation is conducted in four phases:

- **Phase One:** It is the initial rapid growth phase of the organism. During this phase glucose is utilized and pyruvic acid accumulates. The pH level decreases. Carbon source of the medium get exhausted. The growth ceases.
- **Phase Two:** It is the production phase. During this sporulation occurs. The level of the pyruvate reduces. Ammonia in the medium accumulates due to deaminase activity. This results into rise in pH level.
- **Phase Three:** The phase is characterized by the synthesis of cell bound riboflavin in the form of FAD and FMN.
- **Phase Four:** It is the last phase and is characterized by autolysis of the cells. This causes release of free riboflavin into the medium.

The fermentation is carried out under aerobic condition. It is important to note that high-level aeration is not suitable for growth because it inhibits mycelial production and reduces the yield. The optimum temperature for production should be between 28-29°C for approximately 4-5 days. The pH of the medium should remain 6-7.5. An yield of 4-2 g/L can be obtained in 5 days.

### *Isolation*

To obtain more impure form of riboflavin, simple concentration method of the broth at pH 4.5 and drying is used. The procedure includes heating of the broth at

60 °C for 3 hrs with bacterial alkaline protease. The insoluble fraction is obtained after cooling the mixture to 25°C. It is then followed by centrifugation at pH 7.

To obtain more pure form of riboflavin the concentration of the broth by heating to 121°C for 1 hr at pH 4.5 is used. The riboflavin is solubilized and the insoluble fraction is discarded. The solution is further treated with titanium chloride as reducing agent it will precipitate riboflavin. It is then reoxidized in the air and dissolved at 60°C in 10% HCl. After cooling, the solution is neutralized to yield purer forms of riboflavin crystals.

## VITAMIN C (ASCORBIC ACID)

Vitamin C or ascorbic acid, dehydroascorbic acid, hexuronic acid, and the antiscorbutic vitamin – is the very important substance first found in citrus fruits, which prevent scurvy, one of the oldest scourges of mankind. All animal species appear to require vitamin C, but a dietary need is limited to humans, guinea pigs, monkeys, bats, certain fish, and perhaps certain reptiles. These species lack the enzyme L-gluconolactone oxidase, which is necessary for vitamin C synthesis from 6-carbon sugars.

**Chemistry:** Vitamin C is one of the simplest vitamins, the L(+)-isomer of which is biologically active. Ascorbic acid is a compound of relatively simple structure, closely related to the monosaccharides sugars. It is synthesized from glucose and other simple sugars by plants and by most animal species. Two forms of vitamin C occur in nature; ascorbic acid (the reduced form), and dehydroascorbic acid (the oxidized form). Although most of the vitamin C exists as ascorbic acid, both forms appear to be utilized similarly by the human. Also, the body efficiently utilizes either synthetic L-ascorbic acid or the vitamin in its natural form as in orange juice.

Ascorbic acid is easily oxidized to dehydroascorbic acid, which is just as easily reduced back to ascorbic acid. However, dehydroascorbic acid may be irreversibly oxidized, particularly in the presence of alkali, to diketogluconic acid, which has no antiscorbutic activity.

Certain derivatives of vitamin C (for example, erythrobic acid and ascorbyl palmitate) are used as antioxidants in food products to prevent rancidity, to prevent browning of fruit, and to cure meat. Erythrobic acid (D-araboascorbic acid) is poorly absorbed and has little antiscorbutic activity.

**Properties:** Ascorbic acid is a white, odorless crystalline power, which is quite stable when dry. Of all the vitamins, ascorbic acid is the most unstable when in solution. It is highly soluble in water, but not in fat. The oxidation (destruction of ascorbic acid is accelerated by air, heat, light, alkalies, oxidative enzymes, and traces of copper and iron. It is markedly destroyed by cooking, particularly where the pH is alkaline. Cooking losses also result because of its solubility. The destruction of ascorbic acid is slowed down by foods that are acidic, by refrigeration, and by protection from exposure to air.

Animals and plant tissues are the source of this vitamin. Plant sources include vegetables – cabbage, spinach, tomato, and fruits such as oranges and lemons. Microorganisms such as fungi, yeasts and algae also produce ascorbic acid but, in small quantities.

Commercially vitamin C is obtained with semi synthesis from D-glucose. All the processes are operated via 2-keto-L-gluonic acid. The lactonization and isomerization of this 2-keto-L-gulonic acid results into the production of L-ascorbic acid. The 2-keto-L-gluconic acid is also obtained using different oxidative or reductive reactions. It is obtained by reduction of D-glucose involving the inversion of the carbon sequence or by oxidation of D-glucose without inversion of carbon sequence.

$$\text{D-glucose} \xrightarrow{\text{catalytic reduction}} \text{D-sorbitol}$$

$$\text{D-sorbitol} \xrightarrow{\text{bacterial (Gluconobacter oxydans) oxidation}} \text{L-sorbose (Poor yield)}$$

$$\text{diacetone-sorbose} \xrightarrow{\text{chemical oxidation}} \text{diacetone 2-keto-L-gulonic acid}$$

$$\text{D-glucose} \xrightarrow[\text{or 2,5-diketo-D-gluconic acid pathway}]{\text{by L-idonic acid pathway;}} \text{2-keto-L-gulonic acid}$$

The 2-keto-L-gulonic acid is efficiently obtained in double fermentation by 2,5-diketo-D-gluconic acid pathway. It eliminates all the chemical stages, involving only two stages.

**2,5-Diketo-D-gluconic Acid Pathway:** It is the fermentation for Ca-2,5-diketo-D-gluconate. *Erwinia* sp. ATCC 31626 is used for this purpose. This first phase is followed second phase of fermentation of Ca-2,5-diketo-D-gluconate using *Corynebacterium* sp. ATCC 31090. It produces Ca-2-keto-L-gulonate. The sequence of conversion is presented in following stages:

D-Glucose
↓
D-Gluconic acid
↓
2-keto-D-gluconic acid
↓
2,5-Diketo-D-gluconic acid
↓
2-keto-L-gulonic acid

### Phases of Fermentation using Erwinia sp.

**Medium:** The media used in shake-flask and seed tank is same. The production medium for Ca-2,5-diketo-D-gluconate includes ( in g/L): glucose (58.0), CSL (11.3), $(NH_4)_2HPO_4$ (5.6), $CaCO_3$ (184.0), Antifoam agent (0.2).

**First Phase of Fermentation Using *Erwinia* sp.:** The starting culture of *Erwinia* sp is maintained in an agar test-tubes. The culture is being done on an agar medium containing glycerol, yeast extract peptone, $KH_2PO_4$ and $MgSO_4$. It is incubated at temperature of 28°C for 24 hrs. A shake-flask culture (20 hrs at 28°C) is inoculated into a seed tank and incubated at 28°C for 10 hrs with agitation and aeration.

**Second Phase of Fermentation Using *Erwinia* sp.:** The fermentation is carried at 28°C for 26 hrs with agitation (160 rpm) and aeration (1.5 v/v, pressure 1.5 Kg/cm$^2$). After 3 hrs the fermentation culture media should be supplied intermittently with 1152 Kg glucose in 120 L of sterile solution. The addition of glucose solution should be stopped after 15 hrs. The final volume should be 4200 L.

### *Fermentation Stage Using Corynebacterium sp.*

**Production Medium:** The medium includes (in g/L): glucose (20.0), CSL (30.0), $NaNO_3$ (3.45), $KH_2PO_4$ (0.67), $ZnSO_4$-$7H_2O$ (4.9 x $10^{-3}$ ), $MnCl_2.7H_2O$ (0.8 x $10^{-3}$ ), Thiamine. HCl (0.22 x $10^{-3}$), Ca-D-patothenate (0.17 x $10^{-3}$). The initial culture is maintained on agar test tubes at 28°C for 40 hrs. The media is inoculated with *Corynebacterium* sp. (ATCC 31090) at 28°C for 66 hrs. Fermentation has to be carried with agitation (160 rpm) and aeration (0.25 v/v/m). At 16 hours, the fermenter should be provided with $NaNO_3$ (1 g/L). The media is inoculated with 6L of the 1st stage seed.

The seed culture can be obtained by culturing *Corynebacterium* sp. in a seed tank (470 L in 1 $m^3$ stainless steel fermenter) at 28°C for 24 hrs with agitation (244 rpm) and aeration (1 v/v m). Media is inoculated with 6L of the 1st stage seed.

The production media is employed in the main fermenter (4-23 $m^3$ in 10 $m^3$ stainless steel fermenter).

### *Harvesting and Isolation*

The 2-keto-L-gulonic acids is isolated from the harvested broth at the end of second stage fermentation by treating with sodium carbonate. The subsequent concentration to crystallization yields 90% 2-keto-L-gulonate.

# 14. Production of Ergot Alkaloids

Ergot is sclerotium of the Pyrenomycetes, *Claviceps purpurea,* which infects rye and, less commonly, other grains. In infected plants, normal development of the grain is suppressed, and a hardened, purple-black sclerotium develops in place of the grain.

Ergot alkaloids (EA) seem to be a never-ending story. Known since the middle ages as drugs, they continue to play an important role in medical therapy. Because the supply of naturally grown ergots is not sufficient, the biotechnological production of EA has gained in importance.

EA constitute the largest known group of nitrogenous fungal metabolites. Most EA can be separated into two basic groups: clavines and derivatives of lysergic acid. The classic peptide EA (ergo-peptines) is the only natural substances with demonstrated cycled structure. New EA have continuously been identified during the past two decades, e.g. 0-12' – methylergocornine and –12' methyl-α-ergokryptine.

EA continue to be of great interest as drugs themselves, or as prototype compounds from which new drugs can be derived. The biosynthesis of EA lies at level of the genetic information apparatus and continues at the level of physiological realization. Genetic manipulation and knowledge of physiological control are the dominant tools of modern EA production.

Scientific research into EA has moved closely to industrial application and become more sophisticated. The production of EA is today largely carried out through submerged *Claviceps* cultures.

Sometimes they are consumed in large quantities, as might happen if the sclerotia are included in milled flour, a disease known as ergotism (a condition of intoxication which follows the ingestion of excessive amounts of ergot alkaloids; symptoms may include vomiting, diarrhea, thirst and convulsions, and gangrenous lesions may subsequently develop at the extremities). Cattle in the field may also be poisoned by the sclerotia if they eat infected pasture grasses. Ergotism in humans is marked by vomiting, feelings of intense heat or cold, pain in the muscles of the calf, a yellow color in the face, lesions on the hands and feet, diarrhea, and an impairment of the mental functions.

Ergot alkaloids are the secondary metabolites of the ergot fungi and the drugs of high potency. These include a family of related compounds with a common tetracycline ergoline ring system. The derivatives include clavines ergoline ring system. Their derivatives include clavines (Chanoclavine I, Festuclavine, Agroclavine, Elymoclavine, Penniclavine, and setoclavine), and ergoline acids [(+)-lysergic acid, (+)-isolysergic acid, and paspalic acid]. These find a variety of therapeutic applications i.e., uterine atonia, migraine and other vascular heasaches, orthostatic circulatory disturbances, cenile cerebral insufficiency, hypertension, hyperprolactimenia, acromegaly and Parkinsonism.

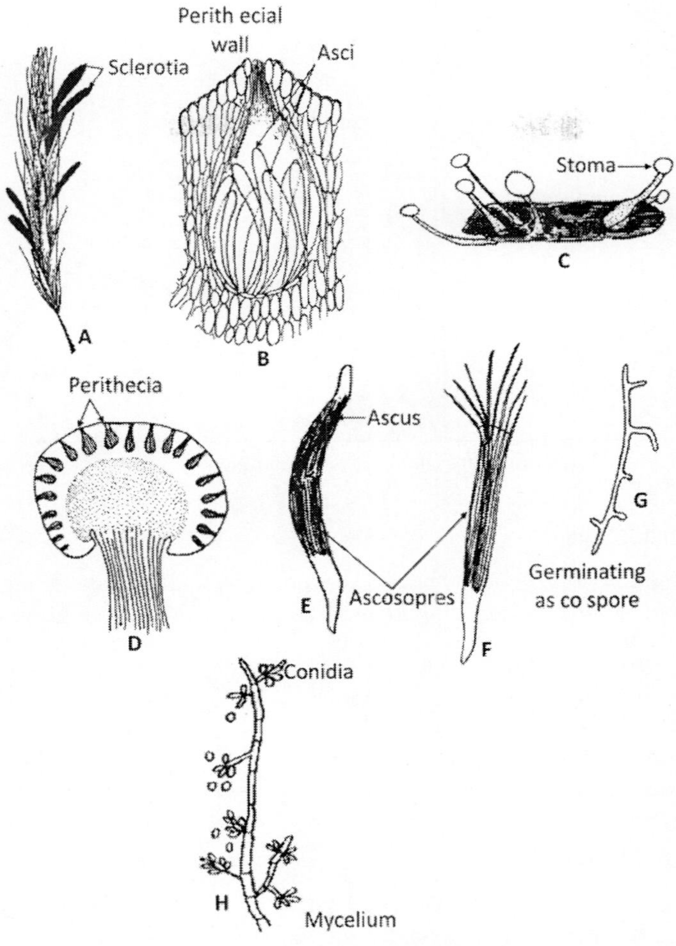


**FIGURE 14.1:** *Claviceps purpurea* ergot of rye. (A) Infected ear with sclerotia; (B) perithecium, (C) germinating sclerotium. It produces several stomata; (D) median longitudinal section of stroma showing perithecia; (E) ascus with needle-like ascospores; (F) ascus discharging ascospores; (G) single needle-like ascospore; (H) germinating ascospore; (I) mycelium with conidiophores and conidia.

## PRODUCING ORGANISMS

EA is produced from genus *Claviceps* of the class Ascocycetes. *C. purpurea* and *C. fusiformis paspali* produce this alkaloids. The *Claviceps* adopts a parasitic way of life. The mycelium grows rapidly with the formation of sclerotium (purple). The sclerotia contain 30% fat and 0.01-1% alkaloids. Other species producing alkaloids include *Balansia*, *Penicillium* and *Aspergillus*. Lysergic acid alkaloids also occur in a higher plants i.e. *Rivea cormybosa*. However, EA have also been found in the *Aspergillus* and *Penicillium* genera, and in the *Convolvulaceae* (morning-glory) family.

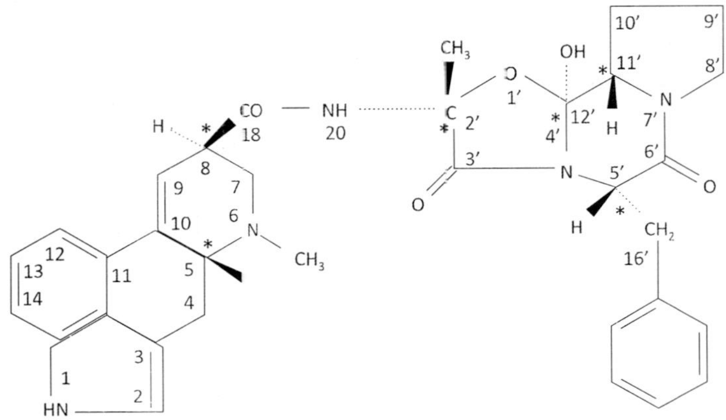

**FIGURE 14.2:** Structure formula of ergotamine – representative of natural ergopeptines.

## Genetic Considerations

The DNA template of *Claviceps* contains sequence elements specifying the start of transcription; other neighbouring sequences are involved in binding of the RNA polymerase complex. In addition, groups of transcription factors are necessary. The RNA Polymerases are suggested to be present in the nuclei in two forms, namely an active and a less active form. The less active form has associated a small peptide inhibitor and has low affinity toward the transcription initiation complex. Addition of stimulatory peptides, as happens when the cells are inoculated on fresh broth medium, competes out the inhibitory peptide. As a result, we observe the high increase in transcription rate in newly diluted cells.

## Physiology of Alkaloid Formation

The biosynthesis of EA lies at level of genetic information apparatus and continues at the level physiological realization. Genes involved in EA synthesis are under rigorous metabolic control. Fungi and plants form ergolenes from the same precursors. The EA formation is more the result of changes in the physiological state than an indication of a new physiological state. Culture growth and EA synthesis are understood as processes, which compete for key metabolic intermediates rather than as mutually exclusive phenomena.

The precursors channeled into ergoline biosynthesis from primary metabolism are L-tryptophan, dimethylallylpyrophosphate and one-carbon donor, i.e., S-adenosyl-L-methionine. Ergopeptines require additional amino acids. In a blind branch of ergopeptine biosynthesis, lactam alkaloids (ergopeptams) with noncylol structure and D-configuration of proline in the peptide moiety of the molecule are produced. The peptide moiety, the biosynthesis of which has been referred to as templete synthesis, originates from a nonribosomal system and is assembled by enzymatic steps. The assembly is not directed by an RNA template, although aminoacyl-RNAs may participate. A template is provided by protein surfaces that originate from mRNA-coded structures. Carboxyl activation is a prereqisite for peptide bond formation. Activation proceeds by formation of phosphate or adenylate intermediates with cleavage of β-γ or α-β-phosphate bonds of triphosphates. A quasilinear o spatial arrangement of activated amino acid substrates directs the peptide sequence. Intermediates remain linked to the enzyme and can be transferred by CoA-like cofactor in intermolecular reactions between multienzymes. The multienzyme structures are directly encoded at the gene level. This type of mechanism has been termed the multi-enzyme thiotemplate mechanism since all activated intermediates formed are thiolesters. Thiolesters are stable at acid pH and thus decompose by hydrolysis. This hydrolysis can be enhanced by thiol reducers, such as dithiothreitol.

## Ergot Alkaloid Enzymes and other factors

The *Claviceps* fungus possesses a great variety of enzymic activity and is versatile in its ability to utilize different compounds as growth substrates. It has also evolved efficient regulatory systems to ensure that the enzymes of specific metabolic pathway are synthesized only when required and in proper amounts for efficient metabolism. The hypothesis appears to be that common regulatory moments may be participating in the synthesis of both alkaloids and lipids.

Enzymes that are repressed or inhibited under conditions of rapid growth generally produce EA. The EA synthesizing enzymes are present in mycelia grown in high levels of phosphate, are activated by decreasing the intracellular phosphate level, and do not undergo rapid turnover, in contrast to both alkaline and acid intracellular phosphatases.

**TABLE 14.1:** Various types of enzymes and other factors involved in ergot alkaloid production

| Enzyme/factor | Functions |
|---|---|
| Isopentenyl Pyrophosphate Isomerase | Isopentenyl pyrophosphate isomerase (EC 5.3.3.2) catalyses the reversible reaction providing dimethyallyl pyrophosphate, the 5-carbon allylic pyrophosphate subunit (isoprene). |
| Dimethylallyltryptophan (DMAT) Synthetase | Dimethylallyltryptophan (DMAT) synthetase catalyses phenylation of tryptophan by dimethylallylpyrophosphate. |
| Agroclavine Hydroxylase | Agroclavine hydroxylase catalyses conversion of agroclavine to elymoclavine. NADPH or liver concentrate (mixture of cofactors) is required for activity. |
| Cytochrome P-450 | Cytochrome P-450-a typical transmembrane highly hydrophobic hemoprotein-is now recognized as an "oxygen-activating enzyme" participating in the hydroxylation of a wide variety of compounds. |
| D-Lysergic Acid-Activating Enzyme | D-Lysergic acid-activating enzyme catalyses both the D-lysergic acid-dependent ATP-pyrophosphate exchange and the formation of ATP from D-lysergic acid adenylate and pyrophosphate. |
| Ergopeptine Synthetase | Condensation of the ergopeptine precursors occurs through the activated intermediates, e.g. CoAthiolesters of lysergic acid. |
| Tricarboxylic Acid and Glyoxylate Cycles | Decrease in the activity of the tricarboxylic acid cycle and rise in that of the glyoxylate cycle precede the intensive EA synthesis. |
| Proteolytic Activity, Turnover | With increased EA formation, the proteolytic activity of the *Claviceps* mycelium decreases. |
| Water | The nature of the water environment carrying chemicals such as organic and inorganic ions, etc. is an important factor in the elucidation of the physiology of EA formation. |
| Tryptophan | Tryptophan is not only a precursor of EA, but is also a factor in |

| Enzyme/factor | Functions |
|---|---|
| | the induction and derepression of enzymes catalyzing EA formation. |
| Mevalonic Acid | Within the EA biosynthesis mevalonic acid is converted to dimethylallyl pyrophosphate, which is a substrate of the enzymic allylation of tryptophan in position C. |
| Hydroxymethylglutaryl-CoA (HMG-CoA) Reductase | Hydroxymethylglutaryl-CoA (HMG-CoA) reductase (EC 1.1.1 88) serves as a regulatory enzyme in the biosynthesis of mevalonic acid. |
| Inorganic Phosphate | Cell differentiation and the onset of the EA synthesis are governed both by the phosphate limitation and the citrate utilization. |
| Citrate | Successful course of the EA biosynthesis requires the presence of higher citrate levels or some other intermediates in the tricarboxylic acid cycle in the cell. |
| Surface-Active Agents | Surface-active agents act by altering the permeability of the cell membrane. They do not change the biosynthetic pathway. The shift of alkaloid production in *Claviceps* cultures, treated with Tween 80, is associated with the shift of organic acids and amino acids in the cell-pool. |
| Vacuoles | Vacuoles contain a number of hydrolytic enzymes, primarily proteases, and also serve as storage pools for amino acids, polyphosphate, and other compounds. Vacuolar enzymes are initially made as inactive proenzymes. |
| Trehalose | Trehalose (α-D-glucopyranosyl-α-D-glucopyranoside) serves mainly as a storage carbohydrate during period of nonproliferation. |
| Fatty Acids | The composition of fatty acids from lipid fraction of high-producing *Claviceps* mutants differs substantially from that of the parent strains. |
| Lipids | Abundant production of lipids is one of indicators of differentiation of the *Claviceps* cells into the sclerotial type. In sclerotial cells, lipids are found in substantially higher amounts than in cells of sphacelial mycelium, and they are unique in that they contain ricinoleic acid. |
| Oleosomes | An extremely large number of oleosomes is typical of the submerged *C. purpurea* cultures. In moment of need, cells are capable of creating *de novo* membranous structures from nonstructured lipid pools (droplets) and protein present in the cell vacuole. |
| Sterols | A total of 15 sterols were found in the lipid fraction of submerged *Claviceps* cultures. Out of them eleven have not yet been described in the *Claviceps* genus. *C.paspali* participates in biosynthesis of almost all sterols under study. |
| Sucrose, Glucans | Extracellular sucrose undergoes complex transformations before its ultimate utilization by *Claviceps* cells. These transformations are species specific *C. purpurea* splits sucrose with β-D-fructofuranosidase, *C. paspali* with endosucrase. |
| Glucanases, Fructofuranosidase | The β-1,3 and β-1,6-D-glucanases are widespread in fungi. They have been found bound tightly to cell walls and extracellular glucans. They are responsible for the hydrolysis of the matrices to which they are bound. |

## COMMERCIAL PRODUCTION IN BIOREACTORS

Selection and screening of potent strains (i.e., sclerotia of *Claviceps* sp.) are required for high yield of suitable alkaloids. The screening may be done in in shake flasks and after mutagenic treatment. Claviceps strains of high producers are unstable. The selected strain should be in heterokaryotic state because homokaryotic parents have rapid growth or abundant sporulation. The stock culture maintenance procedure is important for high yielding strains of *Claviceps purpurea*, because, these high yielding strains proved to be genetically unstable.

### Medium

The media may be of complex natural compounds or synthetic media. The production medium includes a mixture of glucose, 5% malt extract, 2%; a mixture of glycerol, 10% gelatin, 1.25% and NaCl, 0.7%; and milk sucrose are typical adaptive protective media for storage of vegetative hyphae or conidia. Conidia from agar slants or liquid cultures or hyphae developed in a pre-culture medium are suitable for storage. Agar slant cultures can be stored at temperature below 25°C. They are stored under liquid nitrogen or lyophilized. The sealed freeze-dried tubes can be stored at 4°C.

| **Carbon source (non-inhibitory, slowly metabolized carbon sources)** | **Sucrose, sorbitol, malt extract sugars** |
|---|---|
| **Nitrogen source** | Ammonium salts of carboxylic acid, asparagines, tryptone – 1.2-2.0 g/L) |
| **Others** | Phosphates, mineral salts, alkaloids |

A close relationship exists between the production organism and the environment in every fermentation. The adjustment of medium composition and physico-chemical properties helps in the maintenance of a maximum rate and a desirable direction of the required process. Cultivation conditions suitable for the metabolism of primary substances are not suitable for EA metabolism. The EA fermentations are done in the temperature range 20-30°C (favourable 23-26°), in a pH range of 4.6-6.3 (at best 5.2-5.8) in a medium containing slowly metabolized saccharides (as carbon source). **Asparagine** is a superior nitrogen source for saprophytic clavine-alkaloid production in *C. purpurea*. It enters the cell without degradation and its transport into the cell exceeds the cell's biosynthetic need for asparagines. Thus asparagine disturbs the relative pool sizes of various amino acids. This pattern results in a change in the genetically determined ratio at which amino acids are utilized for protein synthesis. The concentration of spores used to seed the inoculum is critical factor. The importance of the inoculum as a factor determining the efficiency of EA production is well recognized industrially.

## PREPARATION OF INOCULUM

Conidia are the propagation units and sporulation occurs on agar media (enriched with grains of cereals and in shaken liquid media). Conidiogenesis (formation of conidia) is enhanced by limited aeration. Media are inoculated with spores or with vegetative cells in the growing phase, incubated at temperatures ranging from 21-27°C for a period of 5-8 days (for liquid cultures) and 10-16 days (for surface cultures). For inoculum development, spores from heavily sporulated working stock (special agar sporulation media with enriched grains of cereals and in shaken liquid media) are used. The starting materials were sclerotia of ergot from rye and different grasses. Sclerotia containing desired alkaloids in relatively high amounts were selected. The media contained in a flask on rotary shaker is inoculated with conidia (2-$10^5$ conidia/mL) of *Claviceps purpurea* strain at 24°C. The germination (up to 5th day) results into production of numerous conidiophores at the end of growth phase before the exhaustion of carbon and nitrogen source and reaches maximum after 7th day (i.e., 12 to 16 days).

## PRODUCTION OF ALKALOID

The hyphae in proliferation stage are transferred to a production in a fermenter containing growth medium. The fermenters are with multiple impellers, suitable for propagation and for the production of *Claviceps* sp. The fermenter should be under strict control of factors such as temperature (24°C), backpressure, pH (5.5), agitator speed, air supply and foam formation. The required osmotic pressure should be between 10-20 bar. It is accomplished by high sugar concentration (200-300 g/L of disaccharides). The high pressure inhibits conidiogenesis and induces differentiation of sclerotial cells. The production medium has to be inoculated with 10-20% v/v pre-culture at the end of growth phase, that allows optimal alkaloid production. The aeration rate of the production culture requires progressive increase from 0.5-1.5 v/v/m and the speed of the impellers from 75-200 rpm. The cultivation lasts up to three weeks – for most strains 12-16 days.

### Peptide Alkaloids

The active principles in ergot that are responsible for ergotism are the alkaloids ergometrine, ergometrinine, ergotamine, and ergotaminine, which occur in the sclerotium.

#### *Ergotamine Production*

For production of ergotamine following medium is used.

**TABLE 14.2:** Amici's production medium for Shake-flask culture

| Chemical | Amount |
|---|---|
| Sucrose | 300 g |
| Citric acid | 15 g |
| $Ca(NO_3)_2.4H_2O$ | 1 g |
| $KH_2PO_4$ | 0.5 g |
| $MgSO_4.7H_2O$ | 0.25 g |
| Yeast extract | 0.1 g |
| KCl | 0.2 g |
| $FeSO_4.7H_2O$ | 7 mg |
| $ZnSO_4.7H_2O$ | 6 mg |
| Aqueous ammonia | |
| pH | 5.2 |
| Tap water add | 1 L |

Total yield after 12 days was. 1.1-1.5 g/L containing 80% of ergotamine is obtained. During the production phase, that lasts 6th/7th days to 12th/14th days, the alkaloids are actively synthesized. The required energy is obtained from oligosaccharides and after 16 days, the autolysis commences.

### Ergotoxine Production

Amici reported the production of ergotoxine (i.e., ergokryptine, ergocornine and ergocristine) in shake-flask culture. The media employed was similar to that employed for ergotamine production. Kobel and Sanglier (1976 and 1978) introduced new medium for submerged cultivation. They isolated the strain that sporulated well on agar media into a sclerotium-like morphological form of the mycelium.

**TABLE 14.3:** Kobel and Sanglier medium for submerged cultivation

| Chemicals | Amount |
|---|---|
| Sucrose | 240 g |
| Ammonium oxalate | 9.6 g |
| Urea | 1.7 g |
| $KH_2PO_4$ | 0.125 g |
| $MgSO_4.7H_2O$ | 0.625 g |
| KCl | 0.31 g |
| $FeSO_4.7H_2O$ | 25 mg |
| $ZnSO_4.7H_2O$ | 10 mg |
| Aqua dest. Add | 1000 ml |
| pH adjusted to | 6.2. |

### Production of Simple Lysergic Acid Derivatives and Paspalic Acid

Lysergic acid is produced by the strain *Claviceps paspali*. It synthesizes (+)-lysergic acid-α-hydroxy-ethylamide in submerged culture. This substance can be converted into (+)-lysergic acid amide, and on alkaline hydrolysis results into (+)-lysergic acid. Lysergic acid is a key substance, in the production of ergot alkaloids and can be employed in the chemical synthesis of their derivatives. Paspalic acid functions as the precursor of (+)-lysergic acid.

**TABLE 14.4:** The production medium

| Chemical | Amount |
|---|---|
| Sorbitol | 50 g |
| Succinic acid | 36g |
| $KH_2PO_4$ | 2 g |
| $MgSO_4.7H_2O$ | 0.3 g |
| NaOH to pH | 5.4 |
| $FeSO_4.7H_2O$ | 10 mg |
| $ZnSO_4.7H_2O$ | 1 mg |
| Water to | 1 ltr. |

The total alkaloid yield in shake flask was 600 mg/L, of which paspalic acid consisted of 86.5%. A mutant strain reported that yields 3g/L of paspalic acid in shake flask. In stainless steel fermenter the production is reduced to 2.3 g/L.

## ISOLATION SEPARATION AND PURIFICATION

### Peptide Alkaloids

Peptide alkaloid is found in the mycelial cells. It is excreted from the medium under alkaline condition and is traced with ether and chloroform. These are then extracted with a water solution of tartaric acid or citric acid. After rendering alkaline they are taken back into organic solvent, which is then evaporated to crystallized form.

## Paspalic Acid

The paspalic acid is a water-soluble compound. It is obtained from the culture filtrate and is extracted by adsorbing on to cation exchange resin (Dowex 50, or Amberlite IR 120). It is subsequently eluted from the resin with dilute ammonia. The elute is evaporated. pH adjusted to 5.5 and finally the substance crystallized.

# 15. Other Primary Metabolites of Pharmaceutical Uses

Microorganisms have the capabilities of synthesizing various chemicals of their own. These chemicals are being used to prepare solvents, lubricants, emollients, demulcents, extractants, adhesives, acidulants, plastics, surface cooling, alternative fuels, dyes cosmetics, antifreeze, digestive aids and flavourings. These organic substances (ethanol, n-butanol, acetone, amino acids, organic acids and enzymes) are produced biologically or by chemical synthesis with the help of microorganisms. A major consideration during the production of these substances is the cost of raw materials.

## FERMENTATION TECHNOLOGY

Microbial fermentation technology utilized in fermentation processes are designed to obtain maximum growth of an organism under the optimum physical conditions in a specific medium for the production of a desired end products. Three general types of growth environments are used: the flask, the shallow tray, and the vat or closed drum. Each of these procedures serves a specific need, and most fermentations use one of these basic procedures.

The chief raw material is the growth substrate, which is usually molasses or starch. The efficiency of the process has to be considered. What fraction of the substrate is converted into product and how long does the conversion take? The other fact to be considered is the cost recovery of the product from the fermentation medium.

The fermentation medium includes nutrients essential to support the growth of microorganisms, and formation of desired end products. The choice of the nutrient should be based on nutritive source and economic aspects. The important sources are carbon, nitrogen, and phosphorus. **TABLE 15.1** presents various nutrient sources used in industrial fermentation.

For fermentation, crude raw materials are usually employed in the medium. There are several factors which should be considered for better fermentation results like trace elements in the medium, quality of water, nature of pipes used for supply of solutions to fermentation reaction centers, aeration (optimum oxygen concentration), pH, required enzymes, optimum temperature, etc. The advent of rDNA technology has further made fermentation more attractive. Industrial chemicals that are aliphatic organic compounds can be distinguished by the absence of benzene rings and similar structures. The aliphatic substances with industrial applications can be broadly divided into two categories, solvents and organic acids. Solvents are not currently made by biological means on an industrial scale, although n-butanol, acetone, and glycerol were made that way. Nevertheless, the solvent industry may return to fermentation because of petrochemicals, the prospect of exploiting thermophilic bacteria and availability of new feed stock.

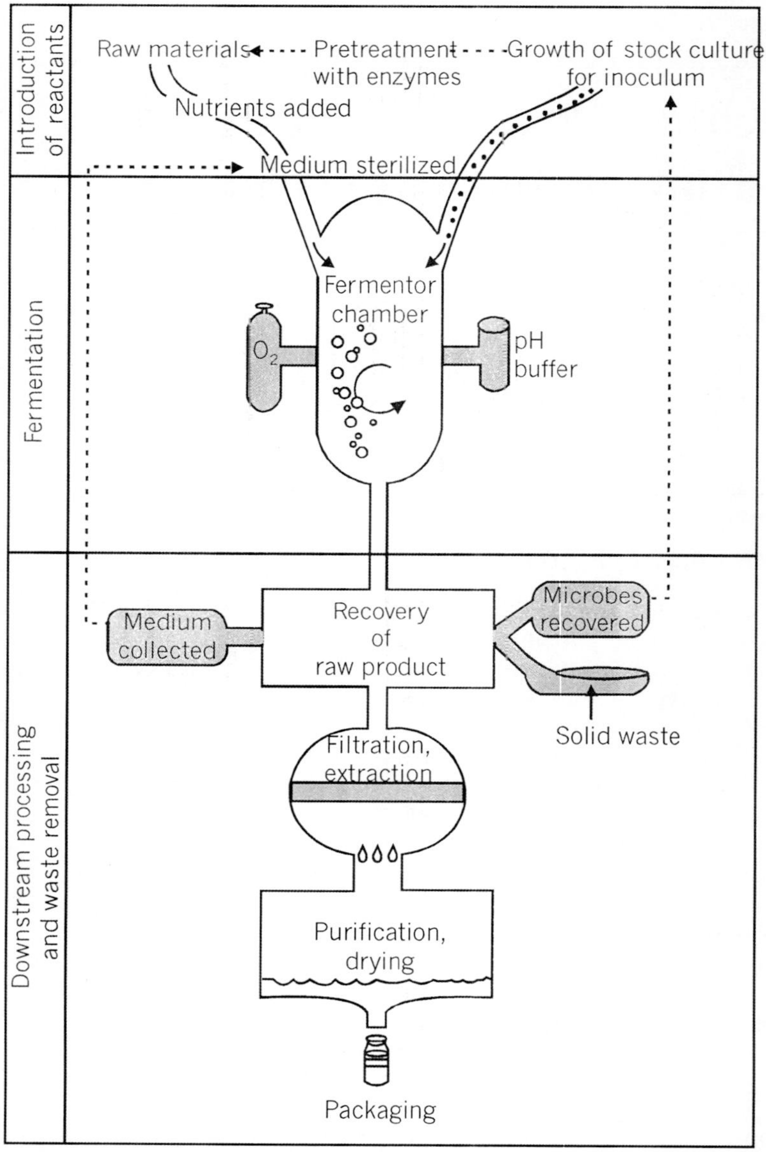


**FIGURE 15.1:** General layout of fermentation plant for industrial production of drugs, enzymes fuels and vitamins.

## Industrial Alcohol

One of the most important and best-known industrial fermentation is that in which ethyl alcohol is produced from sugar and yeasts. Ethyl alcohol ($CH_3CH_2OH$) or ethanol, may be referred to by other names. Ethanol is one of the most important organic chemicals in industry. For example, the alcohol may be named to indicate the source of raw material from which it is manufactured or to indicate the general purpose for which it is to be used, Grain alcohols are alcohols made from grains, such as corn, wheat, or rice. The units of measurement commonly used in reporting ethyl alcohol production are the proof gallon, wine gallon, tax gallon, and barrel.

**Types of Raw Materials:** Types of raw material used for alcohol fermentation may be classified into three principal types:

- the saccharine materials, such as sugar cane, sugar beets, molasses, and fruit juices. Blackstrap molasses is the principal source of industrial alcohol. This material is the syrup that is left after the recovery of crystalline sugar from the concentrated juice of sugar cane. It usually contains 48 to 55 per cent of sugar, mainly sucrose;
- the starchy materials, which include the cereals (corn, malt, barley, oats, rye, wheat, rice, grain, sorghum, and the like), potatoes, sweet potatoes, Jerusalem artichokes (girasol), manioc meal, and other substances; and
- cellulosic materials, such as wood and waster sulphite liquor.

These substrates are used for the synthesis of other compounds that serve as solvents, extracts, dyes, pharmaceuticals, lubricants, adhesives, detergents, plasticizers, surface coatings, cosmetics and resins for the manufacture of fibres. Ethanol is also produced synthetically from petrochemical feedstock or biologically by the yeast.

**TABLE 15.1:** Principle raw materials requited for fermentation

| | |
|---|---|
| Carbon source | Blackstrap molasses or corn but, grains, sulphate waste liquor, whey, potato and wood waste |
| Fats and hydrocarbons | Vegetable oils, petroleum fractions |
| Nitrogen source | Ammonia, ammonium salts, or amino acids. Soybean meal, corn steep liquor (from corn milling) distiller's solubles (from alcoholic beverage manufacture). |
| Ammonia | Pure ammonia or ammonium salts. |
| Potassium salts | Minor quantity |
| Magnesium salts | Minor quantity |
| Calcium salts | Minor quantity |
| Phosphate | As phosphoric acid or orthophosphate salts. |
| Trace minerals | Minor quantity |
| Biotin | For the growth of *S. Cerevisiae* |
| Yeast extract | Provides trace minerals and vitamins |
| pH value | Adjust to 4-5 (4.8-5.0) with sulphuric acid or lactic acid or by allowing lactic acid bacteria to bring about initial lactic acid fermentation. |

**Types of Yeast:** Certain types of yeast (*Saccharomyces cerevisiae*) are the traditional choice for large-scale fermentation. Yeasts are desirable, namely those which are able to produce and tolerate high concentration of alcohol and which possess uniform and stable characteristics or other microorganisms. In the microbiological process, yeasts secrete ethanol as a by-product of fermenting either crude sugar or starch that has been converted into sugar. Strains of *Saccharomyces cerevisiae* are commonly used, but other yeasts, such as *Saccharomyces anamnesis and Schizosaccharomyces pmble* may be employed under certain conditions.

*Zymomonas mobilis* is an important microorganism which ferments sugar twice as fast as yeast does. Thermophilic bacteria are the most efficient fermenters. A major limitation on the fermentative production of ethanol is the capacity of microbe to tolerate the solvent because the growth and the product formation are inhibited by solutions of high osmotic pressure and by high concentration of alcohol. Depending on the type of carbohydrate present in the medium, different microbes were employed in the fermentation viz., *Saccharomyces cerevisiae* (starch and sugar), *S. ellepsoides* (starch and sugar), *Candida pseudotropicalis* (whey, after removal of proteins), *C. utilis* (sulphite waste liquor), *Zymomonas mobilis* (sugars) etc.

**Concentration of Sugar:** A concentration of sugar of 10 to 18 per cent is usually satisfactory, although other concentrations are also used. The initial concentration of sugars (for batch fermentation) should be limited to 12%. When concentration is too high it reacts adversely on yeast or the alcohol produced may inhibit the action of the yeast. The use of low concentration of sugar is uneconomic as it may lead to loss of valuable fermentating space.

Continuous fermentation with sulphite waste liquor can also be employed on commercial scale.

**Nutrient substances:** Although molasses generally contains most of the nutrient substances required for fermentation, ammonium salts, such as ammonium sulphate or phosphates, may be added to the mash to supply deficiencies in nitrogen or phosphorus.

**pH of the Mash:** Fermentation proceeds satisfactorily when the mash has been adjusted to a pH of 4.0 to 4.5. This pH favors the yeast but is sufficiently low to inhibit the development of many types of bacteria. Sulphuric acid is commonly used to adjust the reaction of the mash, although lactic acid is satisfactory.

**Oxygen Tension:** Oxygen in large amounts is necessary in the early stages for the optimum reproduction of yeast cells but is not required for the production of alcohol. During the fermentation, carbon dioxide is evolved and anaerobic conditions are soon established.

**Temperature:** The mash is pitched at a temperature of 60 to 80 °F, depending somewhat on the external temperature. During fermentation, the temperature of the mash rises. The use of cooling coils or sprays on the outside of the tank helps to maintain a suitable temperature. At temperature much above 80 °F, alcohol evaporates rather rapidly. Bacterial growth is also favored.

**Time Required for Fermentation:** A fermentation is usually complete in 50 hr or less, depending on the temperature, sugar concentration, and other factors. The actual period being dependent on the substrate used and by the temperature of incubation.

**Distillation:** The fermented mash (beer) is distilled to separate the ethyl alcohol and fusel oil from the other constituents of the mash. In case there is a shortage of fomenters and the beer cannot at all be distilled immediately, part of the beer is pumped to a storage tank, known as a beer well, where it is held until it can be distilled. During distillation, fractions containing different concentrations of alcohol (wines) and slops are separated. The fractions containing approximately 60 to 90 percent of ethyl alcohol are known as high wines. These fractions are concentrated to 95 per cent ethanol by further distillation or fractionation. The fractions low in alcohol, the low wines, are usually redistilled with new lots of beer. The slops are ordinarily discarded but may be used in a number of ways. Sometimes the slops may be used as a substitute for some of the water in diluting molasses for a new mash. The solids from slops may be concentrated by heat treatment and sold as a fertilizer constituent, for they contain potassium and phosphates in addition to other components. The slops may be used as a core binder (in foundries) or as an adhesive for artificial stone (briquette).

The fermented broth is subjected to successive distillation in continuous stills, to obtain ethanol of 95%.

For further purification special distillation technique is required. Alcohol forms azotropic mixture with water. Therefore, to obtain absolute alcohol, the 5% water has to be removed by forming azotropic mixture of benzene, water and alcohol, which then is distilled by increasing temperature. First the azotropic benzene, water-ethanol mixture is removed, followed by ethanol-benzene azotropic mixtures, so that absolute ethanol remains.

**Yield:** The yield at the end of fermentation will be in the range of 6-9% by volume. The common yield from blackstrap molasses amounts to approximately 90 per cent of the theoretical, on the basis of the fermentable sugars and this yield corresponds to 90-98% theoretical conversion substrate sugar to alcohol.

**Final Treatment:** The 190 proof ethyl alcohol (95 per cent by volume) may be further purified, dehydrated, or denatured.

## Production of Alcoholic Beverages

The alcoholic beverages include beer, wine, whiskey, and other fermented beverages are economically important. The origins of bread and beverage marking are lost in antiquity, but it is known that the Sumerians produced a beer consisting of moistened, fermented bread prior to 7,000 B.C. The Egyptians also made this **bread beer**, and regarded it as a holy gift from Osiris, the God of the death. They were also the first to make leavened bread. Both bread making and brewing were later refined by the Greeks and Romans, and numerous changes in the manner of making beer have been made since these early times.

### *Wine*

The fermentation of the juice of grapes, cherries, and berries to produce wine is an old and well-established procedure. The characteristic and distinctive flavours of wine are partially determined by the variety of fruit used, the environmental conditions under which the fruits were grown, the parts of the fruit used, and the ability of the yeast to form flavoured by-products (particularly esters). In this instance the sugar from these juices may be converted to alcohol via the Embden-Meyerhof pathway. Flavours and aromas occur in the wine as a result of the activities of chemical reactions other than sugars that proceed at the same time. The wind industry is basically the same the world over. Some of the differences that are found in wine originate in the type and quality of the grapes that are used. The physical environment of the grape plant, such as amounts of moisture and climatic temperature changes, as well as the general health of the plants, all reflect in the quality of the grape used in wine-making. Some have larger quantities of sugar, while others contain materials that are conducive in producing more desirable aroma, flavour, or other qualities in the wine.

The organisms used in the production of wine are also important. The type of organism used will influence the quality of the wine, and a variety of different species are used, some to produce especially dry or sweet wine. Traditionally, the yeasts used to ferment the juices were those occurring naturally on the surface of the fruit. Recently the trend has been utilize laboratory cultures of strains of

*Saccharomyces cerevisiae* var. *ellipsoideus*. These strains are chosen for their ability to properly ferment the variety of grapes used and for their ability to impart flavours by formation of by-products which are characteristic for a particular wine.

The wine maker grow the yeasts in the winery by first inoculating about 1 to 5 litre of fruit pulp (must) from the laboratory culture slant and then serially transferring the inoculum to successively larger volumes of fresh must until a quantity suitable for bulk fermentation is obtained. The grape juice is treated with bisulfite to kill undesirable yeast and bacteria.

Fermentation is done in vats. They are usually of wood or stone, and vary in capacity from a small keg to one that holds 190,000 litres or more. The vat is covered to maintain anaerobic conditions that encourage fermentation and also discourage growth of acetic acid bacteria which could convert the vine into vinegar. The course of fermentation (particularly the nature of by-products) and the concentration of alcohol produced are influenced by the temperature, sugar concentration, acidity, and tannin content.

Temperature is especially important as the yield of alcohol is higher at the lower temperatures and it also encourages the formation of pleasant flavours. At higher temperatures the alcohol resistance of the yeast is decreased. A favourable temperature for fermentation is in the vicinity of 10 °C.

The sugar content is over 30%, the fermentation is inefficient because alcohol accumulates and arrests the activity of the yeasts before all the sugar is converted to alcohol. A low acid content favours production of acetaldehyde, glycerol, and volatile and fixed acids, which have a detrimental effect on the flavour, colour, and stability of the wine. In addition, a low acid content gives a lower yield of aromatic principles, which give the wine its bouquet.

After the partial fermentation, the racking operation begins and continues throughout the remainder of the fermentation period. In racking, wine is siphoned off from the sediment, which consists of dead yeast cells and other materials, which settle at the bottom of the vat and is transferred to a fresh vat. This procedure is repeated for several times until the desired clarity of the wine is obtained. Before the last racking, a final clarification is performed by adding some substance (like casein, or bentonite) that carries additional sediment to the bottom. Fermentation continues until the alcohol content is about 12%; and above this point the alcohol kills the yeast cells so that no more fermentation takes place. The wine is then stored in completely filled and sealed tanks for aging. Additional blending, fortification with additional alcohol, heat treatments, refrigeration, or filtration may be required before the wine is bottled.

## Beer

Beer is an alcoholic beverage prepared from fermented grains, usually barley. Several different starting materials may be used in the production of beer, but they all achieve the same end- the production of a carbonated alcoholic drink. The top fermenting yeast, *Saccharomyces cerevisiae*, is the most widely used among all the yeasts. Strains should be chosen that are low-temperature-tolerant varieties. In this manner the low temperature favours the growth of the yeast and not bacteria, which may enter as contaminants on the starting materials. Not all beer is made from *Saccharomyces cerevisiae*. For example, *Saccharomyces carlbergensis* and *Saccharomyces monacensis* are used, especially in Holland. These yeasts grow at the bottom and are known as bottom fermenters.

**Step I:** In initial steps of beer production, barley is **malted** through a controlled germination and grains are washed and soaked for 2 to 3 days to stimulate development of the embryo; the water is then drained off and the seeds germinated. The embryo is allowed to grow until the plumule attains a length equal to approximately three-fourths of the kernel length and is then dried to halt growth. During the short period of germination, the embryo produces a number of digestive enzymes, which begin to hydrolyze food reserves in the seed. From the brewer's standpoint, the most important of these are the carbohydrates, which attack starch and sugars, and of these the most important are especially the amylases, which convert the starch into dextrins, maltose, and a little glucose. The maltose is the principal substratum for yeast fermentation.

**Step II:** Second step in beer production is known as **mashing**. The dried malt is ground and mixed with hot water. Most of the enzymatic conversion of starch to maltose by amylases takes place during mashing, which takes about 2 hours. The aqueous extract is separated from insoluble materials and husks, and is then boiled with hops, the female inflorescence of the hop plant containing essential oils and resins, which impart characteristic flavours to the beer.

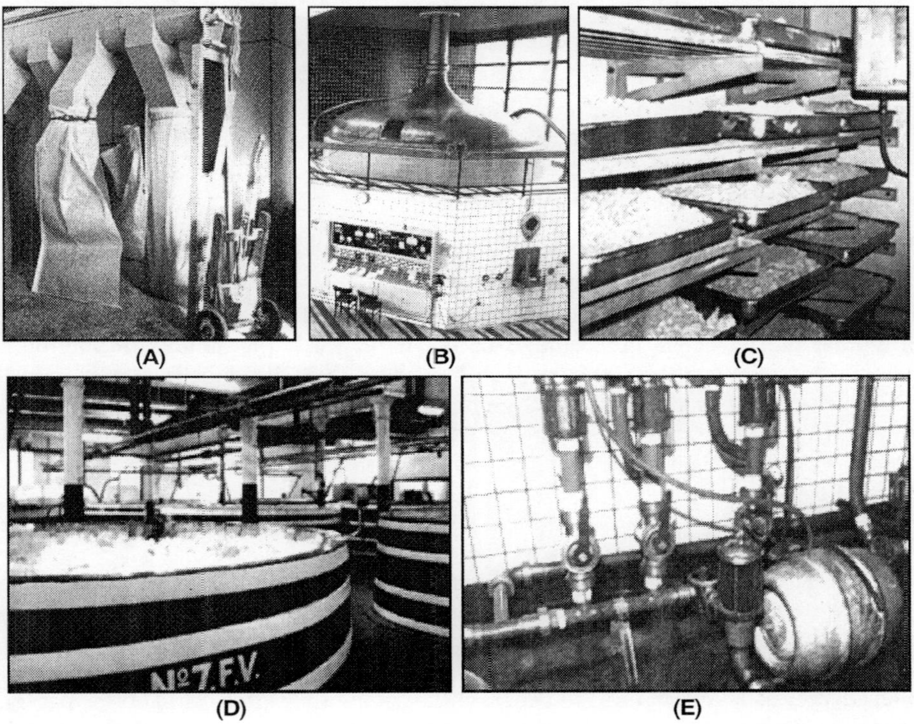


**FIGURE 15.2:** (A to E): Making of beer from barley. (A) It is mixed with warm water to produce an extract called wort. (B) The wart is boiled with hops in a brew kettle. Hopes add flavour to the wort. (C) Trays of drying yeast. Yeast, which ferments the wort, is the other ingredient in beer. (D) Yeast is added to the wort in a fermentation vessel. The froth is caused by bubbles of Carbon dioxide. (E) After filtration and pasteurization, the beer is ready to be put in bottles or barrels.

**Step III:** In this step the aqueous extract is cooled to a temperature between

8°C and 16°C (the temperature is partially determined by the type of beer being produced) and placed in fermentation vats. A selected strain of yeast is added. Traditionally *Saccharomyces carlbergensis* is used to produce larger beer while *S. cerevisiae* is used to produce ales (in the United States, *S. carlbergensis* may be used for both). During a fermentation period of about 5 to 14 days, the sugar is converted to alcohol. The yeasts settle to the bottom of the vat in a flocculent material in the production of the larger type of beer. In contrast, bubbles of carbon dioxide rise to the top of the vat and carry with them the yeast cells and dark flocculent materials from the liquid in the production of the ales. The resultant foamy scum must be periodically removed. When the fermentation period is over, the newly formed beer is allowed to rest for a few days. This period allows the yeast cells to settle out and is then drawn off into weeks. The beer is clarified by addition of gelatinous materials in a manner similar to that in which wines are cleared and may also be filtered through diatomaceous earth. Antioxidants are added since beer changes flavour upon oxidation, and carbon dioxide is added either as the pure gas or by mixing in some freshly fermented beer. The beer is then bottled for sale.

### *Whiskey*

They are made from fermented grains (corn, wheat, barley malt, or rye malt), which are mixed in varying proportions according to the type of whiskey being produced, and fermented by yeasts. The procedures are similar to those involved in beer production, the major exception is that the fermented grain broth is distilled in order to concentrate the alcohol.

### *Miso*

This is an important commodity in Japan, made by fermentation of soybeans. It is sold as pale-brownish paste, from which is made a potage type of soup, serving as a regular morning dish. According to Sakaguchi (1961) the annual production of miso in Japan exceeds 50,000 tons. Cooked soybeans, *koji* made from rice, and 10 to 12 per cent of salt, are ground together along with a little miso from a previous batch. The mixture is packed into closed containers and allowed to ferment at 35 to 45° C, without aeration, for 3 to 5 months. It is then allowed to age from a month or more at room temperature before packaged for sale.

## Amino Acids

Amino acids are produced by biological methods by changing metabolic processes of the cell. These can be obtained by promoting the excretion of the product from the cell. During respiration glucose is converted into pyruvate by the Embden-Meyerhof pathways (glycolysis).

Glycolysis is a term used to describe the sequential series of reactions present in a wide variety of tissues that starts with a hexose number usually glucose and ends with pyruvic acid. The equation of the overall reaction may be written as follows. The reaction takes place in cytoplasm.

$$\underset{\text{Hexose}}{C_6H_{12}O_6} \xrightarrow{?} \underset{\text{Pyruvate}}{2C_3H_4O_3X\ 4H}$$

This is the basic mechanism by which a cell derives energy from glucose.

Fermentation, like glycolysis, is a sequential series of reactions that occurs *in the absence of oxygen.*

In certain species of bacteria, energy metabolism is similar to the processes just described for eukaryotic cells. Anaerobic bacteria such as *Lactobacillus* species (which cause spoilage of milk) ferment glucose to lactic acid by a glycolytic pathway identical to that used in muscle cells. Many aerobic bacteria, such as the spore-forming soil bacterium *Bacillus subtilis*, oxidize glucose to $CO_2$ by a combination of the Embden-Meyerhof pathway and the citric acid cycle. But in two key respects, bacterial energy metabolism is different from that in eukaryotes:

- bacteria lack mitochondria and possess only a plasma membrane; and
- many bacteria are able to grow in unusual environments and do not use glucose or other sugars as a source of energy.

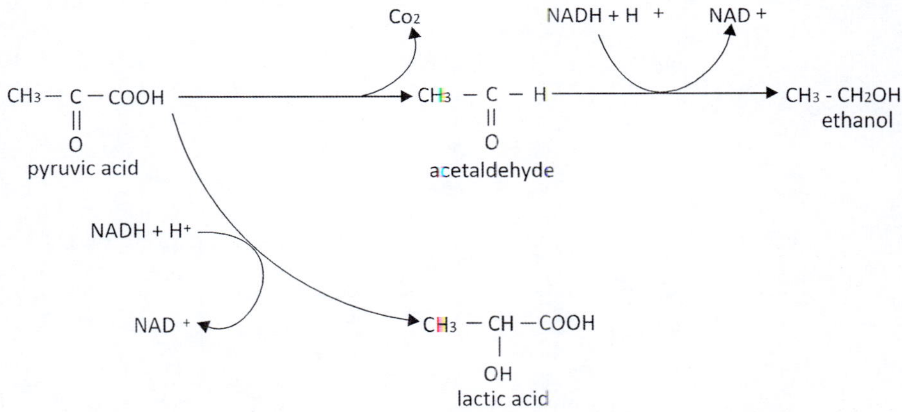


**FIGURE 15.3:** Fermentation of pyruvate to form ethanol or lactic acid. Enzymes involved are (1) carboxylase (2) alcoholdehydrogenase, and (3) lactic acid dehydrogenase.

Although the details of energy metabolism in many bacteria are not well understood, it appears than most, if not all, bacteria utilize a proton-motive force across their plasma membranes. Electron transport chains are often used to generate this proton-motive force, even in certain anaerobic organisms.

The pyruvate then enters the Kreb's cycle (so named because of the cyclic manner in which the starting compound, *oxaloacetate*, is regenerated. It is named after the English biochemist, H.A. Krebs, who played a major role in its discovery) or tricarboxylic acid cycle, in which further oxidation releases more energy. The cycle is short-circuited by a low level of the enzyme alpha-ketoglutaric acid into succinic acid, and by a high level of the enzyme glutamate dehydrogenase, which encourages conversion of $\alpha$-ketoglutaric acid into glutamic acid. At the same time the glyoxylate cycle, is activated to produce more energy. While the cell is growing the glyoxylate cycle competes with the glutamic acid shunt for $\alpha$-ketoglutaric acid. Eventually all the substrate is shunted into glutamic acid.

The excretion of glutamate from a cell depends on many factors, like if the cell is deprived of the vitamin biotin, the membrane develops leaks, allowing more glutamate to pass through. The function of membrane is modified by adding saturated fatty acid or detergent in growth medium. Addition of penicillin to growth medium causes lesions in the cell wall, through which large quantities of glutamic acid can be excreted.

The natural wild isolate of *Corynebacterium glutamicum* excrete glutamic acid into the medium. It is not only deficient in the ability to synthesize biotin but also in has the ability to synthesize α-ketoglutarate dehydrogenase. The mutants of *C. glutamicum* and closely related organisms are suitable for the production of lysine.

This helps in more production of lysine by adding a mutant strain in which regulatory mechanism that ordinary limits the production of an amino acid. In *C. glutamicum* lysine and theonine are both the products of the same sequence of synthetic events, and their simultaneous accumulation inhibits the action of first enzyme in the pathway, aspartate kinase, and hence inhibits further production of lysine.

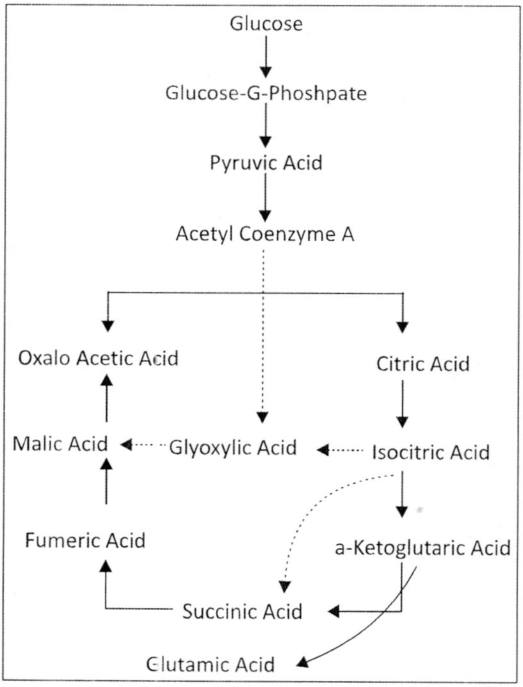


**FIGURE 15.4:** Operation of TCA cycle and glyoxylate cycle in *Corynebacterium glutamicum.*

The mutant strain lacks the enzyme homoserine dehydrogenase, therefore, the metabolic steps leading to the synthesis of threonine are blocked. The mutant bacteria need theonine to grow but it is added slowly so that it and acuminated lysine do not inhibit the further production of lysine.

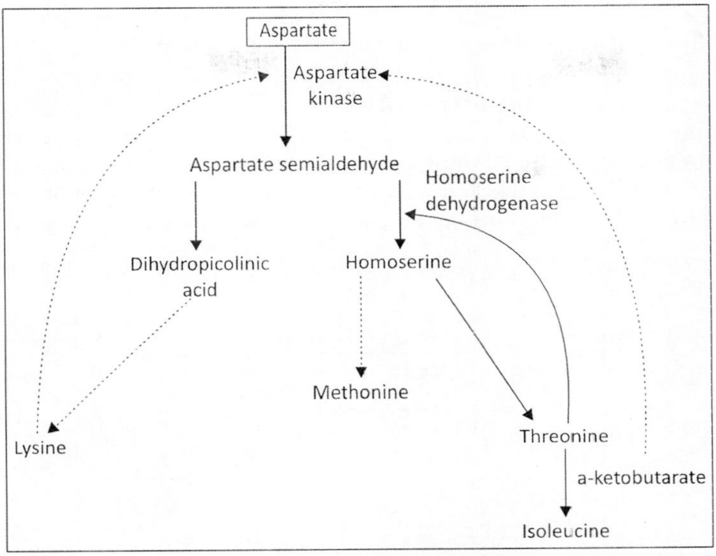


**FIGURE 15.5:** Metabolic pathway for the fermentative production of lysine by *C. glutamicum*-feedback inhibition and its control.

## Glutamic Acid

Glutamic acid is being produced by metabolic activity of various bacteria like, Corynebacterium glutamicum, Brevibacterium flavum, B. lactofermentium, B. divaricatum, B. thiogenitalis, C. allunae, C. herculis, Microbacterium ammoniaphilium etc. They use different kinds of carbohydrates as carbon source. They hydrolyse starch solutions of molasses and beet molasses. If cane molasses contains high level of biotin, then penicillin is added during active growth of the cell. They utilize ammonia, ammonia gas and urea as nitrogen source. In fermentation medium ammonium ions are to be added along with other ions including $K^+$, $Mg^{2+}$, $PO_4^{3-}$, $SO_4^{2-}$, and $Cl^-$. Biotin is added in the fermenting media (added below 5 μg/L). The pH value of the medium is adjusted between 7-8, and optimal temperature range should be 30-35°C.

The *Brevibacterium flavum* utilize acetic acid as a carbon source. The production of glutamic acid depends on the conversion of acetic acid. This gives production up to 98% in 48 hrs.

During fermentation, glucose is oxidized to form citrate and citrate is further oxidized to α-glutarate by TCA cycle.

In process using fed batch culture, a controlled supply of sugar is used as means of overcoming substrate inhibition and of obtaining a high concentration of product in broth, thus reducing product recovery costs.

In order to produce glutamic acid by one step fermentation, initially the fermentation should begin with aerobic oxidation of sugar followed by anaerobic accumulation of α-ketoglutarate, obviously the most efficient pathway of glutamic acid synthesis.

Extraction and purification from concentrated broth is through crystallization of glutamic acid. Ion exchanges resins have been most commonly used for the extraction and purification.

## L-Lysine

Bacteria like *Corynebacterium glutamicum* and *Brevibacterium flavum* are employed for the lysine production. These bacteria are homoserine auxotrophs. The substrate cane molasses is used as carbon source. Other carbon sources include other carbohydrates acetic acid and ethanol. The source of nitrogen includes ammonium salts, ammonium or urea. pH is adjusted at neutrality. For mass production of L-Lysine KL-fermenter is used. The media employed for seed culture and for production has the following composition:

**TABLE 15.2:** Medium employed (seed Culture and for production for *Corynebacterium glutamicum*)

| I Seed culture | | II Seed culture | | Production | |
|---|---|---|---|---|---|
| **Ingredients** | **Qty (%)** | **Ingredients** | **Qty (%)** | **Ingredients** | **Qty (%)** |
| Glucose | 2 | Cane molasses | 5 | Cane molasses | 20 |
| Peptone | 1 | $(NH_4)_2SO_4$ | 2 | Soyabean meal hydrolysate | 1.8 |
| Meat extract | 0.5 | Corn steep liquor | 5 | | |
| Sodium chloride (in tap water) | 0.25 | $CaCO_3$ (in tap water) | 1 | | |

The fermentation carried at 28°C for 6 hrs yield 44 g/L of lysine. The biotin concentration must be greater than 30 μg/L, otherwise L-glutamate accumulates in the medium.

The media employed for *Brevibacterium flavum* yields 75 g/L of lysine utilizing acetic acid and glucose as carbon source.

**TABLE 15.3:** Medium for *Brevibacterium flavum* fermentation for the Production of Lysine

| Ingredients | Quantity (%) |
|---|---|
| Acetic acid | 0.7 |
| $KH_2PO_4$ | 0.2 |
| $MgSO_4.7H_2O$ | 0.04 |
| $FeSO_4.7H_2O$ | 0.001 |
| $MnSO_4. H_2O$ | 0.001 |
| Soyabean protein hydrolysate | 3.5 |
| Glucose | 3.0 |
| Biotin | 50 μg/L |
| Thiamine.HCl | 40 μg |
| pH adjusted | 6.0 |

Fermentation has to be carried out at 33°C with feeding of a solution of acetic acid (60%) until the end of fermentation.

## Citric Acid

Citric acid was originally produced from calcium citrate obtained from cut lemons. Methods for producing citric acid from fungus metabolism were introduced in the United States in 1923. The important organic acid produced by fungi is citric acid which is used commercially as a flavouring ingredient in beverages and foods, especially in dry mixtures such as gelatins, and soft drink powders as tablets, and as the principal acid in the preparation of soft drinks, desserts, jams, jellies, candies, wines, and frozen fruits. About 90 % of the supply comes from *Aspergillus niger* fermentation. The acid is rapidly and almost completely metabolized in the human body and has wide pharmaceutical uses. Specially, its incorporation is effervescent products and as citrates in blood transfusion. Citric acid is also used in astringent lotions to adjust the pH, in hair rinses and hair-setting preparations, in electroplating, in leather tanning, and in reactivating clogged with iron.

**Mechanism of Citric Acid Formation:** Citric acid is one of the principal organic acids produced in the citric (CA) cycle. Recent evidences indicate that the probable origin of citric acid is directly from the CA cycle. When mycelium actively grows, citric acid is produced as an intermediate in the CA cycle and is further diverted to growth-promoting biosynthesis or energy release. For commercial production of citric acid, culture conditions which are inhibitory to growth of the fungus are maintained and the CA cycle itself is inhibited or stopped by enzymatic inhibition of the low pH or by specific enzyme inhibitors, e.g., copper ions. During the production of citric acid, the activity of the condensing enzyme (operating in the condensation of acetyl CoA and oxaloacetic acid to citric) is increased, while the activities of the isocitric acid dehydrogenase and aconitase disappear. The enzyme aconitase is responsible to control the biosynthesis of isocitric acid from citric acid, and in turn isocitric dehydrogenase mediates in the hydrogen removal which yields oxalosuccinic acid from isocitric acid. Therefore, it has been suggested that inactivity of these latter enzymes is the reason for citric acid accumulation.

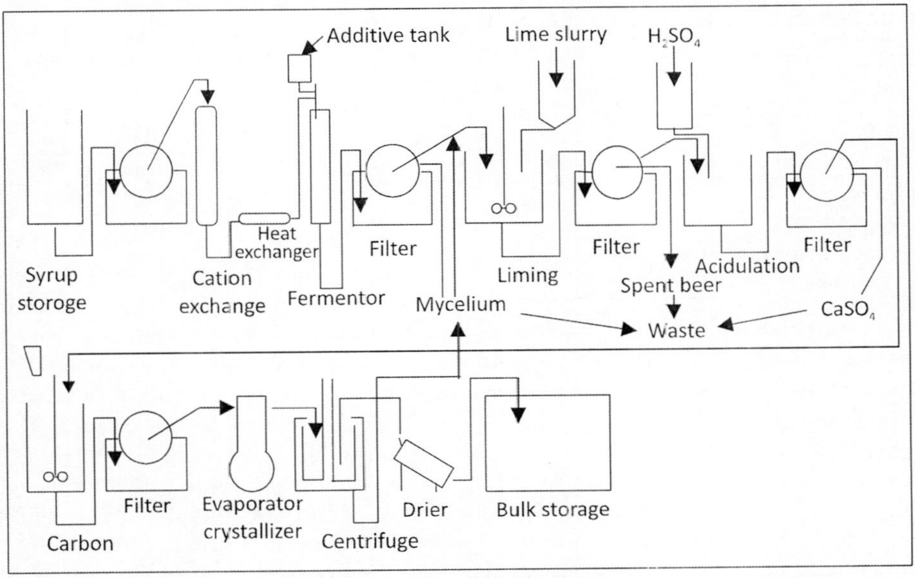


**FIGURE 15.6:** Commercial production of citric acid.

## Commercial Production of Citric Acid

In commercial production of citric acid, *Aspergillus niger* may be cultured on the surface of the shallow pans or as submerged mycelium in aerated vats. Shallow-pan culture has been widely accepted and rapidly added by more efficient vat culture. The requirements and methodology for both the cultures are similar, although in actual plant operation many minute details are different.

For production of citric acid by *A. niger*, the yield is critically determined by the initial sugar concentration (15-18%) and other organic acids produced in the medium. A concentration above 15-18% or lower than this is not only uneconomical, but also amounts to lower yields.

Beet molasses is the least expensive carbohydrate source, and has a high metallic ion content, which may cause low yields of citric acid to be produced. Beet molasses has been widely used in the pan techniques, although pure sucrose or highly refined syrups give higher yields in submerged cultures in vats. Metallic cations are removed from the carbohydrate source by the use of cation-exchange processes and absorbents or (or in the case of beet molasses) by treatment with ferricyanide. The carbohydrate is diluted to a concentration of about 20% to 25%. This high sugar concentration inhibits formation of acids other than citric acid. The manufacture of citric acid needs a low pH, and hydrochloric acid is added to reduce the medium to the range of pH 2 to 5 when the fungal spores in the inoculum germinate. In the vat, the pH is kept below 3.5. Nitrogen is provided by adding ammonium salts. For vat cultures, copper or organic ions may be added as antagonists to iron and aid in the control of mycelial growth which increases citric acid production.

The concentration of other ingredients (i.e., nitrogen, phosphorus, and sulphur in addition to various trace composition of basal medium for citric acid production is presented in following table.

**TABLE 15.4:** Production Media for Citric Acid by *A. niger* Fermentation

| Ingredients | Quantity (%) |
|---|---|
| Total reducing sugars (Molasses, sucrose) | 14-15 |
| Nitrogen source | 0.25 |
| $KH_2PO_4$ | 0.10-0.15 |
| $MgSO_4.7H_2O$ | 0.02-0.025 |
| pH for molasses<br>pH for sucrose | 5.0-6.0<br>20.-3.0 |
| $CuSO_4$ (if molasses is used) | 4.7 mg/L |
| Octadecanol (anti-foam) | 0.75 |

The prepared medium is added with spores from stock cultures. Highly aerobic conditions are needed, and submerged cultures are aerated with sterile air and agitated. Temperature must be maintained in the range of 25° to 30 °C during incubation period (7 to 10 days). Therefore, the harvesting of the citric acid is started. Lime is added to the culture medium to precipitate any oxalic acid, which formed, and the mycelium and calcium oxalate are filtered off. Additional filtration may be required to clarify the medium. A slurry of calcium citrate, which precipitates out. The calcium citrate is filtered out and then treated with sulfuric acid which precipitates the calcium. The dilute citric acid solution is purified by treatment with carbon and is demineralized by successive passage through cation- and anion-exchange resins. This purified citric acid solution is then evaporated, leaving behind citric acid crystals, which are further purified by recrystallization from water.

Unfermented sugar, other acid fermentation products and inorganic impurities are present. In the harvested precipitate the citric acid as calcium citrate from hot neutral aqueous solution, followed by addition of sulphuric acid to remove the calcium as calcium sulphate. Mother liquor is filtered to remove calcium sulphate. Filtrate containing citric acid is treated with charcoal and further concentrated in vacuum at low temperature and crystallized as citric acid monohydrate.

## Lactic Acid

Most of the work on the production of this acid has been done by American investigators. The fungus used is *Rhizopus oryzae* for much better products. The calcium salt of this acid is used as convenient source of calcium for oral administration. It can also be used as a constituent of intraperitoneal dialysis solution. Following a preliminary study of the physiology of the mould, a rapid process was worked out, using a rotary fermenter with forced aeration. The time for the fermentation was reduced to 30-35 hours, and yields of 70-75 per cent were claimed. It is also commercially produced using *Lactobacillus delbruecki* on glucose or lactose. It ferments maltose but not dextrose. Maltose gradually reduced and replaced by dextrose.

The medium for commercial production includes dextrose, 10-15%; $CaCO_3$ 10%; malt sprout in trace quantities (as source for nitrogen and growth factor); and $(NH_4)_2HPO_4$.

Strain of *Lactobacillus delbruecki* (5%) is inoculated in large fermenters at 45°-50 °C temperature at pH 5.5-6.5. At harvest $CaCO_3$ is added to adjust pH up to 10. It is maintained by adding $CaCO_3$. No aeration has to be carried and little agitation is employed to keep $CaCO_3$ in suspension. The fermentation broth is harvested after 5-6 days of incubation. The yield will be in the range of 80-90%.

The broth is heated and then filtered followed by recovery and purification. In one of the procedures, heated and filtered broth is concentrated to allow crystallization of calcium lactate, and acidified with sulphuric acid to remove calcium as calcium sulphate. Lactic acid again recrystallized as calcium lactate and treated with activated carbon to remove colouring impurities. Lactic acid is also extracted by using solvent extraction method with isopropyl ether (from heated and filtered broth). Lactic acid is recovered from isopropyl ether by further counter current washing with water.

## Clinical Dextran

The dextrans are long chain polymers of dextrose (D-glucose) in which the sub-units are joined mainly through $\alpha$ 1:6 glucosidic linkages. The dextran is used as a substrate for plasma. It is commercially produced by *Leuconostoc mesenteroides* and *Leuconostoc dextranicus* (bacteria).

The fermentation medium includes sucrose as a source for carbon, peptones (nitrogen) and yeast extract (source for essential growth factors). The producing organisms hydrolyse sucrose, polymerize the glucose molecule to produce dextran. The fructose so produced is utilized by the organism.

The dextran produced by the fermentation have a polymer of some 2,00,00 glucose units. For clinical use dextran with a relatively homogenous molecular weight of 1,10,000 is used. Therefore, the fermented product is subjected to controlled hydrolysis, followed by fractionation with ethanol or acetone. The end product of dextran has molecular weight above 2,40,000 because this tends to aggregate the erythrocytes in the patient's veins. If the molecular weight is below the prescribed limit, the renal threshold will be excreted rapidly.

The dextran of therapeutically desirable molecular weights should be free from the organic solvents. It should be dissolved in water (6-10%) and should contains 0.9% NaCl or 5% glucose.

# 16. Biotransformation and Steroid Production

Biotransformation (regiospecific and stereo specific bioconversions) is a biological process. In this process the organic compounds are modified into reversible products. These processes are simple but chemically defined. These reactions are catalyzed by enzymes present in microbial, plant and animal cells. The enzymes present in such cells catalyze transformation reactions.

**Microbial cells:** A systematic approach to microbial transformation was initiated by Bertrand (1986). Microbial cells give better results than plant and animals cells. Microbial cells have more surface-volume ratio. They grow fast and have high rate of metabolism. These characteristics help in efficient transformation of the substrate in which they grow. Therefore, growing, resting, or dried microbial cells or enzymes isolated from these cells are added in the substrate for transformation. The transforming agents remain suspended into a medium for use in agitated, aerated fermenter, or packed in columns for continuous operations. They can be immobilized on solid support for increased stability and prolonged operation.

**Plant cells:** The processes of biotransformation are carried out with plant cells for successful production of variety of novel compounds like cardiac glycoside, digoxin, benzyl isoquinoline, etc., involve various types of reactions e.g., hydroxylation of cardenilides. Plant cells can tolerate compounds like steroids, phenolics or alkaloids, etc. Conjugation reactions are the prime aspects of biotransformation carried out by plant cell. These cells act better then microbial cells, if specific enzymatic biosynthetic pathways are needed.

**Animal cells:** Animal cells (cell culture or organ perfusion preparation) are also used in biotransformation of organic compounds. These cells are being used mostly at laboratory-scale.

Out of these cells only use of microbial cells for biotransformations is preferred because they replace several chemical steps, e.g., the 11-hydroxylation of steroids, or when they give abundant access to new complex substrates, which defeat organic manipulation, e.g. the substitution of diosgenin by plant sterols, which are degraded microbiologically into useful intermediates. Similar processes from organic synthesis to biotransformation are more common because steroid bioconversion is applied to particular classes of compounds.

For production of high specific compounds through biotransformation chemical processes are usually preferred, because these processes attack specific site on the substrate and for a single isomer of the product. These reactions are operated at ambient temperature (20-40°C) and pressure (0-34 kPa) in aqueous media. Higher dilutions at which biotransformations are operated create real hazard.

## BIOTRANSFORMATION

Following biotransformation reactions were described by Bertrand (1986).

**TABLE 16.1:** Biotransformation reactions

| | |
|---|---|
| **Oxidation** | **Ethanol to acetic acid**<br>**Glucose to gluconic acid**<br>**Polyhydric alcohols to corresponding sugars**<br>**Isopropanol to acetone**<br>**Glucose to 2-ketogluconic acid**<br>**D-sorbitol to L-sorbose** |
| Hydrolyses | Tannins to gallic acid<br>Di-and trisaccharides to constituent monosaccharides |
| Reduction | Malic acid to succinic acid<br>Fructose to mannitol |
| Resolution of racemic mixtures of | Tartaric acid<br>Lactic acid<br>Glyceric acid |

Present days transformation process employs microbes for the preparation of key intermediates required in organic synthesis i.e., 6-aminopenicillanic acid by microbial deacylation of penicillin G and V, a for commercial production of end products such as steroid hormones.

Many microorganisms specially bacteria are used for biotransformation, e.g., the oxidation of polyhydric alcohols to sugars shows that only polyhydric alcohols with *cis* secondary hydroxyl groups, adjacent to primary alcohol can be oxidized to corresponding sugars by *Acetobacter* species. D-sorbitol is transformed into L-sorbose using *Acetobacter suboxydans*, and glycerol to dihydroxy-acetone, using *A. suboxydans or Gluconobacter melanogenus*.

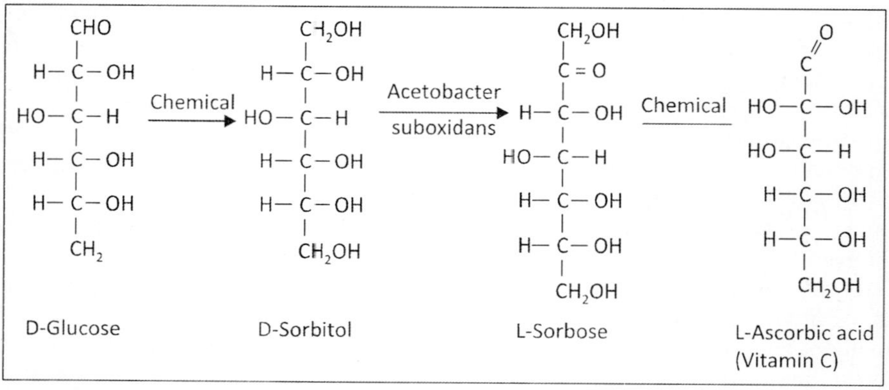


**FIGURE 16.1:** Manufacture of vitamin C. Presentation of transformation of sorbitol to sorbose

## METHODS USED IN BIOTRANSFORMATION PROCESS

Microbial cells serve as a major tool for biotransformation. The prime process is fermentation. The microbial conversion of steroids can also be used for biotransformation of other classes organic compounds.

## Steroid Bioconversion Fermentation

Steroids are a group of organic compounds which have the four membered ring. They are biologically active, similar to hormones, produced by the testies, ovaries, adrenal cortex, and placenta. The steroids differ in the nature of their side groups or side chains, and these differences in structure confer different biological properties on the steroids. Steroids are widely used medically as anti-inflammatory agents, anesthetics, antifertility agents, and in the treatment of sterility.

Steroids are obtained directly from natural sources or can be synthesized. The steroid nucleus produced chemically after many transformations or additions in side chains. Microorganisms help in biotransformations of steroids are as follows:

1. *Rhizopus arrhizus* (fungus) hydroxylates progesterone forming another steroid, 11-α-hydroxyprogesterone by introducing oxygen at position 11.
2. *Cunninghamella blakesleeana* (fungus) hydroxylates cortexolene to form hydroxygen, introduction of oxygen at the number 11 position.
3. *Rhizopus nigricans* (fungus) can also hydroxylate progesterone to produce 11-α-hydroxyprogesterone.
4. *Corynebacterium simplex* dehydrogenates cortisone to produce prednisone.
5. *Corynebacterium simplex* can also bring about dehydrogenation of hydrocortisone or cortisol to produce prednisolone.
6. *Nocardia restrictus* biotransforms $\delta^4$-cholestene-19-hydroxy-3-one into estrone.
7. Androstenodione is converted into testosterone by yeast.

During a typical steroid transformation process, the microorganism, such as *Rhizopus nigricans*, is grown in a fermentation tank, using an appropriate growth medium and incubation conditions to achieve high biomass. In most cases, aeration and agitation are employed to achieve rapid growth. Therefore, steroid (e.g., progesterone) is added to a fermenter containing *R. nigricans*. The product is then recovered by extraction with methylene chloride or various other solvents; purified chromatographically, and recovered by crystallization.

Bioconversion of steroid is performed by using submerged aeration technique in stainless steel tanks with minimal nutritional quantities to allow maximum use of extraction and purification of the transformation product.

The fermentation is carried out in two following phases:

### *Phase - I*

It is the **growth phase** in which maximum growth of culture is obtained. The medium for the growth is simple and of sucrose minimal medium or of complex type. For culture optimum temperature and aeration conditions are needed. The time of incubation period depends on the type of culture (for bacteria, 12-24 hrs; and for fungi, 24-72 hrs) and environmental conditions.

### *Phase - II*

It begins with the addition of steroid at the end of growth phase. Sometimes, the addition may be carried simultaneously with the inoculation. The amount of steroid to be added is variable. It depends on the transforming capacity of the culture, potential toxicity of substrate or type product e.g., deoxycorticosterone have significant antifungal activity. The toxicity problem of the medium is controlled by periodic or continuous addition of substrate in the processing unit. Acetone, ethanol and methanol generally are used for fair solubility of steroids and relatively low toxicity for the conversion enzymes. Dimethylformamide has exceptional steroid solubility properties. It is relatively non-toxic below levels of 2% in the medium. The conversion capacity depends on the charge of the substrate and conversion capacity of the culture. It needs 12 to 72 hrs.

## Analysis and Isolation of Product

For analysis of the product, samples are obtained at regular intervals. These samples are analyzed by using TLC, paper chromatography, gas chromatography or HPLC techniques. It is being done up to the complete course of transformation (1-5 days). The structure of transformed product is elucidated by chemical methods.

The extraction of steroids from broth isolate requires appropriate organic solvents, i.e. methylene chloride, chloroform, ethyl acetate, and methyl isobutylketone. Formation of emultion by broth normally contains transformation products. Products obtained from cell or substrate should be extracted separately.

Successful isolation of product dependents on yield and the similarity of produced product during side reactions. In efficient conversions, the product can generally be crystallized directly from a concentrate of the extract.

## Selection of Organisms

Microorganisms are very convenient enzyme sources, because the enzyme concentrations may be markedly increased by acting on environmental conditions and genetic manipulation. They also have rapid growth rate on inexpensive media. The selection of suitable microbial strain as a source of enzyme is dependent on the fact, that the organism must give high yields of enzyme in shortest possible fermentation time. Extracellular enzymes are usually preferred, because their isolation is relatively simple and expensive processes for disintegrating the cells, can be avoided. The enzyme producing strain should not produce any toxic substances and antibiotics on the medium.

Selection of strains may be done from natural sources, or from that available cultures which should have characters to catalyze desired reaction. Certain microbes perform the desired reactions with a related compound. The plate assay method is used to select organisms that can aromatize 19-nor and 19-substituted steroids and sterols into equilin and related estrogen. These ring A aromatic products reacts specifically with para-nitrobenzene diazonium fluoborate. It develops intense red colour. Colonies developed in solid medium contain suitable steroid substrates which are replicated before the reagent is sprayed. A red ring develops around active colonies.

The selection of cells depends on the following parameters:

1. They should degrade the substrate on which they will grow.
2. They should have the capability to utilize glucose as carbon source.
3. Mutants strain should have the property to degrade steroid substrate at various reaction levels, but can utilize glucose as the carbon source.
4. Mutants strain should be selected which should not accumulate undesirable compounds i.e., *Aspergillus ochraceus*. It blocks the transformations of 11 α-hydroxyprogesterone into 6β, 11α-Di hydroxyprogesterone (undesirable

reaction).

Microbial strains of eubacteria, streptomycetes, yeasts and molds are used for transformation. They can be maintained as per the standard methods (i.e. agar slant, soil cultures, lyophilized culture or frozen cultures at -20°C to –170 °C).

## BIOTRANSFORMATION PROCESS WITH SPECIAL REFERENCE TO STEROIDS

The enzymes present in microbial cells promote reactions at unactivated centers in hydrocarbons. Therefore, any site in the carbon skeleton of a steroid nucleus can be hydroxylated stereo-specifically by microbes.

The other transformation of steroid nucleus by microorganisms include hydrogenations, dehydrogenations epoxidations, and side chain cleavage of –$CH_2OH$=O function of the nucleus.

Deoxycholate is the starting material, but very expensive. The inexpensive sterols sources are corticosteroid cortisone, prednisone, prednisolone, hydrocortisone, dexamethasone, androgen testosterone, estrogen estrodiol etc. Sterols like stigmasterols and sitosterols are obtained as by-products during production of soybean oil. They are produced in vast quantities. Diosgenin is obtained from the roots of Mexican barbisco plant. Sterols are obtained from these plants by removing the side chain beyond $C_{21}$. This can be done by using Mycobacterial species which utilize sterols as carbon and energy source. They bring out degradation in this step. It is economic one.

The steroid nucleus has three **primary** carbon atoms ($C_{18}$, $C_{19}$ and $C_{21}$) and eighteen **secondary** carbon atoms. At each secondary carbon atoms, within the nucleus, there are two sits $\alpha$ or $\beta$- attach –OH group, that can be hydroxylated in either $\alpha$- or $\beta$- configuration. Following are the reactions:

These *reactions* were discussed in brief as follows.

### Hydroxylation

The substrate has specificity of hydroxylase system. This system has advantage for hydroxylation processes which allows the selection of optimal intermediates for hydroxylation in sequence of chemical and microbial steps to obtain desired end products. Microbial hydroxylation is the involvement of direct replacement of the hydrogen atom on a given carbon. The oxygen atom in the hydroxyl group (-OH) is derived from molecular oxygen (gaseous), not from water. The hydroxyl group thus formed always retains the stereo chemical configuration of the hydrogen atom that has been replaced.

The 11 $\alpha$-hydroxylation is high-yielding transformation process. It covers a wide range of types of cultures (*Rhizopus* and *Aspergillus*) and substrates transformed. The starting material is progesterone which is derived from diosgenin and stigmasterol-abundant plant steroids.

The 11 $\alpha$-hydroxylation of progesterone is carried out with *Rhizopus nigricans* which yield more than 85%. *Aspergillus ochraceus* (fungi) is also used for 11$\alpha$-hydroxylate. The conidia of this fungus produce 11 $\alpha$-hydroxy derivative. The 11 $\beta$-hydroxylated into hydrocortisone (yield 60-70%) using *C. lunata*.

There is a slow transformation of steroids without the 3-keto- $\Delta 4$ conjugation. Acylated substrates are hydroxylated as efficiently as the corresponding alcohols, but are often hydrolyzed before they are hydroxylated with *Aspergillus ochraceus*.

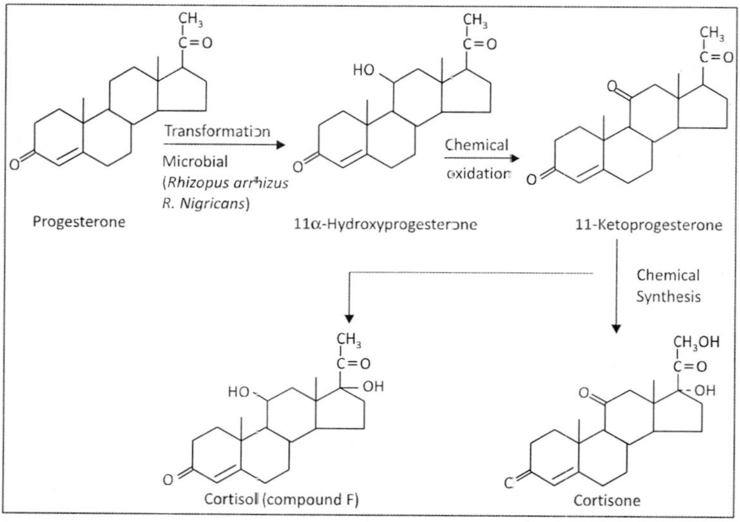


**FIGURE 16.2:** Transformation of progesterone to cortisone and cortisol.

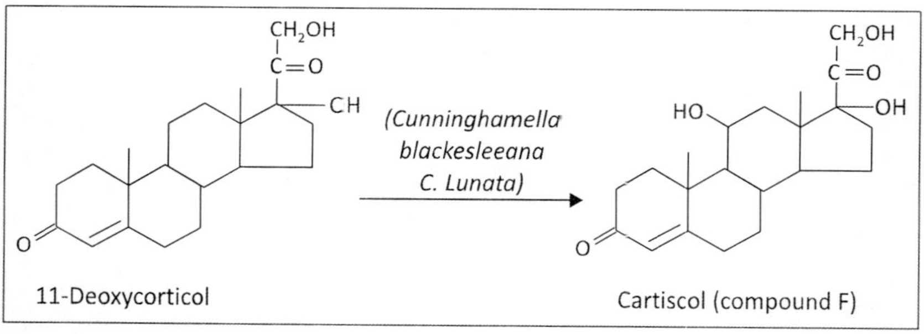


**FIGURE 16.3:** Conversion of 11-deoxycorticol to cortisol (F compound).

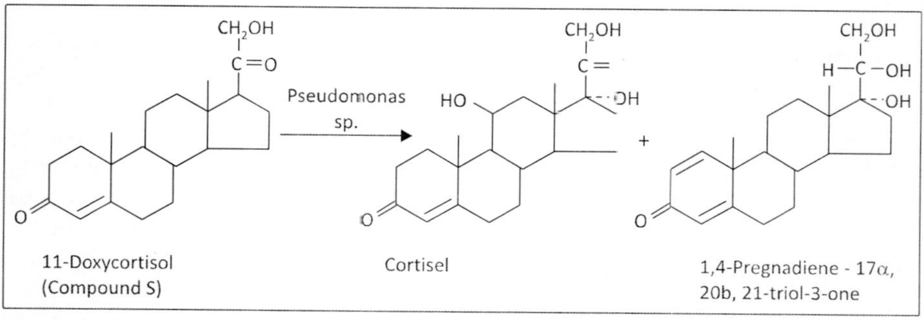


**FIGURE 16.4:** Conversion of 11-deoxycortisol (compound S) into cortisol and 1,4-pergnadiene-17 V, 20 β, 21-triol-3-one.

## Dehydrogenation

Bacteria and fungi are capable of dehydrogenising the secondary alcohol groups of steroids. They produce the corresponding carbon derivative.

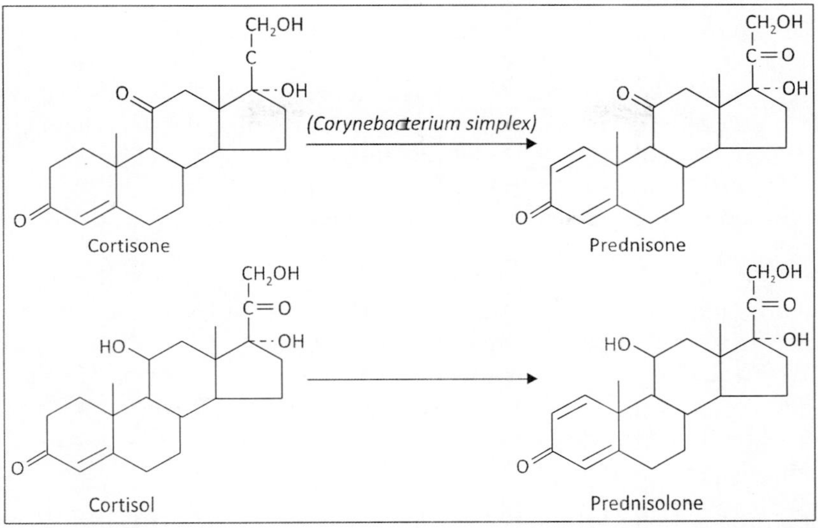


**FIGURE 16.5:** Conversion of cortisone to prednisone in presence of *Corynebacterium simplex* which is converted into cortisol to prednisolone.

C-1-Dehydrogenation is the introduction of a double bond. The degradation of the side chain is carried out by *Cylindrocarpon radicicola, Streptomyces lavendulae, S. tradiae* and by *Fusarium solani* and *F. caucasicum.* At present, microorganisms like *Septomyxa affinis* and *Arthrobacter* (*Corynebacterium*) *simplex* are preferred. These organisms give higher yields of C-1.dehydrogenated and smaller amount of side-chain cleavage products. C-1-Dehydorgenases are also inducible enzymes. It is an anaerobic process (oxygen is not needed), but requires hydrogen acceptor (phenazine methosulphate). Phenazine methosulphate is added to the cell free extracts for maximal activity. In *S. affinis* conidia, C-1-dehydrogenase is not inducible, but reaction is repressed or inhibited by glucose.

## Hydrogenation

Ring hydrogenation includes hydrogenations of $\Delta^1$, $\Delta^4$ and $\Delta^{16}$, $\Delta^1$. The process involves the conversion of prednisone to cortisone, and prednisolone to cortisol, with *Bacillus megatherium.*

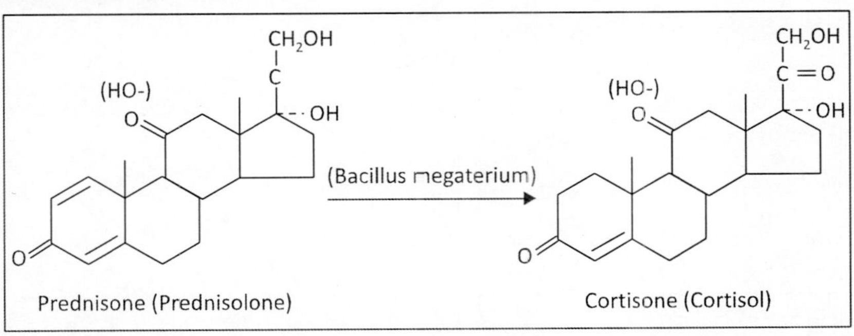


**FIGURE 16.6:** Conversion of prednisone (prednisolone) into cortisone (cortisol).

The conversion of 4-androstane-3,17-dione to androstane-3,17-dions and eiocholane-3, and 17-dione with *Bacillus putrificus* is reported for $\Delta^4$-hydrogenation. $\Delta^{16}$-hydrogenation involves the transformation of $\Delta^{16}$-progesterone 1 by *Rhizopus nigaricus* to 11 α-hydroxy-17 α-progesterone. The microbial reduction of carbonyl groups has been reported for carbonyls at 3,7,17,20 and 22. The reduction of 3-carbonyls results in the production of either 3 α-3β-hydroxyls, while reduction 17- and 20-carbonyl produce only β-configuration.

## Epoxidation

The epoxidation of steroidal double bonds is not common during biological epoxidation. The 9β,11-epoxidation of 9(11)-dehydro-compounds S (I), and the 14α,15-epoxidation of 4(15)-dehydro-compounds S (II), is done by using *Curvalaria lunata* and *Cunninghamella blakesleeana*. These culture normally introduce the 11β-hydroxyl and 14α-hydroxyl into compound S.

## Aromatization of Ring A

C-1-dehydrogenation with substrate lacking methyl group at carbon 10 (19-nor) or suitably substituted at carbon 19 [19-hydroxy, 19(Oxo)] results into ring A aromatization. The C-1-dehydro-derivative formed is unstable. It undergoes spontaneous non-enzymatic rearrangement to ring A aromatic products, e.g., estrone. The side chain of pregnanes and sterols is cleaved by microorganisms. They carry out this dehydrogenation. The 19-hydroxycholesterol and 19-hydroxy-β-sitosterol are converted (by aromatization and side chain cleavage) to estrone with *Nocardia restricta*. The side chain degraded by a stepwise β-oxidation to the 17-keto compound.

With substrate having a 19-methyl group, the steroid nucleus is degraded by *N. restricta* and the other C-1-dehydrogenators through the introduction of a double bond at C1 and a hydroxyl group at 9α. The 9α-hydroxylase reaction is blocked by chelating agents like divalent cholesterol, β-sitosterol, stigmasterol and campesterol into androstadienedione. It is the intermediate for the synthesis of estrogens (estrone).

Biotransformation reactions have been tested in thousands of steroid, sterols and related compounds.

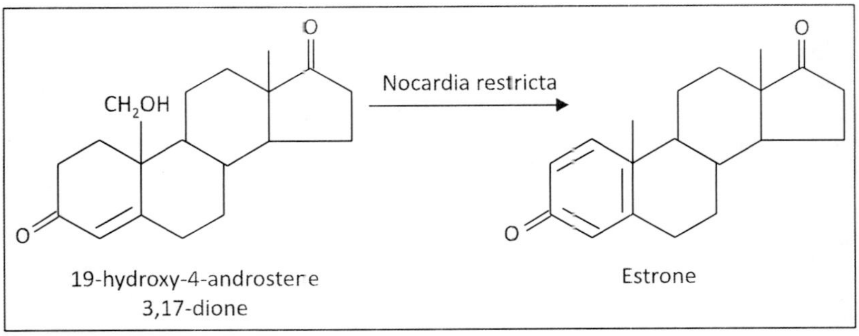


**FIGURE 16.7:** Conversion of 19-hydroxy-4-androstene 3,17-dione in to estrone.

## Synthetic Routes

Progesterone, obtained by the chemical conversion of stigmasterol is transformed by *Rhizopus nigaricus* into 11α-hydroxyprogesterone, which is chemically modified to hydrocortisone and cortisone, or dehydrogenated at C-1 (*S. affinis*), after chemical modification, to prednisolone and prednisone. Alternatively, diosgenin is chemically converted to progesterone or Reich stein's compound S. Compound S is 11β-hydroxylated (*C. lunata*) into hydrocortisone, which is then C-1-dehydrogenataed into prednisolone (*A. simplex*); also cortisone can be C-1 dehydrogenated into prednisone.

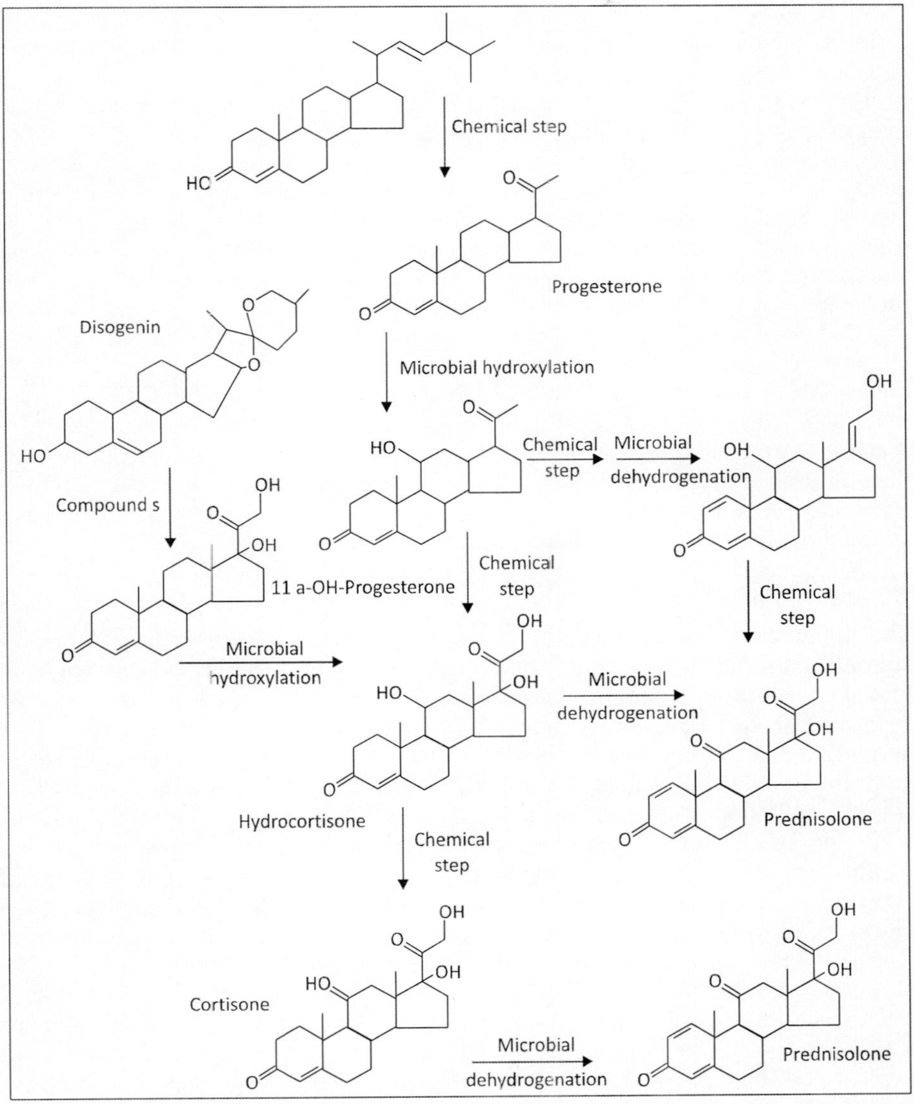


**Fig 16.8:** Production of therapeutical useful steroids. Steps show chemical and microbial transformation.

# 17. Biology of Enzymes

The existence and power of enzymes were first recognized in the nineteenth century when a reaction that has been considered to occur only in the presence of cells and were found also to be mediated by cell-free extracts. It was the Buchner brothers who showed that yeast, a living organism, yielded a non-living cell-free extract capable of fermenting sugars. Subsequent investigators of this field showed that this was due to the presence in the extract of many powerful catalysts, each one is highly specific for one of the steps in the breakdown of glucose. Similar catalysts, or enzymes, are found in all living things. These are organic catalyst found within a living organism which alter the fate of a chemical change, but itself remains unchanged at the end of the reaction; and it is produced by the living organisms but is not itself alive.

Enzymes are catalysts, ubiquitous in nature, responsible for thousands of chemical reactions involved in biological processes of living systems. These are thermolabile proteins produced by living cells catalyzing specific chemical reactions. They are highly specific in their action, promoting only one biochemical reaction with their substrate ensuring synthesis of a bimolecular product under the conditions of mild pressure and temperature and in aqueous solutions. Enzymes are complex organic catalysts, produced by living cells but capable of acting independently of the cells that produce them.

## ENZYMES: THE TAILOR'S SCISSORS

The applications of enzymes (biological catalysts) that can change plants and animals in precise and often remarkably dramatic fashion. In the hands of knowledgeable biotechnologists, *enzymes become the tailor's scissors and the surgeon's scalpel.* There is a strong dependence of advances in enzymology on agriculture. Discovery of enzymes and elucidation of many of their properties were by agricultural chemists. The earliest enzymes studied were primarily from agriculturally important animals, plants and microorganisms. The fast developing wide field of biochemical engineering/biotechnology, and its tremendous promise for the future, have brought basic and applied scientists together in private sector companies, at universities, and in government and industrial laboratories. There are many reasons for the use of microbes in enzyme production. The important characteristics that have made microbes useful in enzyme industries are: (1) their ability to grow on easily available and cheap raw materials, (2) their ability to maintain a physiological constancy, (3) their ability to bring about biochemical transformations under simple culture conditions. Microbes have tremendous capacity of carrying out a variety of reactions (especially secondary metabolism resulting in an inexhaustible supply of secondary metabolites available for commercial exploitation) and (4) high ratio of surface area to volume, which facilitates the rapid uptake of nutrients required to support high rates of metabolism and biosynthesis. These properties of microorganisms are highly preferable over synthetic processes. Presently, industrial microbiology has established as a strong arm of applied microbiology and it is unlikely that, at least in some industries, the microbes could be replaced by synthetic techniques in the near future.

**Enzyme** (Gr. *En* = **in** *zyma* = yeast): Any substance, protein in whole or in part, that regulates the rate of a specific biochemical reaction in living organisms. It is capable of catalyzing a reaction in which substrate(s) are converted to product(s) through the formation of an intermediate enzyme - substrate complex. As with other catalysts, enzymes are responsible for accelerating the rate of a chemical reaction.

**Endo Enzyme.** The enzymes, which act within the cell in which they are synthesized inner core of or substrate molecule, are known as endoenzyme.

**Exoenzyme.** The enzymes, which act outside the cell in which they are synthesized or at the outer ends of substrate molecule, are termed exoenzyme.

The chemical nature of enzymes remained in dispute for a long time. Willstatter, working with peroxidase, had chosen an enzyme of such high catalytic efficiency that he believed that his active preparations were protein-free. By contrast, Sumner's crystalline urease was of such relatively modest activity that his critics attributed the catalysis to a highly active trace contaminant rather than the purified protein itself. Improvements in fractionation procedures have since allowed the purification of many hundreds of enzymes from diverse sources, leading to the realization that all enzymes are proteins. On the basis of chemical composition enzymes are of two types:

1. **Purely proteinaceous enzyme:** These are made up of only proteins e.g., proteases that split protein and amylase that split starch.
2. **Conjugate enzymes:** These are made up of protein molecule, to which a non-protein group (*prosthetic group* or cofactor) is also attached e.g., oxidizing enzymes.

It is essential to understand the difference between the following important terms - substrates, prosthetic group, apoenzyme, coenzymes and cofactor.

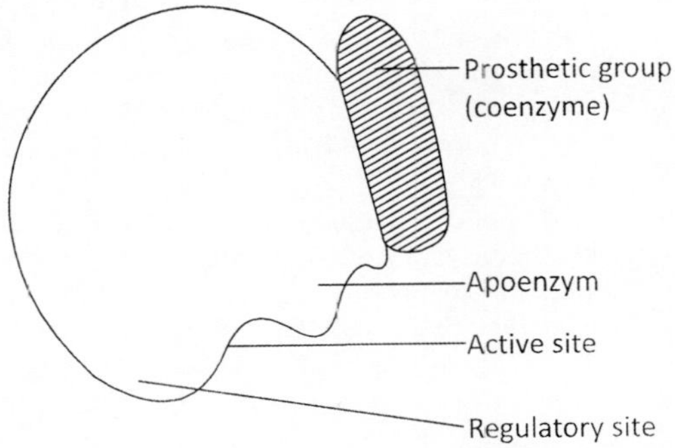


**FIGURE 17.1:** Hypothetical morphology of an enzyme.

**Substrate:** The substance upon which an enzyme act is known as substrate. They are produced by the living protoplast of a cell.

**Prosthetic group:** It is the non-protein group of enzyme. The prosthetic group is firmly bound to protein component of the enzyme by chemical bonds and is not removed by dialysis. Haem, biotin and pyridoxal phosphate, like flavin, usually function as prosthetic groups.

**Apoenzyme:** If, however, the prosthetic group is removed, the remaining protein part of the enzyme is called as apoenzyme and it is inert.

**Coenzyme:** It is the co-worker of prosthetic group and apoenzyme because neither the apoenzyme nor the prosthetic group alone is enzymatically active. The prosthetic group, the coenzymes, as the name specifics, act as co-worker of an apoenzyme. They combine with the enzyme and leave it again in the course of a single catalytic cycle. The coenzymes are vitamins, DPN, NAD, tetrahydrofolate, Coenzyme A and adenosine phosphates, like $NAD(P)^+$.

**TABLE 17.1:** Areas of industrial use of microorganisms, and of application of major enzymes from microbes (main microbial species)

| Industrial uses | Microorganisms used |
|---|---|
| Alcoholic Fermentation | Bear (*Saccharomyces cerevisiae, S. carlsbergensis, S.uvarum*), cider (*S. cidri*), wine (*S. cerevisiae*), sake (*Aspergillus oryzae,* Lactobacillus and *Leuconostoc* spp., *S. cerevisiae*) and other fermentations of fruit juices, distilled sprits, etc. |
| Amino acids, mononucleotides | *Corynebacterium glutamicum.* |
| Baker's yeast | *Saccharomyces cerevisiae, Candida milleri.* |
| Carotenoids | β-carotene: *Blakeslea trispora*; astaxanthin: *Phaffia rhodozyma.* |
| Dextrans | *Leuconostoc mesenteroides*, levans, mannans; xanthans - *Xanthomonas campestris.* |
| Ethanol | *Kluyveromyces fragilis, S. cerevisiae, Zymomonas mobilis*, n-butanol and acetone (*Clostridium acetobutylicum, C. saccharoacetobutylicum*); acetic acid (*Acetobacterium woodii, Clostridium aceticum*) citric acid (*Aspergillus niger, Saccharomycopsis lipolytica*). |
| Fodder yeast | *Candida utilis, Saccharomycopsis lipolytica* |
| Fumaric and lactic acids | *Lactobacillus delbrueckii* |
| Gibberellins | *Gibberella fujikuroi* and other plant growth hormones |
| Production of antibiotics | Penicillins (*Penicillium chrysogenum*), cephalosporins (*Cephalosporium acremonium*), streptomycin, kanamycins, neomycins, tetracyclines, etc. (*Strptomyces* spp.), gramicidin-S (*Bacillus brevis*), polymyxin-B (*Bacillus polymyxa*), bacitracin (*Bacillus subtilis*) |
| Production of Fermented Beverages and Foods | Dried bonito, soyasauce (*Aspergillus oryzae, Pediococcus soyae, Saccharomyces rouxii, Torulopsis* spp.), vinegar (*Gluconobacter suboxidans*), pickled vegetables, cheeses (*Penicillium camembertii, P. roqueforti. Propionibacterium shermanii, Streptococcus* spp.), yoghurt (*Lactobacillus bulgaricus, Streptococcus thermophilus*), lactic acid (sour) drinks. |
| Single-cell proteins | Spirulinas, Chlorellas, and other unicellular algae; (*Methylophilus methylotrophus, Candida tropicalis, C. utilis, Saccharomycopsis lipolytica*). |
| Steroids | *Arthrobacter simplex, Mycobacterium sp., Rhizopus arrhizus, R. nigricans.* |
| Vaccine and microbial bioinsecticides | *Bacillus popilliae, B. thuringiensis* |
| Vitamins | Rriboflavin: *Eremothecium ashbyi*; Vitamin $B_{12}$: *Propionibacterium* sp., Pseudomonas denitrificans ). |
| Abatement of pollution | Treatment of industrial effluents, waste waters, disposal of wastes |

| Industrial uses | Microorganisms used |
|---|---|
| | and garbage; recovery and reuse of biodegradable wastes. Microbial leaching of ores and recovery of mining and colliery wastes. |

**Cofactor:** When a prosthetic group consists of single atom of some metal like $Mg^{++}$, $Fe^{++}$, $Cu^{++}$, $Mo^{+++}$, $Zn^{++}$, then it is known as cofactor and can be easily separated from rest of the protein part. Many enzymes employ either metal ion cofactors or organic cofactors. Thus many oxidoreductases utilize haem, nicotinamide or flavin; the transferases use folate, coenzyme A, pyridoxal phosphate, thiamine and adenosine phosphates; some of the ligases use biotin etc. Among these cofactors, some may be regarded as integral parts of their enzymes; they are known as *prosthetic* groups.

There are about 3000 enzymes known and only few are manufactured at large scale. These enzymes are mainly extracellular hydrolytic enzymes which have the property to degrade naturally occurring polymers for example starch, proteins, pectins, and cellulose. Enzyme glucose isomerase is, however, an exception in this regard.

Enzymes are the directing and controlling biocatalysts present in all living matter, which determine a particular chemical reaction. They accelerate the course of chemical reactions by several orders through a substantial decrease in the activation energy. The catalytic activity of the enzyme secreted by the cells or isolated from them is maintained under suitable conditions and permits the use of these enzymes as catalysts outside the cells.

1. Mutants that produce the enzyme constitutively, that is without an inductor;
2. The production of the desired enzyme in the mutants is not inhibited by repressors; and
3. The yield of enzyme is considerably increased by multiplication of the gene copies, and/or changes in the single sequences preceding the structural gene to bring about high expression.

## PROPERTIES OF ENZYMES

Enzymes are characterized by many biophysical properties. Some common properties are given below:

1. **Proteinaceous nature:** Enzymes are made up of protein. They may be either purely made up of protein or associated with a non-protein part.
2. **Catalytic properties:** Enzymes act as catalysts and influence the speed of chemical reaction but themselves remain unchanged. A small quantity of enzyme can catalyze the transformation of a very large quantity of the substrate into end product e.g. the sucrase can hydrolyse 100,000 times of sucrose as compared with its own weight.
3. **Specificity of enzyme action:** The ability of an enzyme to catalyze one specific reaction and essentially no other is perhaps its most significant property. Close examination reveals that most enzymes can catalyze the same type of reaction.
4. **Reversibility of enzyme action:** Most of the reactions catalyzed by enzymes are reversible and the enzyme can catalyze the reactions in both directions e.g. the lipase can catalyze not only the hydrolysis of fats into fatty acids and glycerol but also can synthesize fats from fatty acids and glycerol as shown

below.

$$\text{Fats} \xrightarrow{\text{Lipase}} \text{Fatty acids + glycerol}$$

5. **Sensitivity of the enzymes:** The enzymes are sensitive to rise of temperature and are destroyed at higher temperature of 60-70° C. However, most of the enzymes act best between 20°C to 35° C. At 0° C or below, they are not destroyed but become inactivated.

## ENZYMES: NOMENCLATURE

A commission has been established for deciding the name of enzyme and is known as **Enzyme nomenclature commission.** The terminology generally agreed for use in relation to enzymes. As in the example of urease, enzyme and classes of enzymes usually are named by appending the suffix *-ase* to the name or abbreviated name of a compound on which the enzyme acts. Enzymes that cause the cleavage of the polypeptide chain by reaction with water are referred to, as a class, as proteinase. Naming of enzymes is known as nomenclature. There are various ways of naming enzymes. The principles of nomenclature are as follows:

- The enzymes are named by adding the *suffix-ase* to the name of the substrate on which they act e.g. Proteinase, sucrase, nucleases, which break up proteins, sucrose and nucleic acids respectively.
- The enzymes can be named according to the type of *function* they are performing e.g. dehydrogenases remove hydrogen, carboxylase help in adding $CO_2$, decarboxylases help in removal of $CO_2$, oxidases helping in oxidation and so on.
- The enzymes are sometimes given *double name*, one after the nature of substrate upon which they act and second according to their function e.g. pyruvic decarboxylases catalyses the removal of $CO_2$ from pyruvic acid, alcoholic dehydrogenase catalyses the removal of hydrogen from alcohol.

## ENZYME CLASSIFICATION

A system of rules by which enzymes are classified on the basis of the substrate they react with and the type of reaction catalyzed. This system provides both a systematic name and a four-part number code. The first number of the code places the enzyme into one of six groups indicating the type of reaction involved. The next two numbers indicate the groups involved in the reaction and the fourth number provides the absolute identification of the enzyme. The enzymes have been classified into following types:

1. Hydrolysing enzymes, 2. Oxidation-reduction enzymes, 3. Ligases, 4. Group transfer enzymes, 5. Desmolases, 6. Isomerizing enzymes, 7. Carboxylation enzymes.

### *Hydrolysing Enzymes*

They catalyse the hydrolysis of complex big molecules into simple, smaller molecules. Depending upon the nature of food substance upon which the enzyme act, they can be further classified into following types.

1. **Carbohydrases:** They hydrolyse the complex polysaccharides into simpler monosaccharides e.g. sucrases, maltases, lactase, cellulase etc.
2. **Esterases:** They hydrolyse the substances containing ester linkage e.g. lipase,

phosphatase.

3. **Proteolytic enzymes:** They hydrolyse the proteins into peptones, and amino acids e.g. pepsin, peptidases.
4. **Amidases:** They hydrolyse amides into ammonia and acids e.g. urease.
5. **Oxidation-Reduction Enzyme:** They catalyse oxidation and reduction reactions e.g. Oxidases, reductases, catalases, dehydrogenases.
6. **Ligases:** They bring linkage between two molecules with the simultaneous breakdown of ATP molecule which supplies energy e.g. synthetases which join amino acids to RNA.
7. **Group transfer enzymes:** The catalyse the transformation of a group from one kind of molecule to another or from one position of molecule to another position in the same molecule and are called as transferases.
8. **Desmolases:** They catalyse the reactions in which long carbon chain is broken or lengthened e.g. aldolase.
9. **Isomerizing enzymes:** They catalyse reactions in which an organic molecule is transformed into its isomeric form and are known as isomerases.
10. **Carboxylation enzymes:** They catalyse the reactions in which $CO_2$ is added or removed and are known as carboxylases or decarboxylases.

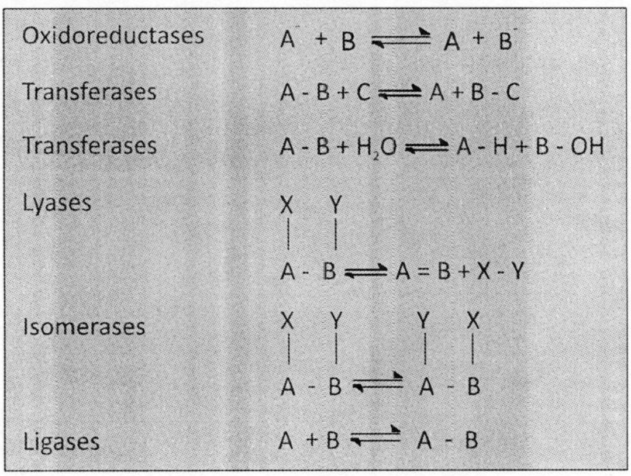


**FIGURE 17.2:** Main classes of enzymes with their reactions.

## ENZYMES AS CHEMO-THERMO REGULATORS

The chemical reactions that take place within the living cell must be precisely controlled and coordinated. These reactions interlock somewhat as a multitude of assembly processes interlock on a modern day production line. For example glucose and other fuel molecules releases energy in a cells by burning. The cells could not tolerate and survive in such high temperatures nor effectively utilize energy that is released in short sudden bursts. Therefore, cells need a slow, steady release of energy that they must be able to regulate to meet metabolic energy requirements. In cellular respiration, fuel molecules are slowly oxidized and energy is extracted in small amounts, which includes sequences of 30 or more reactions (as many as 20 or 30 chemical transformations before it reaches some final state). Such chemical transformations require a system of flexible chemical control. The key elements of this control system are the remarkable enzymes.

## WHAT ENZYMES DO?

An enzyme increases the speed of a chemical reaction without being consumed itself. An enzyme affects the rate of a reaction by lowering the energy needed to activate the reactions. Even a strongly exergonic reaction that releases more than enough energy as it proceeds is prevented from beginning by an energy barrier. This is because before new chemical bonds can be formed, existing ones must be broken. The energy required to overcome this barrier and get the reaction going is called **activation energy.** An enzyme greatly reduces the activation energy necessary to initiate a chemical reaction. Enzyme promotes a chemical reaction. It does not influence the *direction of a chemical reaction* or the final concentrations of the molecules involved. They simply *speed up reactions*. For example carbonic anhydrase is an enzyme which promotes the interconversion of carbonic acid with carbon dioxide and water:

$$CO_2 + H_2O \xrightarrow[\text{Carbonic acid}]{\text{Carbonic anhydrase}} H_2CO_3$$

This reaction occurs in animal cells and tissues (kidneys and red blood cells). Carbonic acid ionizes, forming bicarbonate ions, the forms in which most of the carbon dioxide in the blood is transported. The reaction shown above could proceed without the enzyme but would be very slow. However, in presence of carbonic anhydrase, the reaction proceeds about 10 million times faster. This reveals that a single molecule of this enzyme can promote the conversion of an estimated 600,000 molecules of carbon dioxide into carbonic acid each second.

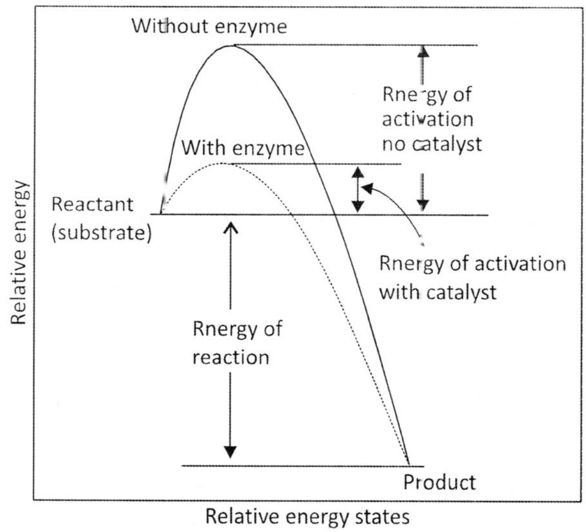


**FIGURE 17.3:** An enzyme accelerates the rate of reaction through lowering activation energy.

## WHAT IS ENZYME KINETICS?

It is the study of enzymes in action. The extremely high rate of enzyme-catalyzed reactions greatly facilitates this study. Consider, for example, the hem-containing proteins, hemoglobin and catalase. Hemoglobin binds oxygen. It may bind and release many oxygen molecules in the course of a minute but they remain oxygen molecules, and at any instant, only one is associated with each hem centre. Catalase, being an enzyme, has a cumulative effect. Again no more than one $H_2O_2$ molecule will be bound per hem, but while it is bound it may react, and one therefore observes a rapid evolution of oxygen - about a million molecules per minute per enzyme molecule.

## HOW ENZYMES WORK

The Chemicals upon which an enzyme operates are referred to as its **substrates**. Enzymes are very **specific**. An enzyme will catalyze only a few closely related chemical reactions, or in many cases only one particular reaction. Enzymes form temporary chemical compounds with their substrates. Theses complexes then break up, releasing the product and regenerating the original enzyme molecules for reuse.

$$\text{Enzyme} + \text{Substrate 1} + \text{Substrate 2} \longrightarrow \text{Enzyme Substrate Complex} \longrightarrow \text{Enzyme} + \text{Product(s)}$$

It is important to note that the enzyme itself is not permanently altered or consumed by the reaction.

Why does the enzyme-substrate complex break up into chemical products different from those that participated in its formation? As shown in **FIGURE 17.4**, each enzyme has one or more regions called **active sites,** which in the case of a few enzymes have been shown to be actual indentations in the enzyme molecule. These active sites are located close to one another on the enzyme's surface, so during the course of a reaction, substrate molecules occupying these sites are temporarily brought together and react with one another. It is thought that when the enzyme and substrate bind together, the shape of the enzyme molecule changes slightly. This produces strain in critical bonds in the substrate molecules so that these bonds break. The new chemical compound thus formed has little affinity for the enzyme and moves away from it. An enzyme can be thought of as a molecular lock into which only specifically shaped molecular keys- the substrates- can fit.

Similar to a lock and key, however, the enzyme and its substrate seem not to be exactly complementary shapes. A recent model of enzyme action, known as the **induced-fit model**, is based on data indicating that the active sites of an enzyme are not rigid. When the substrate binds to the enzyme, it may induce a change in shape in the enzyme molecule, resulting in an optimum fit for the substrate-enzyme interaction. The change in shape of the enzyme molecule can put strain on the substrate. This stress may help bonds to break, thus promoting the reaction.

An organic, nonpolypeptide compound that serves as a cofactor is called a **coenzyme.** Many coenzymes are synthesized from vitamins, particularly from the B vitamins. The coenzyme serves as an adaptor, permitting the enzyme to accept a substrate for which it would, by itself, have little affinity.

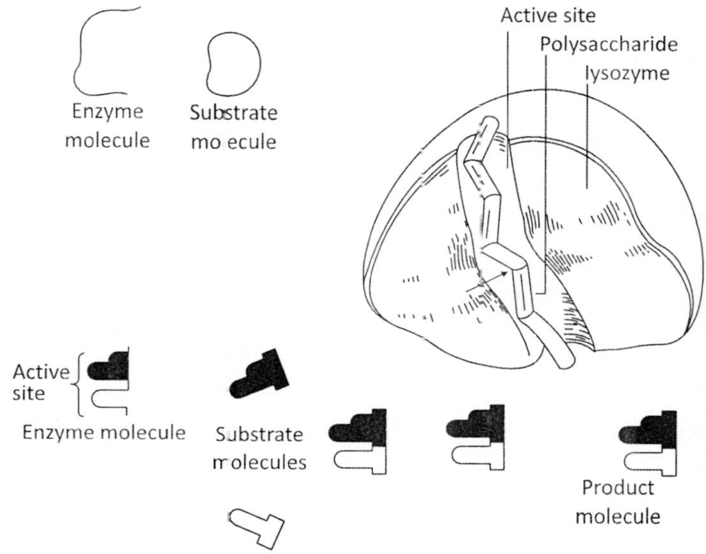


**FIGURE 17.4:** (a) Enzyme-substrate (ES) complex formation, (b) Enzyme in an active site of substrate, (c) Induced-fit hypothesis (Lock as key mechanism).

## ENZYME ACTIVITY

Enzymes perform a wide range of chemical modifications and re-arrangements. For example,

$$\text{Ethanol} \xrightarrow[\text{Dehydrogenase}]{\text{Alcohol}} \text{Acetaldehyde}$$

$$H_2O_2 \xrightarrow[\text{Catalase}]{} O_2 + H_2O$$

$$\text{Xanthine} + H_2O + O_2 \xrightarrow[\text{Oxidase}]{\text{Xanthine}} \text{Urate} + H_2O$$

$$\text{Glucose} + O_2 + H_2O \xrightarrow[\text{Oxidase}]{\text{Glucose}} H_2O_2 + \text{Gluconic acid}$$

$$\text{Starch} \xrightarrow[\text{a-amylase}]{} \text{glucose} + \text{maltose}$$

$$\text{Starch} \xrightarrow[\beta\text{- amylase}]{} \text{maltose} + \text{dextrin}$$

$$\text{Sucrose} \xrightarrow[\text{Rennin}]{} \text{paracasein}$$

$$\text{Lactose} \xrightarrow[\text{Lactase}]{} \text{glucose} + \text{fructose}$$

$$\text{Casein} \xrightarrow[\text{Rennin}]{} \text{paracasein}$$

Enzymes speed up the reaction by lowering their activation energy. They catalyzed it. They are sensitive to environmental conditions. The rate of reaction depends on reactant concentration and the rate increases indefinitely as the concentration increases. The reaction- rate increases linearly with substrate concentration at low concentrations, but then levels off and becomes independent of the concentration at high concentrations.

At high substrate concentration the reaction velocity reaches at limiting value Vmax.

Km is the substrate concentration at which the rate is half maximal.

At high concentrations, most of the enzyme molecules have their active site occupied, and the observed rate depends only on the rate at which the bound reactants are converted into products. Further increases in the substrate concentrations then have little effect. The explanation of this behaviour is known as Michaeli-Menton mechanism.

## *Michaelis-Menten Constant (Km)*

It supposes that an enzyme catalyzed reaction must pass through two or more steps: first enzyme (E) and substrate (S) bind together to form enzyme-substrate complex. The catalysis occurs on account of formation of transient enzyme-substrate (ES) complex. When substrate concentration is high, addition of enzyme can enhance the rate of reaction, till substrate concentration becomes limiting. Then conversion into product (P) occurs within this complex, and the product is released and the original free enzyme regenerated.

Similarly, there is usually a direct proportionality between rate and substrate concentration until the enzyme concentration become limiting. The substrate concentration required to cause half the maximal reaction rate, a value named as Michaelis-Menten constant (Km). Km values are mostly constant and do not depend upon amount of enzymes. Usually the Km values for most of the enzymes studied vary from $10^{-3}$ to $10^{-7}$ molar Km value can be implicated as indicator of substrate concentration, affinity of enzyme with its substrate and it partly indicates enzymes substrate concentration in the cellular compartment, where reaction occurs. It is the affinity part, for which Km values are used most. Km values are inversely proportional to the affinity of enzyme for its substrate. Therefore, higher Km values suffers lower stability of ES complex.

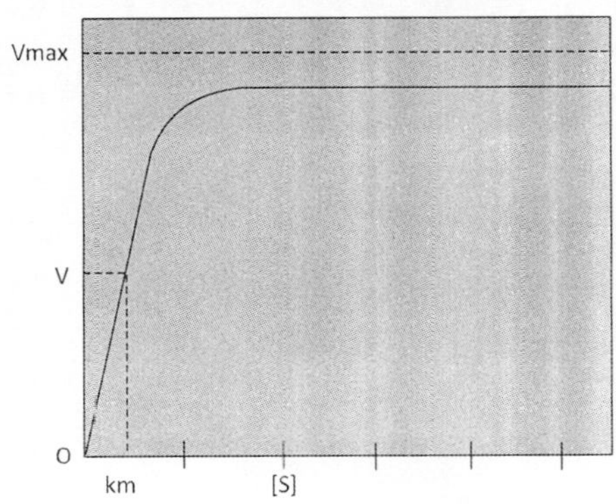


**FIGURE 17.5:** Presentation of reaction velocity V as a function of the substrate concentration [S] for an enzyme catalyzed reaction

$$E + S \underset{k-1}{\overset{\text{fast } k1}{\rightleftarrows}} ES \underset{k_2}{\overset{\text{slow}}{\rightleftarrows}} E + P \quad \text{.................................} (1)$$

Equation 1 shows slow step in the reaction. It is assumed to be formation of E and P from ES, the velocity of the reaction should be

$$V = \frac{d[p]}{dt} k_2[ES] \quad \text{.......... (2)}$$

A maximum velocity (Vmax) is obtained when all of the enzyme is in the form of the Enzyme-Substrate complex. Since the total concentration of enzyme [Et] is equal to [E] + [ES],

$$Vmax = K_2[Et] = k_2([E] + [ES]) \quad \text{.......... (3)}$$

Dividing equation (2) by equation (3) gives:

$$\frac{V}{Vmax} = \frac{[ES]}{[E] - [ES]} \quad \text{.......... (4)}$$

By making use of dissociation constant of the ES complex in equation (1), Ks

$$Ks = \frac{[E][S]}{[ES]} \text{ and } [ES] = \frac{[E][S]}{[Ks]} \quad \text{.......... (5)}$$

Substituting this value of [ES] into equation (4) rearranging gives

$$V = \frac{Vmax[S]}{Ks + [S]} \quad \text{.......... (6)}$$

Equation (6) is the *Henri-Michaelis-Menton* equation. It relates the reaction velocity to the maximum velocity, the substrate concentration , and the dissociation constant for the enzyme-substrate complex.

Shortly after the enzyme and substrate are mixed, [ES] becomes approximately constant and remains so for a period of time termed the steady state. Steady-state kinetic analysis assumes that the concentration of the Enzymes-Substrate complex remains nearly constant. The system following a brief period when the concentration of enzyme-substrate complex [ES] builds up. The ES concentration then remains almost constant for the duration of the steady state, while the concentrations of the substrate and product continue to change.

Therefore, the equation (1) can be presented as:

In this, the ES complex forms from E and S with rate constant $k_1$. ES can either dissociate again with rate constant $k_1$ or go on to P with $k_2$.

For measuring the initial rate of the reaction in the steady state, the rate of formation of ES, Vf, the equation can be expressed as

$$Vf = k_1 [E] [S] \quad \text{.......... (8)}$$

Similarly, the rate of disappearance of ES, Vd, as:

$$Vd = k_{-1} [ES] + k_2 [ES] \quad \text{.......... (9)}$$

If the concentration of ES is virtually constant during the steady state, the rate of formation and disappearance of ES must be nearly equal, Vf = Vd. Therefore, equations, (8) and (9) that describes the situation in the steady state can be represented as

$$k_1 [E] [S] = (k_1 + k_2) [ES] \quad \text{.......... (10)}$$

or

$$\frac{[E]\,[S]}{[ES]} = \frac{k_{-1} + k_2}{k_1} = Km \quad \text{.......... (11)}$$

The constant Km is called the Michaelis constant. If the equation (11) is rearranged, it becomes

$$[ES] = \frac{[E]\,[S]}{Km} \quad \text{.......... (12)}$$

An expression comparable to the Henri-Michaelis-Menten equation (6), with Km in place of Ks, the equation (6) or (13).

$$V = \frac{Vmax\,[S]}{[S] + Km} \quad \text{.......... (13)}$$

By rearranging equation (13) and taking reciprocals of both sides, the equation (13) can be expressed as follows:

$$\frac{1}{V} = \frac{1}{Vmax} + \frac{Km}{Vmax}\,\frac{1}{[S]} \quad \text{.......... (14)}$$

Km is the substrate concentration at which the rate is half maximal. Km and Vmax often can be obtained from a plot of 1/v versus 1/[S].

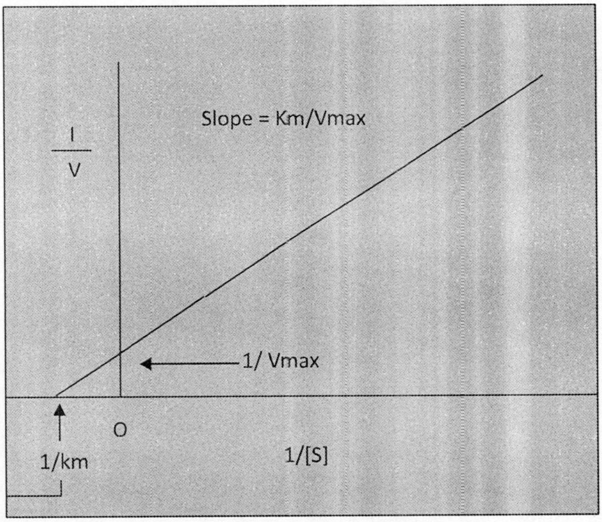


**FIGURE 17.6:** A plot of the reciprocal of the rate (1/v) as a function of the reciprocal of the substrate concentration (1/[S]).

The expression shows that the plot of 1/v versus 1/[S] fits a straight line with the slope of Km/Vmax. The intercept of the line on the ordinate occurs at 1/V = 1/Vmas and the intercept of abscissa occurs at 1/[S] = -1/Km Vmax and Km thus can be determined readily from the graph.

If ES remains in equilibrium with free enzyme and substrate, Km remains equal to dissociation constant for the complex (Ks). Km depends on at least three rate constants and is larger than in Ks.

The turnover number Kcat is the maximum of molecules of substrate converted to product/ unit time/ active site. At high substrate concentrations the rate of reaction remains maximum. At this stage all the active sites are occupied with substrate.

The turnover number is a measure of how rapidly on enzyme can operate once the active site is filled, i.e., at low substrate concentration. This can be expressed as follows:

$$K_{cat} = \frac{Vmax}{[Et]} \quad \ldots\ldots (15)$$

The above equation is the best index to show the effectiveness of an enzyme.

## Multi-Substrate Enzymes

When enzymes catalyze reactions of two or more substrates, they work in a varied ways and can be differentiated by kinetic analysis. Certain enzymes bind their substrates in a fixed order; while some bind in random order. Sometimes bindings of one substrate gives a partial reaction before the second substrate binds.

The only true one-substrate enzymes are the isomerases, which catalyse reactions of the type A= B, and the lyases, which catalyse reactions of the type A - B + C, and are, therefore, one-substrate enzymes in one direction only. Hydrolases, catalysing reactions of the general type A-B + $H_2O$ = A-OH + BH may be regarded as honorary one-substrate enzymes, since one substrate, water, is normally present at constant concentration.

The remaining groups are:

$$\text{oxidant} + \text{reductant} \xrightarrow[\text{product}]{\text{oxidoreductases}} \text{reduced product} + \text{oxidized}$$

$$A + BX \xrightarrow{\text{transferases}} AX + B$$

This large group includes methyl transferases, transaldolase, transketolase, acyl transferases, glycosyl transferases, transaminases, kinases etc.

$$X + Y + ATP \xrightarrow{\text{legases}} XY + ADP + P_i \left(\text{or } AMP + PP_i\right)$$

These three groups of enzymes by definition catalyse reactions involving more than one substrate.

## ENZYME OPTIMA

Enzymes act better under certain narrowly defined conditions known as *optima*. These conditions are optimum at appropriate temperature, pH, and salt concentration, etc. Enzyme pepsin (protein-digesting enzyme of the stomach) works best at the strongly acid pH of 2 whereas amylase (starch-digesting enzyme in saliva and pancreatic juice) works better at pH 8.5 (slightly alkaline). Strong acids or bases irreversibly *in activate* most enzymes. pH permanently changes the molecular conformation of enzymes.

The Michaelis constant Km is affected by ionizations of both free and bound species, and tends to behave in a more complicated way. Enzymes like other proteins, are stable over a limited range of pH. Outside this range, changes in the charges on ionizable amino acid residue results in modifications and eventually lead to denaturation. But within the range of its pH is a characteristic more of the enzyme than of the substrate.

Temperature dependence of the catalytic process can only be meaningful when studied within a range in which thermal denaturation is not significant. It has a narrow range from 5° to 40°C. At low temperatures, enzymatic reactions occur very slowly or not at all, but their activity resumes when the temperature is raised to normal. At lower temperature, the temperature dependence of Kcat can be related to the activation energy of the slowest step in the catalytic pathways. The rates of most enzyme-regulated reactions increase with increasing temperature, within limits. Temperatures greater than 50 °C or 60 °C rapidly inactivate most enzymes by altering their secondary and tertiary level structures. The enzyme is denatured at high temperature. This happens everyday while cooking an egg white; it changes in consistency as the protein is denatured.

## ENZYME INHIBITORS, ACTIVATORS AND INACTIVATORS OF ENZYMES

The rate of an enzyme-catalyzed reaction may sometimes be altered in a specific manner by compounds other than the substrate(s). Activators increase the rate; inhibitors and inactivators decrease it. The study of such agents is of practical importance for several reasons:

1. Inhibition and activation of enzymes by key metabolites provides the normal means of metabolic *fine control* superimposed on the *coarse* control achieved by regulation of the synthesis and breakdown of active enzymes.
2. External interference with metabolism, whether by drugs, pesticides etc. or by undesirable toxic agents, often depends on the inhibition of enzymes.
3. Inhibitors, and especially inactivators, provide a powerful tool for studying the chemical mechanisms of enzyme action.

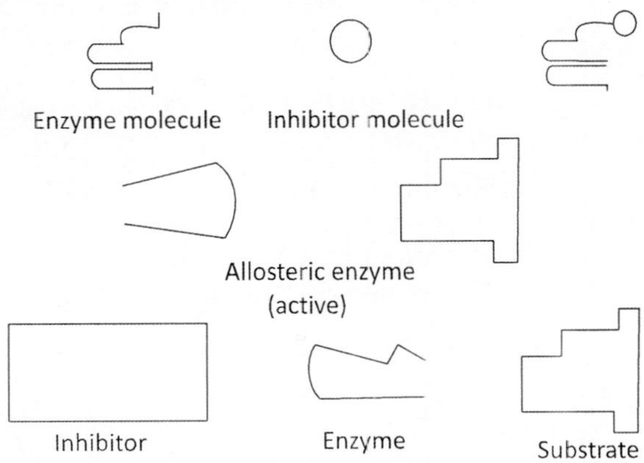


**FIGURE 17.7:** Enzyme inhibition (a) non competitive, (b) Allosteric.

Enzyme inhibition may be reversible or irreversible. Reversible inhibitors can be competitive or noncompetitive. In **competitive inhibition** the inhibitor completes with the normal substrate for the active site of the enzyme. A competitive inhibitor usually is chemically similar to the normal substrate and so fits the active site and binds with the enzyme. However, it is not similar enough to the normal substrate to take its place effectively. The enzyme cannot act upon it to form reaction products. A competitive inhibitor occupies the active site only temporarily and does not damage the enzyme irreversibly.

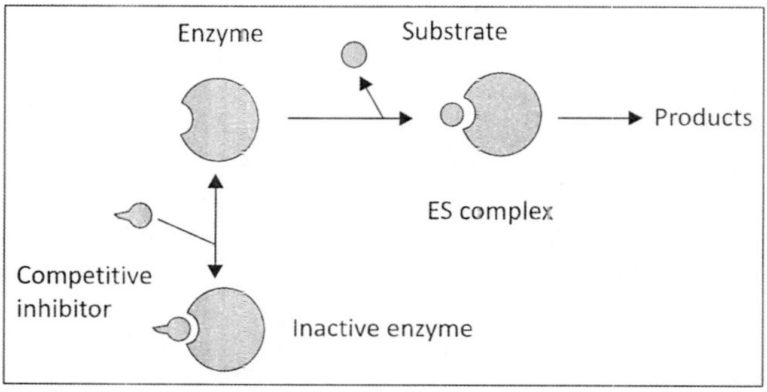


**FIGURE 17.8:** A competitive inhibitor completes with the substrate for binding at the same site on the enzyme.

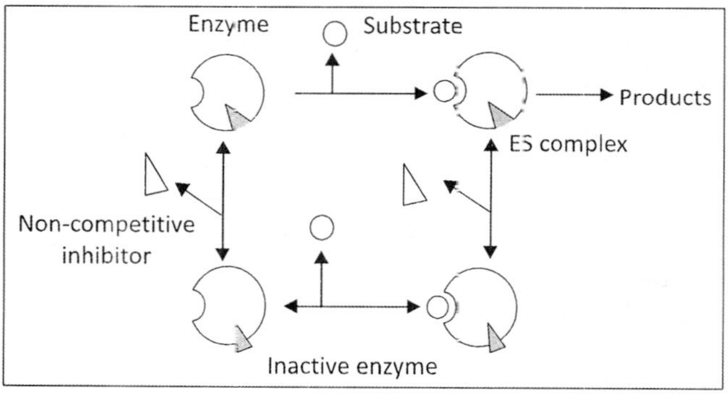


**FIGURE 17.9:** A noncompetitive inhibitor to a different site but block the conversion of the substrate to products.

In **noncompetitive inhibition** the inhibitor binds with the enzyme at a site other than the active site. Such an inhibitor renders the enzyme inactive by altering its shape. Many important noncompetitive inhibitors are metabolic substances that help regulate enzyme activity by combining reversibly with the enzyme.

Many poisons are **irreversible inhibitors** that permanently inactivate or even destroy the enzyme. Nerve gases are irreversible inhibitors that poison the enzyme cholinesterase, essential in the normal function of nerves and muscles. A number of insecticides and drugs are irreversible inhibitors. The antibiotic penicillin and its chemical relatives inhibit a bacterial enzyme necessary for bacterial cell wall construction. Unable to produce new cell walls, susceptible bacteria cannot multiply effectively. Since human body cells do not have cell walls (and so do not employ the susceptible enzyme), penicillin is harmless to humans, except for the occasional allergic patient.

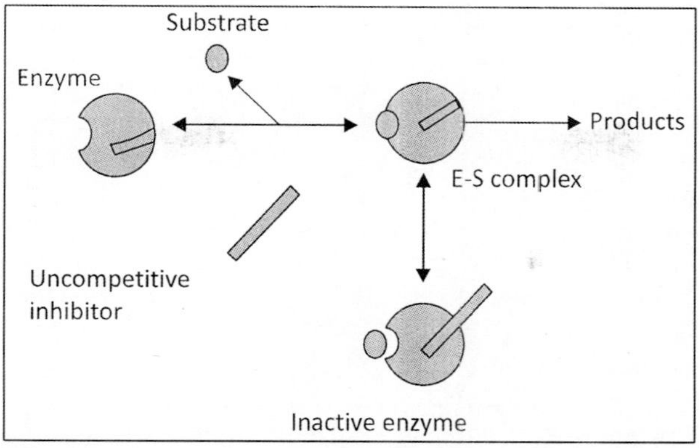


**FIGURE 17.10:** An uncompetitive inhibitor binds only to the substrate-enzyme complex.

### Reversible and Irreversible Inhibition

Enzyme activity can be curtailed by various non-specific agents (acid or alkali, urea, detergents, proteases etc.), which disrupt protein structure. These agents interact with a protein at a small number of loci without markedly disrupting the three-dimensional structure. The two most important considerations in their classification are specificity and reversibility.

## ENZYME SECRETION BY PLANT CELLS

Enzyme secretion is a common feature of most of the living cells. By this process, cells may act on their environment by modifying its chemical composition and ensure their defence againt external aggressions. In pluricellular organisms, groups of cells (glands) become specialized in the secretion of enzymes, which are utilized in a function useful for the whole organism. The control of such a release of enzyme is often mediated by chemical substances, which migrate from one tissue to another one composed of secreting target cells. This coupling between stimulus and secretion is widely distributed in animal organisms.

### The Mechanism of Protein Secretion

The synthesis, intracellular transport, and release of secretary proteins are the basic cellular function common to most eukaryotic cells. In the pancreatic exocrine cell for example, the secretary proteins are the object of six steps or operations, which are: *synthesis*, *segregation*, *intracellular transport*, *concentration*, *intracellular storage* and *discharge*. Proteins for export are generally synthesized on polysomes, attached to the membrane of the rough endoplasmic reticulum. In etiolated radish seedlings submitted to far red light, which is known to induce the synthesis of β-fructosidase, the enzyme activity is found associated to endoplasmic reticulum in a first time, then in Golgi apparatus and in cell wall. There is no evidence for the existence of a single intracellular route leading to enzyme secretion. Although the migration of proteins synthesized in endoplasmic reticulum, through Golgi towards the exterior of cell after exocytosis is likely to occur in some cases, direct transport from reticulum to plasmalemma and transfer of cytoplasmic enzymes across plasmalemma cannot be excluded.

## The Control of Secretion

The secretary process is a complex mechanism. It is known that calcium is essential for the control of the secretary process. Calcium plays a critical role in the exocytosis process, which is a **fusion-fission response** involving the interaction the of the plasma membrane and secretary vesicle membrane. A second requirement for the occurrence of secretion is metabolic energy. An adequate intracellular ionic equilibrium is also necessary. $Ca^{2+}$ is considered as a second messenger in animal cells, able to trigger the discharge of secretary products by regulated cells. In addition, non-regulated cells often requires $Ca^{2+}$ for their secretion. It would be, therefore, necessary to explore the knowledge concerning the regulation of $Ca^{2+}$ in plant cells. It is known that $Ca^{2+}$ is involved in the secretion of some plant enzymes, including peroxidase, α-amylase, and phosphatase. Recently, several articles have been published, which described modes of $Ca^{2+}$ transport across pant membranes. An ATP-dependent $Ca^{2+}$ uptake by isolated membranes vesicles was reported.

Some elements exist for substantiating the hypothesis that phytohormones regulate protein secretion through the mediation of second messenger such as c-AMP or $Ca^{2+}$/calmodulin. However, the existence of an exchange mechanism between $H^+$ and $Ca^{2+}$ suggests that plant growth regulators especially auxin could modify secretary process by a modification of the distribution of protons which indirectly affect $Ca^{2+}$ compartmentation.

## SOURCES OF ENZYMES

Industrial enzymes can be obtained from plants, animals or microorganisms. Various factors such as specificity, pH, thermostability, activation or inhibition, availability and cost, determines the selection from any of the sources.

The important enzymes are:

1. **Glucose Oxidase:** This converts glucose to gluconic acid, which is an industrial chemical and is also used extensively in diagnostic test for measuring blood sugar.
2. **Rennin**, used in cheese production.
3. **Papain**, an enzyme derived from papaya plant and added to animal feeds as a meat tenderizer.
4. **Glucose isomerase** converts glucose to fructose
5. **Pepsin**, an animal protease used in the manufacturing industry.
6. **Urokinase** is used as an anti-thrombotic agent in treatment or pulmonary embolism.
7. **Asperaginase** used in cancer chemotherapy
8. **Tyrosine hydroxylase** which converts tyrosine into DOPA which is specific against Parkinson's disease.

**TABLE 17.2:** Sources of animal enzymes

| Type of enzyme | Source |
|---|---|
| α-amylase | Pancreatic gland of bovine and porcine |
| catalase | Liver |
| chymotrypsin | Pancreatic gland |
| esterase | Pancreatic gland |
| glucose oxidase | Liver |

| Type of enzyme | Source |
|---|---|
| lipase | Pancreatic gland |
| pepsin | Fundus portion of hog stomach |
| rennet | Calf stomach |
| trypsin | Pancreatic gland of bovine and porcine |

Animal gland e.g., the pancreas, is source for hydrolyzing enzymes used as digestive aids. The pancreas, a very rich source contain about 23% trypsinogen and 10-14% chymotrypsinogen. Pancreatin contains several enzymes such as amylase, lipase and protease.

**TABLE 17.3:** Sources of animal enzymes

| Type of enzyme | Source |
|---|---|
| proteases (bromelain and papain) | pineapple and papaya |
| amylolytic | tissues of barley, wheat, rye, potatoes, sweet potatoes, beans and soybeans. |
| α-amylase and β-amylase | Seeds |

The extraction from plants and animals sources is much simpler and less costly. The toxicity and contamination is not a problem.

## MICROBIAL ENZYMES

The enzymes used in the industry are obtained from bacteria, fungi and yeasts. Efforts have been made to replace animal or plant enzymes with enzymes from microbes e.g., amylases and endo-β-glucanases from *Bacillus* and *Aspergillus*, plant and animal proteases by *Aspergillus* and *Thermoactinomyces*, calf chymosin by *Mucor*. The application of genetic engineering gained enormous potential in the use of microbes to produce animal or plant enzyme. The strains constructed by gene cloning procedure will be able to produce abnormal amounts of foreign proteins i.e., synthesis of calf prochymosin by *E. coli/Saccharomyces cerevisiae* and urokinase normally occurring in human urine by *E. coli* cells. The various enzymes and their microbial sources are given in following.

**TABLE 17.4:** Enzymes produced by bacteria

| Type of enzyme | Bacteria |
|---|---|
| α-Amylase | *Bacillus spp.* |
| β-Amylase | *Bacillus cereus* |
| Dextran sucrose | *Leuconostoc mesenteroides* |
| -β-1,3-Gluconase | *Bacillus subtilis* |
| Glucose isomerase | *Arthrobactor globiformis Bacillus coagulans. Streptomycin spp.* |
| Penicillin acylase | *Bacillius megaterium* |
| Proteinase | *Escherichia coli, Bacillus subtilis* |
| Pullulanase | *Klebsiella aerogens* |
| Streptokinase | *Streptococcus spp.* |
| Urokinase | *Bacillus fastidions* |

**TABLE 17.5:** Enzymes produced by fungi

| Types of enzyme | Fungi |
|---|---|
| α-Amylase | *Aspergillus oryzae* |
| Amyloglucosidase | *Aspergillus niger* |
| Catalase | *Aspergillus niger* |
| Cellulase | *Trichoderma reesi, Penicillium funiculsosum* |
| Dextranse | *Penicillium* spp. |
| Glucoamylase | *Aspergillus niger, A. awamori and Rhizopus spp.* |
| Glucose oxidase | *Aspergillus niger* |
| Invertase | *Saccharomyces spp.* |
| Lactose | *Kluveromyces fragilis, Aspergillus niger* |
| Lipase | *Candida lipoytica, Rhizopus delemar* |
| Pectinase | *Aspergillus niger* |
| Proteinase | *Aspergills oryzae* |
| Raffinase | *Mortierella vinacea* |
| Renin | *Mucor miehes* |

Microorganisms have biochemical diversity. The enzyme concentration in these microorganisms is dependent on environmental and genetic manipulations. Microbial enzymes may have varied stability and different pH and temperature optima. Microbial cells require short fermentation times, inexpensive media and simple screening procedures. Microbes can be geneticically manipulated.

## MECHANISM OF ENZYME BIOSYNTHESIS IN MICROBIAL CELLS

Metabolism is mainly regulated by change in the rate of enzyme reaction. The enzyme concentration can be varied by controlled protein synthesis and controlled protein degradation.

Enzymes synthesis by microbial cells may be influenced by various mechanisms. These include-

### Induction

Biosynthesis requires the presence of substrates in the medium. The structural genes coding for the production of many catabolic enzymes are repressed in the absence of substrates of the enzymes. By the addition of substrate, enzyme synthesis is induced or derepressed. In the case of lac operon, the inducer (lactose) binds the repressor and inactivates it. Transcription thus begins and enzyme synthesis occurs. This can be seen in case of the synthesis of sugar or amino acid degrading enzymes.

### Repression

The repressed enzymes can be of the constitutive or of the inducible types. Enzymes are often repressed by the presence of intermediates or end products of the enzyme reactions. This type of regulation is known as end product repression or feedback repression.

### Catabolic Repression

The biosynthesis of the most of extra-cellular depolymerases is influenced by the level of catabolites present in the growth medium. It occurs when cells grow rapidly on readily metabolizable carbon sources i.e., repression of β-galactosidase in *E. coli* by glucose.

### Transcription control during the Growth Cycle

The different behaviour of enzyme biosynthetic in relation to the growth is caused by different regulatory mechanisms. In case of *B. subtilis* transition of cells from vegetative growth to sporulation is accompanied by changes in the predominant forms of RNA polymerase. Four forms of RNA polymerase have been identified in vegetative and early stationary phase cells. Sporulation forms of RNA polymerase recognizes the specific promotor sequence that control expression of genes during growth cycle and sporulation. Genes coding for extracellular proteases indicates overlapping promoters, ensuring gene expression during both growth and stationary phase. Therefore, altered promoter specificity is one of the cellular tools to influence gene expression on the level of transcription during growth. Further, translation efficiency of a gene is influenced by several factors. Rate of translation efficiency of a gene is influenced by several factors. Rate of translation efficiency is affected by the ribosomal binding site, stability of mRNA, the codon usage by the translation machinery, occurrence of repression by attenuation, effectiveness of the secretion etc.

## GENETICS OF ENZYME FORMATION

Modern biotechnology was born because scientists learned to manipulate (engineer) DNA. This capability was derived largely from the field of nucleic acid enzymology. Prior to 1970, there was simply no method available for cutting a duplex DNA molecule into discrete fragments. Discovery of DNA metabolizing enzymes allowed scientists to contemplate and initiate genetic engineering. With the advent of genetic engineering, barriers to the exchange of genetic material were crossed to improve the formation of desired enzyme in various organisms.

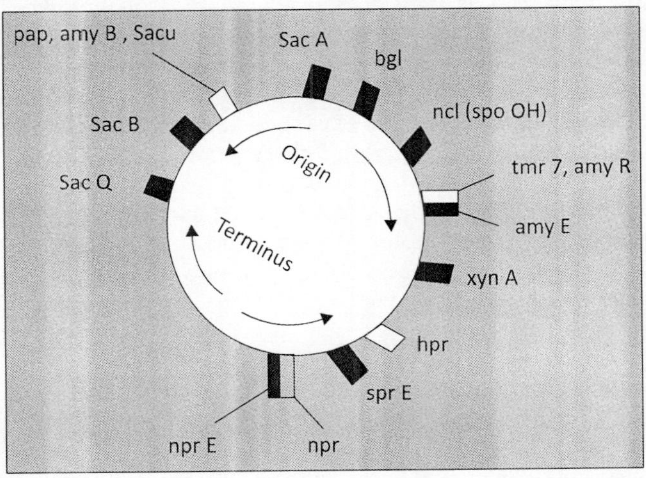


**FIGURE 17.11:** Presentation of location of genes coding for extra cellular enzymes on the linkage map of *B. subtilis*.

Genes coding for intra- or extra- cellular enzymes in Gram negative bacteria are organized in operons while in gram positive bacteria, however, the occurrence of genes for exoenzyme synthesis in clusters in uncertain. Linkage of regulating genes with related structural genes are known to be occurring i.e., amylase (amy R, amy E), levansucrase (sac Q, sac B). Besides regulatory genes, there are genes enhancing the level of simultaneous secretion of different genes (pap, amy B, sac U, in *B. subtilis*.

Classical genetic methods have also been used for the improvement of enzyme formation, e.g., mutation. Through mutations overproduction of enzymes can be obtained by altering mechanisms of regulation and by increasing copies of responsible genes for the production of enzymes. The method of site directed mutageneis of cloned gene is also being used for genetic alteration of enzyme. Chemical mutation is also helpful for the isolation of natural enzyme and modification to its structure.

In vitro methods is also being used to overcome the barriers to genetic exchange between unrelated organisms. Genes encoding enzymes with industrial importance have been cloned. The potential hosts expressing cloned genes are *E. coli*, *B. sublitis* and *S. cereviiae.* Cloning of the structural genes for α-amylase from *B. amyloliquefaciens* in plasmid pUB 110, and transformation of this hybrid plasmid into *B. subtilis* has enhanced the productivity of α-amylase to about 2000 fold. Thus, it becomes possible to produce industrial enzymes of very high quality and purity. Enzymes can be modified by evaluation of their performance. The main objective is to enhance the activity, stability change in pH or temperature optima, change the specificity to catalyze the conversion of different substrate and to perform function in a changed environment.

## ISOENZYMES

Isoenzymes were discovered about 30 years ago and were at first regarded as interesting but of rare occurrences. Since then a wealth of information on enzyme heterogeneity has accrued and it now seems likely that at least half of all enzymes exist as isoenzymes. This is important in many areas of biological and medical science. Thus isoenzyme studies have provided the main experimental substance for the neutral drift controversy in genetics and evolution. Isoenzymes have greatly extended our understanding of metabolic regulation not only in animals but also in bacteria and plants; their existence has made available a multitude of highly sensitive markers for the study of differentiation and development, as well as providing indices of aberrant gene expression in carcinogenesis and other pathological processes. Isoenzymes are also being used increasingly in diagnostic clinical biochemistry. Isoenzymology crosses the traditional boundaries of the biological disciplines and it seems that advances and new applications in this field, are likly to occur at a rapid pace.

### The Cause of Enzyme Multiplicity

There are various causes of enzyme multiplicity and they may be divided into two categories. These are (a) **genetic** or **primary causes**, whereby the organism carries multiple genes each one encoding a different type of enzyme subunit; and (b) post-translational or secondary causes, whereby homogeneous enzyme subunits are modified differentially so as to produce a range of subunits from a single gene. There are, in turn, two types of genetic multiplicity; firstly, multiple alleles at a single genetic locus and secondly, multiple genetic loci.

## Primary or Genetic Isoenzymes

### *Isoenzymes due to multiple alleles at a single genetic locus*

In the diploid genome, each genetic locus is represented twice. For each locus, the individual will either be homozygous, possessing two identical alleles, or heterozygous, possessing two different alleles. Where the genetic locus encodes an enzyme subunit, the homozygous individual can only produce one type of subunit. However, the heterozygous individual with two different allelic variants will produce two different types of enzyme subunits. Within the individual the degree of enzyme multiplicity produced by multiple alleles is limited, as two different alleles per diploid locus is the maximum possible genetic variation of this type. However, from one individual to another, there may be considerable variation in the range of enzyme subunit types produced since there may be a variety of different alleles for the locus in the gene pool of the species.

The enzyme subunit types produced as a result of multiple alleles are likely to differ from each other only in minor ways, such as by individual amino acid substitutions caused by point mutations in the DNA nucleotide sequences. Where a genetic locus is active, both alleles will usually be expressed. Thus in the same individual, although the total activity of the enzyme may vary considerably between different kinds of cell, the isoenzyme profile will be constant throughout, homozygous individuals displaying only one subunit type, and heterozygotes possessing two different subunit types.

### Isoenzymes due to Multiple Genetic Loci

Many enzymes are encoded at more than one genetic locus and, where this the case, each locus will produce a different type of enzyme subunit. As the expression of each genetic locus can be controlled independently, the organism may synthesize one type of enzyme subunit in a particular cell, and another enzyme subunit elsewhere. Furthermore, the expression of genetic loci may alter during the course of development, and therefore the type of enzyme subunit produced in a tissue may change. Thus multiple genetic loci permit **differences in isoenzyme profile** both from one tissue to another, and from one developmental stage to another even within the same tissue. Such variation in isoenzyme profile are not possible where the enzyme multiplicity is due to multiple alleles operating at only a single locus. However, as all members of the same species will possess the same genetic loci, multiple enzyme-encoding loci in the absence of multiple alleles cannot account for isoenzyme differences between one member of the species and the next. The isoenzyme subunits produced by multiple loci are likely to differ extensively as a result of numerous amino acid substitutions, and also deletions and additions of residues may have occurred resulting in small size differences between the subunits.

## Secondary or Post-Translational Isoenzymes

Proteins may be modified in numerous ways following their synthesis. Such possible modifications include the addition of carbohydrate, limited proteolysis, and the covalent modification of amino acid side chains. Post-synthetic alterations affecting only part of the enzyme subunit population, so that modified and unmodified subunits are found in the same organism, will result in isoenzymes. Where the post-translational modification process is very active in some tissues but not in others, the result will be a tissue-specific distribution pattern of the secondary isoenzymes, mimicking the effect of multiple genetic loci. An example of this is one particular pyruvate kinase subunit type. There are three major types of pyruvate kinase subunit in mammals, each coded for by an independent genetic locus.

**TABLE 17.6:** Enzymes, coenzymes, and mineral activating substances

| **Enzyme** | **Coenzyme or co-factor In processes of glycolysis** | **Activating mineral elements** |
|---|---|---|
| Phosphatase | Phosphate | $Mg^{2+}$, |
| Phosphoglutamase | | $Mg^{2+}$ |
| Aldolase | | $Fe^{2+}$, $Zn^{2+}$, $Co^{2+}$ |
| Enolase | | $Mg^{2+}$, $Mn^{2+}$, $Zn^{2+}$ |
| Pyruvate kinase | ADP In the tricarboxylic cycle | $Mg^{2+}$, $K^{+}$, $NH_4^{+}$ |
| Decarboxylase and pyruvate dehydrase | Thiamine pyrophosphyte fatty acids, CoA | $Mg^{2+}$, $Mn^{2+}$ |
| Isocitrate dehydrogenase | NADP | $Mn^{2+}$, $Mg^{2+}$ |
| Succinate dehydrogenase | FAD | $Ca^{2+}$, $Cr^{3+}$, $Al^{3+}$ |
| Aconitase | | $Fe^{2+}$, -SH |
| Dehydrogenase | | $Mn^{2+}$, $K^{+}$, $Rb^{+}$, $NH_4^{+}$ |
| Pyruvic oxidase | | $Mg^{2+}$, $Mn^{2+}$ |

Among the three most important functions namely, the mechanical, osmotic and chemical, the later depends solely upon the bio acceleration of metabolic reactions caused by enzymes.

# 18. Production of Enzymes

Microorganisms, plants and animals are the major sources of natural enzymes, but much focus has been diverted on microorganisms as a source of enzymes. Scientists and industrialists rely on enzymes of microbial source. Recombinant DNA technology has greatly influenced the possibilities for optimizing enzyme yield and type through strain selection, mutation, induction and selection of growth condition etc. As discussed in the chapter *Biology of enzymes*, enzymes differ greatly in their stability to temperature and to extremes of pH. For example, bacterial proteases are relatively heat stable and remain active under alkaline conditions and is being used as soap powder additives. On the other hand, fungal amylases have their greater sensitivity to heat and are being used in the baking industry. Microbial enzymes are used in fermentation processes commercial level and substrate is required in large quantities. These substrates are starch hydrolysate, molasses, corn steep liquor, whey and cereals. For industrial production of enzymes suitable and low-prices substrates are required. Enzymes at industrial level are produced by solid substrate process using fungal sources for commercial extracellular enzymes production. Other technique is conventional batch submerged culture for bacterial source and is commonly used.

## SOLID SUBSTRATE FERMENTATION

Solid substrate methods for production of fungal enzymes includes growth of filamentous fungus on a moist wheat or rice bran with added nutrient salts as substrates to produce amylases, proteases, pectinases and cellulases. During the process, the control of pH, temperature and humidity is very difficult.

### Fermenters

The fermenters are rectangular or circular trays of a few centimeter depth, mounted on racks and air is blown over the surface and beneath the mesh bottom. Another type is of two stage rotary system in which partially fermented bran from an upper chamber is conveyed to a lower rotary tray where fermentation process is completed. Substrates composed of longer beds can be used. These can also be of fixed bed types in which air is forced though a bed of inoculated substrate which is periodically agitated by some mechanical device, or they can be of conveyed type. The other type of fermenter has been developed in which the rotating drum used horizontally, disposed cylinders, mounted on rollers and provides both support. It acts as rotation device.

### Medium

The media comprise of cereal, which provide nutritional requirements to the microorganisms. The cereals are cracked and grinded to make them in utilizable form. They are sterilized using steaming technique at low pH or by chemical methods, or a combination of both. The mass of the moistened bran is inoculated with a suspension of spores or vegetative inoculum of the selected strain.

Incubation is done for 1-10 days at temperature 40° to 84 °C. Moisture content and temperature of the fermenting media should be controlled properly, though it is a difficult task. During the process, several metabolic reactions are under operation. Such reactions generate metabolic heat generated, which causes loss of water loss from cultures.

**TABLE 18.1:** Types of media employed for the commercial production of enzymes

| Source | Enzyme type | Medium |
|---|---|---|
| *Aspergillus niger* | Lactase | Wheat bran 100 parts), Zn, Fe and Cu (traces in 60 parts) 0.2 N.HCl |
| *Aspergillus niger* | Amyloglucosidase | Cassava meal (10 parts), (NH4)2 SO4 (0.97 parts), urea (0.24 parts), tap water (10 parts) |
| *Aspergillus oryzae* | Proteinase | Wheat bran (10 parts) and 0.2 N.HCl (5 parts) |
| *Coniothyrium diplodiella* | Pectinases | Defatted rice bran (10 parts) and tap water (6 parts). |
| *Mucor pusillus* | Rennin | Wheat bran (10 parts), water (7 parts). |

## Advantages

The volume of fermenter is needed in less quantity as compare to submerged fermenter system. Inoculum stages are generally unnecessary. Spore seeding can be done directly. Recovery by extraction of the harvest with small volume of water yields relatively highly concentrated solutions of enzymes. The sterilization process is not expensive.

## Disadvantages

Solid substrate fermentation process has several disadvantages and limitations. The process is limited to fungal organisms. The important one to maintain and control the environmental conditions of the fermenter, this includes temperature, pH, moisture content and oxygen transfer. It is a difficult task. Pretreatment of moist media except in rice, is usually required. Control of the progress of the fermentation is difficult because of in-homogenous condition of the culture.

## SUBMERGED FERMENTATION

In submerged fermentation process bioreactor are used similar in deign and function to those used in antibiotic production processes. It is constructed from stainless steel and has a capacity of 10-50 $m^3$ and includes systems for mixing by mechanical internal agitator, external pumps, bubble columns, and air lift loop. Cultivation involves the suspension growth in liquid medium. Sterilization and process control of submerged fermentation are easy.

The medium is prepared after adjusting optimum or required pH value. The medium is pumped to previously stem sterilized fermenter vessel through HTST sterilization units. Thermolabile additives are added separately to the sterile medium. Inoculum buildup should be before transfer to the production stage vessels. Sterile compressed air is introduced into the fermentation broth through a sparger system of a ring type or a single or multiple inlet type. Agitation is done using multiple, flat bladed disc turbines mounted on a vertical shaft. Often types include axial flow propellers or counter current stirrers. Temperature of the fermenters is maintained by using internal cooling coils or external heat exchange surface.

In situ probes and instruments are used to monitor and control varied physical conditions during development of the inoculum during production stages. pH value, dissolved oxygen, $CO_2$, redox potential, etc should be controlled properly.

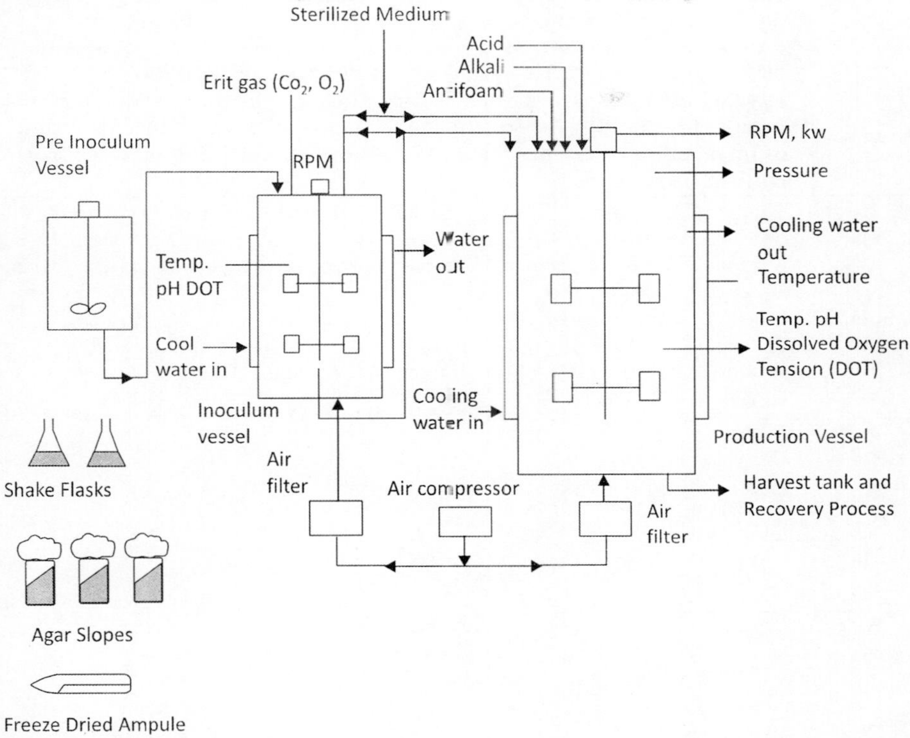


**FIGURE 18.1:** Schematic representation of a typical fermentation plant.

## Steps of Enzyme Production

### *Development of Inoculum*

The stock cultures obtained from a freeze dried ampoules used for the development of inoculum. These cultures after rehydration are used to prepare master cultures on agar slants. Which is further used for preparation of subsequent liquid culture inoculum. Growth from the shake-flask stage for inoculum build up is carried in stirred, aerated fermenters. At first stage, harvesting of the cells or spores is done aseptically from solid medium to obtain suspension as inoculum. The inoculum is then transferred by pumping or air pressure to the next stage of inoculum build up or to the production fermenter stage.

### *Fermenter Stage*

The fermentation is done in either batch, fed batch or continuous cultivation.

**TABLE 18.2:**

| | |
|---|---|
| **Batch Culture** | **In this culture growth proceeds, nutrients are consumed cell mass increases, and the enzyme yield rises to a maximum level before conditions become deleterious. The deleterious conditions develop due to exhaustion of essential nutrients or toxic/inhibitory end product accumulation. At the termination stage of the fermentation may be judged by various parameters like changes in dissolved oxygen, pH. $CO_2$ output, etc. At final stage the broth the transferred from the fermenter for harvesting of the cells followed by extraction of intracellular enzymes from biomass or extracellular enzymes from culture filtrate.** |
| Fed Batch Culture | This fermentation technology is used when the substrate causes catabolite repression. The process needs feeding of additional nutrients, to extend the production phase at particular times and in response to specific physiological state of culture. |
| Continuous Culture | It needs continuous feeding of fresh sterile medium to the culture, while at the same time removing the same volume of broth. It maintains constant volume of culture for indefinite time in a steady state. The fermentation is carried in a chemostat. |

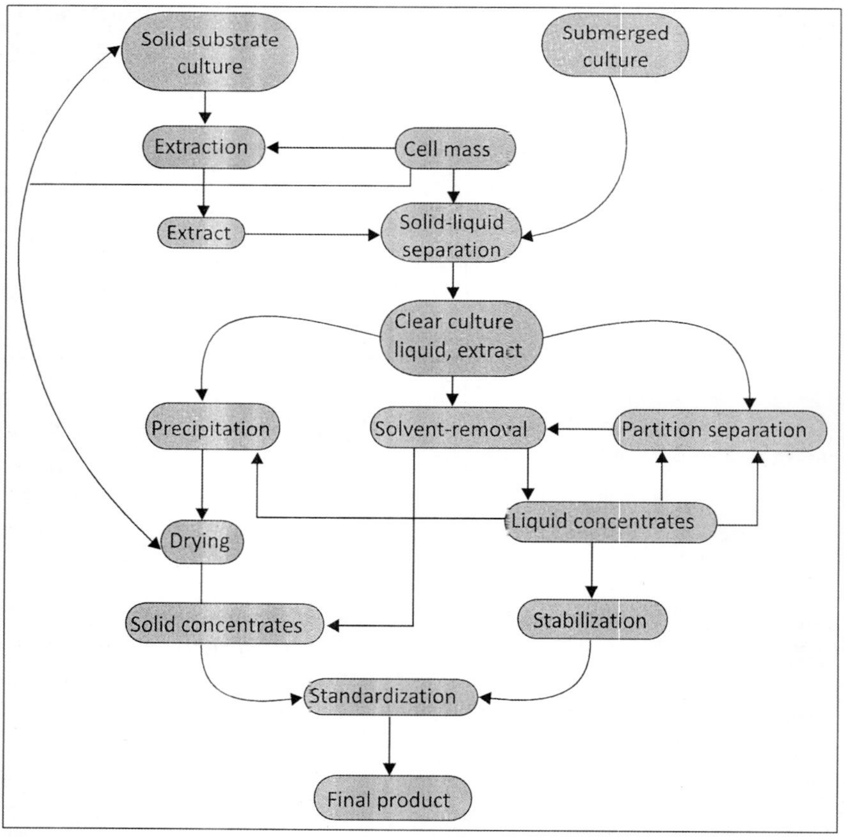


**FIGURE 18.2:** Presentation of outline of the extraction and purification of an enzyme.

## Factors affecting Submerged Culture

### *Temperature*

Optimum temperature is required for the production of a particular enzyme and growth of the microorganism. It varies. The activity of an enzyme towards a used substrate is also influenced by temperature. Therefore temperature during fermentation should range in such a way that cell growth is properly encouraged and subsequently changes with enzyme accumulation.

### *pH*

The pH optima is required for stability of an enzyme. It varies. For getting maximum growth rate of the organism it is essential to maintain pH optima. Therefore, pH profile during fermentation and enzyme production phase should be imposed in such a way that the maximum growth and enzyme production could be obtained.

### *Dissolved Oxygen Tension (DOT)*

Proper aeration rate, agitation ratio and gas phase pressure are required to maintain dissolved oxygen tension. The oxygen below its critical concentration affects the growth of organisms. Adequate aerobic condition in growth culture is an essential requirement. The level of oxygen in submerged microbial cultures can be achieved by either increasing the mass of oxygen supplied to the fermenter/unit time by increasing the overpressure in the fermenter head space or by increasing the rte of air supply to the fermenter. Oxygen enriched air may effect higher oxygen transfer rates.

## EXTRACTION OF ENZYME

Various techniques are available for enzyme extraction. Therefore, the choice of extraction method depends on factors like nature of the source (extracellular or intracellular scale of operation) and stability of the enzyme. For enzyme extraction cell disintegration is required which releases major proportion of enzyme. Following are the methods for cellular disintegration which depends on characteristics of cells.

## Physical Disruption Method

It is done by pressure homogenization or grinding in the pressure of an abrasive material, and chemical or enzymic lysis.

**TABLE 18.3:**

| **Grinding with Abrasives** | **In includes**<br>**Grinding with alumina using mortar and pestle for small scale production.**<br>**Agitation of cells with Dynomill instrument is carried out with glass or metal heads in grinding mills for large scale.** |
|---|---|
| Liquid Shear | This method is useful for disruption of bacterial, yeast and fungal cells. The process involves passage of suspensions of cells at high pressure through office into a chamber at atmospheric pressure. In case of filamentous organisms, it requires to macerate material prior to disruption. |
| Solid Shear | It is used on a small scale production and involves extrusion of cell |

| | |
|---|---|
| | material at high pressure through a narrow office at low outlet temperature (-20°C) resulting in ice crystal formation aiding disruption. |

### Chemical Treatment Method

The method involves treatment with alkali, detergents, osmotic shock, organic solvents, lysozyme and EDTA.

**TABLE 18.4:**

| **Treatment with Alkali** | **The cell are disrupted at a pH within range of 10.5 to 12.5. No live cells remain to be carried over into final product. For example for production of L-asparaginase from *Erwinia* at a pH 11.5 using NaOH. In this treatment the enzyme remains stable for 20-30 minutes.** |
|---|---|
| Detergents | Detergents acts on both on cell wall and cytoplasmic membrane under low stringent conditions. This makes the cell permeable to certain protein materials. The method is useful for the extraction of membrane bound enzymes like cholesterol oxidase from *Nocardia* sps. Detergents like cholates, sodium dodecylsulphate, ceytilmethyl ammonium bromide (CTAB). Tween and Triton (X-100 and X-450) are in common use. |
| Osmotic Shock | This technique is suitable for Gram negative organisms. The osmotic shocks release hydrolytic enzymes present in the perilasmic spaces. It is done by rapid change in the solute concentration of the suspension buffer. |
| Organic Solvents | Isopropyl alcohol, ethanol, and ethyl acetate are used for extraction of enzymes from yeast cells. Organic solvents are costly, toxic, cause protein denaturation and flammability, therefore, they are not used for large-scale enzyme extraction. |
| Lysozyme and EDTA | It is a delicate and specific method of cell disruption. It is suitable for small-scale production. Lysozyme is effective with gram positive bacteria. |

## PURIFICATION OF ENZYME

Enzyme purification is a complex process and various steps are to be conducted in sequence for the purification of any enzyme. For enzyme purification such method should be used which can give high degree of purity; high overall recovery of activity; reproducibility. Therefore, the purification steps have to be selected and linked accordingly.

After the homogenization of the original raw material, the cell debris are separated using centrifugation or filtration method. This will give more or less clear extract. With microorganisms the extract may contain nucleic acids and polysaccharides besides proteins, therefore, nucleic acids, cell debris, preliminary purification and concentration steps should be followed first then final purification and is done. Finally concentrated and purified enzyme is obtained.

### Removal of Nucleic Acids and Cell Debris

The release of nucleic acid makes the solution more viscous because it contains broken cell parts as major contaminant. The removal of nucleic acids and cell debris is done either by differential sedimentation or precipitation, and it can be affected by centrifugation or filtration. Protamine sulphate streptomycin, ceytyltrimethylammonium bromide are used as precipitants. Digestion with nuclease may reduce the viscosity of the solution and can be removed in later stages of purification. Thus, soluble proteins may be separated from organelle-sequestered proteins. At second step, the supernatant which contains the enzyme be further subjected to process for the removal of small organic and inorganic molecules, other proteins, unwanted contaminant and water.

## Preliminary Purification

Success of purification depends on picking a number of procedures and combing them in an effective manner. Procedure is combined to purify the specific enzymes from crude whole cell extracts. Following are the methods:

### *Separation Based on Solubility*

Solubility based separation is mainly based on precipitation. Once a crude extract of protein is made the mixtures of different fractions are separated involving precipitation steps. The solvent should be selected in such a way that it should not the solubility and the structure of protein. It is also necessary to control pH, temperature and the composition of solvent.

At isoelectric point proteins show a minimum solubility. The decrease in solubility at the isoelectric point shows that the individual protein molecules (which all have similar charges at pH values away from their isoelectric points) cease to repel each other. Contrarily, they coalesce into soluble aggregates.

**Precipitation:** The solubility of proteins depends on the concentration of salts in the medium. The solvent properties are changed by specific interaction between charged side chains and the solution ions, or particularly at high salt concentrations. Ammonium sulphate is commonly used salt. A salt is the weak acid. Salt together with the possibility of a slow loss of ammonia, effects the change in the pH. Sodium sulphate is an effective salt. It has low solubility-Phosphates and used as buffers in the biological pH range are good separators.

Precipitation of protein depends on salt concentrations. A salt is added to an aqueous buffered solution of a protein to a concentration just below that at which the required component get precipitated. The initial precipitate is removed by centrifugation. Further salt is then added to the clear supernatant to a level just high enough to precipitate the enzyme. This helps in removal of unwanted material in the first precipitate. Further unwanted materials are left in the solution and are removed in the second step. Organic solvents (polyethylene glycol or polyacrylic acid) are not used for this purpose due to price factor. Such solvents are only used for low scale production.

As the solubility of protein molecule is lowest at its isoelectric point, successive precipitation of different enzymes from a solution can be achieved by changing the pH. These precipitants get settled easily and can easily be separated by using centrifugation technique.

**Adsorption:** Crude extract processing is performed by batch adsorption and elution in initial stages of purification and concentration. Resins, substituted dextrans or agarose, and substituted cellulose are used as adsorbents on small scale are. Hydroxyapatite is also used because of its low cost. Differential elution of many bound proteins can be performed with increasing concentration of phosphate. A common feature of all adsorption techniques is the principle of adsorption, followed by adsorption or elution. Separation is achieved by adsorption or elution of either enzymes or impurities.

**Aqueous two phase separation:** Removal of polysaccharides and separation of cell debris is done by using an aqueous liquid-liquid two phase extraction method. The process permits processing of large volumes and which decreases the starting volume in two parts. Aqueous two phase systems is prepared by mixing solutions of polyethylene glycol and dextran, or polyethylene glycol and specific salts (i.e. potassium phosphate or ammonium sulphate).

1. **First Step:** It is a two-phase separation process of enzyme in which enzyme remains in the upper PEG-rich phase.
2. **Second step:** In this stage the enzyme is found in the lower phosphate rich phase. The partitioning depends on the molecular weight and charge, the concentration and molecular weight of the polymers, the temperature, pH and ionic strength of the mixture, and the presence of polyvalent salts. This is an easier method to separate liquids of different density, than solids from liquids on large-scale. It is advantageous for large-scale enzyme purification.

## Final Purification

Chromatography technique is the most widely used for the isolation of proteins. For large-scale isolation, all of the commonly used techniques may be employed. Low volume, adequate purity and high priced products of biotech pharmaceuticals are obtained by using this technique. The products should be isolated from a complex mixture of other protein.

Ultrafiltration method is also used for separation of abundant phosphate and PEG from enzyme. It lowers the ionic strength for the following batch wise adsorption to an ion-exchanger. Even after adsorption from ion-exchanger the enzyme solution have a relatively high ionic strength. For this purpose, hydrophobic chromatography is suitable.

After desorption by decreasing salt gradient the enzyme solution may be too diluted and must be concentrated before following get-filtration. Concentration may again be done by ultrafiltration or by precipitation with ammonium sulphate or an organic solvent, which permits further fractionation. Dye or affinity chromatography for may also be applied. For highest purity preparations electro focusing may followed as the final step.

# 19. Production of Enzymes by Fermentation

Fermentation-derived enzymes are produced mainly by variants of the solid substrate techniques which is suitable for filamentous fungi, and submerged fermentation in aerated stirred tanks for bacteria. The enzymes of microbial origin are the oxidoreductases and kinase. But majority of enzymes used on large-scale are simple hydrolases.

## OXIDOREDUCTASES

### Pyridine Nucleotide Dehydrogenases

It is mainly used for clinical diagnostics and other analytical applications. Conversion of oxidized and reduced forms of NAD and NADP could be detected by monitoring a change in the absorption at 340 nm. Redox enzymes with pyridine nucleotides as cofactors are used in chemical synthetic steps. The analytical applications used the enzymes from mammalian organs, although they may be supplemented by enzyme from microbial origin.

### NAD/NADP Dehydrogenase

Most microbial enzymes are species for either NAD OR NADP. This class includes lactate dehydrogenases, malate dehydrogenases glutamate dehydrogenases and diaphoresis. The commercial lactate dehydrogenase is prepared from yeast and *Leuconostoc* species. The enzyme is used for determination of ADP, alanine aminotransferase, lactate, lecithin, creatine, pyruvate, pyruvate kinase and triglycerides. The commercial preparations for malate dehydrogenase is obtained from yeast, *Bacillus stearothermophilus*, and *Thermos thermophilus*. The enzyme is used for determination of asperate aminotransferase, and $CO_2$ in blood sample. The strain is cultivated on a glucose salt-yeast extract. Glutamate dehydrogenase enzyme is used in coupling reaction of urease with NAD/NADH system for determination of urea in blood samples.

#### *Production of Glycerol Dehydrogenase*

The enzyme is used for estimating triglycerides in blood samples. It is being produced from *Cellulomonas flavigena* and is cultivated on glycerol-polyptone-yeast extract-salts medium while *Candida valida* is cultured on glycerol and dihydroxyacetone as carbon source. Strains of *Geotrichum* and *Endomyces* utilize propane 1,2-diol, butan-1,2-diol, and acetone as carbon source. The carbon source is exhausted within 24 hrs.

#### *Production of Glucose Dehydrogenase*

This enzyme is used for determination of blood glucose. Fermentation is done by bacteria - *Bacillus megatherium*. Production of enzyme is during stationary phase in batch fermentation using glucose as substrate. The maximum rate of enzyme production occurs after initial utilization (5%) of glucose (i.e., after 10 hrs) and persists for further 10 hrs.

## OXIDASES

Commercially, oxidases are used in analytical methods.

### Production of Glucose Oxidase

Fungi like *Aspergillus niger* and *Penicillium* spp. are used for production of glucose oxidase through fermentation processes are developed to produce high levels of glucose oxidase and catalase. Concurrent production of catalase is advantageous to keep hydrogen peroxide concentration at low under conditions of use.

#### *Small-scale Production*

Presence of succinate, malate, asparate, glutamate or ammonium ions affects the rate of enzyme production. Ammonium ion inhibition can be reduce by production of asparate. The enzyme is extracted from mycelial cell by disruption. *Penicillium purpurogenum*, under optimum conditions secrete the enzyme in to the medium and gives maximum yield and requires a longer period of three days for production.

**TABLE 19.1:** Growth medium for small scale production of Glucose Oxidase

| **Sucrose (as carbon source) Beet molasses are also used** | 5% |
|---|---|
| Corn steep liquor | 2% |
| Nitrogen source | Nitrates and urea |
| Temperature for inoculated media | 28°C |
| Time | 48 hours (Maximum production can be achieved within 16-20 hours) |

### Cholesterol Oxidase

The enzyme is used for determination of cholesterol in the blood sample. It is obtained from *Nocardia rhodochrous*. Cholesterol is required for the organism as an inducer for the synthesis of this enzyme. A poorer contact between cholesterol in the medium and the cell may result into a loss in the enzyme yield.

**TABLE 19.2:** Growth medium for small scale production (800 L fermenter) of cholesterol oxidase

| **Glycerol** | **6.g L$^{-1}$** |
|---|---|
| Yeast extract | 12 g.L$^{-1}$ |
| Salts | |
| PH | 6.7 |
| Incubation temperature | 29°C |
| dissolved oxygen tension (DOT) | within the range of 35-50% saturation |

The enzyme is harvested and can be released without cell disruption by washing with Triton X-100 (isoctylphenoxy-polythoxyethanol).

## HYDROLASES

### Glycosidases

#### *Bacterial α-Amylases (α-1, 4-Glucan Glucanohydrolase)*

The enzyme is obtained from *Bacillus* species, *Pseudomonas saccharohiba* and *Clostridium* spp. using fermentation technique. It is carried out by the aerated submerged cultivation technique. *Bacillus licheniformis* and by *Bacillus stearothermophilus* produce highly thermostable enzymes. In stationary culture *B. subtilis* gives high yield. A careful balance of carbohydrate and nitrogen ingredients of the medium is vital. Concentration of nitrogen source (organic/inorganic) should be higher than necessarily required for growth of the organism. However, it should be combined with small amount of proteinaceous compounds.

**Effect of Sugar (carbon source):** Readily metabolizable sugar in the medium causes poor production during the growth phase and causes catabolite repression. Higher yields can be obtained by using glucose as carbon source than that of maltose. A high carbohydrate level in the medium stimulates protease production and decreases amylase production.

The pH of the medium should be maintained near neutrality. The incubation at 25-35°C has to be carried for 3-4 days.

*Bacillus subtilis* in stationary culture produces surface-pellicle growth and is associated with higher yield of amylase, therefore, sterile air should be circulated over the pellicle. At the time of harvest the culture can either be filtered or centrifuged and aqueous portion concentrated by evaporation, or amylase can precipitated from aqueous solution by addition of acetone, ethanol, isopropanol or ammonium sulphate.

**TABLE 19.3:** Medium constituents, temperature, time duration, organisms and equipments used for bacterial α-amylases production

| Media constituents | Temp. | Duration | Species | Type/Scale of equipment |
|---|---|---|---|---|
| Yeast extract $Na_2HPO_4$, NaCl, $CaCO_3$, and glycerol | 55°C | 20 hrs | *B. coagulans* | 500 L. fermenter batch |
| Corn starch, peptone corn steep, liquor, salts | 55-65°C | 40-72 hr | *B. stearothermophilus* | Shake-flasks 50 ml culture |
| Barley, malt sprout, ammonium phosphate | 30°C | 40 hrs | *B. subtilis* | 6-L. fermentation batch. |
| Potato starch, ground barley, soybean flour, $Na_2SO_4$, $CaCO_3$ | 30°C | 144 hrs | *Bacillus licheniformis* | Shake-flask 100 hr culture |

#### *Fungal α-Amylases*

The enzyme produced from fungal source is low in thermal stability and the spectrum of sugars produced from the starch differs for fungal amylases. They yield maltose and some glucose.

It is mainly produced from *Aspergillus oryzae*, and *A. niger*. *A. oryzae* is universally used as the source for commercial fungal α-amylase. For commercial production of α-amylases from *A. oryzae* solid state fermentation is used while submerged fermentation technique is used for *A. niger*. This fungus grows on starchy-salt medium and produce other enzymes like amyloglucosidase. Wheat bran serves as basic component of the medium, and is spread in relatively thin layers in trays. It is moistened with water or dilute acid, sterilized, and inoculated with spores. After growth, the harvest can be dried at 50°C or less and then ground to liberate amylase. Amylase can also be extracted by water from wheat bran culture, precipitated by the addition of alcohol, and dried at 55°C or less.

### *β-amylase*

The enzyme is produced from barley, malt or Soyabean seeds. Bacteria like *Bacillus polymyxa* (starch-peptone-salt medium), *B. circulans*, *B. cereus* (in shake flasks), *B. cereus* var. *mycoides* (in 10 L fermenter) are used.

The gene responsible for β-amylase obtained from *B. cereus* has been cloned in *E. coli*. Recombinant DNA techniques have also been used to obtain the gene for enzyme on a plasmid suitable for cloning in *B. subtilis* using *E. coli* as intermediate host. A transformed *B. subtilis* strain can be grown in shake flasks at 30°C using a thinned starch, soya meal, sodium phosphate medium to produce 10-20 fold more.

### *Amyloglycosidase (Glucoamylase, α-1,4-Glucan Glucohydrolase)*

It is used for production of glucose and high fructose syrups from starch in conjunction with α-amylase and glucose isomerase. The enzyme occurs in various microorganisms like starch degrading molds (*Aspergillus*, *Mucor*, *Rhizopus*, *Endomyces* etc.) and in bacteria (*Aerobacter* and *Clostridium*). For commercial scale production of the enzyme utilizes fungi like *A. niger*, *A. awamori* and *A. foetidus*.

*A. niger* is used as the production strain in submerged cultivation. *Mucor* and *Rhizopus* are used in solid-substrate fermentation.

The submerged fermentation medium contains 12-20% organic matter. Maize, wheat, barley, rye or sorghum are used as starchy raw ingredients. The inorganic nitrogen source is ammonium salts. Removal of transglucosidases is important and can be achieved with range of techniques. Co-precipitation with maleic anhydride co-polymers and heteropoly acids are commercially employed for removal of transglucosidases.

### Cellulase

Microorganisms like bacteria and fungi are cellulolytic in nature. Cellulase is commercially prepared from *Aspergillus niger*, *Trichoderma viridae*, *Neurospora* and some other organisms. *Trichoderma viridae* produces an enzyme which degrades insoluble cellulose.

Cellulase in fungi is an inducible enzyme and they produce cellulase when fungal cells are grown on cellulose, on glucan of mixed linkages including β-1,4 bond, and on few oligosaccharides. The inducing effect is due to cellobiose. It is the soluble hydrolysis product of cellulose. Other inducers are β-1,4 galactoside, saphorose, and β-1,2-glucoside. At low concentration (0.1%), cellobiose serves as an inducer of cellulose while at high concentration (0.5-1.0%) it represses cellulase formation, and it can also act as an inhibitor of cellulose action.

High yields of cellulase is obtained from *A. niger* using wheat bran tray method.

*Trichoderma reesei* produces cellulases when grown on a glucose carbon source, provided that appropriate trace elements and calcium are present. *Trichoderma reesei* in solid substrate cultures are able to produce more concentrated extracts than were submerged cultures, at 57% moisture. Continuous submerged cultures using *T. reesei* with a glucose medium in a 10 L fermenter, yields were higher at pH 3.0. Trace elements like iron, manganese, zinc and cobalt are needed for enzyme production but not for growth of *Trichoderma reesei*. The peptone, Tween, and cellulose concentrations are occasionally varied to a minor degree. Use of protein hydrolysate such as peptone enhances the production of cellulase. Tween 80 enhances the production of soluble cellulose.

During fermentation the pH value lower down from 5.7 to around 2.7. This occurs because ammonium ions uptake is followed by consumption of cellulose. For optimum growth and enzyme production pH regime should be at 3.5. For production inoculum needs 4-5 days. Concentration and purification of the enzyme is done using precipitation, adsorption or gel filtration techniques.

## Proteolytic Enzymes

Proteinase enzymes are commercially important. The important proteinase enzymes are as follows:

**TABLE 19.4:**

| Type | Source |
|---|---|
| Carboxyl proteinases (acid) | *Aspergillus* spp., *Mucor* spp., Calf stomach (rennin), Pig stomach (pepsin) |
| Metalloproteinases (Neutral) | *Bacillus thermoproteolyticus* (thermolysins), *Bacillus amloliquefaciens, Bacillus subtilis, Aspergillus* spp. |
| Serine proteinases (alkaline) | Animal pancreas (trypsin and chymotrypsin), *Bacillus licheniformis, Bacillus subtilis and Bacillus amyloliquefaciens* |
| Thiol proteinases | Papaya (papain), Pineapple (Bromelin) |

### *Proteinases of Fungal Origin*

The important protease producing fungi are *Aspergillus oryzae. A. sojae, A. niger. A wentii, Mucor pusillus, Mucor miehei, M. delemar* and *Amylomyces rouxii*. The fungus is grown on wheat bran under fermentation conditions similar to those for amylase production. At sporulation, proteolytic enzymes are present in the medium.

### *Microbial Rennin*

Under laboratory condition and at low scale (2L-capacity fermenter) microbial rennin is obtained from *Endothia parasitica* strain. A 5% inoculum obtained from 96 hrs- shake flask stage is used for the production in a stirred, aerated (0.5 vvm) vessel for 48 hrs at 28°C.

**TABLE 19.5:** Composition of the medium for Inoculum development and for production of rennin by *E. parasitica*

| Inoculum preparation | (per litre) | Main fermentation | (per litre) |
|---|---|---|---|
| Casein hydrolysate | 5.0 g | Glucose | 10.0 g |
| D. Mannitol | 5.0 g | $KH_2PO_4$ | 0.5 g |
| $FeSO_4.7H_2O$ | 15.0 g | $MgSO_4.7H2O$ | 0.25 g |
| Soluble starch | 10.0 g | $NaNO_3$ | 3.0 g |
| Soya protease digest | 2.0 g | Skin milk | 10.0 g |

| Inoculum preparation | (per litre) | Main fermentation | (per litre) |
|---|---|---|---|
| Trace elements | 1.0 g | Soyabean meal | 30.0 g |
| Yeast extract | 2.0 g | | |

Solid substrate cultivation of *Aspergillus oryzae* and *A. sojae* are used for proteinases production which produce a mixture of enzymes containing acid aspartyl proteinase, neutral Metalloproteinases, and serine proteinase are produced. The optimum pH for activity should remain 4.0 to 5.0. In submerged culture, with nutrient broth, *A. oryzae* produce only serine proteinase. *A. niger* and closely related species with black spores often produce acid proteinase of the pepsin type. Proteases are recovered by procedures similar to those for mold amylases.

### Proteases of Origin

Bacterial strains like *Bacillus licheniformis*, *B. amyloliquefaciens* and *B. stearothermophilus* produce proteases enzyme. High yields could be obtained from *B. subtilis* if grown satisfactorily in culture. High carbohydrate level in the medium stimulate protease production and depresses amylase production. During fermentation, insoluble protein of the medium is first partially hydrolyzed by boiling in dilute acid/enzymatic treatment. pH of the medium should be kept neutral. Fermentation proceeds for 3-5 days at an incubation temperature of 37°C.

The biosynthesis of this enzyme is correlated with particular growth phases of the microbial culture. *Bacillus* spp. produces extracellular protease during post-experimental growth phase. The production of enzyme depends on exponential growth phase and medium composition. Catabolic repression by glucose declines proteinase production. Carbon source in the media is ground barley while source of nitrogen may be protein hydrolysates or sodium glutamate. The level of the starch level is limited. are used as.

For harvesting of enzyme, culture is filtered or centrifuged. The aqueous portion is concentrated using evaporation technique at reduced pressure and at temperature (not less than 40°C). Recovery can also be carried by precipitation.

## PENICILLIN AMIDASES (PENICILLIN ACYLASE)

The enzyme is used for preparation of 6-amino penicillanic acid (6APA) from naturally produced penicillin G or V. This compound is an important intermediate in the synthesis of semisynthetic penicillins.

**TABLE 19.6:** Types of penicillins produced by industrial fermentation

| Source | Product | Fermenter | Grown on | Medium/L | Temperature |
|---|---|---|---|---|---|
| Bacteria (*E. coli*) | Penicillin G (phenyl acetyl 6-APA) | 100 L aerated fermenter vessel | phenyl acetic acid and glutamate | $KH_2PO_4$, 3g; $K_2HPO_4$, 7g; $(NH_4)SO_4$. 1g; $MgSO_4.7H_2O$, 0.2 g; Phenyl acetic acid, 2 g and Sodium glutamate, 5 g. | 30°C (8-10 hrs) |
| *E. coli* and Molds | penicillin V-amidases | 25 L | | Glucose, 30 g; $KH_2PO_4$, 1g; Yeast extract, 10 g; $MgCl_2.4H_2O$, 05 g; $Ca(NO_3)_2.4H_2O$, 0.5 g; and $FeSO_4$, 0.5 g. | 27°C |

# TRANSFERASES

## Glycosyl Transferases

Glycosyl transferases enzyme is used for manufacture of dextrans. It is used for medical and laboratory purpose.

## Cyclodextrin Glycosyl Transferase

The sources of this enzyme are *Bacillus* spp., *Klebsiella pneumonia* and *Micrococcus varians.* The enzyme produces cyclic oligosaccharides from starch. A mixture of cyclodextrin containing hexamer, heptamer, octamer is produced. The preparations for these depends on the source of enzyme and the condition of the reaction.

**TABLE 19.7:** Medium for production of cyclodextrin glycosyl transferase

| Compound | Concentration/L |
|---|---|
| $CaCO_3$ | 1.5 g. |
| Caseinhydrolysate | 6 g |
| $FeSO_4.7H_2O$ | 0.25 g |
| KCl | 0.5 g |
| $KH_2P_4$ | 1 g |
| $MgSO_4.7H_2O$ | 0.5 g |
| $NH_4Cl$ | 3g |
| Sodium citrate | 0.5 g |
| Starch | 16 g |
| Yeast extract | 6 g |

Strains in 200 ml culture were incubated for 12-24 hrs at 25°C. Static culture gave higher yields of activity than shaken cultures.

# GLYCEROL KINASE

The enzyme is used for determination of triglycerides levels in blood samples. *E. coli* $K_{12}$. *Candida mycoderma* are used for production of this enzyme at commercial level.

### *Medium*

The medium for seed culture for production of glycerol kinase includes (in per L) Biotin (5 μg), $CaCl_2$ (0.1 g), $CuSO_4.5H_2O$ (1 mg), $FeSO_4.7H_2O$ (1 mg), Glycerol (90%) (9g), KCl (0.4g), $MgSO_4$ (0.1 g), $NH_4H_2PO_4$ (0.85 g), Vitamin $B_{12}$ (0.1 mg), $ZnSO4.7H_2O$ (1 mg), and pH (adjusted with ammonia solution, 7.0. It should be held for 30-40 hrs at room temperature with aeration.

# GLUCOSE ISOMERASE

The enzyme is used to catalyses the isomerization of D-xylose to D-xylulose. Bacterial glucose isomerase reacts with D-glucose to product D(-)-fructose which is commercially used in the preparation of sweet syrups from starch hydrolysates. The enzyme is produced suing strains of *Streptomyces flavogriseus, Streptomyces violaceus, S. olivochromogens, S. wedmorensis, Bacillus coagulans, B. stearothermophilus, B. licheniformis, Arthrobacter* spp. and *Actinoplanes missouriensis* etc. It is produced by submerged fermentation either in batch culture or in continuous culture. Most of the wild strains requires xylose as an inducer, and glucose cannot act as inducer.

## Production of Glucose Isomerase using *Bacillus coagulans*

For laboratory scale production, continuous fermentation with different limiting nutrients are used with dual carbon-oxygen limitation. Catabolite repression regulates the production of glucose isomerase. Production in continuous culture is repressed in the presence of surplus carbon and inorganic growth limiting factors. In batch fermentation, growth on xylose, this strain produces enzyme during cell growth.

### *Medium*

The medium for Glucose Isomerase production by *Bacillus coagulans* includes (in g/L) $(NH_4)_2SO_4$ (3.0), $K_2HPO_4$ (1.0), maize meal (80.0), $MgSO_4.H_2O$ (0.2), $MnSO_4.H_2O$ (0.05), xylose (4.0), and yeast extract (5.0). Initial pH should be maintained at 7.0. Culture in shake-flask (single stage) is incubated at 50°C for 40 hrs.

## Production by *Streptomyces wedmorensis*

The enzyme is produced by *Streptomyces wedmorensis* when an inert suspended solid is added to the medium. Production-scale process with two 10% inoculum stages uses 5 $m^3$ of the medium in the main fermentation. Addition of 0.1% histidine to casein hydrolysate and yeast extract promotes enzyme production.

### *Medium*

The medium for the production of glucose isomerase by *Streptomyces wedmorensis* (25 L fermenter for 2 stage inoculum) includes (in g/L) $COCl_2.6H_2O$ (0.24), corn steep liquor (40.0), wheat bran (40.0), time required (48 h), temperature (30 °C), and pH (7.0).

### *Purification*

For purification of enzyme cell rapture (enzyme is intracellular) and solubilizing of preparations techniques are used. After removing the cellular material, the enzyme is then adsorbed on DEAE-cellulose, recovered and washed.

# 20. Applications of Enzymes

Enzymes are protein biological or organic catalysts and are associated with metabolic reactions, respiration, the conversion and transfer of energy within living systems, and the formation of various macromolecules and cellular components. They exhibit specificity, therefore, used in pharmaceutical industries, in therapeutics and in clinical systems. At least 2,000 different enzymes are now known, each of them capable of catalyzing a specific reaction. Moreover, different cell types synthesize different types of enzymes – no cell contains all of the known enzymes. The specific enzymes produced by a cell are the major factors determining the biological activities and functions performed by that cell.

For thousands of years, processes such as brewing, bread making and production of cheeses have involved the use of enzymes. Enzymes of fungi and bacteria have a wide variety of industrial applications, including alcoholic beverages, food, detergents, and pharmaceuticals. In the production of alcoholic beverages, microbial amylase may be used instead of the enzyme from barley malt to prepare grains for fermentation. Protease can also chill proof beer by degrading the proteins that cause cloudiness in such beverages upon refrigeration. Enzymes are used in varied industrial sectors, consumer products, and scientific pursuits.

In past few years interest in various applications enzymes has been developed because of greater understanding of the function and nature of enzymes and also advances involved in handling of these, particularly their immobilization. Most of the enzymes catalyze the reactions different from their inherent reactions as well as under special conditions, i.e., hydrolases catalyze, transferase reactions and semi synthesis or transformation of antibiotics etc., carried out based on their non-physiological characteristics.

## APPLICATIONS IN PHARMACEUTICAL INDUSTRIES

Enzymes possess multicatalytic functions. They are used for the synthesis of optically active amino acids. Enzymes demonstrate stereo specificity in catalysis.

Enzymes decarboxylases catalyze decarboxylation of L-amino acids and also stereo specific protonation of the formed amines. The synthetic route for preparation of cortisone is form stigmasterol, which chemically converts it into progesterone. Hydroxylation of progesterone is carried out by *Rhizopus nigaricus.* It yields II-α-hydroxy progesterone. Use of microbes in transformation of steroid is an important aspect of pharmaceutical industries. All the enzymatic reactions are based on fermentation technology.

### Synthesis of Amino Acids

Enzymes are used for synthesis of optically active amino acids e.g., D-amino acids, and their derivatives. A variety of procedures for the synthesis of amino acids and their derivatives with microbial enzymes have been developed. Mainly hydrolytic enzymes [e.g., pyridoxal 5'-phosphate (pyridoxal-p)] are used. Pyridoxal-p dependent enzymes play the important role in amino acid metabolism because they catabolize transamination, decarboxylation, racemization, replacement, addition, elimination and a few other reactions of amino acids.

## Synthesis of Nucleosides and Nucleotides

Enzymatic procedure for the production of inosine 5'-monosphate (IMP) and guanosine monophosphates (GMP) has been developed with 5'-phosphodiesterase nucleases. Similarly, a number of antibiotics, antiviral compounds, antineoplastic compound, immunosuppressors, coenzymes, and sugar containing nucleosides or nucleotides have also been synthesized. Currently, these nucleosides and nucleotides are produced by a salvage synthesis i.e., microbial conversion of an intermediate synthesized chemically or enzymatically, by an enzymatic or chemical conversion of an intermediate produced by fermentation, or by direct fermentation.

## Synthesis of other Compounds

### *Antibiotics*

Penicillin acylase catalyzes the deacylation of benzyl penicillin to yield 6-aminopenicillanic acid phenyl acetic acid. The reaction is based on pH. At pH 7.5 the reaction proceeds while at pH 6.00 it becomes reverse. The industrial semi-synthesis or transformation of penicillins and cephalosporins is with an immobilized acylase. The enzyme is also used for the stereo specific acylation of intermediates in the chemical synthesis of β-lactam antibiotics.

### *Peptides and Proteins*

**Peptides:** Different proteases and carboxypeptidases are employed for the peptide bonds i.e., the conversion of porcine insulin to human insulin. Only carboxyl terminal amino acid is different in human and porcine insulin (i.e., alanine for porcine and theonine for human insulins). The synthesis of human insulin from porcine insulin is carried out with a lysine specific protease enzyme of *Achromobacter lyticus*. Different proteases and carboxy peptidases are also used for the synthesis of pharmaceutically important peptides like trypsin (human insulin), *Achromobacter* protease I (human insulin), chymotrypsin (encephalin, caerulein) subtilisin (angiotensin), carboxypeptidases Y (human insulin, encephalin, caerulein), thermolysins (angiotensin, caerulein), and pepsin and fungal carboxypeptidases (eldocin, aerulein). Aspartame (L-aspartyl-L-phenylalanine methyl ester) is a dipeptide sweetener. The synthesis of the dipeptide with thermolysins immobilized on an ion exchange resin or by a microbial exopeptidase has also been developed.

**Proteins:** Peptidoglutaminases hydrolyze r-amide group of glutamine in glutaminyl-peptide converts it in to glutamate residue.

**Triglycerols (triglycerides):** Enzyme lipase produced by *Rhizopus delemar* converts olive oil to cacao butter-like oil.

**Cyclodextrin:** Cyclomaltodextrin glucanotransferase in starch medium synthesizes cyclodextrins in the presence of complex forming reagents such as acetone, sodium dodecyl sulphate, I-decanol.

**Dentifrice enzymes:** Dextranase hydrolyses dextran. This reaction is effective in reduction and protection of dental plaques and caries. They hydrolyze dextran to glucose and isomaltose at pH 5.5 with 55% hydrolytic efficiency.

## APPLICATIONS IN THERAPEUTICS

In some heritable human genetic disorders there remains deficiency or even a total absence of one or more enzymes in the diseased tissues, and under abnormal conditions there may be an excessive activity of a specific enzyme. These abnormalities are sometimes may be controlled by a drug designed to inhibit its catalytic activity. Some body enzymes are misses due to inborn errors of metabolism. Therefore, missed enzymes are replaced.

These enzymes are in solubilized form or immobilized on some substrate and may be given orally or injected or implicated in soft tissues as prosthesis.

### For Treatment of Gaucher's, Tay-sach's, and Fabry's Diseases

Gaucher's, Tay-sach's, and Fabry's are hereditary diseases and are caused due to deficiency of enzymes and result into abnormal accumulation of materials that would otherwise broken down by missing enzymes.

Gaucher's disease (a lysosomal storage disease) occurs due to absence of enzyme glucocerebrosidase and leads to accumulation of glucocerebrosides in various tissues. If the enzyme is replaced in body, the symptoms may be relived in an individual.

Tay sach's disease is caused due to absence of enzyme hexose deaminase. The enzyme is usually present in the lysosomes of ganglion brain cells and removes the carbohydrate residues of the ganglioside lipids. If the ganglioside concentration in these cells accumulates, it affects the brain function and ultimately leads to the death of an individual.

Fauvism disease is caused due to lack of the enzyme *glucose-6-phosphate dehydrogenase* in the red blood cells. This affects the ability of cells to regenerate glutathione, which leads to membrane fragility and haemolytic anemia.

Not all the enzyme deficiency diseases require enzyme replacement. For treatment of such diseases offending substance should be removed from diet. For example, phenylketouria (PKU) is caused due to inability of body cell to metabolize phenylalanine, which results in the accumulation of toxic levels of phenylalanine and its abnormal metabolites. Other example are galactosemia (*galactose or galactose urydyltransferase*), Orotic aciduria (*orotidine-1-phosphate decarboxylases*) and arginosuccinuria (arginosuccine lyase) etc.

Uses of hydrolases or oxidase enzyme regenerating systems are also common for some enzyme replacement therapy. The inherited deficiency of *galactokinase* is cured by injecting coencapsulated galactokinase with ATP/ADP and pyruvate kinase. It initiates ATP generation system.

The treatment of malabsorption secondary to pancreatic insufficiency is caused by chronic pancreatitis, pancreatectomy and cystic fibrosis. The manifestation of pancreatic malabsorption is steatorrhea a voluminous malodorous stool laden with fat that is not being adsorbed. If the output of pancreatic lipase is less than 10% of normal, clinical steatorrhea will not occur, therefore, the treatment of steatorrhea Therefore, replacement of preparation lipase enzyme is essential.

For treatment of chronic pancreatitis and pancreatic fistulae, pancreatic enzymes (a mixture of *lipase*, *amylase* and *protease*) are used. Pancreatic enzyme preparation consists

Treatment of acute lymphocytic leukemia, T-cell leukemias and lymphomas, is depended totally or partially upon exogenous asparagines, which are enzymatic destructed enzyme asparaginase. It is injected into plasma. It arrests their growth and sometimes result in reduction of tumour mass.

Blood clots are dissolved in tissue by plasminogen activator (TPA). They convert plasminogen to plasmin. Plasmin enzyme breaks down protein fibrin which is the main component of blood clot.

Blood clots are dissolved by enzymes urokinase and streptokinase and both soluble fibrinogen and insoluble fibrin clots are destroyed. These enzymes activate plasmin.

Enzyme hyaluronidase hydrolyse hyaluronic acid and accelerates the subcutaneous spread of both particulate matter and solutions of depolymerising the hyaluronic acid. The enzyme is injected through drug.

*Subtilins* are proteolytic enzymes. They digests the denatured proteins present in necrotic tissues. These enzymes are used as adjunct for biochemical debridement of lesions of burns, decutibus ulcers incisional traumatic and pyrogenic wounds, and ulcers secondary to peripheral vascular disease.

Burned areas and dermal ulcers are cured with the help of enzyme collagenase. Papain is used in the debridement of necrotic tissues. Solution of crystallized trypsin is inhaled to liquefy viscous sputum.

Fibrin and liquefy pus are being cured with a mixture of fibrinolysin of bovine plasma and deoxy ribonuclease when used topically. It is obtained from bovine pancreas. The mixture is also used as debriding agent in surgical wound, ulcerative lesion and, second and third degree burns. It is also used intravaginally in severe cervicitis and vaginitis.

## CLINICAL ANALYSIS

Disease are diagnosed by clinical analysis, which includes the measurement of activity of certain enzymes in the blood plasma, red blood cells, or tissues samples. Serum enzymes have their importance in clinical analysis. They are intracellular and elevated in hyperactivity diseases, malignancy or injury to cardiac, hepatic, pancreatic, muscle, bone and tissues. For identification of specific tissue involved in the disease is determined by the type of enzyme elevated. It is helpful for determinations and differentiation of various pathological conditions.

### Transferases

Transfer of amino or phosphate group to another is catalyzed by enzymes transferases like aspartate aminotransferase and alanine aminotransferase. They are helpful in clinical diagnosis. Aspartate aminotransferase is found in liver, cardiac and skeletal muscle. It is elevated in myocardial infraction and Duchenne muscular dystrophy. Alanine aminotransferase is found in the liver tissues. The levels of both the enzymes increased during liver diseases, acute toxic or viral hepatitis, obstructive jaundice and hepatic cirrhosis. These enzymes catalyze the transfer of amino group from glutamic acid to keto acids (oxaloacetic or pyruvic) to form aspartic and α-ketoglutaric acid with Aspartate aminotransferase while alanine and α-ketoglutaric acid with alanine aminotransferase. These products are analyzed using colorimetric and spectrophotometric methods.

### Ornithine Transcarbamylase

Ornithine transcarbamylase enzyme is involved in urea cycle in serum. It is used in the clinical investigation of the liver disease. The enzyme catalyses the conversion of ornithine to citruline. The normal serum value is 0 to 0.4 U/L.

## Creatine Kinase

Creatine kinase ( a transferase enzyme) present in the muscle and brain tissue. The enzyme catalyses the transfer of phosphate groups from creatine phosphate to adenosine diphosphate (ADP) and converted into adenosine triphosphate (ATP). Activated CK activity is observed by increase of ATP in the creatinine phosphate-ADP reaction in the presence of glutathione or cysteine thiol activators. Fluorimetric method is used for ATP determination. The enzyme is elevated in myocardial infraction and Duchenne muscular dystrophy. The level of enzyme remains normal levels in liver disease.

## Oxidoreductases or Dehydrogenases

This group of enzyme catalyzes hydrogen transfer in cellular oxidation process. Lactic, α-hydroxybutyric, Malic, glutamic, isocitric, and sorbitol dehydrogenases are of diagnostic importance. Lactic dehydrogenases catalyzes the reversible conversion of pyruvic to lactic acid in the presence of NADH. The level of lactic dehydrogenases increases in hepatic disease, pernicious anemia and muscular dystrophy. Lactic dehydrogenases and Malic dehydrogenases phosphatases and leucine aminopeptidases are found in isoenzyme forms. Isoenzymes have similar substrate specificity but different biophysical properties. Malic and sorbitol dehydrogenases in the presence of NAD, catalyze the conversion of malate or sorbitol to oxaloacetate or fructose respectively. They are used for diagnostic importance in MI and acute liver injury.

## Hydrolases

These are enzymes catalyse the addition of elements water across the bond, which is cleaved. Hydrolases like amylases, lipases, phosphatases, 5-nucloetidase, γ-glutamyl transferase and leucine aminopeptidases are used for clinical purposes.

### *Salivary and Pancreatic Enzymes*

These enzyme hydrolyze starch to maltose and dextrins. The activity of amylase activity is measured using method based on the loss in certain properties of starch or by the generation of reducing substances. Elevation in concentration is observed in acute pancreatic, acute abdominal conditions and salivary gland diseases.

### *Lipases*

They catalyze the conversion of triglycerides to glycerol and fatty aids. Determination of serum lipase is used for the diagnosis of acute pancreatitis (10 fold increase), pancreatic injury or inflammation of organs contiguous to pancreas.

### *Phosphatases*

They catalyze the hydrolysis of orthophosphoric esters, and are grouped into acid oralkalie phosphatases. Acid phosphatase is of first diagnostic value in metastatic carcinoma of prostate. Acid phosphatase activity (prostate enzyme) is investigated by using tartrate inhibitor.

### 5'-nucleotidase

It is an alkaline phosphomonoesterase. It hydrolyses nucleotides with a phosphate radical attached to 5'-position of the pentose (i.e. AMP). It is elevated in liver disorder.

### Serum Lysozyme (muramidase)

The activity of this enzyme is increased in certain kinds of leukemia. Enzyme serum arginasea hydrolyses arginine to ornithine and urea, and serum guanase. They are sensitive indicators of hepatic necrosis.

### Serum γ-Glutamyl Transferase

The activity and level of this enzyme is increased during diseases of liver, bile ducts and pancreas. Together with alkaline phosphatase, leucine aminopeptidases, and 5'-nucleopeptidase, serum γ-glutamyl transferase can be tested in the group of cholastasis-indicating enzymes. This method is based on the hydrolysis of γ-glutamyl-p-nitroanilide.

# 21. Enzyme Immobilization

Microbes are, in many ways, packages of enzymes. The use of enzymes in industrial applications is limited because most of the enzymes are relatively unstable, of high cost isolation, purification, and recovery of active enzymes from the reaction mixtures after the completion of catalytic process. Due to these reasons, industries have been making limited use of free enzymes to manufacture many products such as high-fructose syrups or semisynthetic antibiotics such as ampicillin. To do this, the *enzyme must be immobilized on the surface of some solid support or otherwise manipulated so that it can convert a continuous flow of substrate to product without being lost.* These techniques have also been adapted to live whole cells, and sometimes even to dead cells. Whole cell systems are difficult to aerate, and they lack the single-enzyme specificity of immobilized enzymes. However, whole cells are advantageous if the process requires a series of steps that can be carried out by one microbe's enzymes. They also have the advantage of allowing continuously running processes with large cell populations operating at high reaction rates. Immobilized cells, which are usually anchored to microscopically small spheres or fibers, are presently used to make high-fructose syrups, aspartic acid, and numerous other products of biotechnology.

## IMPORTANT FEATURES ENZYME IMMOBILIZATION

1. Enzyme immobilization involves confining soluble protein catalysts, or individual buoyancy of cells in a reactor system, which the charge is allowed to pass continuously.
2. Enzyme immobilization is used in development of biosensor, bioelectronics sensor in fermentation technology as well as enzyme therapeutics.
3. Enzyme immobilization is utilized in anchoring on the surface phase however, may be in the form of fine particulate, membranous or monolithic spheres.
4. Enzyme in turn may be bound to another enzyme via cross-linking. Thus, rendering the overall composite to be an insoluble, but polymeric enzyme system.
5. It implies for restrictive localization of enzymes.
6. The immobilization is obtained via covalent coupling employing adsorption or physical entrapment of enzymes within the enzymes phase (polymeric matrix).
7. The nature of immobilized enzyme is characteristically dependent upon the character of matrix and enzyme phase.

## DISADVANTAGES OF ENZYMES

1. The use of enzymes in industrial applications is limited.
2. The enzymes are relatively unstable and are rapidly inactivated by heat.
3. One has to pay high cost for isolation, purification, and recovery of active enzymes from the reaction mixtures after the completion of catalytic process.
4. For soluble enzymes in batch operations is uneconomical because active

enzyme is lost after each reaction.

## ADVANTAGES OF IMMOBILIZED ENZYMES

1. The above disadvantages could be overcome using immobilized enzymes.
2. Immobilized enzymes are physically confined during a continuous catalytic process.
3. Immobilized enzymes could be recovered from the reaction mixture and used over and over again, thus improving the economy of the process.
4. Immobilized enzymes have advantages over their soluble counterparts and can be reused, process.
5. They can be operated continuously and can be readily controlled.
6. The products can be easily separated and enzyme properties (activity and stability) can be altered favourably.

## PROPERTIES OF IMMOBILIZED ENZYMES

Enzymes can be used in best form if we best understand the changes, which occur after immobilization, mainly in their physical and chemical properties. In such enzymes changes occur in their stability and kinetic properties because of the microenvironment imposed upon them by the supporting matrix and by the products of their own action.

### Stability

During insolubilization enzymes coupled to inorganic carriers were generally more stable than those attached to organic polymers when stored at 4 or 23 ° centigrade. The stability of the enzymes might be expected to either increase or decrease on insolubilization, depending on whether the carrier provides a microenvironment capable of denaturing the enzymic protein or of stabilizing it. Isolating the enzyme molecules from mutual attack by immobilizing them on a matrix reduces inactivation due to autodigestion of proteolytic enzymes. Stability to denaturing agents sometimes changes upon insolubilization.

### Kinetic Properties

Kinetic properties generally decrease in specific activity of an enzyme upon insolubilization. It can be attributed to denaturation of the enzymic protein caused by the coupling process. After insolubilization of an enzyme it finds itself in a microenvironment, which may be drastically different from that existing in free solution. The new microenvironment may be a result of the physical and chemical character of the support matrix alone, or it may result from interactions of the matrix with substrates or products involved in the enzymatic reaction.

The Michaelis constant is decrease by more than one order of magnitude when substrate of opposite charge to the carrier matrix is used. This happens at low ionic strengths, and when neutral substrates are used.

The diffusion of substrate from the bulk solution to the microenvironment of an immobilized enzyme can limit the rate of the enzyme reaction. The rate at which substrate passes over the insoluble particle affects the thickness of the diffusion film, which in turn determines the concentration of substrate in the vicinity of the enzyme and hence the rate of reaction. Diffusion of large molecules is also limited by stearic interactions with the matrix. This reflects the relative activity of bound enzymes towards high molecular weight substrates. It is usually found to be lower than towards low molecular weight substrates. This, however, may be an advantage in some cases, since the immobilized enzymes may be protected from attack by large inhibitor molecules.

## TYPES/METHODS OF IMMOBILIZATION

The methods of immobilization fall into four main categories (1) Physical adsorption onto an inert carrier, (2) Inclusion in the lattices of a polymerized gel, (3) Cross-linking of the protein with a bifunctional reagent, and (4) Covalent binding to a reactive insoluble support.

These methods depend upon physical relationship of the catalyst to the matrix (carrier). The selected matrix must enhance the operational stability of the immobilized enzyme purification. The carriers used for this purpose may be porous or nonporous materials with organic (natural or synthetic) or inorganic nature. The catalyst may be covalently bonded to the polymer, physically adsorbed onto the polymer, cross-linked with itself and possibly another inert protein, entrapped inside a polymer matrix or encapsulated in a polymer bag.

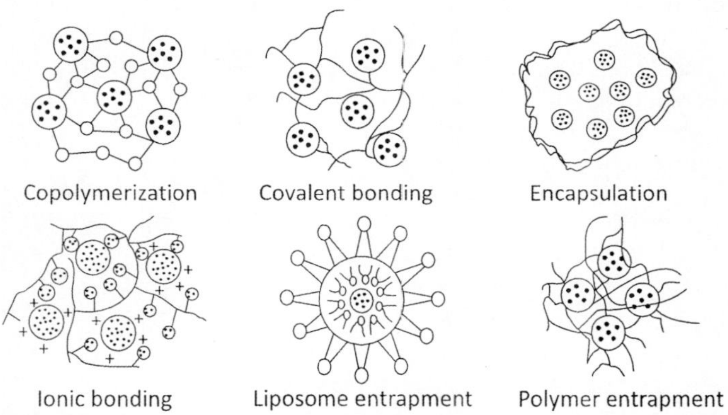


**FIGURE 21.1:** Types of enzyme immobilization binding techniques

### *Adsorption*

Adsorption is the oldest of techniques used to immobilize enzymes, dating back to 1916 when Nelson and Griffin used both charcoal and aluminium hydroxide to adsorb invertase. Since that time wide range of organic and inorganic substances have been utilized as supporters fro absorbed enzyme. Both organic materials such as charcoal, various cellulose derivatives, and ion exchange resins and inorganic materials including silica, alumina, titania, glass and various naturally occurring minerals have been used.

Physical adsorption of an enzyme onto a solid is the simplest way of preparing immobilized enzymes. The method depends on non-specific physical interaction between the enzyme protein and the surface of the matrix, brought about by mixing a concentrated solution of enzyme with the solid. Various types of compounds are used for adsorption like calcium phosphate ($\alpha$-amylase), charcoal (catalase), cellophane - followed cross-linking with glutaraldehyde, inorganic adsorbents- (glucose oxidase), charcoal and DEAE-sephadex (invertase), agarose gel, DEAE-sephadex (amyloglucosidase) and cellulose (subtilisin).

The process of adsorption of an enzyme is done by contacting the enzyme and polymer support in a stirred reactor or by percolating the enzyme through a packed bed, tube or membrane formed from a support material. During this process different specific and non-specific forces are applied e.g., electrostatic, hydrophobic interactions or affinity bonding to specific legends attached to the matrix. The amount of enzyme, which is adsorbed is dependent on the concentration to the unit surface of the carrier and the activity increases with increasing concentration of enzyme which approaches to a saturation value, with decrease of the specific activity. The procedure includes the mixing of enzyme under suitable conditions like pH, nature of the solvent, ionic strength, concentration of the enzyme and adsorbent, and temperature.

Since no reactive forces are involved in the process therefore, there are no conformational changes in the enzymes on immobilization. Alumina, activated carbon (charcoal), clays, collagen, glass diatomaceous earth, Hydroxyapatite, cellulose metal oxides and various siliceous materials are commonly used adsorbents.

Physical adsorption of enzymes is the hydrophobic interactions. The adsorption is based on both ionic and hydrophobic attractive forces. Because of the weak bonds involved, desorption of the protein is dependent on changes in temperature, pH, ionic strength or even the mere presence of substrate.

**Advantage of adsorption:** For insolubilizing enzymes no reagents are required. Only a minimum of activation steps are needed. Adsorption is cheap, easily carried out. It tends to be less disruptive to the enzymic protein than chemical means of attachment, the binding being mainly by hydrogen bonds, multiple salt linkages, and Van der Wall's forces.

### *Occlusion*

Occlusion involves no disruption of the protein molecules. Confining enzymes within the lattices of polymerized gels method is used for enzyme immobilization. The free radicals generated on the course of the polymerization may affect the activity of entrapped enzymes. In this process, only low molecular weight substrates can diffuse rapidly to the enzyme. Therefore, this method is unsuitable for enzymes, which act on macromolecular substrates, such as ribonuclease, trypsin, Dextranase, etc. This is the disadvantage.

Occlusion allows the free diffusion of low molecular weight substrates and reaction products. The method is based on polymerization of hydrophilic matrix in an aqueous solution of the enzyme and break up the polymeric mass to the desired particle size. In this process, there is no bond formation between the enzyme and the polymer matrix.

Polyacrylamide gel has pores and the broad distribution in pore size of synthetic gels of the polyacrylamide type inevitably results in leakage of the entrapped enzyme, even after prolonged washing. This disadvantage may be overcome by cross-linking the entrapped protein with glutaraldehyde. Sometimes, ultrafiltration membranes of well-defined pore size may be used to occlude the enzyme is also used.

### *Cross-Linking*

Intermolecular cross-linking of the protein is another method for enzyme immobilization. Since the enzyme is covalently linked to the support matrix, very little desorption is likely using this method. It is achieved by intermolecular cross-linking of the protein, either to other protein molecules or to functional groups on an insoluble support matrix. Cross-linking an enzyme is expensive and insufficient, because as some of the protein material inevitably acts mainly as a support which results in relatively low enzymic activity. The method of cross-linking is best used in conjunction with one of the other methods. This helps to prevent the leakage from polyacrylamide gels as mentioned earlier. It is used more widely as a means of stabilizing adsorbed enzymes. For example, reported that carbamyl phosphokinase cross-linked to alkyl amine glass with glutaraldehyde lost only 16% of its activity after continuous use in a column at room temperature for fourteen days.

### Covalent Binding

It is the common method used for enzyme insolubilization. Enzymes are made of chains of amino acids and have number of reactive side chains, which can be utilized for forming chemical bonds with appropriately active support. The technique includes the formation of covalent bonds between the enzyme and the support matrix. For selection of type of reaction through which a given protein is insolubilized, the choice remains limited because the binding reaction is performed under conditions, which do not cause loss of activity of enzyme. In this method the active site of the enzyme should be unaffected by the reagents used.

Covalent bonding has an advantage of an attachment not reversed by pH, ionic strength or substrate. However, the active site of an enzyme may be blocked through the chemical reaction and the enzyme rendered inactive.

The various groups of enzyme that take part in the formation of chemical bonds are amino, imino, amide hydroxyl, carboxyl, thiol, methylthiol, guanidyl, imidazole and phenol ring. The functional groups of proteins are used for covalent binding under mild conditions. The important functional groups are: (i) the alpha amino groups of the chain and the epsilon amino groups of lysine and arginine, (ii) the alpha carboxyl group of the chain end and the beta and gamma carboxyl groups of aspartic and glutamic acids, (iii) the phenol ring of tyrosine, (iv) the thiol group of cysteine, (v) the hydroxyl groups of serine and theonine, (vi) the imidazole group of histidine, and (vii) the indole group of tryptophan.

The activated polymers used are hydro gels incorporated with diazo, carbodimide or azide groups.

**TABLE 21.1:** Covalent coupling. Binding agent or reagents, enzyme and carrier matrix used for enzymes immobilization

| Binding agent/Reaction | Enzyme | Carrier Matrix |
|---|---|---|
| Direct coupling | α-Amylase | DEAE-cellulose |
| Cyanuric chloride | Amyloglycosidase | DEAE-cellulose |
| Isocyanate | Cellulose | Polyurethane |
| Isocyanate | Glucose isomerase | Polyurethane |
| Isothiocyanate | Glucose oxidase | Porous glass |
| Isocyanate | Pectinase | Polyurethane |
| Carbodiie activation | Pronase | CM-Sephadex |

The above table shows that the process needs number of reactions to couple with functional groups on the protein other than the amino and phenolic residues. Coupling of amino ethyl cellulose to the carboxylic acid residues of enzymic protein in carried out in presence of carbodiimide and thiol residues of a protein.

It is oxidatively coupled to the thiol groups of a cross-linked copolymer of acrylamide and N-acryloyl-cystein. Increased number of reactive residues of an enzyme may increase yield of insolubilized enzyme. The alternative reaction sites are essential for activity of an enzyme. The wide variety of binding reactions, and insoluble carriers with functional groups capable of covalent coupling, or being activated to give such groups, makes this a generally applicable method of insolubilization, even if very little is known about the protein structure or active site of the enzyme to be coupled.

### Entrapment

The term entrapment has been used to cover a range of immobilization techniques. O'Driscoll (1976) has chosen to distinguish between those methods, where a gel or membrane is formed around enzyme by polymerization and those methods where the enzyme is retained in a preformed polymer by subsequent formation of a cross-linked network. Microencapsulation and fiber entrapment are clearly other examples of this immobilization technique. Hollow fiber and membrane devices have also been used to retain enzymes in given location without actually binding enzyme to a support material.

The technique is based on coupling enzymes to the lattice of a polymer matrix or enclosing them in semipermeable membranes It prevents the release of proteins while allowing the diffusion of substrates and products. In this method the enzyme itself does not bind to the membrane or gel matrix, and there is a little inactivation of the biological activity as compared with the coupling method. The active sites of enzyme molecules are not blocked by chemical bonds and enzyme remains in its native state.

It method is easy to perform. The enzymes are dissolved in a solution of the polymers precursors. The polymers may be polyacrylamide gels or natural gels (cellulose triacetate, agar, gelatin or alginate). To prevent the leakage of the low molecular weight enzymes from the gel, the average pore size of the gel should be as high as possible. The excessive diffusion limitation variability of pore size has to be controlled.

**Advantages:** That allows high local concentration of enzymes. Enzymes can contained without any chemical modification or bonding that may lead to activity loss. Multiple enzyme systems can be handled readily since entrapment is essentially the method living cells to retain their enzyme systems.

**Disadvantages:** (1) Polyacrylamide gel is used with sensitive enzymes of viable cells. (2) The free radicals generated during the course of polymerization are toxic. (3) Agar and carragenin can be depolymerized by mild heat. (4) For viable cells calcium alginate is the suitable gel for immobilization.

Penicillin acylase is fibre entrapped enzymes. It is immobilized by entrapment in the micro cavities of synthetic fibres. The physical entrapment is carried out by dissolving a fibre forming polymer (cellulose triacetate) in an organic solvent immiscible in water and emulsifying this solution with aqueous solution of enzyme. The emulsion is extruded through a spinneret into liquid coagulant (toluene or petroleum ether) that precipitate the polymer in a filamentous form, with a micro droplet of enzyme solution entrapped in the fibre. This method is helpful in reducing the diffusion limitation.

### Encapsulation

Encapsulation provides large surface area to enzyme to contact with the substrate. This method is useful because several enzymes can be immobilized in a single step. Enzymes are immobilized within microcapsules. These are prepared from organic polymers. The membranes encloses the enzyme and remain semipermeable to the substrate and the products. This is an easy and economic technique. The stability of the biocatalyst in solution is an important aspect of this technique. This method is not applicable for high molecular weight substrates. Occasionally inactivation of enzyme occurs during immobilization procedure, and the enzyme may become incorporated into membrane wall. Leakage of enzyme takes place through microcapsule. Using this technique, capsules of suitable size can be obtained. The diffusion limitation is also controlled. Capsules are made from ethyl cellulose nitrocellulose, polystyrene, polyvinyl acetate, polyamine glycol, polyethylene etc. A variant of this technique is the entrapment of enzyme solution in the lumen of hollow, semipermeable fibres.

## CHOICE OF IMMOBILIZATION METHOD

Immobilization is the method in which surface of the enzyme are attached with reactive groups outside the active catalytic and binding site. It is necessary to know the active sites of particular enzymes. It helps to select and avoid reaction with the essential groups therein. Alternatively, these active sites can be protected during attachment as long as the protective groups can be removed without loss of enzyme activity. Sometimes, this protective function can be fulfilled by a substrate of the enzyme or a competitive inhibitor; this also contributes towards retention of tertiary structure of the enzyme.

The surface on which the enzyme is immobilized plays several following vital roles:

1. Retaining of tertiary structure in the enzyme by hydrogen bonding or the formation of electron transition complexes.
2. Retention of tertiary structure may also be a vital factor in maximizing thermal stability in the immobilized state. Therefore, it is essential to have knowledge about new findings in the chemical nature of soluble thermostable enzymes.
3. The microenvironment of surface and the immobilized enzyme has an anionic or cationic nature of the surface, which causes displacement in the optimum pH of the enzyme up to 2 pH units. This may be accompanied by general broadening of the pH region in which the enzyme can work effectively.

## COMPARISON AND EVALUATION OF IMMOBILIZATION TECHNIQUES

The choice of the methods depends on the characteristics of the enzymes and their substrates including. It should be of low cost and effective. Efficient immobilization of enzymes can be achieved by combination of methods i.e., adsorption followed by cross-linkage. Each technique has specific limitations.

**TABLE 21.2:**

| Type of method | Characteristics |
|---|---|
| Physical adsorption and ionic bonding | It is prepared easily under mild conditions. Any variations in the ionic strength or pH of the solution, due to weak binding forces between the enzyme and matrix, results into leakage of the enzyme from the matrix. Helpful to retain initial high activity and low operational stability. |

| Type of method | Characteristics |
|---|---|
| Intermolecular cross-linkage and covalent binding | It is the chemical modification of the enzyme. It results into conformational changes of its structure and partial loss in activity. To minimize these problems, reactions should be carried out under suitable possible conditions. The operational stability of such immobilized enzymes high. |
| Entrapment method possess | Enzymes immobilized by this method have high retention of specific activity because there is no binding between enzyme and carrier. The technique is limited to the use of low molecular weight substrates and product molecules because strong mass transfer resistance is imposed by the entrapping polymer or membrane. |

## EFFECTS OF IMMOBILIZATION ON KINETICS AND PROPERTIES OF ENZYME

The immobilization process changes catalysts from homogenous (soluble enzyme) nature to heterogeneous one, in which enzyme is associated with a supporting or enveloping matrix. During immobilization, the activity of the enzyme is usually lost because of the reactions involved in the process or by occlusion of active sites in the enzyme support complex.

After immobilization of the enzyme there occurs many changes due to the physical and chemical nature of the carrier used, and the conformational changes in the enzyme structure. These changes result into conversion of catalysts to heterogenous nature with significant change in its properties. The reactivity of enzyme is also changes because of actively stability, pH and temperature optimum, kinetic constants (resulting from a change in activation energy) and substrate specificity.

### Michaelis Constant

Michaelis constant (Km) is increased or decreased after enzyme immobilization. The determination of Km value reflects the affinity between enzyme and substrate. A decrease in the Km value of an immobilized enzyme increases rate of reaction than its soluble native form. Conversely, an increased in Km value implies the use of higher substrate concentration to accomplish the same rate of reaction obtained with the free enzymes. The apparent Km value lowers down when the charges on the support and substrate are opposite. The decrease can be attributed to partition effects due to an increase in the electrostatic attractive forces between the carrier and the substrate. It leads to an increase in the concentration of the substrate in the microenvironment of the immobilized enzyme. Km value also decreases when internal and external diffusional effects are present under the operational conditions applied.

**TABLE 21.3:** Types of diffusion effects

| External diffusion | Limited due to an unstirred layer (the Nernst layer) around the particles of an enzyme. |
|---|---|
| Internal or pore diffusion | It is caused by substrate having to diffuse through the support matrix. It is significant when enzyme is immobilized by binding on porous supports and in entrapping techniques. |

Km value increases when conformational changes of the enzymes protein molecule and the interaction of the substrate with the enzyme occur. It is affected by stearic hindrance. These effects are pronounced in the covalent binding techniques because enzyme molecules are chemically modified.

## Maximum Reaction Velocity (Vmax)

The maximum reaction velocity of an immobilized and normal enzyme remains almost same. The Vmax value for an immobilized glucoamylase in the presence of maltose (as substrate) is 10 times more than the native enzyme. For gel-entrapped β-D-fructofuranocidase, the Vmax is 10 times lower than the free enzyme. The effectiveness depends on enzyme activity and diffusion limitation during immobilization.

### *pH Activity*

Immobilization of enzyme causes changes in pH profile and depends on the charges of the enzyme molecule. In the support binding method, when the matrix is charged, the kinetic behaviour of the immobilized enzyme vary from that of the free enzyme. This difference occurs due to partition effects that cause different concentrations of charged species (e.g., substrates, products hydrogen ion, hydroxyl ions etc.) in the microenvironment of the immobilized enzyme and in the bulk solution, due to electrostatic interactions with fixed charges on the support.

Presence of $H^+$ ions has pronounced effect on the enzyme activity. It depends on whether the initial pH is above or below the pH optimum. Substrate diffusion limitation also affects the response of an enzyme reaction rate and the response of an enzyme to change in pH.

### *Temperature*

Enzyme has a temperature optimum. It is active only in narrow range of temperatures. Beyond this optimum enzyme rapidly denatured and inactivated. The enzymes remain viable between 30°C and 80°C. Enzyme activity as a function of temperature may be expressed using Arrhenius equation.

$$k = A_e \, H/RT$$

Where, k = reaction rate, T = absolute temperature, A = arrhenius constant, H = activation energy and R = gas constant.

Diffusional effects in case of entrapment and binding on porous supports, protects enzyme against heats denaturation. Therefore, the optimum temperature of the enzyme becomes higher than that of the native enzymes. The diffusional effects, which protect the enzyme against heat denaturation lead to both lower activation and deactivation energies.

### *Stability*

Stability is the retention of activity for a longer periods under suitable conditions of storage and depends on the nature of the carrier surface to which it is bound. The prevention of conformational inactivation or protection of active groups on enzymes from reactive groups in solutions enhances the stability of immobilized enzymes. During operating conditions, the activity of enzymes is lost due to enzyme denaturation and physical loss from the carrier. Loss from matrix may be either as a result of carrier erosion or from breaking of bonds between enzyme and carrier or from gel entrapment matrix. Covalent coupling leads to stable preparations, while with weakly bound, physical adsorbed, or ionic bound enzymes the activity is lost into solution depending on temperature, pH, ionic strength and physical operating conditions.

Stability is also dependent on the type of reactor used. Abrasion or attrition may cause loss of activity for enzymes immobilized on organic support. Microbial contamination also causes loss of enzyme activity under operating conditions. Loss of activity of immobilized enzymes is diffusion limited.

Stability of immobilized enzyme is also affected by pH. Partition effects enhance pH stability. Stability enhances on alkaline side, when immobilized on polyanionic supports and on acid side if it were on polycationic support.

Heat stability is enhanced by covalent-coupling and entrapment methods, while it is decline in physical adsorption or ionic linking methods. An increase in the storage stability for enzyme by covalent coupling is usually at low temperatures (40°C).

## IMMOBILIZATION TECHNIQUES FOR CELLS

Cell immobilization methods are the extensions and modifications of techniques developed for immobilization of enzymes. These include cross-linking, copolymerization, covalent binding and particularly entrapment. Following are the methods of cell immobilization.

### Cross-linking

Cells walls of microbes contain free amino or carboxyl groups, which can be cross-linked by a bi or multifunctional reagents. These are glutaraldehyde or toluene diisocyanate. Cells are also immobilized by ionic cross-linking through a flocculation mechanism by addition of poly-electrolyte. The method is also employed in combination with entrapment techniques as an additional cell fixing and gel stabilizing procedure.

### Copolymerization

This method is the extension of cross-linking method. The method involves the use of an inert *spacer* (gelatin, albumin or polyethylenimine), which is added to the cell suspension before cross-linking with glutaraldehyde. This results in biocatalyst with cells entrapped in a covalent network.

### Covalent Binding

Covalent binding is the linkage of biocatalyst to the carrier through reactive groups on the matrix. The coupling agents like glutaraldehyde, diisocyanate, aminosilane, carbodiimide are used to introduce a space group on the carrier surface. This subsequently can interact with reactive groups on cell surface.

### Adsorption

A suitable adsorbents displaying high affinity towards biocatalyst causes minimal denaturation are used for this purpose. The adsorption is an electrostatic interaction between the charged support and charged cell and is dependent on the chemical nature cell wall. It provides the ionic sites for attachment to a charged support. Carrier, and biocatalyst remain together through van-der-walls bonding forces, inorganic bonds or H-bridges.

### Entrapment

For physical entrapment techniques plastic polymers are used. Pectin and carrageenan additional partial covalent bonds are formed between the cell and the support. Polyacrylamide involving a free radical polymerization results in damage to the cells. The technique is associated with a diffusion limitation. Rate of cellular enzyme system is declined. The permeability barriers imposed by cell membrane can be reduced by controlled autolysis, lyophilization, drying or freezing of cells, or by treatment of the cells with toluene or surfactants. Physical entrapment affects the cell viability, however, it conserves enzyme activity.

## BIOREACTORS

The technology utilized in fermentation processes are designed to obtain maximum growth of an organism under the optimum physical conditions in a specific medium for the production of a desired end products. Three general types of growth environments are used: the flask, the shallow tray, and the vat or closed drum. Each of these procedures serves a specific need, and most fermentations use one of these basic procedures. A fermenter or **bioreactor** is a container designed to provide an optimum environment in which microorganisms or enzymes can interact with a substrate and form the desired products. The fermenters are of two types:

**TABLE 21.4:**

| **Open bioreactor** | **It allows continuous processing with substrate entering at one and end products leaving at the another.** |
|---|---|
| Closed bioreactor | In this type, the processing is done in the batches. This type of fermenter is used for the production of antibiotics. The microbes are grown on nutrients placed in the fermentation vessel at the start of the fermentation. The vessel is cooled by a water jacket. Air is pumped into bottom of the liquid, and acid or alkali added as necessary. A stirrer keeps the contents well mixed. Steam lines are provided so that the vessel can be sterilized after each fermentation batch. |

During fermentation, it is necessary to regulate many factors within predetermined values: oxygen and carbon dioxide, pH, temperature and media concentration etc. It is also essential to maintain high degree of sterility within fermenter. The fermenter should be made of stainless steel or copper because such fermenters are resistant to steam sterilization.

### *Pre-treatment and Purification*

Pre-treatment is always given to the raw substrate in addition to sterilization. For example molasses are treated for removal of iron salts from it. Molasses are the byproducts of sugarcane processing consisting of 50% sucrose. Similarly, starch in corn syrup is hydrolyzed to sugar before yeast can convert either to ethanol. Both the products are acidified and diluted.

### *Downstream Processing*

At the end of the fermentation, the desired products may present in a very small quantities (just a few milligrams per cubic decimeter in case of pharmaceuticals). Therefore, it is necessary to give treatment to such undesirable waste products. The treatments of such products is collectively called downstream processing.

### *Purifying the Products*

When fermentation process is completed, the vessel is full of a thick broth microbial cells, some unconsumed nutrients and dissolved products. It is necessary to clean the vessel after every operation.

## Fermenter Design Considerations

Modern fermentation processes require a fermenter that should provide an environment suitable for the growth of a pure culture and/or a defined mixed culture, which can run free from contamination under controlled conditions. A well-designed vessel will also ensure that culture is contained with no aerosol leaks of the vessel contents, since repeated exposure to even a pathogen can in certain circumstances, be hazardous.

The design must incorporate a device for mixing the contents, an air supply for aerobic processes, probes to monitor the environment and regulators to control it. There must be provision for inoculation and sampling, as well as for charging and discharging the vessel. In continuous culture, it is necessary to monitor and control the flow rate of the medium as well as the culture volume and mass.

Incorporating all these features means that the construction has many potential sources for the entry of contamination. Good aseptic design at this point is crucial. The following design rules will apply:

1. There should be no direct connections between sterile and non-sterile parts of a system.
2. Minimize flange connections. These can move under vibration and heat and provide entry for contaminates.
3. Use all-welded construction if possible.
4. Avoid dead spaces and crevices, etc.
5. Various systems should be independently sterilizable.

## Immobilized plant Cell Fermenters

The immobilized enzyme reactors are of different types. These include the stirred tank reactor, continuous flow stirred tank reactor (CSTR), plug flow column reactor (PFR), fluidized bed reactors, ultrafiltration reactors, electrochemical reactors, membrane bioreactors, airlift fermenter, sparged tank fermenter and deep jet fermenter. Few of these are described below. Immobilization of plant cells has recently been developed as an alternative technique to suspension cultures for production of secondary metabolites. Immobilized plant cell reactors offers many advantages. However, immobilized cell systems can be applied when the product of interest is released in the growth medium. A modification to the conventional stirred tank is the inclusion of a support matrix to **immobilize** the roots. While this modification is simple in form it introduces a large difference in operation. However, there can be problems of sufficient mixing in such systems upon scale-up.

## Design and Applications of Bioreactors

A bioreactor or fermenter is a vessel constructed of glass, polycarbonate or steel in which cell cultures or organ cultures are cultivated. It has to fulfill the following requirements: (1) A sterile environment. (2) Access for environmental monitoring e.g. pH, temperature, dissolved oxygen. (3) Provision for sampling. (4) Provision for addition of antifoam, acid, base, fresh medium etc. (5) Supply of air. (6) mixing, and temperature control, addition or removal of heat.

A wide variety of shapes and sizes of vessels have now been used. Major types of bioreactors are:

Stirred tank, (2) Air-lift column, (3) Trickling film, (4) Mist, (5) Rotating drum,

(6) Bubble column, (7) Immobilized plant cell fermenter.

### Stirred Tank Bioreactor

Stirred tank bioreactor is a classical representative of bioreactor types where air is dispersed by mechanical agitation. Various types of stirrers, i.e. spin, helix, bladed, paddle, etc. have been designed for this purpose. Its behavior has been very well studied in a number of biological systems. Temperature, pH, amount of dissolved oxygen and nutrient concentration can be controlled better in this type of reactor.

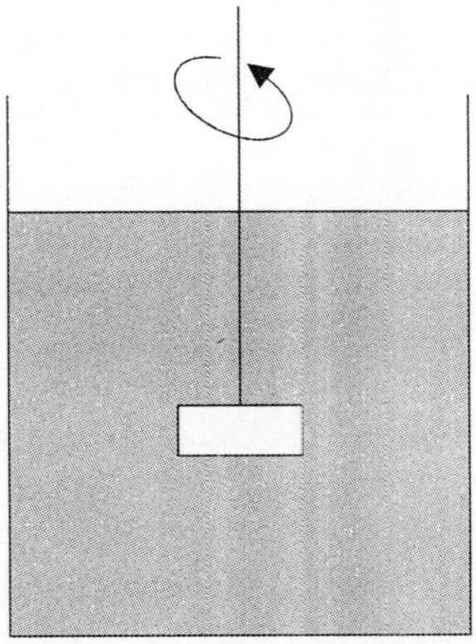

**FIGURE 21.2:** Stirred tank reactor.

### Continuous Flow Stirred Tank Reactor

Continuous flow stirred tank reactors are of large size, therefore, they are less productive and percent conversion of substrate to produce is low but the concentration of product is high. The high concentration of product inhibits the activity of the catalyst which may reduce productivity to a greater extent. For this reason, the rate of conversion remains low. It is uneconomic for conversion factor of more than 90%. Continuous flow reactors are basically of two types: the continuous flow stirred tank reactor with complete mixing, and the plug flow reactor with no mixing. Conditions within continuous flow stirred tank reactor remain identical to those at its outlet.

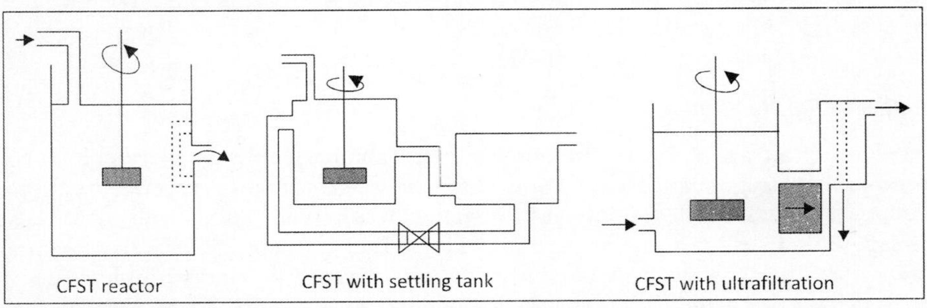


**FIGURE 21.3:** Various types of continuous flow stirred tank reactors

The catalyst is suspended in a large tank through which substrate flows and retained within the reactor by subsequent sedimentation, filtration or being attached to the stirrer paddles. Reactants are mixed well. Diffusional limitations are reduced to the minimum. Control of pH, temperature and replacement of spent catalyst is easy. High power is required for efficient agitation.

### *Packed Bed Reactor*

These reactors are available in a various designs. These are of small size and have high productive output. High viscous substrates tend to block these reactors. The mixing of the substrate with enzyme if poor this causes diffusional limitations. The catalyst is compressible and therefore, blocking of the flow through the catalyst may occur. During industrial production the flow direction of the substrate solution is important.

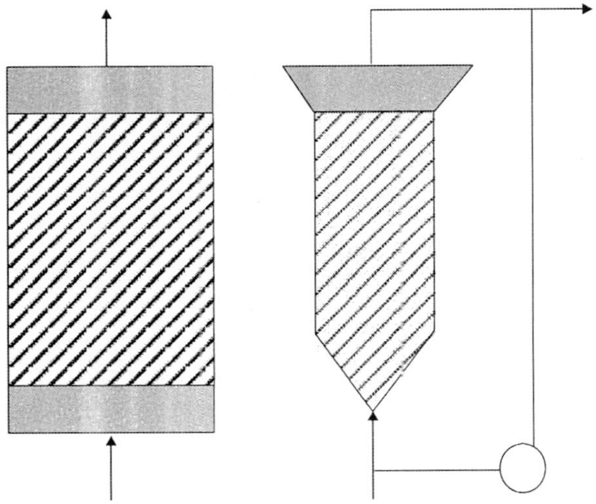

**FIGURE 21.4:** Bioreactors: (A) Packed bed reactor (B) Fluidized bed recycle reactor

The downward flow causes compression of the bed of enzyme columns, therefore, flow in upward direction is preferred, particularly when gas is produced during enzyme reaction. Spirally rolled up sheets of immobilized enzymes can be introduced into the reactor or immobilized enzyme fibre skins can be used. They are easy to handle because have automatic control and operation. They are of low cost and easy quality control of products.

### *Fluidized Bed Reactor*

These reactors are appropriate for substrate solution of high viscosity and a gaseous substrate or product are used in a continuous reaction system. Biocatalyst articles are maintained in suspension relative to each other by upward passage of substrate or gas at high flow rates. High-pressure substrate stream is forced into the base of the reactor. It lifts and mixes the catalyst. Fluidized bed reactor has characteristics of both suspended and supported growth system because both colloidal or even solid substrates can be used. The reactor has low diffusional limitations, therefore, productivity of the reactor can be increased by high biocatalyst loading. High flow rate may washout the biocatalyst and this problem can be overcome by recycling the substrate sing tapered fluidized bed or by using several fluidized bed in series. The reactor gives high degree of conversion using high concentration of enzyme. The reactor is advantageous because there is no blocking of biomass is washout.

## *Membrane Bioreactors (Hollow Fibre and Rotating Membrane)*

These bioreactors are made from cellulose acetate with a uniform matrix. Fibres with acrylic polymers or polysulphose are having asymmetric wall configuration. The bioreactor of hallow fibres have highly porous surface walls which provides surface for cell growth. Free diffusion of ions and molecules is through ultrafiltration layers. Catalyst is separated from the substrate stream by the semipermeable wall of the ultrafiltration hollow fibre. These reactors are used for soluble enzymes, which can be easily replaced.

## *Air-Lift Fermenter*

It is a gas liquid bioreactor where compressed air is used for aeration and agitation is based on the draught tube principle. Air lift reactors upto 100 litre capacity have been used extensively for the cultivation of *Cathranthus roseus*. Pneumatically agitated fermenters are more suitable for the production of secondary metabolites than stirred tank reactor. Eight fold higher yield of tripdiodide from cells of *Tripterginum wildordii* was obtained in an air lift reactor when compared to a stirred tank reactor.

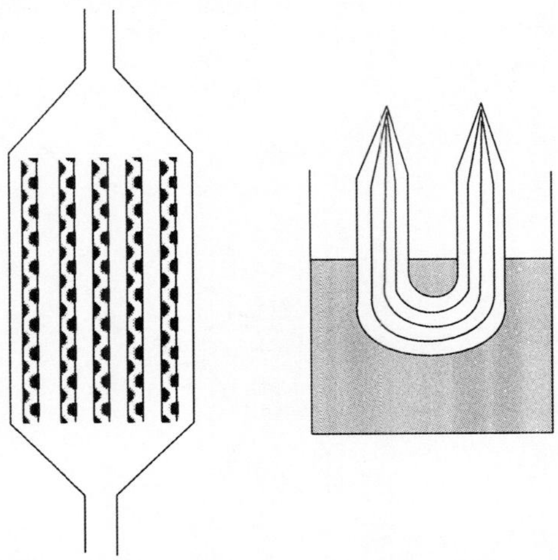

**FIGURE 21.5:** Hollow fibre reactors.

Oxygen transfer at low shear, less contamination due to non-movable parts, low operation cost because of simple design are a few merits of air- lift type reactors. The demerits are the development of dead zones inside the vessel due to insufficient mixing and nutrient supply caused by high cell densities The air lift column with cell immobilization is a superior cultivation system for hairy root cultures.

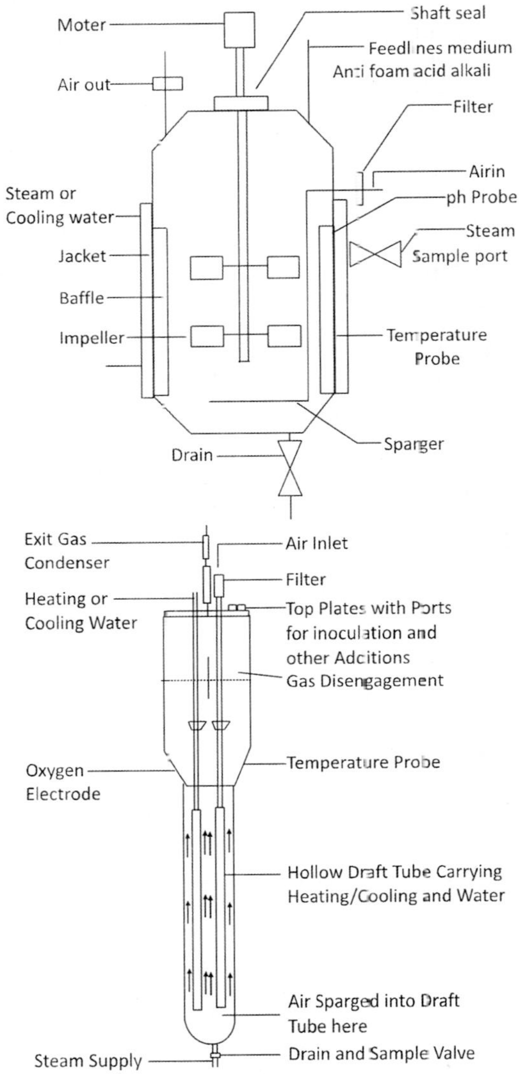


**FIGURE 21.6:** Bioreactors Upper: Stirred tank bioreactor, Below: Air-lift fermenter.

### *Trickling Film Bioreactor*

In this type of bioreactor the medium trickles over glass beads. The media is recirculated from a reservoir and sprayed over the fixed bed of roots and permitted to flow down through the root bed. The spray can be either a fine mist or a coarser spray. Roots are inoculated on top of the glass beads. These grow over the surface of the bed and down between beads in the film of circulating medium. The length of the pipe between the P.T.F.E. filter and the airlift is kept as short as possible to minimize oscillation of culture medium within the airlift tube. Despite this, a screw clamp construction is required in the pipe returning the culture medium to the bottom of the air lift to act as a throttle valve to further damp oscillations which otherwise could, on occasions, lead to a reversal of air flow up through the base of the reactor.

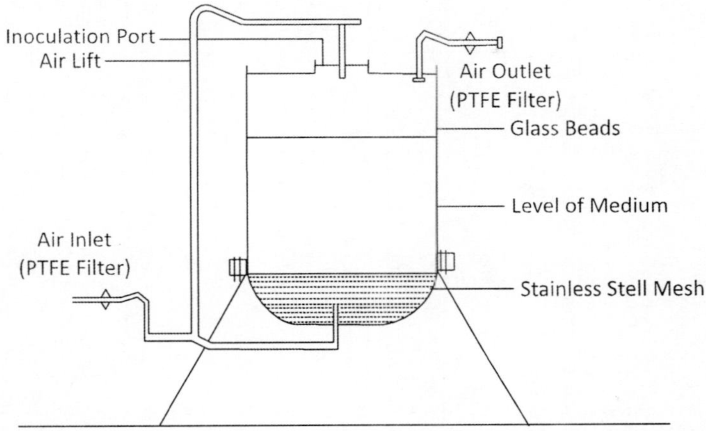


**FIGURE 21.7:** Trickling film bioreactor

This reactor configuration has promise in terms of scale-up since the flow patterns are strongly influenced by gravity, which acts uniformly over the bed in contrast to localized power input for mechanical agitation but harvesting of the roots is difficult because of their adherence to the glass beads.

### *Mist Bioreactor*

In this type of reactor medium is pumped through mist head using peristaltic pump fitted with timer. Glass fiber is inserted at the base of the bioreactor vessel to filter out cells sloughed from the root cap and any other debris, which could clog the jet in the mist head. A fine stainless steel mesh filter is also incorporated into the mist head itself to further guard against blocking of jet by cell debris. These type of bioreactors have the advantage that the medium can be drained off and the wet weight can be determined directly. Wet weight can thus be recorded at frequent intervals without jeopardizing the sterility of cultures.

### *Rotating Drum Reactor*

A reactor configuration which operates on principles similar to a fill and drain reactor is the rotating drum (Kondo *et al.*, 1989). In this configuration, the 'filling' takes place as the tissue rotates below the media surface; draining occurs as the tissue rotates up out of the media. This configuration minimizes the problems associated with the timing of the fill and drain cycle, but may have some limitations in scale.

It consists of a horizontal rotating drum on rollers. The rotating motion of the drum facilitates proper mixing of gas-liquid in reactors, thereby promoting efficient oxygen transfer to cells at high densities. The rotating drum reactor imparts less hydrodynamic stress. The major disadvantage of this type of reactor is its dependence on high energy requirement for large scale operation. Two other reactor systems are shaken bottle bioreactor and aerated bottle bioreactor.

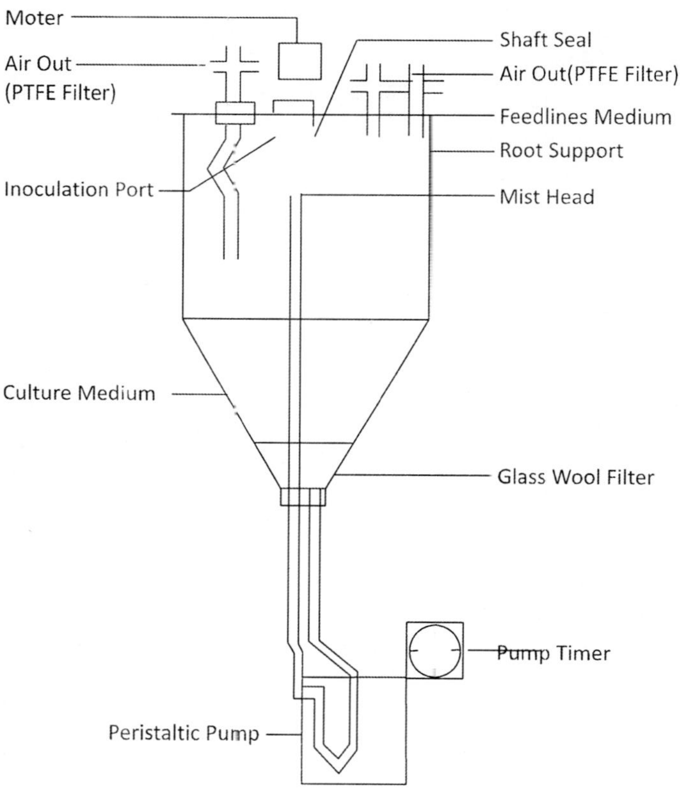


**FIGURE 21.8:** Mist bioreactor

### The Shaken Bottle Bioreactor

It is 1 litre Duran bottle shaken, on an orbital shaker. Aeration is by passive diffusion through a bacteria proof (0.2 micron) PTFE filter.

### The Aerated Bottle Bioreactor

It is the same as the Shaken flask except that two PTFE filters are used as an inlet and outlet. Air is bubbled through the medium without a sparger head.

### Ultrafiltration Reactors

A continuous ultrafiltration membrane device is used for high molecular weight substrate and a low molecular weight product. They help in carrying out depolymerisation reactions especially when using soluble enzyme.

### Bubble Column Bioreactor

The development of air sparged stirred reactors proved highly successful and they are widely adopted. While the majority of the recent bioreactor designs are quite sophisticated and rather complicated, bubble column reactors are gaining an important place in both chemical and biochemical industries. Unlike mechanically agitated reactors, bubble columns are simple to construct and operate. They consist of vessel usually cylindrical in which gas is sparged into liquid.

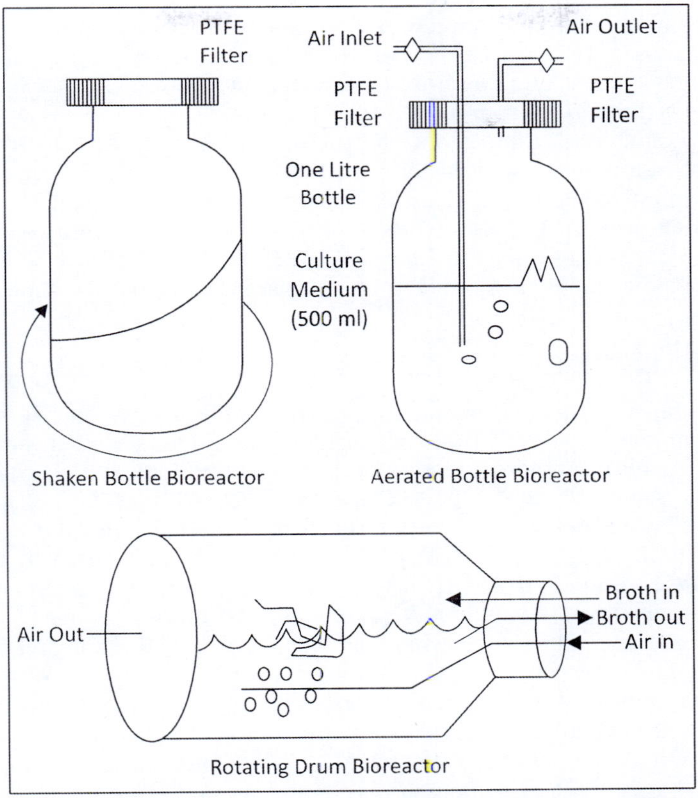


**FIGURE 21.9:** Rotating drum reactor.

They have no moving parts, as adequate levels of mixing can be achieved with the sparged gas. In bubble column reactors, all energy needed for agitation, as well as the oxygen required for the culture is provided by the sparged air. The most important advantage of bubble columns, results from their simplicity. As there is no need to introduce energy by mechanical means, the sealing of the stirrer shaft assembly, which is most difficult problem to overcome in the mechanical design of a bioreactor, is eliminated. In addition, the absence of shafts in the head space of the vessel gives more room for entry ports, an important factor in vessels with relatively small dimensions. Mechanical simplicity is particularly important in biological processes where sterility has to be maintained over extended periods. The absence of shafts eliminates a very expensive and equally vulnerable feature of mechanically stirred bioreactors thereby increasing process reliability.

It is generally believed that bubble columns are less suited to processes involving highly viscous liquids. However, there is evidence to the contrary. In an analogous situation however the use of a bubble column resulted in more homogeneous mixing the bubbles released at the sparger coalesced immediately to form very large bubbles known as slugs which rose rapidly along the axis of the column setting the entire liquid into circulation with upward movement near the cylinder axis and downward movement near the walls. The sparger determines the initial bubble size and shape in a given liquid. A sparger with small diameter holes, such as perforated or sintered plate, generates smaller bubbles than a single orifice sparger, and therefore provide a higher interfacial gas-liquid contact area in the vicinity of the sparger.

Many commercially important bioprocesses involve shear sensitive cultures. Hairy root cultures are also shear sensitive, have been grown successfully in bubble column reactors. Experimental observations showed that the growth of plant cell cultures in bubble column reactors, stirred tank reactors and shake flasks was similar. Oxygen transfer at low shear, low operation cost and low hydrodynamic stress makes bubble column reactor an attractive choice for scale up studies.

## CHOICE OF ENZYME REACTOR

In enzyme reactor, the highest specific activity, in terms of weight of enzyme and support employed, is desirable. It is considered an added bonus if the support is employed to fulfill another function. The most important of these functions is separation since, if this occurs simultaneously with reaction, unfavorable equilibria may be displaced. One approach is to use a molecular sieve as the support and, in packed reactors to pulse the reactor bed alternatively by passage of substrate solution and water, so that bands of unused substrate and product are progressing down the column. It so happens that these enzymes for which this would be useful are also those, which in some cases benefit in having the enzyme immobilized on a porous support.

For an industrial reactor it is preferable to use supports that are non-biodegradable such as glass, silica, Celite, Bentonite, alumina, or titanium oxide, if possible. Even the linkages between enzyme and support can be non-biodegradable, as with the titanium procedure. As far as the biodegradability of the enzyme, in some cases, when in active use, enzymes will partially protect themselves, notably all those that effect oxidation with production of hydrogen peroxide as a by product.

Various types of reactors have been proposed: packed beds with downward flow, suspended particles in a fluid been with upward flow of substrate, a simple stirred reactor, tubular reactors, membrane reactors, and many others. In some of these the physical nature of the surface becomes a major problem. Thus some supports that form excellent packed beds fail to do so when coated with enzyme and particles which ideally self-suspend in a fluid bed then aggregate during use requiring more power to pump through substrate allied to a lowered catalytic activity from a decreased surface area. Certain problems were encountered with porous glass supports until it was realized that the enzyme was not simply losing activity but the glass itself could dissolve. This has now been overcome by the zirconium treatment of the glass surface. Additives to enzymes intended to preserve enzyme activity are also rarely fully disclosed. This often causes confusion as to the weight of protein in the enzyme product and can even interfere with the coupling procedure or decrease its efficiency.

# 22. Applications of Immobilized Enzymes

Immobilized enzymes and cells are of use industrial, clinical analytical processes, in novel drug delivery system design, in extra corporeal hemoperfusion, in analytical techniques, in secondary metabolite production, in designing of bioreactors and biosensors, in immunoadsorptive and affinity chromatographic separations, in aphaeresis, in drug targeting, in bio-environmental modification, in fabrication of surrogate organs and cellular systems, in blood detoxifiers designing, and in fundamental biochemistry and microbiology. The application potentials are mainly based on the concept that *in-vitro* simulation and culturing of biocomponents, cells and plants for biological activity under external manual control could be concluded for evaluations. Following are the important uses:

**Uses of immobilized Biocatalyst:** The potential uses of immobilized biocatalysts as retention of catalyst, high catalyst concentration, micro-environmental control, fine process control and separating biocatalyst and process product. Some of the representative examples and process operations are discussed in this chapter.

- **Retention of catalyst:** Retention of catalyst provided by immobilization process offers two types of distinctive benefits. **Trigonopsis variable** whole cells were immobilized in calcium alginate beaded gel and used in bioconversion of amino acids to α-ketoacids. In clinically applicable bioproducts it is sometimes desirable to remove biocatalyst totally for its antigenic nature, or systemic toxicity.
- **High concentration of catalyst:** The high catalytic density results in a more compact system, reducing capital cost but also reduces the volume of reactor, which has to be specifically treated of kept sterile. The high concentration of catalyst leads to two major benefits. One is the extreme rapidly of the enzyme reaction in the matrix and the other is the ability to generate specific compounds at high local concentrations.
- **Micro-environmental control:** The micro-environmental in the vicinity of an immobilized catalyst distinctively differs from the homogenous equivalent. Surface charge or hydrophobic regions on the matrix interact specifically with the catalyst, and even where this does not occur, effects on ionic balance, pH, water activity and osmolarity will directly contribute to and have an impact on the aqueous environment. These effects collectively can increase the thermostability of the enzymes by reducing the degree of freedom of the protein chain.
- **Delinking growth and production:** It has application in the production of cell metabolites, by separating growth and production phases.
- **Use of Immobilized Cells:** The major use of immobilized cells is in the synthesis of secondary metabolites. Most of these are small to medium molecular weight biochemical and can arise either by *de novo* synthesis from

primary metabolites or by bio-transformations of similar compounds from either from plant or animal kingdom.

- **Use of Monoclonal Antibodies in Active Immobilized Enzyme Preparation:** Antibody based immobilization produces highly active enzyme system. The method essentially involves binding of enzymes t a suitable carrier/support using monoclonal antibodies.
- **Immobilized Enzyme and Industrial Processes:** Immobilization technology has gained wide range of applications in the industrial field as well as in the medical and analytical field.

**TABLE 22.1:** Immobilized enzymes in industrial application

| Industrial | Applications |
|---|---|
| Analytical | Enzyme electrodes based on changes in pH, redox potential, oxygen tension<br>Affinity chromatography and purification<br>Immobilization in affinity purification<br>Immobilized avidin-biotin based systems<br>Enzyme thermistors based on heat of reaction<br>Bioprobes based on cellular metabolism |
| Applications of Immunoadsorption Technique | Immobilized enzyme based immunoassay<br>Immobilized haptens as immunopurifier<br>Immobilized protein A/G in IgM purification, Radioallergosorbent test (RAST)<br>Immobilization based novel drug delivery systems (Urea-urease modulated system, Glucose oxidase-glucose modulated system, Triggered drug delivery system)<br>Immobilization of artificial cell as artificial organs<br>Artificial cells immobilized enzymes and multienzymes<br>Artificial Pancreas, Blood detoxifier<br>Artificial liver, Artificial Kidney<br>Artificial Cells Used as blood substitutes (surrogates)<br>Gene Probes, Immobilization and Drug Targeting Using Artificial Cellular Systems<br>Bilayer Vesicular Systems as Cellular Carriers (Liposome appended with antibody (IgG) or their fragments and conjugates)<br>Immunoliposome<br>Immobilized legends as pilot molecule. |
| Auto analysis | Repeated auto analysis of small samples (e.g., blood samples)<br>Continuous stream monitoring of large volume |
| Biosensors | Electrochemical sensors<br>Potentiometric sensors<br>Thermal and mass calorimetric biosensor<br>Optical biosensor<br>Acoustic biosensors<br>Surface plasmon<br>Glucose sensor |
| Enzyme reactors | Batch reactors<br>Continuous flow reactors<br>Continuous flow stirred tank<br>Cell-free system and bioreactors (packed reactor, hollow fiber reactor). |
| Fine chemicals | Optical isomer resolution by hydrolysis of ester and amides |

| Industrial | Applications |
|---|---|
| | ∝-keto acid production<br>Redox reactions using dehydrogenases<br>Cyanide detoxification<br>Iodination of proteins<br>Phenol degradation<br>Radioactive chemical |
| Fundamental biochemistry | Study of enzyme interactions with other solutes<br>Selective cleavage of biopolymers |
| Industrial processes | Production of high-fructose syrups, Production of L-amino acid by resolution |
| Medical | In vivo devices to treat failure of kidney, pancreas (Islets)<br>Detoxification following drug overdose<br>Body monitors for key electrolytes and glucose |
| Pharmaceuticals | Selective hydrolysis of penicillin G<br>Steroid modification<br>Production of monoclonal antibodies<br>Animal vaccines |

As discussed in previous chapter that immobilized enzymes should have operational stability. It determines the success or failure of an industrial process because it affects the costs of the biocatalysts ad regenerating deteriorating columns. Use of immobilized enzymes for industrial production reduces the cost to half as compare to soluble enzyme processes.

## PHARMACEUTICAL APPLICATIONS

Immobilized biocatalysts are used for manufacturing of drugs in pharmaceutical industries, specially in production of antibiotics, steroids, amino acids, etc.

### Uses in Antibiotics Production

#### *Penicillins*

Immobilized enzymes are used for commercial production of 6-amino penicillanic acid, penicillins, cephalosporins etc. Production can be carried in column, stirred tank or batch membrane reactors.

**TABLE 22.2:**

| Enzyme | For production of | Source | Process |
|---|---|---|---|
| Penicillin amidases | 6-aminopenicillanic acid (6-APA): A-Penicillins | G (benzyl penicillin) or penicillin V (phenoxymethyl penicillin) | It is produced penicillin by deacylation with penicillin amidase. Semisynthetic penicillins are also produced as an important intermediate. These antibiotics are produced by fermentation. Immobilized enzyme process replaced the conventional processes based on soluble enzyme. |
| penicillin amidase from *E. coli* | 6-APA from penicillin G | Penicillin amidase from *E. coli* trapped into cellulose | It is produced in a column as a part of recirculation batch reactor. 5% substrate solution |

| Enzyme | For production of | Source | Process |
|---|---|---|---|
| trapped into cellulose triacetate fibres | | triacetate fibres from penicillin G | hydrolyses to 90%. Retention time required is 1.5-2.0 hrs. |

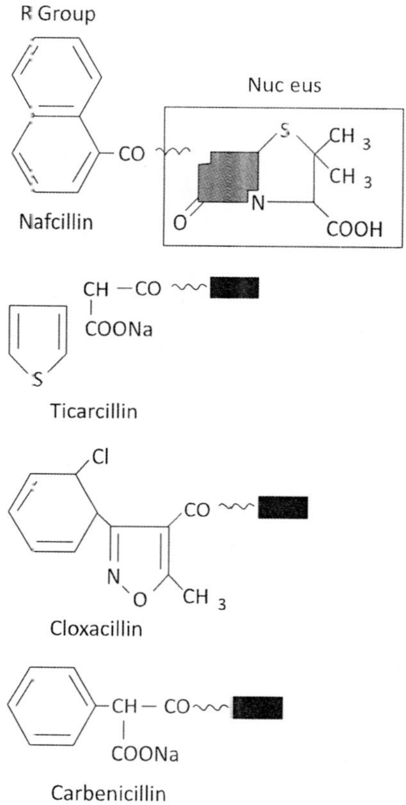


**FIGURE 22.1:** Chemical structure of penicillins. All contain a thiazolidine ring (yellow) and a beta-lactam ring (red), but each differs in the nature of the side chain (R group), which is also responsible for differences in biologic activity.

Penicillin amidase immobilized by covalent binding to Amberlite XDA-7 with glutaraldehyde, by physical adsorption to bentonite, or by ionic binding to DEAE-sephadex, and by covalent binding to a copolymer of acrylamide and maleic anhydride can be used for the production of 6-APA from penicillin V. Continuous method from penicillin G using *E. coli* cells immobilized with polyacrylamide is also employed for this process and pure 6-APA yield of about 78% can be obtained. The half-life of column at 37°C is 42 days and at 40°C it will be 17 days.

For laboratory-scale production of Penicillin G under from glucose involves the use of conidia, mycelium, and protoplasts of *Penicillium chrysogenum* immobilized with K-carrageenan, polyacrylamide, and calcium alginate. A continuous flow bioreactor (150 ML) can be used. Conidia of *P. chrysogenum* immobilized, with K-carrageenan operates for up to 16 days. 1.2 mg/g cell/hr yield can be obtained from glucose (7.0 mg/g glucose). The half-life of production is > 15 days with a media containing 10.0 g of glucose/L. The duration of residence is 20 hrs.

### Ampicillin and Amoxicillin

Penicillin amidase from *E. coli* entrapped in cellulose triacetate fibre. It is employed in continuous flow reaction for production of ampicillin or amoxicillin from 6-APA and D-phenylglycine methyl ester respectively. About 50% conversion is obtained. Succinolyated penicillin amidase adsorbed on DEAE-sephadex is used for the production of ampicillin. About 67% yield is obtained. in stirred batch reactor. Whole cells of *Achromobacter* sp. of *Bacillus megatherium* immobilized by ionic binding on DEAE-cellulose in a column reactor gives an yield of about 54% of ampicillin from 6-APA.

### Cephalosporins

Cephalosporins are obtained using fermentation method with *Cephalosporium acremonium*. Cephalosporin amidase obtained from different microorganisms can be immobilized by various methods. The immobilized enzyme systems are used for the production of various cephalosporin derivatives.

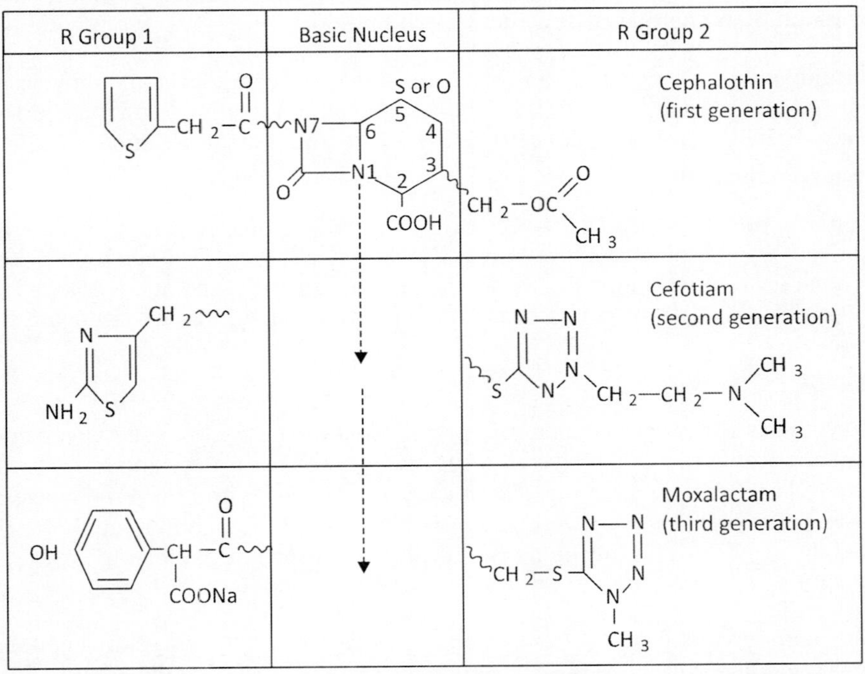


**FIGURE 22.2:** The structure of cephalosporins. Like penicillin, they have a beta-lactam ring (red), but a different main ring (yellow). However, unlike penicillins, they have two sites for placement of R groups (at positions 3 and 7). This makes possible several generations of molecules with greater versatility in function and complexity in structure.

Enzymatic deacylation of cephalosporin is carried out with compounds which contain 7-ADCA nucleus. 7-phenylacetamidodesacetoxy-cephalosporanic acid (phenyl acetyl) and 7-phenoxyacetamidcdesacetoxy-cephalosporinic acid (phenoxyacetyl-7-ADCA) are the substrate which are obtained by ring expansion reaction from penicillin G and penicillin V. The enzyme from *Bacillus megatherium* adsorbed on celite or alumina can be employed to carry out reverse reaction i.e., the acylation of 7-ACA or 7-ADCA. The cephaloglycine and cephalothin are synthesized from 7-ACA and D-phenylglycine methyl ester or 2-thiophene acetic acid by continuous passage of substrate solution through immobilized enzyme column for commercial purpose.

### Cephalexin

It is produced from 7-ADCA and D-phenylglycine methyl ester which is obtained from *E. coli* entrapped in cellulose triacetate fibres. About 75% of the 7-ADCA can be converted to cephalexin in 1 hr at 25°C. Immobilized acetone dried whole cells of *Achromobacter* sp. adsorbed on DEAE-cellulose or Hydroxyapatite can also be used.

### Cephalosporin C

Cephalosporin C can be synthesized from 3-(-N-morpholino) propane sulphuric acid by whole cells of *Streptomyces clavuligerus* immobilized with polyacrylamide.

### Tylosin and Nikkomycin

Macrolide antibiotic tylosin and the nucleoside peptide antibiotic nikkomycin are produced using living cells of *Streptomycin* spp. and *S. tendae*, immobilized with calcium alginate. Air bubbled reactor is used for continuous production.

## Steroids Production

Transformation of steroid is the biocatalytic conversions. Several enzymes are employed for steroid transformation, which are required as cofactors. Whole microbial cells containing systems of cofactors are immobilized and employed for large-scale steroid transformations. Following three methods are used for this purpose.

### Method 1

The synthesis of hydrocortisone and prednisolone from cortexolene is obtained by immobilizing the biocatalyst by entrapment with polyacrylamide. For this purpose, 11 β-hydroxylation mycelia of *Curvularia lunata* and for $\Delta^1$-dehydration (whole cells of *Corynebacterium simplex*) is used. By combining both the immobilized systems prednisolone can be obtained from cortexolene.

### Method 2

In this method whole cells of *Corynebacterium simplex* are entrapped in collagen membrane and are used for the production of prednisolone from hydroxycortisone. Chips of collagen cell complex packed in column are used for continuous operation.

### Method 3

In this method *Corynebacterium lunata* mycelia and immobilized *Arthrobacter simplex* cells are used. It is the two steps transformation (cortexolene to its Δ‘-dehydro-11-hydroxy derivative).

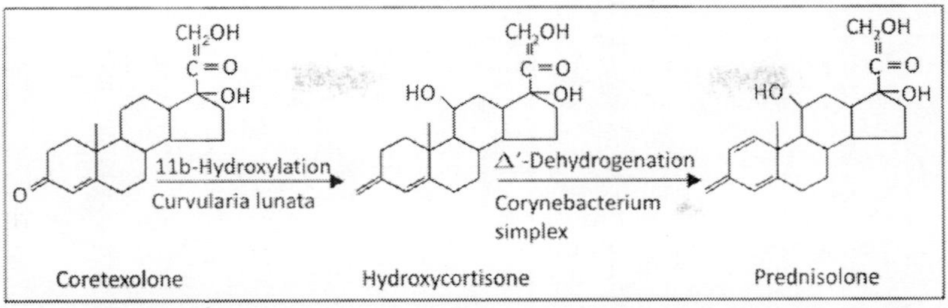


**FIGURE 22.3:** Presentation of conversion of cortexolene to hydrooxycortisone for production of prednisolone.

## Amino Acids Production

Amino acids are of common use in pharmaceutical industries. For production various large-scale processes are based on immobilized enzyme. Amino acid production by enzymic resolution using amino acids acylase and fructose syrups from glucose isomerase are few among them.

### *Optical Resolution of DL-Amino Acids*

**Acyl-DL-amino acid:** Synthesized acyl-DL-amino acid is asymmetrically hydrolyzed by aminoacylase. It gives L-amino acid and unhydrolyzed acyl-D-amino acid. Both products are easily separated using solubilisation method. Unhydrolyzed acyl-D-amino acid is racemized and reused for the resolution procedure.

Immobilized aminoacylase is obtained from *Aspergillus oryzae* and is used for the continuous optical resolution of DL-amino acids. The enzyme is obtained by ionic binding to DEAE-sephadex covalent binding to iodoacetyl cellulose, and entrapment of polyacrylamide gel. Ionic binding to DEAE-sephadex is used for industrial process. Acetyl-DL-amino acid solution is regularly charged into the enzyme column through a filter and heat exchanger. The enzyme reaction occurs while passing through the column. L-amino acid crystallizes from the concentrated effluent. Acyl-D-amino acid contained in the mother liquor is racemized by heating and can be reused for optical resolution.

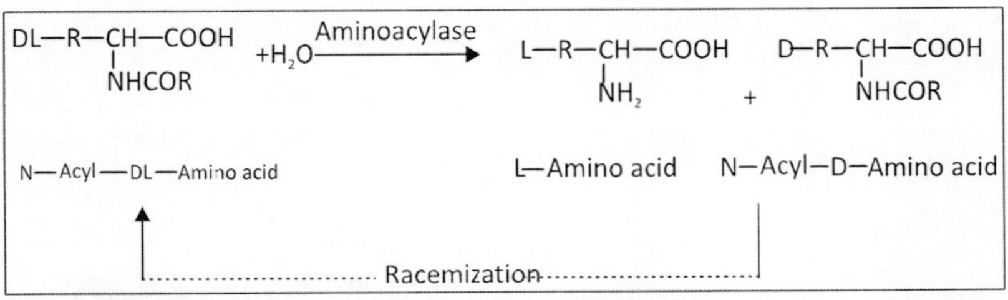


**FIGURE 22.4:** Acyl-DL-amino acid is asymmetrically hydrolyzed by enzyme aminoacylase. It produces L-amino acid and unhydrolyzed acyl-D-amino acid.

**L-Tryptophan:** Acetyl-DL-tryptophan can be hydrolyzed asymmetrically by entrapping the mold aminoacylase in cellulose triacetate fibres to produce L-tryptophan and acetyl-D-tryptophan on pilot scale.

## *Optically Active Amino Acids*

**L-aspartic acid:** It is produced industrially by fermentation/enzymatic batch process from fumaric acid and ammonia using aspartase.

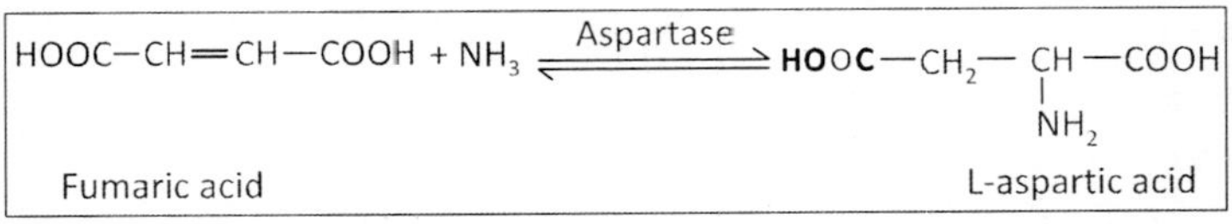


**FIGURE 22.5:** Conversion fumaric acid to L-aspartic acid using enzyme aspartase.

Whole microbial cells of *E. coli* are immobilized with K-carrageenan. They are treated with glutaraldehyde and hexamethylenediamine. Column packed with immobilized cells of *E. coli* with highest productivity has a half-life period of 680 days at 37°C.

**L-alanine:** L-alanine production is obtained from L-aspartic acid by using *Pseudomonas dacunhae* cells which are immobilized with K-carrageenan.

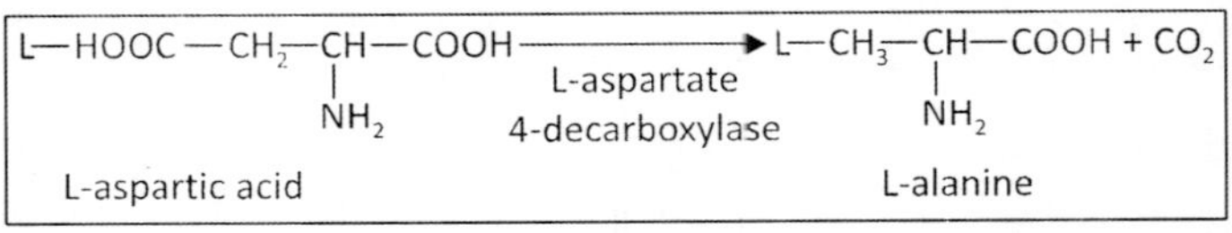


**FIGURE 22.6:** Conversion of L-aspartic acid in to L-alanine using enzyme L-aspartate 4-decarbozylase.

For this purpose a closed column reactor is used under high pressure. This completes plug flow of the substrate solution. Carbon-dioxide gas is evolved during the and is dissolved into reaction mixture. pH of the reaction mixture is not a limiting factor.

**L-DOPA (3,4-dihydroxy-L-phenylalanine):** L-DOPA is useful for treatment of Parkinson's disease

(i) β-tyrosinase obtained from cells of *Erwinia herbicola* is immobilized by cross-linking with collagen and glutaraldehyde or dialdehyde starch. It is used for the production of L-DOPA.

(ii) β-tyrosinase obtained from cells of *Escherichia intermedia* is immobilized by covalent binding to CNB-activated Sepharose. The immobilized enzyme is used for the production of D-α-phenylglycine and D-α-hydroxy phenylglycine from phenyl hydantoin and P-hydroxyphenyl hydantoin respectively.

## Acids Production

Processes have been developed for production of Acetic acid is produced using enzymes immobilized obtained from *Acetobacter aceti*. These enzymes are immobilized with porous ceramics, citric acid by *A. niger* with calcium alginate. Lactic acid is obtained by *L. casei* with polyacrylamide, *L-Malic* acid by *Brevibacterium flavum* with carrageenan, 12-ketochenodeoxycholic acid *B. fuscum* with carrageenan, 2-ketogluconic acid by *Serratia marcescens* with collagen etc.

## Other Organic Compounds Production

### *Coenzyme A*

Coenzyme A obtained from pantothenic acid, L-cysteine, and ATP. For continuous production is carried with *Brevibacterium ammoniagenes* cells immobilized on polyacrylamide gel.

### *Flavin Adenine Dinucleotide (FAD)*

This is produced from flavin mononucleotides and ATP by whole cells of *Arthrobacter oxydans*. The cells possessed high FAD pyrophosphorylase activity and were immobilized in a film polyvinyl alcohol cross-linked with tetraethyl silicate.

### *Pyridoxal 5'-Phosphate*

It is produced with high enzyme activity from pyridoxine 5'-phosphate by whole cells of *Pseudomonas fluorescens* immobilized on a film of polyvinyl alcohol cross-linked with tetraethyl silicate.

### *Adenine Arabinoside and Vitamin $B_{12}$*

**Adenine Arabinoside:** Cells of *Enterobacter aerogenes* with transglycosylation activity are immobilized with a hydrophilic photocrops linkable resin. They are used for production of adenine arabinoside from uracil arabinoside and adenine in appropriate water-organic solvent system.

**Vitamin $B_{12}$:** Similar immobilization procedure for cells of *Propionibacterium* sp. were used for the production of vitamin $B_{12}$.

### *Dihydroxy Acetone*

Dihydroxy acetone is a pharmaceutical intermediate product obtained after oxidation of glycerol with *Acetobacter xylinum* cells immobilized with polyacrylamide gel.

### *Proinsulin*

*Bacillus subtilis* cells carrying plasmids encoding for rat proinsulin are immobilized in agarose beads. The immobilized cells are used for continuous production of proinsulin in a small continuous stirred tank reactor.

### *Interleukin-2 and Monoclonal Antibodies*

Animal cells (lymphoblastoid MLA 144 cells and hybridoma cells) immobilized with agarose beads are used for the production of interleukin-2 and monoclonal antibodies.

### *Prostaglandin*

Prostaglandin synthesis using arachidonic acid is carried out utilizing ram seminal microsomes immobilized on photocrosslinkable resin.

### *Ajmalicine Isomers*

The cells of *Catharanthus roseus* (an angiosperm plant) are immobilized with calcium alginate were used for the synthesis of ajmalicine isomers from tryptamine.

### *Anthraquinones*

De novo synthesis of anthraquinones is performed by immobilizing plant cells of *Morinda citrifolia* with calcium alginate from secologanin.

For the *hydroxylation of alkaloids*, *Daucus carota* and *Digitalis lunata* immobilized with calcium alginate were used to produce 5-β-hydroxygitoxigenin from gitoxigenin and digoxin from digitoxin.

## APPLICATIONS IN CLINICAL ANALYSIS

For medical diagnosis and for industrial monitoring programs enzymes are extensively used. They are also used for carrying out analysis because they have specificity and the range of reactions that they catalyze. One of the advantages of using immobilized enzymes system based on the flow through autoanalyser units are being developed. In clinical analysis, enzymes are used as regents to measure specific metabolites, and are measured as such in body fluids and tissues, as an indicator of pathogenic conditions or metabolic disturbances.

### Enzyme Electrodes (Biosensors)

The natural biosensors are the *sense organs*, primarily the chemical sensors of smell and taste. Biosensors technically in their various forms share a reliance of biological materials as sensing elements. Technical biosensors have been under intense development since the middle of 1960 with the prospects of commercial potential offered by biotechnology. These are the combinations of biologically active material displaying characteristic specifically with chemical or electronic sensor to convert the response into electrical signals.

Biosensors are, in fact, biocatalysts. They are used as purified enzyme, antibody or as whole microbial cell or as an organelle. A biosensor used as an **immobilized** biological molecule (usually an enzyme or an antibody) or a whole microbial cell to detect or *sense* a particular substance. The biosensor does this by reacting specifically with the substance to be detected (hence the use of enzymes or antibodies) to give a product which is used to generate an electrical **signal** by means of a device called **transducer**.

The response of biosensor is measured in terms of substrate used or product form. They are of different types like, carbon electrode, glucose electrode, ion sensitive electrode, photocell, oxygen electrode, adenosine electrode, etc.

For example a glucose electrode is constructed by immobilizing a layer of glucose oxidase in polyacrylamide gel around a platinum oxygen electrode. When a solution of glucose is brought into contact with electrode, glucose and oxygen diffuses into an enzyme layer and are converted into gluconolactone and hydrogen peroxide lowering the oxygen concentration around the electrode. $O_2$ concentration read by electrode is proportional to glucose concentration in the sample. The effective control of the rate of reaction is ensured by high enzyme loading and limited diffusion of substrate.

Biosensor technology couples our knowledge of biology with advances in microelectronics. A biosensor is composed of a biological component, such as a cell or antibody, linked to a tiny transducer. Biosensors are detecting devices that rely on the specificity of cells and molecules to identify and measure substances at extremely low concentrations. When the substance of interest collides with the biological component, the transducer produces a digital electronic signal proportional to the concentration of the substance. Biosensors are used for following purposes:

- measure the nutritional value, freshness and safety of food
- provide emergency room physicians with bedside measures of vital blood components
- locate and measure environmental pollutants.

Biosensor technology is progressing very fast on the front of techniques, their applications specially in the fields of analytic medicine, industry and environment. It is sometimes also useful in monitoring the presence of specific chemicals both accurately and rapidly.

Several enzyme electrode probes have been described in the literature. Enzyme is immobilized to the inner wall of tubing (nylon preferred) and enzyme reactor is placed in series with appropriate detector. Sugars, amino acids, urea, cholesterol, creatinine, penicillin and acetylcholine, etc., can be measured.

Microbial electrode sensors also have also been developed. The whole cell is immobilized and a product of the relation is measured. i.e., oxygen and $CO_2$ probes.

Biosensors are used for glucose estimation. A nylon coil is used to which hexokinase and glucose-6-phosphate dehydrogenase are immobilized. Glucose is estimated by the rate of formation of NADH. It may be used effectively with few samples to analyze daily. In other analyzer, glucose oxidase immobilized to porous inorganic support and held in a column is used to measure glucose.

The device is based on enzyme immobilized to the inner surface of a nylon tube linked to an automatic pipette has been developed. The system has been used for urea, glucose, uric acid and amino acids.

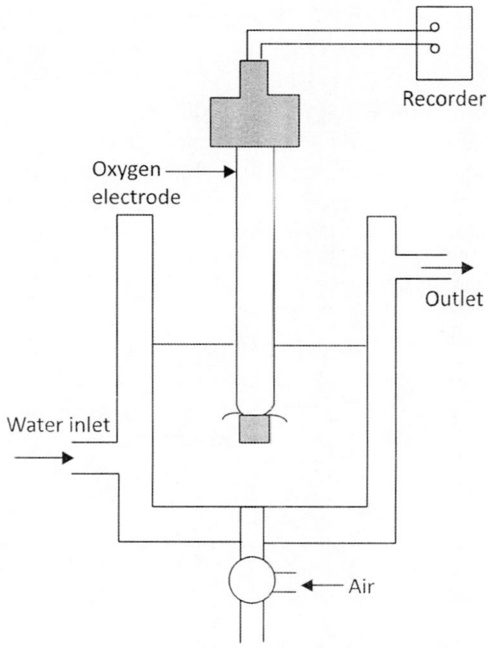


**FIGURE 22.7:** Enzyme electrode

**Penicillin electrode:** These electrodes are based on β-lactamase entrapped within semi-permeable membrane at the base of a pH electrode is also developed for in vessel monitoring of fermentations.

**L-lysine electrodes:** An enzymes sensor with lysine decarboxylase immobilized on carbon dioxide gas sensing electrode has been developed. It is used for assaying L-lysine in an amino-acid mixture.

## Biosensor for Analysis of Organic Compounds

Analysis of organic compounds in fermentation and other samples can be done with the help of biosensor. Compounds such as glucose, acetic acid, lactic acid, formic acid, alcohol, methane, glutamic acid, cephalosporin antibiotic, nystatin antibiotic, nicotinic acid, vitamin B1 etc., can be analyzed.

### *Medical Sciences*

Hepatitis antigens found in blood during infection by the virus can be detected. Abnormal amounts of urea in blood or urine in kidney diseases can be measured. Hormone gonadotropin produced during pregnancy can be tested and measured. High concentrations of creatinine produced after heart attack can be analyzed.

### *Industrial Uses*

They are used to know about the nature of industrial products of acids, alcohols, phenols, pollutants, etc. The industrial workers can know the presence and concentrations of hazardous chemicals in the environment surrounding them. Biosensors can detect various chemical warfare agents, nerve gas, etc.

### *Environmental Analysis (Protection)*

They are used in analysis of BOD requirements, ammonia, nitrite, sulfite measurements, etc.

Biosensor technology is changing so rapidly that potentially commercializable devices quickly become obsolete as new technology emerges. Very few biosensor devices are likely to be universally applicable and thus the most appropriate sensor technology has to be linked to the market need. There has been some developments in biosensor research in the country among the institutions such as NPA, New Delhi; CECRI, Karaikud; IACS, Calcutta; TIFR, Bombay; IIT, New Delhi; CSIO, Chandigarh; CEERI, Pilani.

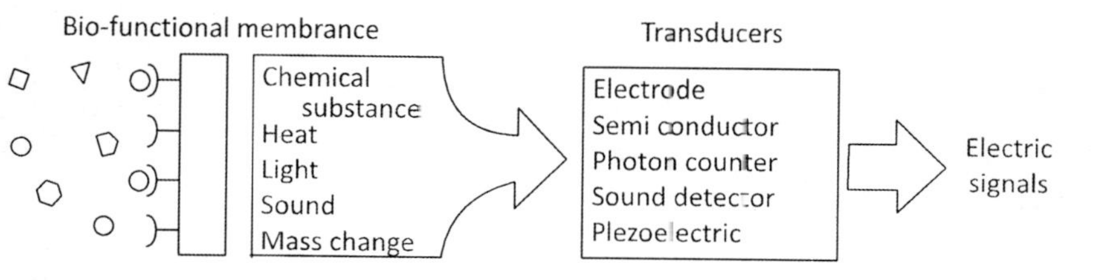


**FIGURE 22.8:** Biosensor: basic principles.

## Conventional Biosensor

Biosensors are composed of a bifunctional material and a transducer and have been developed and applied to analytical fields, clinical analysis, food industry and environmental measurements. Biosensors have their roots in military research, as means of detecting nerve gases and other chemical warfare toxins. Their applications have branched out to include simple to use *alternate site* diagnostic devices for home, doctor's office, or drug use screening; medical and surgical monitors (small enough to fit inside a blood vessel) and environmental quality monitors.

Immobilized enzymes, microorganisms and antibodies are used as molecular recognition materials. Electrochemical devices have often been used for transducers. Various enzymes have been used as molecular recognition elements. An enzyme electrode is composed of an enzyme immobilized membrane and an electrode. The principle of an enzyme electrode is based on the detection of electro active species produced or consumed by the enzyme reaction. For example, a conventional glucose sensor is composed of **glucose oxidase** (GOD) and an electrode. GOD oxidizes glucose with the consumption of oxygen and produces gluconolactone and hydrogen peroxide. Measuring the consumption of oxygen with an oxygen electrode or production of hydrogen peroxide with hydrogen peroxide electrode, the concentration of glucose can be determined. This type of glucose sensor is in commercial use for the diagnosis of diabetes. There are many kinds of biosensors using the same principle and devices, and which are developed and used in the fields of clinical analysis and measurement of foodstuffs.

## Microbial Biosensor

Microorganisms have also been utilized as molecular recognition elements. A microbial sensor consists of a microorganism immobilized membrane and an electrode. Various kinds of microbial sensors have been developed and applied to the measurement of biological compounds. The principle of a microbial sensor is based on either the change of respiration or the amount of produced metabolites as the result of assimilation of substrates by microorganisms. Furthermore, the use of auxotrophic mutants can selectively determine many kinds of substances. For example, the vitamin $B_{12}$ sensor was constructed by using immobilized *Escherichia coli* 215. The *E. coli* 215 strain requires vitamin B12 for its growth. The linear relationship was obtained in the range between 5 x 109 and 25 x 109 g/ml. Within 25 days, the decrease in the response was approximately only 8 per cent.

Recently, microbial sensors using thermophilic bacteria have been developed. The use of thermophilic bacteria can possibly reduce contamination of other microorganisms by the use of high temperatures to obtain long term stability. For example, BOD and carbon dioxide sensors are constructed by using thermophilic bacteria isolated from a hot spring, Good linear correlation was observed between the BOD sensor response and BOD value in the range 1 to 10 mg/1 BOD (JIS) at 50°C. The Sensor signal was stable and reproducible for more than 40 days. For the carbon dioxide sensor, a linear relationship was obtained in $NaHCO_3$ concentration between 1 and 8 m M at 50°C and the response time was 5 to 10 min. The linear relationship was also observed in the $CO_2$ concentration range 3 to 8 per cent.

Micro biosensors have many following advantages

1. Implantation in the human body and are suitable to *in vivo* measurement.
2. Can be integrated on one chip and are useful for measuring various substrates in a small amount of sample solution simultaneously.

3. Since semiconductor fabrication technology is applied to micro biosensors, it is possible to develop disposable transducers for biosensors through mass production.

## Development of Micro biosensor

Micro biosensors are based on **ion sensitive field effect transistor** (ISFET) and were first reported by Bergveld (1970). Matsuo et al. (1974) improved the ISFET using silicon nitride as the gate insulator to construct micro pH sensitive devices. They show rapid response, low power consumption, low noise and no need of a high impedance amplifier. The general circuit diagram for measuring the gate output voltage is shown in below.

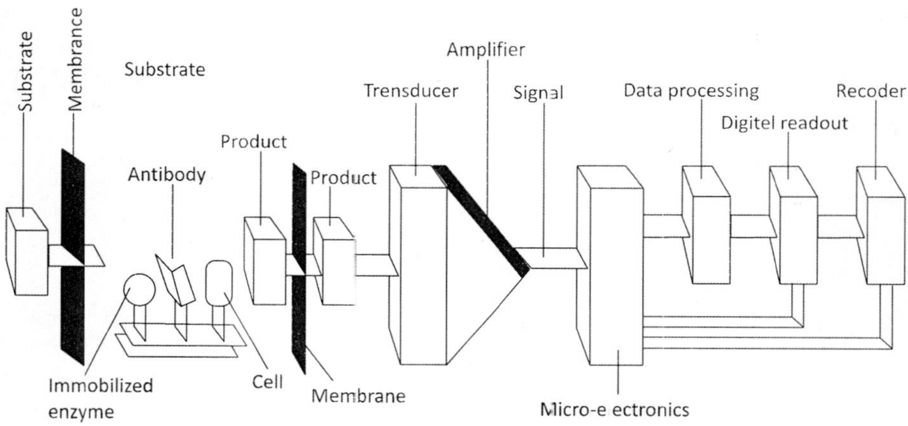


**FIGURE 22.9:** Schematic outline of biosensor

In this circuit, the voltage between the source and drain is controlled constantly and the current between source and drain is also held constantly. The Ag/AgC1 electrode is immersed in the same solution. The surface potential on the silicon nitride of the ISFET is affected by pH of the solution, with concomitant change in the gate voltage, which is proportional to the change in surface potential. Therefore, the surface potential change in the ISFET, caused by the change of pH, can be measured as the change in the gate output voltage.

### *Fabrication*

ISFETs are fabricated by using semiconductor technology. Hence, it is easy to miniaturize and integrate ISFETs on one chip. ISFET is used as a Potentiometric transducer, therefore, enzymes which cause pH changes in its reaction, such as urease and oxidase, can be used.

### *Urea Sensor*

A urea sensor consisted of a urease immobilized membrane and a pH electrode. Urease catalyzed reaction cause pH changes, so that ISFET can be used as a transducer. A micro urease sensor is fabricated as follows:

An ISFET was laid inside a vacuum chamber and 3 amino propyltriethoxysilane (3Aptes) is vaporized at 80°C and 0.5 Torr for 30 min, followed by glutaraldehyde (GA) treatment under the same conditions. The chemically modified ISFET was covered with cellulose acetate membrane containing 1,8 diamino 4 amino methyl octane and GA. The ISFET was immersed in urease solution. The urea sensor gives the linear relationship between the initial rate of the output gate voltage and the logarithm value of urea concentration in the range 16.7 to 167 mM and can be used for 20 days with slight degradation of the enzyme activity.

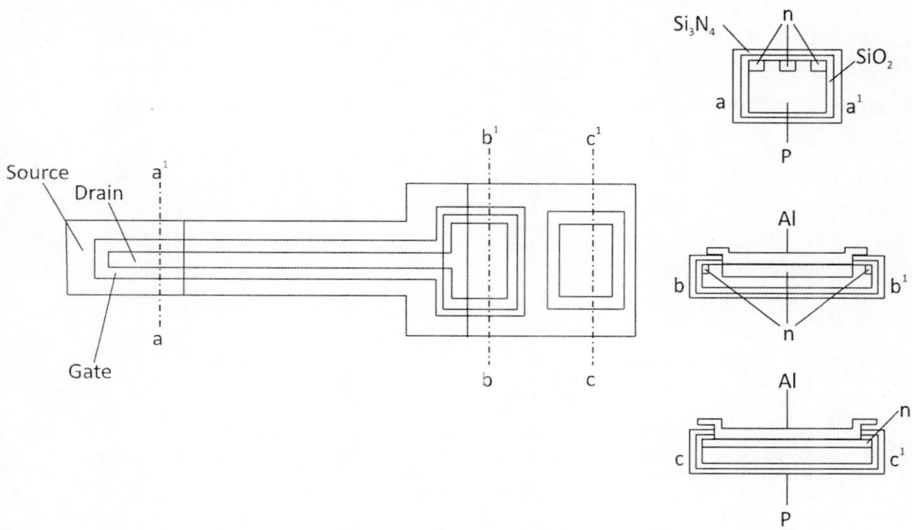


**FIGURE 22.10:** Structure of ISFET

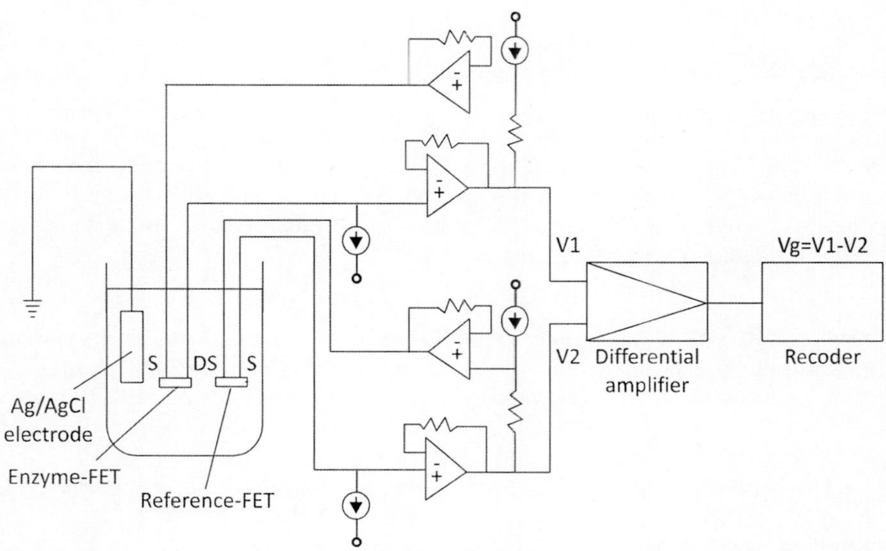


**FIGURE 22.11:** Circuit diagram of the measuring system.

### *Alcohol Sensor*

The study of an alcohol-sensitive micro biosensor using an ISFET and the enzyme system existing in the cell membrane is reported. The cell membrane of acetic acid producing bacteria has a complex enzyme system oxidizing ethanol to acetic acid via acetaldehyde. This system consists of membrane bound alcohol dehydrogenase (ADH), aldehyde dehydrogenase (ALDH) and an electron transfer system. This complex enzyme system, therefore, can be used for application with an ISFET.

### Biosensors using Amorphous Silicon ISFET

The ISFET device can only be manufactured by using a silicon wafer for the substrate. In recent years, devices made from amorphous silicon have received widespread attention because of their great applications potential. Various substrates such as glass and plastics can be used for preparing amorphous silicon and transistors can be fabricated with a number of different structures such as a needle of a syringe.

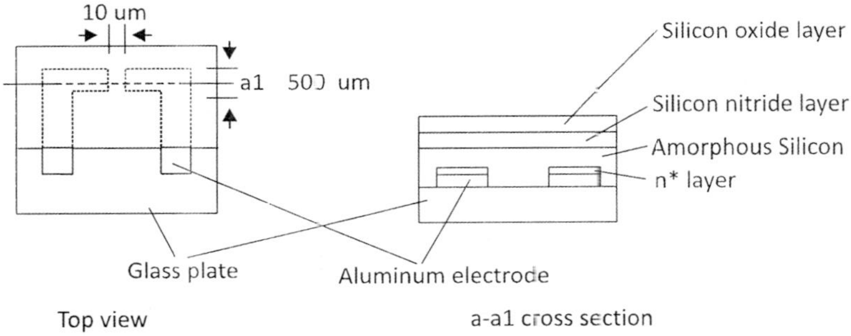


**FIGURE 22.12:** Structure of amorphous silicon ISFET.

### *Hypoxanthine and Inosine Sensor*

To maintain quality, evaluation of freshness is important in the fish industry. When a fish dies, adenosine 5' triphosphate (ATP) decomposition in the fish meat occurs and adenosine 5' diphosphate (ADP) and adenosine 5' monophosphates (AMP) and related compounds are generated

$$\text{ATP} \rightarrow \text{ADP} \rightarrow \text{AMP} \rightarrow \text{IMP} \rightarrow \text{HxR} \rightarrow \text{Hx} \rightarrow \text{X} \rightarrow \text{U}$$

where IMP, HxR, Hx, X, and U stands for inosine 5' monophosphate, inosine, Hypoxanthine, xanthine, and uric acid, respectively, Consequently, Hx accumulation with an increase in storage time can be used as an indicator of fish meat freshness. Therefore, simple and rapid methods for the determination of Hx and HxR are required in the seafood industry.

**Hypoxanthine Sensor:** Hypoxanthine is measured on the basis of the reaction catalyzed by Xanthine oxidase (XO). The pH change caused by uric acid is detected by using aSi ISFET.

**Inosine Sensor:** Inosine sensor is fabricated similarly to the Hypoxanthine sensor by using nucleoside phosphorylase and XO co-immobilized on aSi ISFET simultaneously.

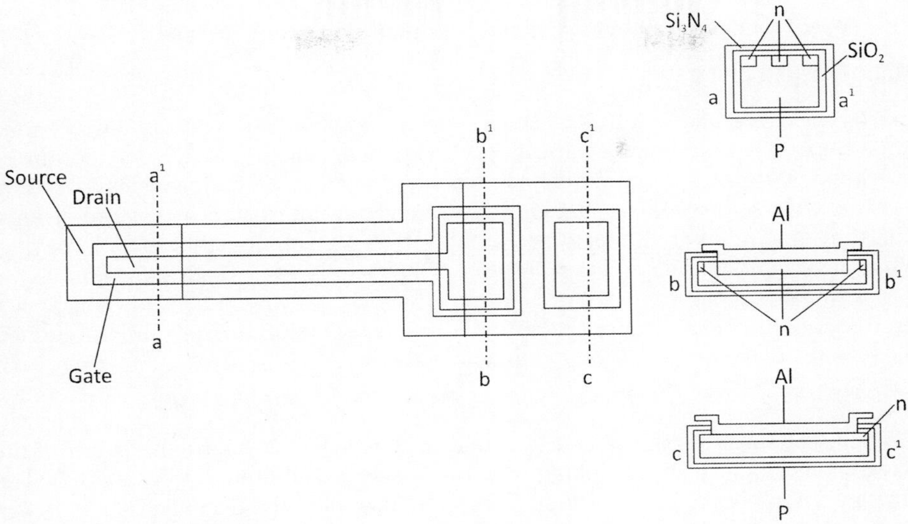


**FIGURE 22.13:** Fabrication process of micro-oxygen electrode.

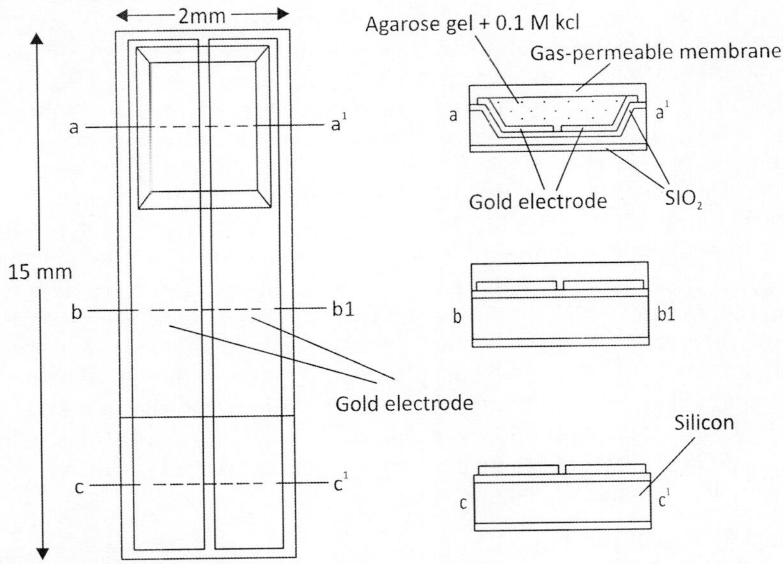


**FIGURE 22.14:** Structure of micro-oxygen electrode

After 90 seconds from injection of inosine solution, the gate voltage gradually increases and reaches a steady state in approximately 7 min. Xanthine formed by the decomposition of inosine catalyzed by nucleoside phosphorylase is subsequently oxidized to uric acid by XO. The linear relationship was obtained in the range 0.02 to 0.1 mM by plotting the initial rate of the gate voltage change with respect to the logarithm of inosine concentration. The oxidation of hypoxanthine to uric acid by xanthine oxidase is initiated immediately after injection. The response to inosine, however, has a time lag of 90 sec after injection, This phenomenon is attributed to the three step reaction On the basis of this time lay. This sensor can determine inosine and hypoxanthine simultaneously.

## Micro-Oxygen Electrode

Clark-type oxygen electrodes have been applied to various biosensors, immobilizing either enzymes or microorganisms, which catalyze oxidation of biochemical organic compounds. At present, several oxygen electrodes, based on conventional semiconductor technology, have been fabricated by several groups, but they are not yet in mass production. The reason for this is that oxygen electrodes contain electrolyte solution, which results in difficult adhesion of the gas permeable membrane to the substrate, even if epoxy resin is used. Thus, mass production of such devices is difficult. Recently, miniaturized and integrated biosensors have been required in clinical analysis.

### *Glucose and Carbon Dioxide Sensors using Micro-Oxygen Electrode*

The glucose sensor is fabricated by immobilizing GOD on the sensitive part of the oxygen electrode by cross linking with bovine serum albumin (BSA) and GA. The enzyme immobilized membrane is formed by dropping the sensitive part into the mixture containing 2 mg of GOD. 20 μl of 10% BSA solution and 10 μl of 25% GA solution. The glucose sensor responds as soon as the glucose solution is injected into the buffer solution and reaches a steady state in 5 to 10 min. The sensor responded almost linearly for glucose concentrations between 0.2 and 2 mM, which is comparable to conventional glucose sensors.

A microbial $CO_2$ sensor using this oxygen electrode was constructed by Suzuki et al. (1988). Autotrophic bacterium named as S17, which can grow with only carbonate as the carbon source is used. (available at Fermentation Research Institute, Japan). Bacterial whole cells are immobilized on a micro-oxygen electrode. The sensitive area of the oxygen electrode is immersed in 0.2% sodium alginate solution containing S17 whole cells, then removed and immediately immersed in 5% $CaCl_2$ solution to form bacteria immobilized calcium alginate gel. The negative photo-resist as the gas permeable membrane is formed over the bacteria immobilized gel. The photo-resist in only exposed to UV light for a few minutes. The response time is 2 to 3 min. Carbon dioxide was supplied by acidification of $NaHCO_3$, the concentration of which can be related to $CO_2$ concentration. The linear relationship is obtained between the current decrease and $NaHCO_3$ concentration in the range 0.5 to 3.5 mM. The lowest detection limit was 0.5 mM $NaHCO_3$ within the margin of the noise amplitude. Above 3.5 mM, no significant increase in response was observed.

## Integrated Multi-biosensor

In clinical analysis, about 20 constituent elements are analyzed at the same time. To detect various substances in a small amount of sample solution simultaneously, it is important and necessary to develop micro multi-biosensors. Recently, various kinds of integrated biosensors using ISFET and microelectrodes have been reported. These biosensors are based on the ISFETs and electrodes, which were coated with enzyme immobilized membranes. It is necessary to develop an enzyme immobilized membrane fabrication method that meets the following requirements:

1. An enzyme immobilized membrane should be precisely deposited onto a gate

region or small working electrode.

2. A deposited membrane should not peel off the sensitive surface area in practical use.
3. Different enzyme membranes can be prepared without mixing.
4. Fabrication processes are applicable to a wafer and compatible with the IC process.

## Novel Biosensor Based on New Transducers

Conventional biosensors have consisted of electrochemical devices. On the other hand, other transducers such as optical fibers, image sensors, piezoelectric devices, and SAW devices are currently used for biosensors.

### *Image Sensor*

Most clinical analyses are based on the determination of soluble marker substances in body fluids such as blood and urine. The direct analysis in cell or tissue level is greatly important in clinical diagnosis. In the cancer detection, highly sensitive and rapid detection methods for abnormal cells are required. Cell diagnosis is carried out mainly by the visual inspection of trained experts or the use of a flow cytometer.

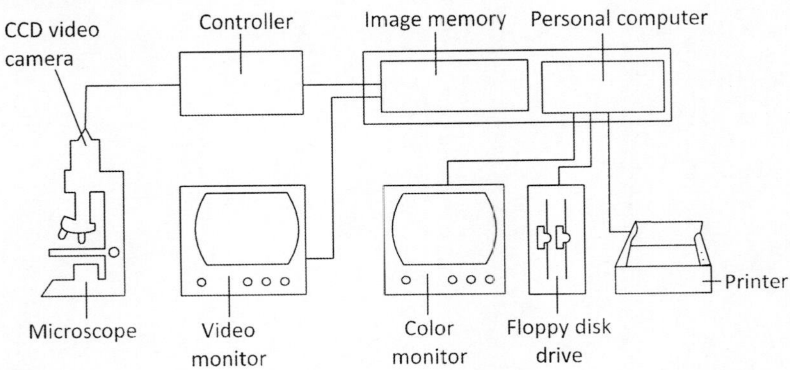


**FIGURE 22.15:** Schematic diagram of the imaging sensor system.

Recently, much attention has been focused on image analyzing systems composed of an image sensor and a microcomputer system. Image sensors are classified into XY address methods and the charge transfer method. In the XY address method, optical signals of each address are read by switching on the corresponding circuit. The charge transfer method was first demonstrated by using a bucket bridge device and presently by the more advanced charge coupled device (CCD). Most image sensors in practical use are now of the CCD type.

The CCD is an integrated semiconductor chip composed of photodiode arrays and charge transfer circuits. Electric charges accumulated at each photodiode are transferred systematically to the output terminal by controlling the electric potential in the chip. Output pulse height correlates with the brightness at the corresponding photodiode. Thus, a visual image focused on the CCD can be converted to a succession of analog pulses, since, the photodiodes are arranged approximately by 10μm separation, the same degree of image resolution can be obtained. There are many advantages to the solid state CCD image sensor as compared with a conventional vidicon. For example, they are compact in size, have high sensitivity are distortion less, have no afterimage, have low power consumption, and they have a long operational life.

# 23. Plant Tissue Culture: Basic Concepts

*Plant tissue culture is the technique of growing plant cells, tissues and organs in an artificial prepared nutrient medium static or liquid, under aseptic conditions.* It has advanced the knowledge of fundamental botany especially in the field of agriculture, horticulture, plant breeding, forestry, somatic cell hybridization, phytopathology and industrial production of plant metabolites, etc.

## IMPORTANCE OF TISSUE CULTURE

All the cells in an organism carry the same genetic information yet show variations in expression. Our knowledge of cell and tissue cultures has been developing with full swing, specially in biotransformation, forestry, genetic engineering, morphogenesis, somatic hybridization, secondary metabolite production, hybridization, variety development and their conservation, maintaining pathogen free plants and rapid clonal propagation, totipotency, differentiation, cell division, cell nutrition, metabolism, radio biology, cell preservation, etc. It is now possible to cultivate cells in quantity, or as clones from single cells; to grow whole plant from isolated meristem and to induce callus or even single cell to develop into complete plant either by organogenesis or directly by embryogenesis in vitro.

The production of haploid through tissue culture from anthers or isolated microspores and of protoplasts from higher plant cells has served as the basic tools for genetic engineering and somatic hybridization. Tissue culture technique helps to propagate plants of economic importance such as orchids and other ornamental plants in large numbers by their meristem culture or by other in vitro methods. This provides them virus-free plantlets. Propagation of valuable economic plants through tissue culture based on the principle of **totipotency** (*every cell within the plant has potential to regenerate into a whole plant*).

In plant breeding, embryo, ovary and ovule culture as well as in vitro pollination have been employed to overcome morphological and physiological sterility and incompatibility. In recent years, plant tissue culture technique is in increasing use for producing haploids from anthers or isolated microspores, and of protoplasts from higher plant cells and the recognition of the potential of these materials in genetics and plant breeding. One of the most significant developments in the field of plant tissues culture during recent years are the isolation, culture and fusion techniques which have their special importance in studies of plant improvement by cell modification and somatic hybridization.

Plant tissue culture technique is a boon in the studies of the biosynthesis of secondary metabolites and provides an efficient means of producing economically important plant products (fine chemicals). Plant tissue culture also provides raw material to pharmaceutical, cosmetic and confectionary industries. Examples are berberine, ginseng, shikonin and vanilla.

## Global Picture of Plant Tissue Culture

The potential value of plant tissue culture technology is being commercially exploited by various organizations all over the world. The Horticulture Industry responded very quickly to the micro propagation research. At present more than 400 million plants are produced through tissue culture in different parts of the world. There is a potential market for billions of US dollars per year worldwide for tissue culture products. There are more than 65 laboratories all over the world producing more than a million plants per year. The total production capacity of North American laboratories probably exceeds 150 million plants per year. The production in Asian countries is nearing 100 million plants, which includes orchids, temperate and tropical flowering crops, foliage plants and plantation crops.

**TABLE 23.1:** Land marks in plant tissue culture research*

| Worker | Year | Contribution |
|---|---|---|
| G. Haberlandt | 1902 | Was first to culture isolated plant cells in vitro on artificial medium. |
| W.J. Robbins, W. Kotte | 1922 | Culture of isolated roots-organ culture. |
| P.R. White | 1934 | Demonstrated capability of indefinite culture of tomato roots (long periods). |
| R.J. Gauhtheret, P. Nobecourt | 1939 | First culture cambial tissues isolated from carrot. |
| P.R. White | 1939 | Callus culture of tobacco tumour tissue from interspecific hybrid of *Nicotiana glauca* x *N. langsdorfii*. |
| J. Van Overbeck | 1941 | Discovered nutritional value of liquid endosperm of coconut (coconut milk) for culture. |
| P.R. White, A.C. Braum | 1942 | Experiments on crown-gall and tumour formation in plant growth of bacteria-free crown-gall tissue |
| A. Caplan, F.C. Steward | 1948 | Use of coconut milk and (2,4-Dichlorophenolacetic acid) for proliferation of cultured carrot and potato tissues. |
| G. Morel | 1950 | Culture of monocot tissues using coconut milk. |
| W.H. Muir | 1953 | Developed technique for culture of single isolated cells (nurse culture method). |
| W. Tubecke | 1953 | Haploid cultures from pollen of gymnosperm (*Ginkgo* sps.) |
| C.O. Miller, F. Skoog | 1955 | Discovery of cytokinins, e.g. kinetin, as potent cell division factor |
| E. Ball | 1955 | Culture of gymnosperm tissue (*Sequoia*) |
| F. Skoog, C.O. Miller | 1957 | Predicted hypothesis that shoot and root initiation in cultured callus is regulated by the proportion of auxin and cytokinin in the culture medium |
| E.C. Cocking | 1960 | Enzymatic isolation and culture of protoplasts |
| G. Morel | 1960 | Development of shoot apex culture technique |
| G. Morel | 1964 | Use of modified shoot apex technique for orchid propagation |
| S.G. Guha, S.C. Maheshwari | 1966 | Cultured pollen and anthers can produce haploid embryos |
| J.P. Nitsch | 1974 | Culture of microspores of *Datura* and *Nicotiana*, to double the chromosome number, and to harvest seeds, from the homozygous diploid plants just within 5 months |
| R. R. Hendre | 1975 | Established technique for obtaining virus free sugarcane, |

| Worker | Year | Contribution |
|---|---|---|
| and coworkers | | citrus, potato and cassava. |
| G. Melchers | 1978 | Production of somatic hybrid from fusion of protoplasts isolated from potato and tomato. |
| S. K. Mukherjee, S. Bhaskaran | 1980 | Isolated auxotrophic mutants for vitamin B components in *Nicotiana*. |
| A.Mascarenhas, V. Jagannathan | 1982 | Isolated spontaneous variants from wheat callus cultures. |
| K.A. Barton, W.J. Brill, | 1983 | Insertion of foreign genes attached to a plasmid vector into naked plant protoplasts |
| M.D. Chilton | 1983 | Production of transformed tobacco plants following single cell transformation or gene insertion |
| H. Kohn and coworkers | 1985 | Somatic hybrids in tobacco mediated by electro fusion. |
| E. Sundberg and K. Glemelius | 1986 | Somatic hybrids in Brassicaceae |
| R. Nadgauda, A. Mascarenhas | 1986 | Isolated and planted variants of sugarcane resistance to mosaic virus and turmeric plants of the variety 'Tekurpeta' in the field. |

## PLANT TISSUE CULTURE: PRINCIPLES

The technique has developed around the concept that a cell is **totipotent** that is has the capacity and ability to develop into whole organism. The principles involved in plant tissue culture are very simple and primarily an attempt, whereby an explant can be to some extent freed from inter-organ, inter-tissue and inter-cellular interactions and subjected to direct experimental control.

The most common culture in plant tissue is **callus**, which is wound tissue composed of undifferentiated, highly vacuolated and unorganized cells.

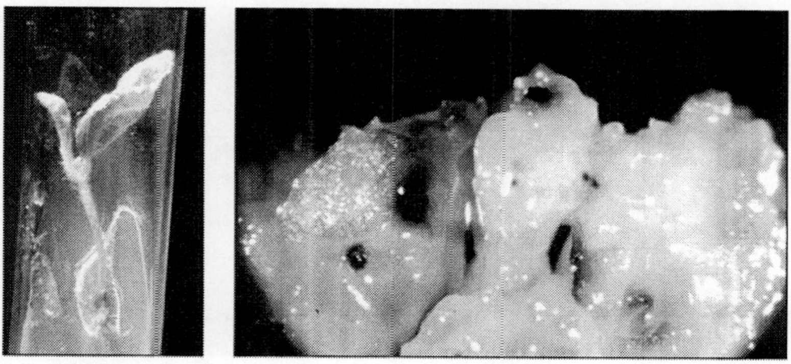

(A) (B)

**FIGURE 23.1:** (A) In vitro cultured plant material, (B) A callous.

### Callus Culture

For raising the callus tissues, a tissue culturist must have clear understanding of some basic principles. A cell from any part of the plant like shoot apex, bud, leaf, mesophyll cells, epidermis, cambium, anthers, pollen, fruit etc., when inoculated in a suitable medium under aseptic laboratory conditions can able to differentiate and multiply. This results into the formation of an amorphous mass of cells known as **callus**, which can be induced to re-differentiate on appropriate medium to develop **embryoids** which directly develop into the plantlets, eventually giving rise to a whole viable plant.

The term **clone** (from the Greek *klon*, meaning: a slip or twig suitable for plant propagation) was suggested by Webber (U.S A.) in 1903 to explain those plants which were obtained by a sexual reproduction; it is even applied to DNA multiplication (cloning of genes in bacteria). In strict scientific sense, cloning means *an organism obtained from a single cell through mitotic divisions.*

## Meristem Culture

When a meristem is cultured in vitro, then it produces a small plant bearing 5 or 6 leaves. This could be obtained within a few weeks. Then the stem is cut into 5-6 small micro cuttings, which under favourable conditions, become fully grown plants.

## Organ Culture

A body of higher plants has complex inter-relationships between different organs like root, shoot, apical meristem, leaf primordia, floral buds, ovary, ovule, anther lobs, pollen grains, fruit, seed, etc. In this method a particular organ is isolated and cultured under laboratory conditions in a chemically defined medium where they retain their characteristic structures and other features and continue to grow as usual. In organ culture, organs are not induced to form callous, therefore, it differs from the callus culture where the organization of the intact tissues is lost.

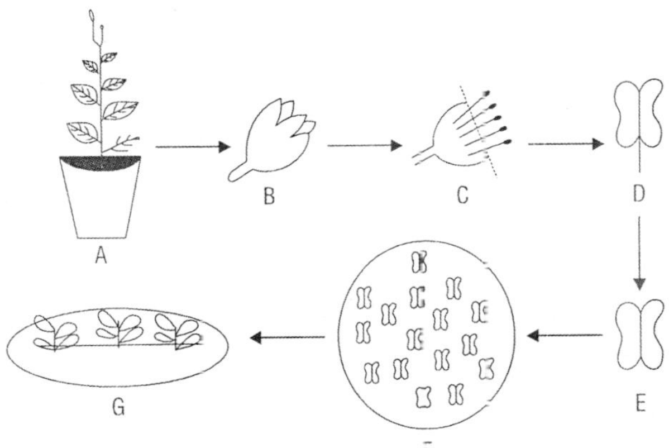


**FIGURE 23.2:** Diagrammatic representation of the anther culture protocol for tobacco. (A) Flowering diploid plant. (B) Floral bud excised from plant. Calyx and corolla re of approximately equal lengths. (C) Floral bud with calyx and corolla removed. Buds normally contain five anthers, which are shown still attached to filaments. (D) Isolated stamen with anther and filament. (E) Anther removed from filament. (F) Petri dish with cultured anthers. (G) Plantlets emerging from anthers.

This technique provides an experimental system to define the nutrients and growth factors that are usually received by the organ from other organs of the plant body and from surrounding environment. It also helps us in understanding the inter-dependence of organs with respect to various physical and chemical growth factors including growth hormones. Organ culture technique also provides the knowledge about the various problems of morphogenesis and the sites of biosynthesis of specific metabolites and growth compounds. It may be used as a tool for improvement of various economically important crops.

Organ culture may be grouped into two major categories: **vegetative organs** (root culture, leaf culture, and shoot tip culture) and **reproductive organs** (complete flower culture, isolated ovary culture, isolated ovule and embryo culture, pollen mother cell culture, seed and fruit culture).

## THE CONCEPT OF TOTIPOTENCY OF CELLS

Some concepts in science become inherently acceptable long before their practicality is demonstrable. This was so in the concept of the totipotency of cells of higher plants. Even in the mid-twenties one encountered the tacit view that, apart from inherent practical difficulties, there was no theoretical reason why one should not rear a begonia plant from a single leaf hair cell. This view was traceable **first** to the then well recognized principle that as cells divide mitotically, they do so equationally to produce daughter cells in facsimile.

**Secondly**, the concept was due to G. Haberlandt's historic but then unsupported claim, expressed in 1902, that one day it should be possible to rear plants from isolated surviving cells of flowering plants. Haberlandt's insight was, in fact, the more remarkable because he even stated that out of surviving somatic cells artificial embryos would be reared asexually. In higher plants, embryos do develop in situ from appropriately stimulated somatic cells but without any prior sexual act. At the outset, the special significance of the fertilized egg in plants should be seen as restricted to its role as the genetically unique product of a fusion of male and female gametes. Having established the genetic constitution of a given individual, the zygote (or fertilized egg) really behaves developmentally as a very general kind of living cell- a cell with a built-in capacity to grow in an organ, the ovule, that fosters that growth. Moreover, as will be shown, and in a satisfying number of cases, isolated somatic cells may return to a simulated zygotic state and grow into embryos and to plants under the appropriately applied conditions which furnish the requisite nutrients and stimuli.

In higher plants the provision of nutrients to the embryos of higher plants in the embryo sac and ovule, via the endosperm for their attachment in the ovule via a suspensor serves a mere mechanical purpose, while the pro-embryo and its later formed cotyledons absorb nutrients and stimuli directly over their outer surfaces. Thus, if somatic plant cells can be induced to form immature embryos, it is very much simpler to provide for their nourishment externally than for embryos of higher animals, which need a blood supply.

This may well account for the fact that testicular teratomas of the mouse, lacking a blood supply, fail to develop. However, these structures do embark upon a simulated embryogeny and indeed similar asexual forms have long been known to occur in other animals. By contrast, and as will be shown below, proembryos of several plants which originate from free somatic cells can now be caused to form embryos, and plantlets by the thousands; this occurs even in isolation and in a limited volume of an appropriate aqueous solution which behaves, in this respect, like an artificial ovule or embryo sac.

There is, however, another obvious reason why totipotent development from somatic cells is more feasible in plants than in higher animals. Such animals develop their organs early, they assign to them highly specialized functions and then integrate them by means of the circulating tissue of the blood, by a nervous system spinal cord, and brain, and by highly specialized hormones which are the product of discrete glands.

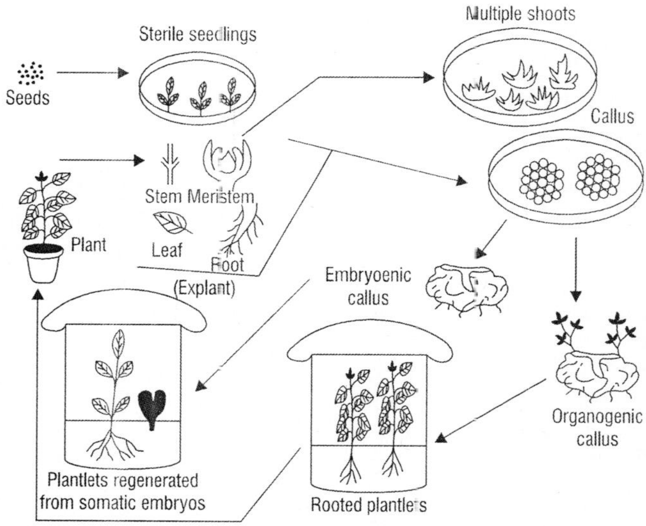


**FIGURE 23.3:** General presentation of steps used for plant tissue culture.

Moreover, these developmental events in animals essentially occur but once. By contrast, higher plants create their organs repetitively and, often, virtually indefinitely. Plants do all this by arranging, in their growing regions, living cells which retain the essential characteristics of the plant and which, during development, depart from, or return to, the situation to be seen in the primary growing apices of shoot and root by relatively small and often easily reversible steps.

# 24. Plant Tissue Culture: Methods

The setting up of a tissue culture laboratory needs proper planning. It depends upon space availability, volume of work to be carried out and funds. Usually, the available laboratory space is being divided into five distinct laboratory areas. These are: *Media preparation area/room*, *Aseptic transfer chamber area*, *environmentally controlled culture room*, *Analytical room and Acclimatization room*.

## MEDIA PREPARATION: AREA/ROOM

### Requirements

Laboratory tables or benches and revolving stools with adjustable height, Hot plate and magnetic stirrers, analytical loading single pan balance with precision of ± 0.001 g, weighing range 0.1 mg-180g, digital read out, Top pan loading balance for quick weighing, range 100 mg-500g capacity, sensitivity 0.1g, Refrigerator and freezer, Water purification and storage system, Glassware washing facility with proper drainage, Gas outlet, Electric hot-air oven range upto 250 ± 2°C and microwave oven, Digital pH meter, Range 0-14 pH, accuracy 0.1 pH temperature compensation 0-100°C, Autoclave preferably horizontal., Continuous supply of single and double distilled water

### Cabinets or Shelves

For storing glassware, plastic wares, chemicals, plugs and appliances required for media preparation are as follows:

**Requirements:** Culture tubes/conical flasks/petri dishes of various capacities, Measuring cylinders 25 ml, 100 ml, 500 ml, 1000 ml, Cotton for plugs/plastic caps (autoclavable), General glassware's/plastic wares of various capacities such as volumetric flasks, beakers, reagent bottles, pipettes, vacuum filtration system and glass rods.

### Culture Media, Washing Powder/ Liquid Detergent, Disinfectants

#### *Requirements*

Powder or liquid detergents or wetting agents such as Tween-20, 70% alcohol and absolute alcohol. Glass distilled water, Stock solutions of nutrients of tissue culture media or ready-made/pre-mixed powdered media, Sucrose, Agar (tissue culture grade), Sterile culture vessels with distilled water, Chemicals of analytical grades (Inorganic, organic salts, vitamins, amino acids, growth regulators/hormones and activated charcoal), Coconut milk, yeast extract, malt extract, casein hydrolysate and extracts of potato, carrot and tomato.

### Other Requirements

Spatulas, weighing butter paper/boats, stirring bars (magnetic) for magnetic stirrers and stirring retrievers, Brush (flask and test tubes), gloves (disposable 23 cm), mop (household), scoop, towels (household), wastebasket (large and medium), Service lines such as gas, water, electricity, vacuum pump and generator, A low bench/table/desk for culture evaluation and data recording.

## Aseptic Transfer Chamber Area

### *Requirements*

Laminar air-flow cabinets to provide constant flow of air across working area, Dissecting and stereo zoom microscopes and lens magnifier, Micro-dissecting scissors, scalpel handles with blades, forceps (various sizes), catheter trays, needles and inoculating loops, Gas outlet, vacuum facility, Tissue paper/filter paper, Sterilizer (dry heat with glass beads) and pipette dispenser (automatic).

## Environmentally Controlled Culture Room

### *Requirements*

Racks with light arrangements on timers and controlled temperature (25 ± 2°C) maintained with air cooler or by window air conditioners fitted with temperature indicators, Incubators having dark and light/dark photoperiod with controlled temperature, Rotary shakers of variable speed from 80-220 rpm to take 100 ml or 250 ml Erlenmeyer flasks with arrangement of lighting to provide an intensity of 2000 to 4000 lux, Lux meter to measure intensity of light in culture room, incubator and shaker.

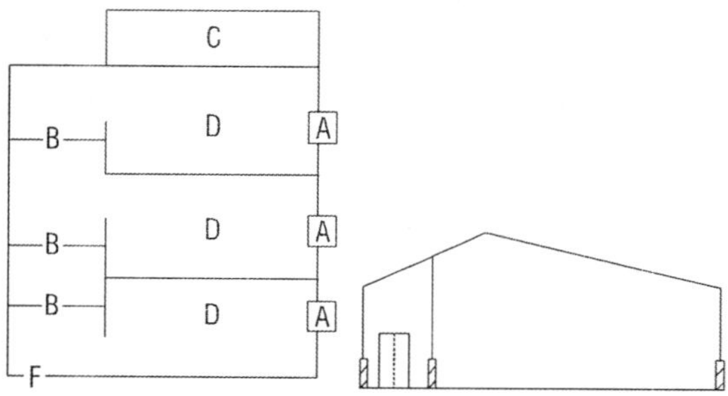


**FIGURE 24.1:** Design and elevation of greenhouse with three compartments having desert coolers. (A) Desert-cooler (B) Air-vents (C) Water tank (D) Greenhouse chamber (E) Entrance.

The cultures are usually incubated at 25 ± 2°C under 16:8 lights: dark photoperiod. The source of light for the cultures in racks should be make available with cool-day-light (fluorescent tube lights of 40 watts, 2000 lux). It is advisable to connect the controlled culture room with power generator for emergency power supply. Some space should also be kept for incubating cultures in continuous darkness.

## Analytical Room

### *Requirements*

Inverted microscope for bright field, dark field with compensating wide-angle eyepieces, preferably with photo-micro graphic attachment, Colorimeter for chemical estimation such as chlorophyll, starch, nucleic acid, phenols, oxidizing enzymes etc. Low speed centrifuge with continuous variable electronic speed control, Chemical reagent racks for qualitative and quantitative chemical analysis. Viscosity meter, Gas outlet.

## Acclimatization Room

The hardening chamber needs high illumination (4,000-10,000 lux) and high humidity (90-100%, through mist and fog systems). Humidity is required for conditioning tissue culture plants after taken out from rooting media and transfer to pots under greenhouse.

## Miscellaneous items

### *Requirements*

Air conditioners, uninterrupted power supply (UPS) and emergency light, Bunsen burners, Permanent markers, tapes (autoclave indicator), tape label (self-adhesive), aluminium foil and Para film, Fluorescent lamps/tubes, Trays and baskets for cultures, Plastic carboys to store water and other solvents, UV germicidal lamp, Metal racks to keep test-tubes in culture room, Gas lighter/match box, Fire saving, equipment and first-aid box etc., Filter paper, culture trays, culture boxes and culture tube racks.

To ensure the growth and development of an explant, it has to be provided with a suitable nutrient medium and proper laboratory conditions for culturing. These operations have to be carried out under aseptic conditions.

## Specifications of Laboratory Equipments

The following laboratory equipments are required for tissue culture.

- **pH Meter (digital):** 230V 50Hz, single phase supply with combined pH electrode, range 0-14pH/0-1400mv, temperature compensation 0-100°C.
- **Balances:** Manual - single pan, capacity 100-200g, sensitivity 0.1mg. Electronic - top loading, precision €0.005g, range - 0 to 1200g.
- **Electric Hot Air Oven**: Thermostatically controlled, range 5°C to 250 ± 2°
- Microscopes
- **Dissecting:** two lenses (eyepieces) 10X, 20X.
- **Laboratory:** with facilities for bright field, dark field, phase contrast, wide angle eyepiece with photomicrographic attachment.
- **Inverted:** Wide angle eyepieces, quintuple objective turret adjustable specimen 55 mm petridish holder and photomicrographic attachment.
- **Centrifuge** (low speed): Variable electronic speed control, speed indicator, Amp. meter, timer, dynamic break, 0 starting switch, 230V 50Hz.
- Electric Autoclaves
  - **Vertical:** with safety valves, pressure gauge, and steam-release cock.
  - **Horizontal:** mounted to tubular stand, heavy hinges
- **Steamers:** With immersion heaters of ejection safety device, size: 31cm x 1 3cm x 10cm to 61cm x 20cm x 15cm.
- Filter Sterilization Equipment: Syringe filter holder: 2.5cm
- Pressure Filter Holder: 4.7cm
- **Manestry Stills:** Electrical water stills, 2-lit capacity/hr of 1.5 kw.
- **Double Distilled Water Equipment**: Built-in energy regulator all glass water stills, heat resistant boiling flask with heater.

- **Environmental Growth Cabinets:** Cabinet with controlled temperature, light and humidity, temperature range 4° to 45°C with timer to regulate photoperiods.
- **Gyratory Shakers:** Capacity: 50 to 1000 ml flasks, speed: 80 to 200 rpm.
- **Laminar Airflow Cabinets:** Constant flow of purified air, sizes: 0.6m, 1.2 m and 1.8m.

The other miscellaneous equipments which are required for tissue culture are: air conditioners, arrow heads, Bunsen burners, deep-freeze, dissecting needles, glassware, forceps, florescent lamps/tubes, heaters, hot plates with magnetic stirrer, inoculation cabinets, metal trays and bowls (for transport of cultures), tubes, refrigerators, UV germicidal lamps, wooden or metal racks, etc.

## ASEPTIC TECHNIQUE

In *in vitro* condition plant cells and microbes have basically same requirements. When the culture medium contains sugar (as carbon source) it attracts a variety of microorganisms, which grow fast than that of the cultured tissue in medium, and they ultimately kill the plant cells. It is, therefore, necessary to have complete aseptic condition around the culture equipments, which prevents contamination of the culture medium. Following are the three main sources of contamination of the medium and the subsequent methods to check them:

**FIGURE 24.2:** Portable autoclave.

The microorganisms may be present in the nutrient medium at the time of its preparation. Proper plugging and autoclaving the culture tubes/flask can destroy these microorganisms. The medium can be completely sterilized by maintaining it at 120° C for about 20 minutes at 15 lb pressure in the autoclave.

1. The explant (plant part to be cultured) may carry microorganisms with it, therefore, the plant part should be surface-sterilized by mercuric chloride (1 to 2%) or by sodium hypo chloride solution for 30 minutes.
2. Precautions must be observed to prevent the entry of microorganisms when the plug of a culture is removed during transfer of the plant material to the medium or from one medium to another. The inoculation chamber may be sterilized by UV-radiations.

3. Correct pH of the medium is important. Highly alkaline or acidic pH affects the nutrient uptake in culture tissues. Therefore, the tissue culture medium is adjusted to a pH of 5.6 to 6.0 before autoclaving.
4. Semi-solid and liquid media are most commonly used for growing plant cells. A high concentration of gelling agent (agar-agar, gelatin, silica gel) makes the medium very hard and decreases the nutrient uptake by the tissues. Agar at 0.8% to 1.0% concentration is widely used.

## STERILIZATION OF PLANT TISSUES

It is essential to remove dirt and debris from the plant tissue and should be washed in a weak detergent solution and rinsed several times with distilled water prior to sterilization. Immersing them briefly in a 70% ethanol solution, which wets and spreads over the tissue surfaces more effectively than a higher concentration alcohol solution, cleans some woody tissues, such as buds and twigs.

- **Sodium Hypo chlorite (NaOCl):** It is the most common chemical agent used to sterilize plant tissues (0.025%-0. 25% NaOCl). Diluted household bleach can also be used for this purpose, which normally contains 5.25% NaOCl. It is equally effective and considerably less expensive.
- **Calcium hypo chlorite (CaOCl):** It can be used as a substitute for NaOCl. CaOCl causes slightly less damage to plant tissues but tend to precipitate out of solution. To avoid the accumulation of CaOCl on the plant tissue surfaces, sterilization solutions should be filtered or decanted prior to use.
- **Hydrogen Peroxide ($H_2O_2$) Solution:** Plant tissues can also be surface sterilized using a $H_2O_2$ (3%-10%). It is much easier to remove from tissues than NaOCl and CaOCl.
- **Other Substances:** Plant tissue can also be surface-sterilized **by bromine water** (1%-2%), **silver nitrate** ($AgNO_3$ 1%) and **mercuric chloride** ($MgCl_2$ 0.1%-1%).

### Cleaning (Preparation of Glasswares/Plastic wares (Autoclavable)

1. Clean the glasswares/plastic wares in 10% commercial detergent liquid or powder for 1 hr and then in HCl for 2 hr.
2. Remove the traces of detergent and acid by thorough washing with tap water
3. Rinse vessels with double distilled water and allow them to dry over night at room temperature.

### Sterilization

1. Plug glasswares such as conical flasks or test tubes with non-absorbent cotton or cover by the plastic caps. Wrap the petri dishes with aluminium foil. Place forceps and scalpels in test tubes, plug the tubes with cotton or cover with aluminium foil. Plug the mouth end of the pipettes with cotton. Wrap them individually in aluminium foil.
2. Autoclave glasswares and instruments at 121°C for 1 hr.
3. For dry-heat sterilization, metal instruments should be sterilized in an oven at 140-160°C for 2 hr.
4. Filter sterilization for heat labile amino acids, vitamins, growth regulators, antibiotics, natural complexes should be through Millipore filtration assembly using filter membranes of 0.45 or 0.22 μM porosity. Plug the receiver flask

with cotton. Assemble the filtration assembly and wrap the filtration unit with paper or foil. Autoclave the receiver and filtration unit at 121°C for 1 hr. Attach the filtration unit with receiver flask with vacuum pump in a laminar flow bench pour solution to be sterilized into the filtration unit. Apply slight air pressure to start filtration. Transfer the desired volume to sterile flasks under laminar air-flow bench. Use a sterile pipette for drawing filter sterilized solution to autoclaved medium.

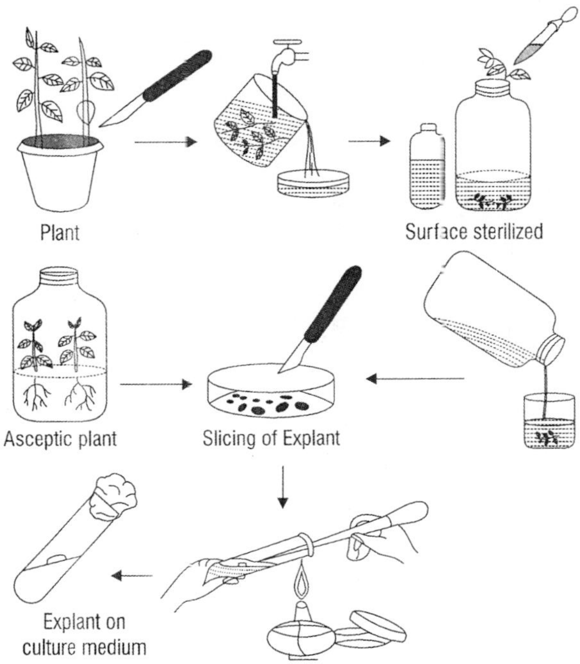


**FIGURE 24.3:** Diagrammatic presentation of the procedure for surface sterilization of plant material and inoculation of explant for culture.

## Surfactants

- **Tween 20 or Triton X-100:** These are scientific reagent-grade surfactants and are often added in low concentrations (0.05%) to chemical sterilization solutions. Their use ensure that the sterilizing agent come in contact with the entire plant tissue surface.
- **Stirring of the Tissues**: stirring the tissues during sterilization also facilitates good surface contact.
- **Ultrasonic Bath:** It is an effective method to ensure good surface contact during sterilization treatment in an ultrasonic bath like those used to clean dentures.

This technique is particularly useful for sterilizing buds and woody tissues that have many small surface crevices and cracks.

After surface sterilization, a minimum of three sequential rinses with sterile distilled water is recommended to remove any remaining chemical sterilizing agent.

## *Control of Bacterial and Fungal Contaminants by Antibiotics in Plant Tissue Culture*

Antibiotics are used to reduce, control, or eliminate contaminants in plant tissue cultures.

It is suggested that antibiotics should not replace careful aseptic technique. Most antibiotics are heat-sensitive and cannot be autoclaved. They are usually dissolved in water or another suitable solvent, filter-sterilized, and added to autoclaved culture media that has cooled to 45°- 50° C. Falkiner (1990) suggested the list of antibiotics which are suitable for plant tissue culture.

**FIGURE 24.4:** Tissue culture / growth room with culture vessels.

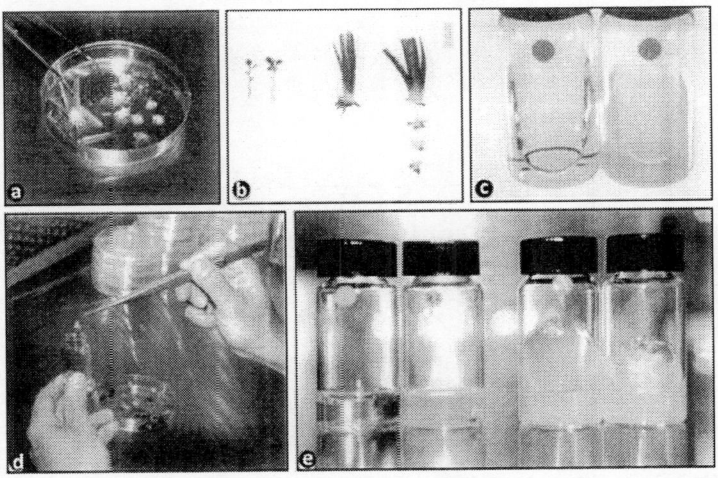

**FIGURE 24.5:** Culture indexing for cultivable contamination. (a) Aseptic technique is use to obtain tissue sections from the Stage I shoot culture. (b) Shoots with upright stems (left) are sectioned into nodal segments. The shoot tip is either cultured in the indexing medium or transferred to fresh Stage I medium. Cross-sectional stem disks are made from shoots with basal meristems (right). Scale base = 1 cm. (c) Indexing medium is prepared as a liquid and agar-solidified slants. (d) Several tissue samples are directly placed in the liquid indexing medium. (e) Clouding of inoculated liquid indexing medium is a positive indication of the presence of cultivable contaminants in the tissue sample. Development of colonies on the surface of the medium and/or halo in the medium where the tissue sample was stabbed indicated culture contamination.

## IN VITRO ENVIRONMENT

A piece of plant tissue taken out from original site of plant and transferred to an artificial tissue culture media for the growth or maintenance is called **explant material**. The choice of tissue depends upon on ultimate goal of the tissue culture project. Any piece of the plant tissue can be used as an explant material.

Various factors of an explant tissue source influence the culture on tissue culture media. These are: (1) physiological and ontogenetic age of organ or tissue, (2) quality of source plant, (3) season in which explant tissue is obtained, (4) size of the explant and (5) aim of the culture.

## PRETREATMENT TO EXPLANT TISSUES PRIOR TO CULTURE

Many explant tissues especially woody tissues produce phenolics into tissues culture medium, which cause browning of explants material. Antioxidants such as L-ascorbic acid, free acid 0.1 g/l; citric acid, free acid anhydrous 0.15 g/l; L-cysteine HCl 0.05 g/l are commonly used for checking browning problem. The browning pigments are toxic to the plant growth in cultures. Best results are obtained when explants are treated with freshly prepared antioxidant solutions.

### *Procedure*

1. Prepare the mixture of L-ascorbic acid (100 mg/l) and citric acid (150 mg/l) in double distilled water
2. Filter sterilize the solution through a 0.22 μm filter unit
3. Store the explant material in a cold antioxidant mixture and incubate explants in refrigerator at 0°C for 5-30 min. to allow the tissue to soak in antioxidant solution
4. Commercial bleach contains about 5% sodium hypochlorite and thus may be used at a concentration at 10-20%, which is equivalent to 0.5-1.0% sodium hypochlorite.

**TABLE 24.1:** Commonly used disinfectants for plant tissue culture

| Disinfectant | Concentration (%) | Exposure (min) |
|---|---|---|
| Calcium hypochlorite | 9-10 | 5-30 |
| Sodium hypochlorite | 0.5-5 | 5-30 |
| Hydrogen peroxide | 3-12 | 5-15 |
| Ethyl alcohol | 70-95 | 0.1-5.0 |
| Silver nitrate | 1 | 5-30 |
| Mercuric chloride | 0.1-1.0 | 2-10 |
| Benzalkonium chloride | 0.01-0.1 | 5-20 |

### *Sterilization procedures may be enhanced by*

1. Place the material in a 70% ethyl alcohol solution before treating with another disinfectant solution. The use of a two step (two source) sterilization procedure is beneficial.
2. Use of a wetting agent (Tween 20 or 80 respectively) is beneficial. It reduces surface tension and allows better surface contact.
3. Conduct the sterilization process under vacuum. This helps in the removing air bubbles and provides efficient sterilization process.

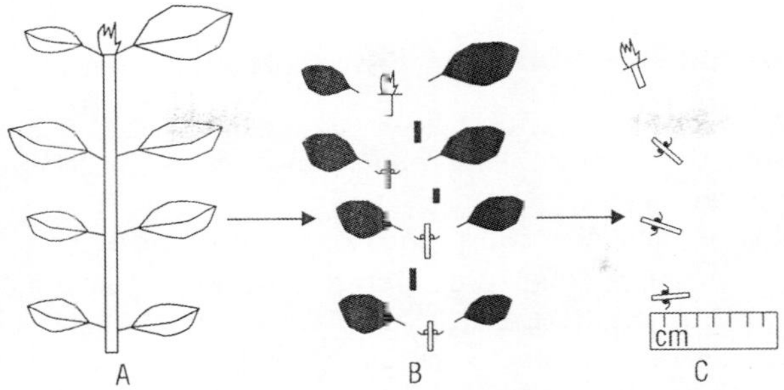


**FIGURE 24.6:** Diagram illustrating how to cut shoots (A) into single-node segments 2 to 3 cm long, removing the leaves (B), and keeping a small part of the petiole subtending the bud (C). Shaded sections are discarded.

## Procedure

1. Wash explants in a mild detergent before treatment with disinfecting solution (exclude herbaceous material which may not need this treatment).
2. Within a few days, cells proliferate and a callus culture is obtained. In this, dividing cells form a layer of meristem and build a globular mass of non-dividing parenchyma. Alternatively, they may form small meristematic and build a globular mass of non-dividing parenchyma. Alternatively, they may form small meristematic zones interspersed in non-meristematic regions, yielding a sort of nodulated callus.

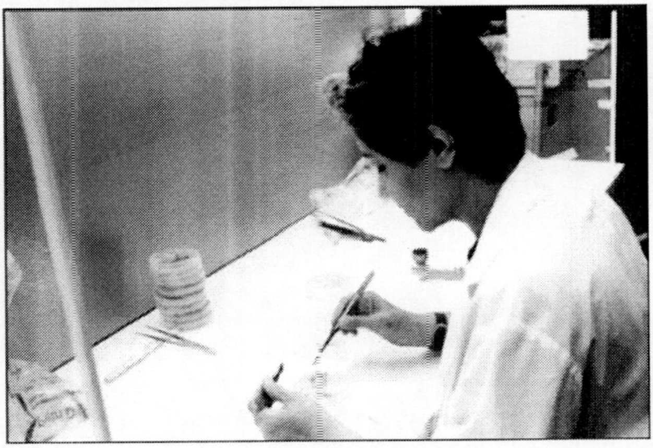

**Fig: 24.7:** Inoculation of explants on a laminar airflow bench.

**Organogenesis** is the development of adventitious organs or primordia (embryoids) from undifferentiated cell mass (callus) in tissue culture. Mostly a balance between cytokinin and auxin controls it. A relatively high ratio of auxin: cytokinin induces root formation in callus tissues whereas; a low ratio induces shoot formation. **Caulogenesis** is a type of organogenesis by which only adventitious shoot bud initiation takes place in the callus tissue. When it is applicable for root, it is known as **rhizogenesis**. Anomalous structures when develop during organogenesis is called **organoids**. The localized meristematic cells on a callous, which give rise to shoots and/or roots, is termed as

**meristemoids**.

## MICROPROPAGATION THROUGH ORGANOGENESIS

The principle of micro propagation is based on the phenomenon that shoot tip when cultured on tissue culture medium, can develop large number of shoots identical to the parent plant.

The technology provides many advantages over conventional propagation methods such as product development in highly speedy manner, uniforming in product development, large population can be produced in small growing space, quick multiplication of elite and genetically engineered products, and in vitro storage of germplasm. The technique also helps in enhancing tissue culture products via specific product format, rooted or unrooted micro cutting, bulblets, and embryos, certified plant material free of pathogens and production can be obtained throughout the year as most of the multiplication phase is completed under artificially controlled conditions.

Through the use of auxin and cytokinin, hormonal control of organogenesis became feasible. Skoog and Miller (1957) demonstrated that the differentiation of root, shoot, and both root and shoot in tobacco pith tissue is the result of auxin-cytokinin ratio (i.e., the differential regulatory role is concentration-dependent). A high level of cytokinin promoted shoot, and auxin root formation. An equal concentration of cytokinin: auxin resulted in callusing. However, the requirement of exogenous supply of growth hormones for differentiation is dependent upon the endogenous level of these substances. To prove the efficacy of somatic cells to behave as zygote, Reinert (1958) and Steward (1959) reported the first somatic embryogenesis from the tissues of carrot.

By culturing shoot tips of *Lupinus* and *Tropaeolum*, Ball (1946) raised complete plants, which could be transplanted. Shoot tip culture helped Morel and Martin (1952) to recover virus-free plants of *Dahlia* and, subsequently, in rapid multiplication of orchids.

### MAJOR STAGES OF MICROPROGATION

Selection of an elite mother plant

⇓

Explant pieces ⇒ Trimming

⇓

Surface sterilization ⇒ Washing

⇓

Establishment of cultures on appropriate growth medium

⇓

Transfer of multiplication medium

⇓

Rapid shoot or embryo formation

⇓

Embryo shoots or plantlets transferred to sterilized soil or artificial medium by different gradual weaning processes (gradual reduction in humidity and nutrient levels)

**FIGURE 24.7:** Procedure of removing plant parts.

## CULTURE MEDIA AND PREPARATION

The success in cell, tissue and organ culture technology is related to the selection or development of the culture medium. As no single medium will support the growth of all tissue

Various media compositions are frequently used for tissue culture. A literature search is useful for selecting the appropriate culture medium as a starting points in developing a medium for specific purpose such as callus induction, axillary bud proliferation, organogenesis, somatic embryogenesis, anther culture etc. A nutrient medium generally contains inorganic salts, vitamins, growth regulators, a carbon source and gelling agent. Other components added for specific purposes include organic nitrogen compounds, hexitols, amino acids, antibiotics and plant extracts. The Murashige-Skoog medium (MS) (1962), Revised Murashige-Skoog medium (Raj Bhansali and Arya, 1978), White's medium (1963), Linsmaier and Skoog (LS) (1965), B5 (Gamborg et al, 1968), Nitsch and Nitsch (1969), Woody plant medium (Llyod and McCown, (1981), Somatic embryogenesis medium (Raj Bhansali, 1988, 1990) and derivatives of these media have wide applications for different plant species and for different culture objectives. The decision on using type of media for the metabolic needs of the cultured cells and tissues, is a major factor of success in plant regeneration process.

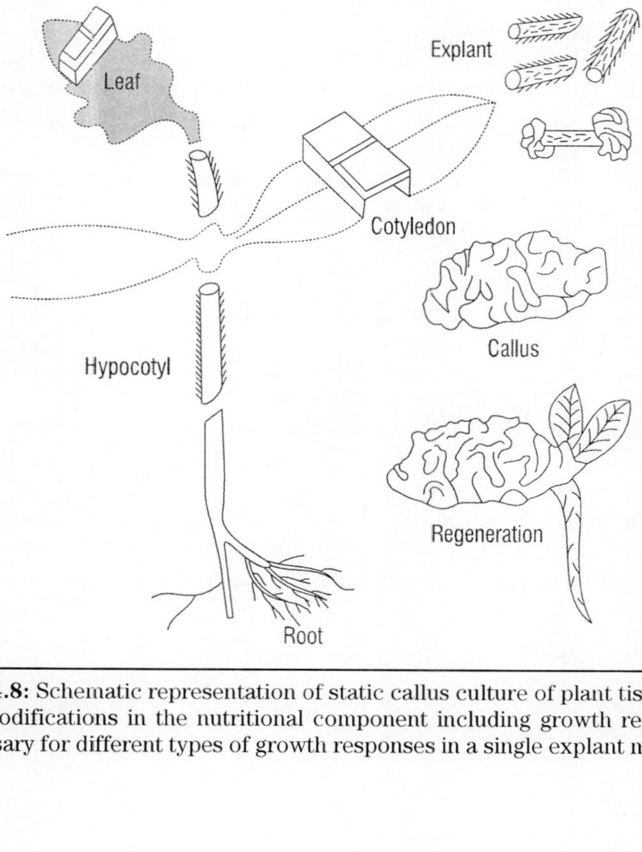


**FIGURE 24.8:** Schematic representation of static callus culture of plant tissue cultures therefore modifications in the nutritional component including growth regulators are often necessary for different types of growth responses in a single explant material.

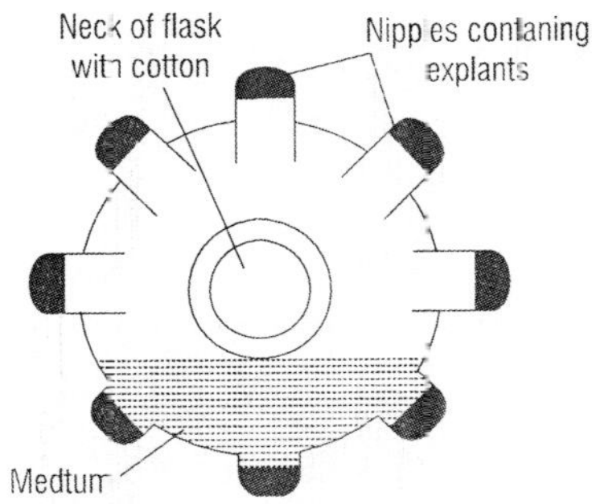


**FIGURE 24.9:** Tissue culture techniques are now practiced on a grand scale. Special nippled flasks have proved very useful for the liquid culture of single cells and cell clumps.

## MEDIA COMPONENTS

### Inorganic Salts

A relatively small number of mineral salts are used as component of media for plant tissue culture. The inorganic salt formulations can vary in various reported media, however MS formulation is most widely used with or without modifications. The distinguishing feature of MS inorganic salts is their high content of nitrate, potassium and ammonium in comparison to other salt formulations. The stocks are prepared at 100 X (times) the final medium concentration. Each stock is added at the rate of 10 ml per 1000 ml of medium prepared.

The Na-Fe EDTA stock should be protected from light stored in bottle that is amber coloured or wrapped in aluminium foil. Concentrated salt stocks enhance the accuracy and speed of media preparation. Guidelines for maintaining stock solutions:

1. All salt stocks should be stored in the refrigerator and are stable for several months
2. Always prepare stocks with glass distilled or demineralized water
3. Label the stock solutions clearly with date
4. Reagent grade chemicals should be used to ensure maximum purity
5. The nitrate stock usually precipitate out and must be heated until crystals are completely dissolved before using
6. Any stock showing cloudy or has bacterial or fungal growth should be discarded
7. Do not combine the stock to other stocks unless they are stable and compatible.

### Plant Growth Regulators

The four classes of growth regulators are commonly used in tissue culture media i.e., auxins, cytokinins, gibberellins and abssisic acid. The type of growth regulators and concentration used will vary according to the cell culture purpose.

An auxin (IAA, NAA, 2,4-D or IBA) is required for the induction of cell division and root initiation in cultured tissues. The auxins are mostly used in combination with cytokinins. The 2,4-D is used for callus induction where as IAA, IBA, and

NAA are used for root induction. Auxin stocks are usually prepared by dissolving in ethanol, I N NaOH or 1 N KOH until crystals dissolved (not more than 0.3 ml/10 mg of auxin), rapidly adding 90 ml of distilled water and increasing the volume to 100 ml in a volumetric flasks. Though the auxins are thermostable, however IAA is destroyed by low pH, light, oxygen and peroxidases. The NAA and 2,4-D are most stable form of auxin.

The cytokinins (KN, BAP, 2iP and Zeatin) are adenine derivatives; promote cell division, and shoot proliferation, organogenesis and somatic embryogenesis. They have essential role in differentiation and micropropagation of most plant species.

The cytokinin stocks are prepared in a few drops of In HCl and water to dissolve crystals. Gentle heating is usually required for complete dissolving of crystals. Bring the stock up to the desired volume by adding double distilled water in volumetric flasks. Cytokinin stocks can be stored for several months in the refrigerator. Cytokinins are thermostable during autoclaving in media.

The gibberellins are infrequently used in plant tissue cultures as it can inhibit callus growth but for meristem culture after shoot primordia formation are used in plant regeneration and elongation.

The stock solution of $GA_3$ can be prepared by dissolving in water and adjusting the pH 5.7. They are not thermostable therefore should be filter sterilized.

The abssisic acid is useful in embryo culture and somatic embryogenesis. Abssisic acid is heat stable but light sensitive. Stock solutions can be prepared in double distilled water and stored in coloured bottle in refrigerator. Dilution of stock solutions may be as per the requirement.

*Vitamins*: Vitamins have catalytic functions in enzyme reactions. The most commonly used vitamins in tissue culture media are nicotinic acid ($B_3$), thiamine ($B_1$) and pyridoxine ($B_6$). They are added in medium before autoclaving. The stocks are usually prepared in water at 100X or 1000X (10 ml per 1000 ml medium or 1 ml per 1000 ml medium) and stored in a freezer.

## Carbon Source

The carbohydrates in form of sucrose or glucose (2-5% W/V), as a carbon source are essentially required in tissue culture as cells or tissues are generally not photosynthetically active. Lower levels of a carbohydrate may be used in protoplast culture but higher levels are required for embryo or anther culture.

**Sugars** undergo caramelization prolonged on autoclaving (too long period) and will react with amino acid compounds. Sugars are degraded and form melanoidin, which are brown, high molecular weight compounds that can inhibit cell growth.

**Hexitols:** Among hexitols, myo-inositol has been found very important ingredient in tissue cultures, it is considered as growth promoter in tissue cultures. This has an action like carbon source as well as vitamin like. Mannitol or sorbitol is good osmotical for protoplast isolation. It is water-soluble and stock can be made up at the strength of 100 X (10 ml aliquots are used for 1000 ml medium).

## Gelling Agent

In tissue cultures, washed or purified agar of TC grade or Difco-bacto agar grade is used. The agar must be kept in motion while dissolving otherwise it will burn on the bottom of the flask. The agar must be completely dissolved before it is dispensed into the culture vessels.

The agar can also be melted in a autoclave or in a foil capped Erlenmeyer flask for 15 min at 121°C and dispensed aseptically into sterile containers by using laminar air-low bench before solidification of agar.

### Amino Acids and Amides

The amino acids and amides are very important in tissue cultures especially for the morphogenesis. All L forms of amino acids are commonly used, as L-tyrosine can contribute to shoot initiation, L-arginine can facilitate rooting, and L-serine can be used in haploid embryos induction in microspore cultures. L-cysteine is used for controlling phenol leaching from explant tissues. Amides such as L-glutamine and L-asparagine can induce somatic embryogenesis.

### Antibiotics

The various fungicides and bactericides are used in case plant explants on cultures excessively contaminated. These chemicals are toxic not only to contaminants but also to cultures or explant materials so restricted use should be made for additions into the culture medium. The antibiotics are soluble in water should be made fresh and be added to the medium. The antibiotics are soluble in water should be made fresh and be added to the medium after autoclaving by filter sterilization.

### Natural Complexes

The natural complexes such as coconut (endosperm) milk (CM), yeast extract (YE), malt extract (ME), tomato juice, potato extract, casein hydrolysate (use enzyme digest) and fish emulsion are used in tissue cultures for various purposes. Addition of these complexes in the medium make the medium undefined, since variation in growth promoting or inhibiting compounds in these complexes.

*Antioxidants*: The antioxidants such as citric acid, ascorbic acid, pyrogallol, phloroglucinol and L-cysteine are used in tissue culture to reduce excessive browning of the explants. Adsorbents like PVP and activated charcoal are also used for checking excessive browning.

## ADDITIONAL REQUIREMENTS

**Quality of water, chemicals and natural complexes:** Demineralized or double distilled water of high purity are used in making stocks and medium. Glass distilled water is most desirable and stored in clean containers.

*Callus-induction medium*

| | |
|---|---|
| Murashige and Skoog medium | 1.01 |
| 2,4-D | 1.0 mg |
| Agar | 8.0 g |

Prepare 1 litre of standard MS medium. Add the 2,4-D and adjust the medium pH to 5.5 using 0.1 M NaOH. Dissolve the agar in the medium in a steam bath. Dispense the medium into culture tubes or vessels and autoclave for 20 min at 121° C.

*Chlorate Selection Medium*

| | |
|---|---|
| $MClO_3$ | 600.0 mg |
| $Ca(NO_3)_2\ 4H_2O$ | 118.0 mg |
| $MgSO_4\ 7H_2O$ | 19.5 mg |
| $K_2HPO_4$ | 19.7 mg |
| P-IV trace metal solution | 3.0 ml |

Dissolve the salts in 900 ml of distilled water. Add the P-IV trace metal solution and adjust the medium pH to 7.0. Add additional distilled water to make 1 litre of medium.

*Chlorate Selection Medium Overlay*

| | |
|---|---|
| Chlorate selection medium | 500.0 ml |
| Gel-rite ™ gelling agent | 4.0 g |

Slowly add the Gel-rite a little at a time to the chlorate selection medium while stirring the mixture with a magnetic stirrer. Set the mixture in a steam bath to dissolve the Gel-rite. Dispense 4 ml of the overlay medium into each culture tube; 4 ml should spread out as a very thin layer over the surface of the media in plates.

*Embryo Culture Medium*

| | |
|---|---|
| Murashige and Skoog medium | 1.01 |
| Agar | 8.0 g |

Prepare 1 litre of standard MS medium. Adjust the medium pH to 5.5 using 0.1 M NaOH. Dissolve the agar in the medium in a steam bath. Autoclave for 25 min at 121°C. Allow the medium to cool to 50°C in a temperature controlled water bath. Pour the medium into sterile 100 mm petri plates.

*L/10 Green Algae Medium*

| | |
|---|---|
| $Ca\ (NO_3)_2\ 4H_2O$ | 0.118/ g |
| $MgSO_4\ 7H_2O$ | 0.0195 g |
| $K_2HPO_4$ | 0.0197 g |
| P-IV trace metal stock | 3.0 ml |

Dissolve all of the salts in 1 litre of distilled water. Adjust the medium to pH 7.0 by adding 1 M HCl or 1 M NaOH.

*Micropropagation Medium*

| | |
|---|---|
| Murashige and Skoog medium | 1.01 |
| Indolebutyric acid (IBA) | 1.0 mg |
| Benzylaminopurine (BAP) | 3.0 mg |

Prepare 1 litre of standard MS medium. Add the IBA and BAP and adjust the medium pH to 5.5 using 0.1 M NaOH. Dissolve the agar in the medium in a steam bath. Dispense the medium into culture tubes or vessels and autoclave for 20 min at 121°C.

*Murashige and Skoog (MS) Medium*
*Macro salts*

| | |
|---|---|
| $NH_4NO_3$ | 1.65 g |
| $KNO_3$ | 1.90 g |
| $CaCl_2\ 2H_2O$ | 0.44 g |
| $MgSO_4\ 7H_2O$ | 0.37 g |

$KH_2PO_4$ .................... 0.17 g

*Micro salts*

$FeSO_4\ 7H_2O$ .................... 27.80 mg
$Na_2EDTA\ 2H_2O$ .................... 33.60 mg
Kl .................... 0.83 mg
$H_3BO_4$ .................... 6.20 mg
$MnSO_4\ 4H_2O$ .................... 22.30 mg
$ZnSO_4\ 7H_2O$ .................... 8.60 mg
$Na_2MoO_4\ 2H_2O$ .................... 0.25 mg
$CuSO_4\ 5H_2O$ .................... 0.025 mg
$CoCl_2\ 6H_2O$ .................... 0.025 mg
Organic supplements
Myo inositol .................... 100.00 mg
Nicotinic acid .................... 0.05 mg
Pyridoxine HCl .................... 0.05 mg
Thiamine HCl .................... 0.05 mg
Glycine .................... 0.20 mg
Sucrose .................... 20.00 g

Dissolve the salts and organics in 800 ml of distilled water. Adjust the medium pH to 5.7 by adding 1 M NaOH. Add additional distilled water to adjust the final volume to 1 litre.

*MS/C Medium*

MS salts and organic supplements ....................
IAA solution .................... 8.0 ml
Kinetin solution .................... 2.5 ml
Agar .................... 8.0 g

Dissolve the salts and organics in 800 ml of distilled water. Add the IAA and kinetin solutions. Adjust the medium pH to 5.7 by adding 1 M NaOH or 1M HCl. Add additional distilled water to adjust the final volume to 1 litre. Add the agar and heat the medium on a hot plate or in a steam bath until the agar melts. Stir the medium occasionally until the entire agar is dissolved and the solution is clear. Do not let the medium boil. Dispense 8 ml aliquots in 20 X 150 mm culture tubes (approximately 120 tubes of medium).

*Myriophyllum aquaticum Shoot-Induction Medium*

Murashige and Skoog medium .................... 1.0 litre
[2-isopentenyl] adenine (2iP) .................... 2.0 mg
Agar .................... 8.0 g

Prepare 1 litre of standard MS medium. Add the 2iP and adjust the medium pH to 5.7 using 0.1 M NaOH. Dissolve the agar in the medium in a steam bath. Dispense the medium into culture tubes or vessels and autoclave for 20 min at 121°C.

*M. aquaticum Stock Plant Medium*

Murashige and Skoog medium .................... 1.01
Agar .................... 8.0 g

Prepare 1 litre of standard MS medium. Adjust the medium pH to 5.7 using 0.1 M NaOH. Dissolve the agar in the medium in a steam bath. Dispense the medium into culture tubes or vessels and autoclave for 20 min at 121°C.

*P-IV Trace Metal Solution*

| | |
|---|---|
| $Na_2EDTA$ | 0.750 g |
| $FeCl_3$ $6H_2O$ | 97.0 mg |
| $MnCl_2$ $4H_2O$ | 41.0 mg |
| $ZnCl_2$ | 5.0 mg |
| $CoCl_2$ $6H_2O$ | 2.0 mg |
| $Na_2MoO_4$ | 4.0 mg |

First dissolve the $Na_2EDTA$ in 500 ml of distilled water, then dissolve the remaining metal salts. For greater accuracy, it may be easier to prepare 10 X concentration stock solutions of the Zn, Co, and Mo salts and add 1/10 of these stocks to the P-IV stock solution.

*Pandorina Ammonium Medium*

| | |
|---|---|
| $NH_4Cl$ | 27.0 mg |
| $CaCl_2$ $2H_2O$ | 100.0 mg |
| $MgSO_4$ $7H_2O$ | 19.5 mg |
| $K_2HPO_4$ | 19.7 mg |
| P-IV trace metal solution | 3.0 ml |

Dissolve the salts in 900 ml of distilled water. Add the P-IV trace metal solution and adjust the medium pH to 7.0. Add additional distilled water to make 1 litre of medium.

*Pandorina Nitrate Medium*

| | |
|---|---|
| $NaNO_3$ | 35.0 mg |
| $CaCl_2$ $2H_2O$ | 100.0 mg |
| $MgSO_4$ $7H_2O$ | 19.5 mg |
| $K_2HPO_4$ | 19.7 mg |
| P-IV trace metal solution | 3.0 ml |

Dissolve the salts in 900 ml of distilled water. Add the P-IV trace metal solution and adjust the medium pH to 7.0. Add additional distilled water to make 1 L of medium.

*Pandorina Hypoxanthine Medium*

| | |
|---|---|
| Hypoxanthine | 68.0 mg |
| $CaCl_2$ $2H_2O$ | 100.0 mg |
| $MgSO_4$ $7H_2O$ | 19.5 mg |
| $K_2HPO_4$ | 19.7 mg |
| P-IV trace metal solution | 3.0 ml |

Dissolve the salts in 900 ml of distilled water. Add the P-IV trace metal solution and adjust the media pH to 7.0. Add additional distilled water to make 1 litre of medium.

*Pandorina Uric Acid Medium*

| | |
|---|---|
| Uric acid | 84.0 mg |
| $CaCl_2$ $2H_2O$ | 100.0 mg |
| $MgSO_4$ $7H_2O$ | 19.5 mg |
| $K_2HPO_4$ | 19.7 mg |
| P-IV trace metal solution | 3.0 ml |

Dissolve the salts in 900 ml of distilled water. Add the P-IV trace metal solution and adjust the medium pH to 7.0. Add additional distilled water to make 1 litre of medium.

*Potato Dextrose Agar*

| | |
|---|---|
| White potatoes, sliced | 250 g |
| Dextrose | 20 g |
| Agar | 15 g |

Boil the potatoes in 500 ml of distilled water for 15 min until soft. Filter this mixture through cotton to remove most of the particulate matter. Dissolve the dextrose in 200 ml of the potato infusion. Add 800 ml of distilled water. Adjust the final pH to 3.5-4.0. Dissolve the agar in a steam bath or on a hot plate. Autoclave at 121°C for 25 min.

*Trypticase-Soy Broth Medium*

| | |
|---|---|
| Trypticase | 17.0 g |
| Phytone | 3.0 g |
| NaCl | 5.0 g |
| $K_2PO_4$ | 2.5 g |
| Glucose | 2.5 g |

Dissolve the ingredients in 1 litre of distilled water. Adjust the medium pH to 7.3 by adding 1 M NaOH. Dispense 10 ml aliquots in 20 X 150 mm culture tubes. (Approximately 100 tubes of medium).

*Yeast Extract Broth (YEB)*

| | |
|---|---|
| Yeast extract | 1.0 g |
| Beef extract | 5.0 g |
| Peptone | 5.0 g |
| Sucrose | 5.0 g |
| $MgSO_4$ $7H_2O$ | 0.5 g |

Dissolve all the ingredients in 1 litre of distilled water. Adjust the medium pH to 7.0 by adding 1 M NaOH. Dispense and autoclave for 25 min at 121° C.

*Yeast Extract Indicator Medium (YI)*

| | |
|---|---|
| Yeast extract | 1.0 g |
| Lactose | 10.0 g |
| Agar | 20.0 g |

Dissolve the yeast extract and lactose in 1 litre of distilled water. Adjust the medium pH to 7.0 by adding 1 M NaOH. Add the agar and dissolve it by heating the mixture in a steam bath or on a hot plate. Autoclave for 25 min at 121°C.

## Sterilization of Media

Tissue culture media are generally sterilized by autoclaving at 121°C and 1.05 kg/cm$^2$ (15-20 psi). The time required for sterilization depends upon the volume of medium in the vessel. Dispense medium in small aliquots whenever possible as many media components are broken down on prolonged exposure to heat. Medium exposed to temperatures in excess of 121°C may not properly get or may result in poor cell growth.

**TABLE 24.2:** Minimum Autoclaving Time

| Volume of Medium per Vessel (ml) | Minimum Autoclaving (min) | Volume of Medium per Vessel (ml) | Minimum Autoclaving (min) |
|---|---|---|---|
| 25 | 20 | 500 | 35 |
| 50 | 25 | 1000 | 40 |
| 100 | 28 | 2000 | 48 |
| 250 | 31 | 4000 | 63 |

Minimum autoclaving time includes the time required for the liquid volume to reach the sterilizing temperature (121° C) and 16 min at 121° C. Times may vary due to differences in autoclaves. Validation with system is recommended. Several medium components are considered thermolabile and should not be autoclaved. Stock solutions of the heat labile components are prepared and filter sterilized through a 0.22 mm filter to sterile container. The filtered solution is aseptically added to the culture medium, which has been autoclaved and allowed to cool to approximately 35-45°C°. The medium is then dispensed under sterile conditions.

### Use and Storage of Coconut Water

Coconut water has been shown to stimulate shoot proliferation in many species of plants. Coconut water is prepared from selected coconuts by filter sterilized and frozen prior to use. This precipitation should not affect the growth of the plant tissue. Coconut water can be divided into smaller aliquots, corresponding to standard medium batch size, and refrozen until needed. Coconut water should be used at a concentration of 5-20% (v/v).

## RELATED PROCEDURES

### Ultraviolet Light

UV light may be divided into three wavelength groupings near UV (315-400 nm), mid range UV (280-315 nm) and far UV (200-280 nm). Maximal sensitivity in humans is at about 280 nm. Exposure to direct or indirect mid-range or for UV can cause acute eye irritation after a latent period of 2-24 hrs. Because retina is not sensitive to UV eye damage may result without the subject being aware of the exposure. Skin is also sensitive to UV, which may cause for skin cancer. Hence protect your eyes and skin from the effects of UV irradiation by wearing goggles with side shields by clothing, and by limiting exposure.

### Preparation of Phenol

All crystalline phenol must be redistilled at 160° C to remove contaminants that cause or cross linking of DNA or RNA. Soon after distillation add 0.1% hydroxyquinoline. The melted phenol is extracted several times with an equal volume of 1.0 M Tris pH 8.0 followed by 0.1 M Tris pH 8.0 and 0.2% ß-mercaptoethanol, until pH of the aqueous phase is 7.6. Phenol is stored in aliquots at 4° C under equilibration buffer for periods upto 1 month.

Phenol is widely used as a disinfectant and germicide. It is a dangerously toxic material that can produce poisoning when ingested, inhaled or absorbed trough the skin. The toxic effect includes headache, dizzines, nausea, weakness, and difficulty in breathing, unconsciousness and death. Phenol is corrosive to skin, initially producing a softened area followed by severe burns.

1. If phenol is spilled on the skin, flush immediately with large amounts of water. Do not use ethanol.
2. If eyes are contaminated, wash them with running water for about 15 min, call for medical help.

## Working with $^{32}$P Labelled Compounds

ß-particles with energy of 1.71 MeV (6.1 meter range in air) is emitted by $^{32}$P. Hence $^{32}$P labelled compounds must be handled carefully with much caution using shields. When ß particles hit targets, electromagnetic radiation known as Bremsstrahlung is produced, the yield of which is directly proportional to the density of material used for shielding. Therefore, a low-density material may be added to absorb the Bremsstrahlung emitted. Always ear gloves (two pairs if necessary), protect eyes and use dosimeters. Always cover the work area with absorbant papers and use survey meter to check spillage. Eating, smoking or drinking while handling radioactive compounds should be banned. Use special tape to label containers and tubes in which radioactive materials are kept. The maximum permissible burden of $^{32}$P is 30 uCi but the maximum permissible burden for bone is only 6 μCi.

## Silanization of Plastic and Glassware

Place the plastic and glasswares to be silanized in a desiccator. Add about 1 ml of dichlorodimethyl silane in a small container in the desiccator. Pull vacuum on the system until dichlorodimethyl silane boils. Close the system and allow sitting for about 2 hrs. Open the desiccator in a fume hood and allow to air out several hours. Rinse the plastic and glassware with water and autoclave.

**TABLE 24.3:** Problems and Possible Solutions in Plant Tissue Culture Work

| Symptoms | Possible Causes | Possible Solution |
|---|---|---|
| Culture contamination | Source heavily infested/ infected with micro organisms Poor sterilization | Improve sterilization method (70% Ethanol, drop of detergent, antimicrobial agents) clean plant material/ select unexposed tissue. |
| Explant dies | Strong disinfectants. | Use weaker disinfectant/ change sterilizing agents |
| | Media too strong | Use 1/2 or 1/4 strength |
| | Wrong stage of growth | Obtain explants at different stage of growth |
| Culture blackens and dies | Contaminated by micro-organisms | Discard with care. Review sterile technique and sanitation |
| | Bleeding | Subculture immediately. Transfer more frequently Try different agar concentrations |
| | Agar problem | Check water purity Try different formula |
| | Water problem | |
| | Wrong formula media constituents | |
| Explant live but no growth | Dormant | Chill for a month |
| | Media too harsh | Use explant at different stage of |

| Symptoms | Possible Causes | Possible Solution |
|---|---|---|
| | | growth |
| | Wrong formula | Lower salts and hormones try different formula |
| Culture live but no growth | Too cold | Increase temperature |
| | Wrong medium | Try different medium |
| Shoots tool long (leggy), and poor multiplication leaves are yellow, watery | Too little cytokinin | Increase cytokinin Change cytokinin/ auxin ratio |
| Leafy shoots too short | Hormones too strong | Decrease or omit hormones |
| No multiplication | Too little cytokinin | Increase cytokinin |
| | Needs chilling | Cold store 4-8 weeks |
| | Too cold | Run cytokinin/ auxin grid |
| | Requires dormancy period | Increase temperature Cold treat 3-8 weeks |
| Fat stems, small and pale leaves | Too much cytokinin | Decrease cytokinin/ increase auxin |
| Unwanted callus | Wrong hormones | Decrease or omit hormones Run cytokinin/ auxin grid |
| Leaves chlorotic | Contaminant | Index for contaminants |
| | Too hot | Decrease temperature |
| | Wrong formula | Try different medium |
| Leaves succulent (watery), virtrification, abnormal stem, embryos | Osmotic potential upset | Decrease temperature |
| | Too high cytokinin | Increase agar strength |
| | Wrong agar | Decrease hormones |
| | Culture too old | Try different agar |
| | | Transfer more often |
| Premature rooting | Wrong hormone balance | Transfer more often |
| | | Run cytokinin/ auxin grid |
| | | Increase cytokinin/ decrease auxin |
| Red stems/ embryos/ cells | Stress | Change light and/ or temperature |
| | | Lower sucrose content |
| | | Increase nitrate ($NO_3$) |
| | | Transfer more frequently |
| Non-friability of callus | Too much sugar Not enough $NO_3$ Culture too old Media composition Plant source | Using number of media having range of hormones. Select most rapidly growing callus. Change to liquid culture and after friable to agar medium. |

Source: Bhansali, R. R. (1995).

# 25. Protoplast Fusion Technology

The field of tissue culture has expanded at such a rate that the time has now arrived for individual techniques such as protoplast culture to break off as their own individual areas.

Protoplasts are plant cells with a plasma membrane but without the cell wall. The absence of cellulosic cell wall permits advantages to fuse protoplast of similar or on different species and the fused product can generate into whole plants. This is called somatic fertilization or fusion of two vegetative cells to form hybrids.

## IMPORTANCE OF PROTOPLAST ISOLATION AND CULTURE

The isolation, culture and fusion of protoplast are one of the most fascinating fields of research. The techniques are important for the following reasons:

1. To develop novel hybrid plant through protoplast fusion, genetic engineering would continued to be an exciting area of research in modern plant biotechnology This technology holds great promises to synthesize a plant of desired characteristics.
2. This helps in crop improvement by somatic hybridization and cell modification.
3. The protoplast in culture can be regenerated into an entire plant.
4. The technique in future will be one of the most frequently used research tools for tissue culturists, physiologists, pathologists, molecular biologists, cytogeneticists, and biotechnologists.
5. It provides a tool for isolating protoplasts and exploring the possibilities of genetic engineering.

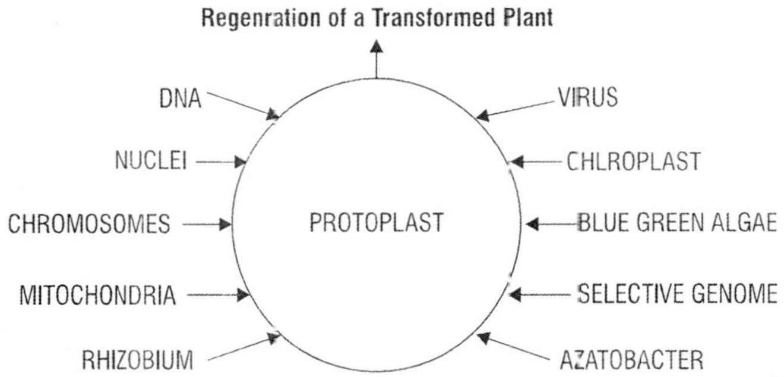


**FIGURE 25.1:** Hypothetical illustration for the uptake of different cell organelles and introduction of genetic material into plant protoplasts. This technique results into the regeneration of new plant bearing desired characters.

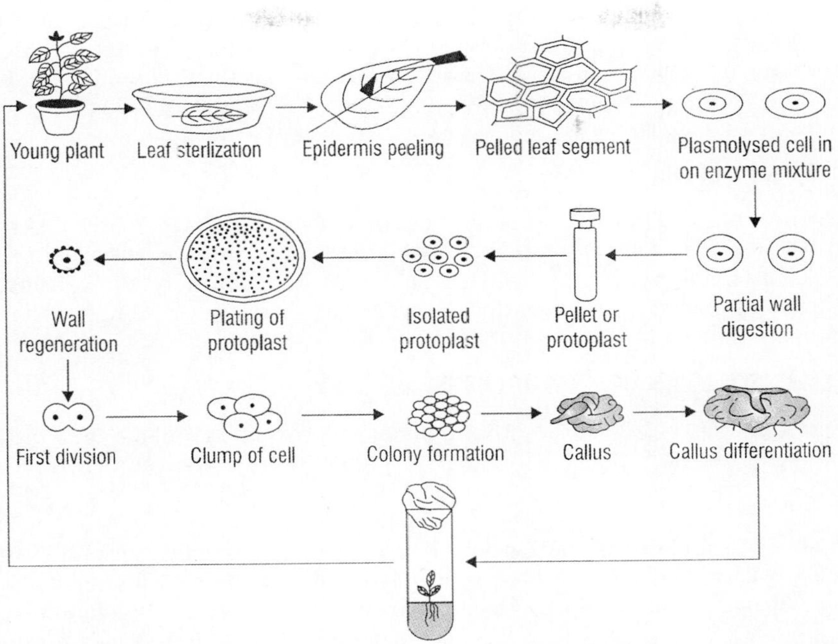


**FIGURE 25.2:** Systematic illustration of the technique used for isolation, culture and regeneration of plants from leaf protoplasts.

**TABLE 25.1:** Isolation of Protoplast from Various Plant Parts

| **Plant tissues** | **Young leaves, roots, stems** |
|---|---|
| Flower parts | Petal, reproductive organs |
| Callus/cells | Friable callus tissues/cells |
| Suspension cultures | Fast growing cells |
| Aseptically grown plantlets | Leaves, stems, roots |

Protoplast is the living material of the cell where as an isolated protoplast is the cell from which the cell wall is removed. Protoplasts can be isolated from almost all plant parts viz., roots, leaves, fruits, tubers, root nodules, endosperms, crown gall tissues, pollen mother cells, pollens, pollen tetrads and cells of callus tissues grown in vitro. Protoplasts are isolated from cells by two methods: (1) Mechanical isolation, and (2) Enzymatic isolation.

**TABLE 25.2:** Application of plant protoplast in fundamental research of plant sciences

| **1.** | **Structure** | **Cell wall, Plasma membrane, Organelle- structures and biosynthesis** |
|---|---|---|
| 2. | **Function** | Biochemistry, Photosynthesis, Absorption |
| 3. | **Genetics** | Fusion, Organelle and DNA uptake |
| 4. | **Plant Pathology** | Virus, RNA uptake, Pathotoxin |
| 5. | **Differentiation** | Morphogenesis, Embryogenesis |

## Enzymatic Method

Enzymatic method is most often used in isolation of protoplasts from various plant parts. Enzymatic methods generate very large numbers of protoplasts in comparison to mechanical methods, but in some instances, the enzymes have deleterious effects on plant metabolism. However, the effects of enzymes used on protoplasts (which is unknown) are overlooked in this method. Enzymatic isolation of protoplast has some advantages over mechanical method such as: (i) cells is not damaged, (ii) Osmotic shrinkage of protoplast is much less.

### Mechanical Method

Mechanical method, though not most often is in use, still has its merit. It is rarely used because it is an extremely tedious process that results in the yield of only very small numbers of protoplasts. The cells are plasmolysed, causing the protoplast to shrink away from the cell wall. The protoplasts obtained from this method are then cultured on suitable culture medium.

## ENZYMIC ISOLATION OF PROTOPLASTS: METHOD

The starting material is sterile in vitro shoot cultures or any plant part. In these cases protoplasts are enzymically isolated in a two-step or a one-step method.

### *First*

Isolated cells are prepared (maceration) by treating the leaf or other tissue segment with macro enzyme (a pectinase) in 13% mannitol. The isolated cells obtained from this incubation are purified by filtration through a nylon mesh.

### *Second*

The protoplasts are prepared by incubating these isolated cells in 2 per cent cellulase for about 90 minutes. In the *one-step* method both enzymes (cellulase + pectinase) are used simultaneously. In *one-step* method 2 to 3 hours at 20 to 22° C is adequate to release significant numbers of protoplasts.

### *Isolation of Protoplast from Leaves*

The process of isolating protoplasts from leaves involves a number of steps; these include surface sterilization, removal of epidermis, enzymic digestion and cleaning of the protoplasts.

For isolation, the physiological state of the leaves is extremely important. The leaves normally chosen for experimental purposes should be fully expanded and from young plants (vegetative state, e.g., 10-14 weeks in tobacco). The lower epidermis is removed and is used to isolate epidermal and guard cell protoplasts. The remaining peeled leaf is used for isolation of mesophyll protoplasts. The isolation of protoplast involves four main steps, which are,: (1) Surface sterilization of leaves. (2) Peeling off the epidermis. (3) Enzymatic treatment.

### *Method*

1. Fully expanded young leaves are surface sterilized by dipping them into 70% ethanol for one minute and then in 2% solution of sodium hypochlorite for 20-30 minutes.
2. The leaves than washed three times with sterile glass distilled water to remove traces of sodium hypochlorite.
3. The subsequent steps to isolate sterile protoplast are taken up under aseptic conditions.
4. Lower epidermis of sterilized leaves in carefully peeled off and the stripped leaves are cut into small pieces.

5. Peeling off process is easier with flaccid leaves than those with turgid leaves.
6. The protoplast is generally isolated by two methods: (i) Direct Method and Sequential Method.

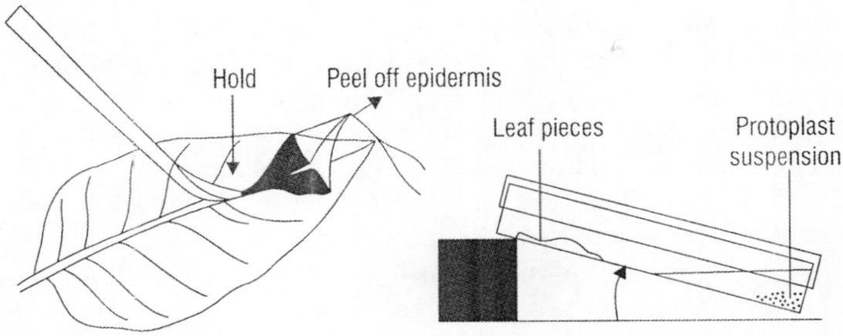


**FIGURE 25.3:** (A) Peeling off leaf epidermis. (B) Release of protoplasts from peeled leaf pieces after enzyme treatment.

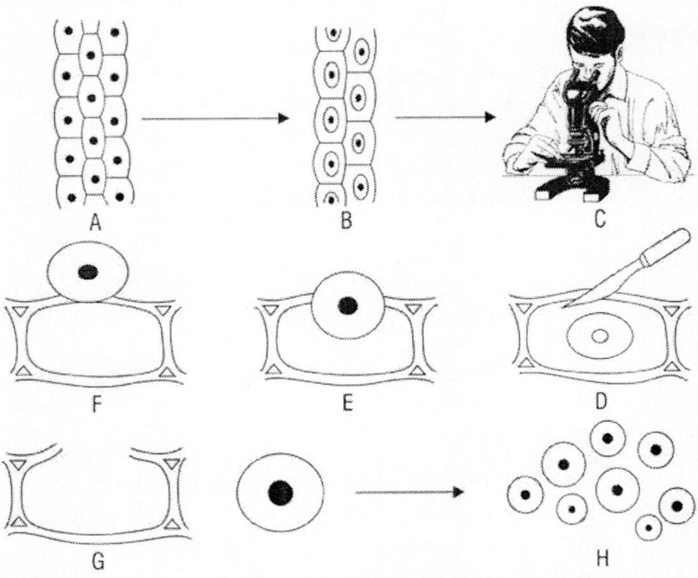


**FIGURE 25.4:** Method of mechanical isolation of protoplasts. (A) A small piece of peeled epidermis, (B) Plasmolysis of the same cells, (C-F) Steps of removal of protoplasts from the cell after incision, (G&H) Free isolated protoplasts.

### *Direct Method*

1. Peeled leaf pieces are placed with lower surface down, into 20 ml solution of 13% mannitol and inorganic salts contained in petridishes (90 mm) for one hour.
2. The mannitol-inorganic salt solution is pipette out with the help of Pasteur pipette and an enzyme mixture (20 ml) in mannitol (0.5% macerozyme + 2% Onozuka cellulase in 13% sorbitol or mannitol at pH 5.4) is added and incubated overnight (15 to 18 hours) at 20 to 22o C.
3. The petri plates are then gently agitate to facilitate the release of the protoplasts, pushing the larger pieces of leaf-material to one side with a sterile

forceps and keeping the petridish at an angle of 15° C.

4. After sometimes protoplasts will settle down in the petridish and then protoplast-enzyme mixture is transferred to 15 ml screw-capped centrifuge tubes using Pasteur pipette.
5. Centrifuge the material at 100g for one minute. This process is repeated for 2 to 3 times and protoplasts are washed with 13% sorbitol solution.
6. For the final washing, 20% sucrose solution is used in place of sorbitol and centrifuged at 200g for one minute.
7. The cleaned protoplasts float and debris settles down.
8. The floating protoplasts are carefully pipette out with a Pasteur pipette leaving behind the remains of mesophyll cells.
9. This method is useful for the isolation of protoplast both from spongy and palisade parenchyma.

### Sequential Method

In this method of protoplast isolation cells are first separated and then transformed into protoplasts. This method is completed into two steps:

#### Step I. Isolation of Cells

1. The peeled leaf segments are placed into a petridish (140 mm) containing an enzyme mixture (macerozyme 0.5%, potassium dextran sulphate 0.3% in 13% mannitol at pH 5.8).
2. The petridishes are put into a desiccator for vaccum infiltration of enzyme mixture for about 4 to 5 minutes, and then transferred to a water bath at 25° C.
3. Agitate them slowly.
4. After 15 to 20 minutes the enzyme mixture is replaced by fresh one and is again incubated for one hour. By this process cells are completely separated from one another.
5. The isolated cells are filtered through a nylon mesh, then centrifuge at 100g for one minute.

#### Step II. Isolation of Protoplasts

1. Isolated cells are now treated with another enzyme mixture containing 2% cellulase in a 13% solution of mannitol at pH 5.4 for 90 minutes at 30°C and then centrifuged at 100g for one minute.
2. The supernatant is decanted off and the protoplasts are washed three times with mannitol.
3. The protoplasts are finally cleaned with 20 to 25% sucrose solution. This method can be used for isolation of protoplasts from palisade cells.

### Isolation of Protoplast from Shoot Apex

The use of sterile shoot cultures as a starting material for leaf and stem protoplast isolation is popular. One key advantage is that the material is already sterile and does not have to be surface sterilized prior to incubation in the digestive enzyme mixture. Sterile shoot cultures are used as starting material in a wide range of plant species, from potato and tobacco, and a wide range of tree species.

### Isolation of Protoplast from Root and storage organs

Protoplasts have been isolated directly from a wide selection of roots and storage organs. These include potato tubers, artichoke tubers, onion roots and root nodules. The enzyme treatments are sometimes slower and often a more complex enzyme mixture is required.

### *Isolation of Protoplast from Pollen Grains*

Pollen protoplasts have some advantages over the somatic cells being found in uniform in chromosome number (ploidy). Protoplast obtained from mature pollen could be used in the study of pollen ontogeny, mutation, somatic hybridization, cell modification and regeneration of haploids. Fusion causes the formation of a normal diploid instead of doubling the chromosome number (to tetraploid) of the new hybrid. It is also possible to regenerate haploid plantlets from isolated pollen protoplasts.

Pollen grain has highly sculptured and chemically complex cell wall. It is covered in a highly resistant coat of sporopollenin, as polymer of carotenoid esters and can be removed by a pretreatment with strong oxidizing solutions.

Pollen grains taken out of the anther lobes are sterilized with 2% sodium hypochlorite solution for 10 to 12 minute and subsequently washed 2 to 3 times with sterile distilled water. The concentration and combinations of the enzyme incubation to pollen grains depend upon their stage of development. Protoplasts are released in many ways described as follows:

Protoplast ooze out through germ pores due to dissolution of pecto-cellulosic membrane by enzyme action. The enzyme mixture used normally contains high concentrations of β-1,3-gluconase, which is capable of digesting both cellulose and callose.

1. Due to mechanical rupture of the exine.
2. By partial dissolution of the exine.

### *Isolation of Protoplast from Cultured Cells*

Actively dividing cells in suspension cultures are proved as most suitable material for isolation of non-germ protoplasts in large amount. Moreover, plant breeders can also use protoplasts isolated from various somaclones as raw material for somatic hybridization and cell manipulation. Protoplast isolation through enzymatic activity from older plant parts is very difficult. Protoplasts have been isolated from suspension culture of many plant species like, carrot, rose, *Atropa* sp., sugarcane, etc. Protoplasts can be easily isolated from such cultures by treating the filtered suspension with 2 to 4% Onozuka cellulase in 0.6 M mannitol, for 4 to 6 hours at 32° C in a gently shaking water bath.

**Other Sources:** Protoplasts have also been isolated from coleoptiles, flower petals, aleurone layers and plant cell tumours or galls. The basic isolation technology is the same in all cases.

## METHODS OF PROTOPLASTS CULTURE

### Laboratory Facilities

Protoplasts can be cultured in liquid as well as solid medium. Concentration of enzyme, intensity of light (300-3000 lux) and temperature (28-45° C) play vital role in protoplasts culture.

## Enzyme Mixture and Osmotic Stabilizer (Osmotica)

The composition of the cell wall indicates that the digestion mixture should be able to degrade cellulose, hemicellulose and pectin and in some cases callose. The number of commercial enzyme preparations available is somewhat limited and very often they are contaminated with a wide range of biologically active impurities of other enzymes- i.e., proteases, nucleases and lipases. They have a degrading effect on the plasma membrane of the isolated protoplasts.

**TABLE 25.3:** Plant species in which protoplast have been successfully cultured

| Source | Enzyme mixture used | Culture medium | Growth response |
|---|---|---|---|
| **MESOPHYLL CELLS** | | | |
| *Brassica napus* L. | Cellulase Onozuka P 1500 (0.5%) + Rhozyme HP 1500 (0.5%) + Driselase (0.5%) + Hemicellulase | B + NAA (10-6M) + BA (10-6 M) | Plants |
| | (0.5%)+ Driselase (0.5%)+ Hemicellulase | | |
| | (0.5%) in sorbitol (4.5%)+ mannitol (4.5%) | | |
| *Cucumis sativus* | Preplasmolysis in a salt medium and cellulase (0.3%) +pectinase (0.4%)+ PDS (0.5%) in 11% mannitol | Harad's medium (1973) | Cell clusters |
| *Datura innoxia* | Macerozyme (1%)+ cellulase Onozuka (3%) in 0.5M mannitol BAP (0.4 mg/l) + NAA (0.4 mg/l) | Durand et al. (1973) | Plants |
| *Hyscyamus niger* | Pectinol (0.02%) + cellulase Onozuka R 10 in KCl (2.5%) + MgSO (1%) + 0.6 M mannitol | Durand et al. (1973) | Callus |
| *Nicotiana tabacum* cv. xanthi (2x) | Macerozyme (0.5%) + cellulase Onozuka (2%) in NAA (3mg/l )+ 6-BAP(1mg/l) + 0.7M mannitol | Nagata & Takebe (1971) | Plants |
| **CALLUS** | | | |
| *Saccharum officinerum* | Cellulase Onozuka 4S (5%) + glusulase (1.5%)+ sorbitol + 0.4M + glucose 5.5 mM | Mod. White's medium + yeast extract | Callus |
| **STEM CALLUS** | | | |
| *Atropa belladonna* | Cellulase R10 (1.5%) + in 0.5M sorbitol + NAA (2 mg/l) + kinetin (0.5mg/l) | Mod. MS + | Embryoids & plants |
| **OVULAR CALLUS** | | | |
| *Citrus sinensis* | Cellulase (1%) + pectinase (1%) + PDS (0.3%) in 0.14 M sucrose + 0.28M mannitol + 0.28M sorbitol | Murashige & Tucker's (1969) basal medium | Embryoids |
| **PERICARP CALLUS** | | | |
| *Vitis vinifera* | Cellulase Onozuka SS (2%) + macerozyme (1%) in 0.1M $CaCl_2$ + 0.14M KCl + (2000 mg/l) kinetin + 0.2 mg/l) +NAA (0.1 mg/l) | B5 + Casein hydrolysate | Callus |
| **CROWN GALL** | | | |
| *Parthenocissus* | Macerozyme R 10 (0.01%)+ cellulase | Nagata & Tabeke | Callus |

| Source | Enzyme mixture used | Culture medium | Growth response |
|---|---|---|---|
| *tricuspidata* | + NAA (0.1 mg/l) + Onozuka R 10 (2%) in 13% mannitol | (1971) | |

Following are the commercial enzyme preparations used for isolation of protoplasts.

1. Onozuka R10 cellulase (Kinki Yakult Biochemical, Nishinomiya, Japan)
2. Cellulase R10 (Kinki Yakult Biochemical, Nishinomiya, Japan)
3. Macerozyme R10, pectinase (Calbiochem-Behring, San Diego, California, USA)
4. Hemicellulase (Rohm and Hass Co. Philadelphia, PA., USA)
5. Driselase, cellulase, (Kyowa Hakko Kogyo Co., Tokyo, Japan)
6. Cellulysin, cellulase (Calbiochem-Behring, San Diego, California, USA)
7. Pectinase (Sigma, St. Louis, MO, USA)
8. Macerase (Calbiochem-Behring, San Diego, California, USA)
9. Rhozyme HP 150, Hemicellulase (Rohm and Hass Co. Philadelphia, PA., USA)

The enzyme mixture is normally dissolved in culture media together with an osmotic stabilizer. Calcium (2 to 10mM) is a necessary component and phosphate (0.5 to 2.0 mM) appears to stabilize the isolated protoplasts.

After the digestion of cell wall the isolated protoplast is subject to osmotic stress. If an osmotic stabilizing agent is not included in the medium the isolated protoplasts would take in water by the process of osmosis and would eventually burst as there is no cell wall to constrain the cells. The osmotic agents may be sugar, alcohols, sorbitol or mannitol (13% w/v). Sucrose can be used for this purpose. The concentration of osmoticum should not be high because it causes protoplast shrinkage.

Protoplast is cultured in liquid medium in Erlenmeyer flasks or in micro chambers. Microchamber is prepared by placing two 22 mm$^2$ coverslips about 18-22 mm apart on a drop of heavy mineral oil on a slide. A drop of medium containing protoplasts is placed in between the two coverslips. A third coverslip is then gently placed on the medium and finally sealed with mineral oil or parafilm. Through this method protoplast can be cultured in a small amount and for a limited time.

## PURIFICATION OF ISOLATED PROTOPLASTS

Protoplasts contain cell debris and broken cell organelles in medium. A number of methods have been used to separate the intact protoplasts from this surrounding cellular debris.

### Sedimentation and Washing

1. The crude protoplast suspension in osmotically adjusted medium is decanted into a conical tip centrifuge tube.
2. It is centrifuged at low speed (50-100 x g/5 min). Under these conditions the intact protoplasts accumulate in the form a soft pellets in the tip of the tube.
3. The supernatant, containing the broken cells debris, cell wall, and cell organelles etc., are then carefully pipetted off.
4. The pellets are then gently resuspended in fresh culture medium containing mannitol, and rewashed. This process is repeated two or three times.

### Flotation

As intact protoplasts have a relatively low density as compared to other cell organelles, many types of gradients have been used that allow the protoplasts to float and for the sedimentation of cell debris.

A concentrated solution of mannitol, sorbitol or sucrose may be combined with the enzyme-protoplast mixture in a Babcock bottle and then centrifuged at low speed.

Protoplasts can be pipetted off from the top of the tube. The concentration of sucrose used for flotation varies from 0.3 to 0.6 M. The flotation method causes less loss due to damage than the palleting and washing method. However, the high osmotic concentrations of solution may cause osmotic stress to the protoplasts, which may delay the wall formation, or leads to eventual loss of the viability.

### Other Purification Methods

Other more complex palleting and flotation systems are also used for protoplast purification. Some workers have layered crude protoplast preparation of *Hyscyamus muticus* onto 20% Percol (polymer of solicos). After centrifugation debris was in the lower half of the gradient and protoplasts floated to the top.

After obtaining a reasonably pure isolated protoplast preparation, it is important to determine what proportion of the protoplasts is alive and to be able to culture.

## PROTOPLAST VIABILITY TESTING

In early days the viability of protoplast is determined by observing its cytoplasmic streaming, exclusion of Evans Blue dye changes in protoplast size induced by changes in the level of osmoticum and measure of photosynthetic and respiratory activity. Most of these methods have proved to be rather unreliable for determining viability.

The most frequently used methods for estimation of viability is the use of fluorescein diacetate (FDA). As the FDA accumulates within the plasma membrane, viable protoplasts fluoresce green or white. Protoplast preparation treated with FDA (normally 0.01%) should be observed within 5 to 15 minutes, because after this period FDA dissociates from the membrane.

Another stain commonly used is **phenosafranine** (0.01%). This stain detects dead protoplasts, which turn red in the presence of the dye. Viable protoplasts remain unstained even after 2 hours of staining.

Isolated protoplast can be cultured in several ways of which agar-embedding technique in small petridish is commonly followed. In petridishes, embedding of protoplasts in solid agar medium is known as **plating of protoplasts**.

### General Steps of Protoplast Culture

Dilute the protoplast culture medium (1 x $10^5$ protoplasts/ml) and transfer 2 ml of culture medium containing protoplasts in an airtight petri dish. Add 1 ml of protoplast culture medium containing double agar concentration (1.6%) to each petri dish containing protoplasts in liquid. For this, melt the medium having agar and store at 30-35° C until required. Higher temperature is injurious to protoplasts. Mix gently but quickly by rotating petri dish. Allow the medium to set and seal petri dishes with parafilm. Incubate petri dishes down in an incubator at 25 ± 2° C with continuous illumination at 2000 lux. Protoplasts can also be cultured by suspension culture (without agar) and droplet culture method by placing of protoplast suspension containing protoplasts plate (1 x $10^4$ or 1 x $10^5$ ml). Incubation procedure is same as described above.

The plated protoplast can be handled very easily and the agar medium gives good support to the protoplast. Though various techniques have been developed for the culture of plant protoplasts however, the style of the culture often depends upon the objective of the experiment.

## CULTURE MEDIA FOR PROTOPLAST CULTURE

### Protoplast Culture Media for PC I Group

**TABLE 25.4:** F.5 Medium (Frearson et al., 1973)

| Constituents | Concentrations (mg/l) | Constituents | Concentrations (mg/l) |
|---|---|---|---|
| $CaCl_2.2H_2O$ | 850.0 | $ZnSO_4.7H_2O$ | 4.3 |
| $CuSO_4.5H_2O$ | 0.025 | BAP | 1.0 |
| $CoSO_4.7H_2O$ | 0.15 | Biotin | 0.05 |
| $FeSO_4.7H_2O$ | 13.9 | Folic acid | 0.5 |
| $H_3BO_3$ | 3.1 | Glycine | 1.0 |
| $KH_2PO_4$ | 353.6 | Mannitol | 130,000.0 |
| KI | 0.498 | Meso-inositol | 100.0 |
| $KNO_3$ | 525.0 | NAA | 2.0 |
| $MgSO_4.7H_2O$ | 739.0 | Nicotinic acid | 5.0 |
| $MnSO_4.4H_2O$ | 11.15 | Pyridoxine HCl | 0.5 |
| $Na_2EDTA$ | 18.75 | Sucrose | 10,000.0 |
| $Na_2MoO_4.2H_2O$ | 0.125 | Thiamine HCl | 1.0 |
| $NH_4NO_3$ | 412.25 | pH | 5.8 |

**TABLE 25.5:** Nagata and Takeba Medium (1971)

| Constituents | Concentrations (mg/l) | Constituents | Concentrations (mg/l) |
|---|---|---|---|
| $CaCl_2.2H_2O$ | 220.0 | $Na_2EDTA$ | 37.3 |
| $CuSO_4.5H_2O$ | 0.025 | $Na_2MoO_4.2H_2O$ | 0.25 |
| $CoSO_4.7H_2O$ | 0.030 | $NH_4NO_3$ | 825.0 |
| $FeSO_4.7H_2O$ | 27.8 | $ZnSO_4.4H_2O$ | 8.6 |
| $H_3BO_3$ | 6.2 | BAP | 1.0 |
| $KH_2PO_4$ | 680.0 | Mannitol | 130,000.0 |
| KI | 0.83 | Meso-inositol | 100.0 |
| $KNO_3$ | 950.0 | NAA | 3.0 |
| $MgSO_4.7H_2O$ | 1233.0 | Sucrose | 10,000.0 |
| $MnSO_4.4H_2O$ | 22.3 | Thiamine HCl | 1.0 |
| | | pH | 5.8 |

**TABLE 25.6:** V 47 (Binding, 1974)

| Constituents | Concentrations (mg/l) | Constituents | Concentrations (mg/l) |
|---|---|---|---|
| $CaCl_2.2H_2O$ | 735.0 | BAP | 0.4 |
| $CuSO_4.5H_2O$ | 0.015 | Biotin | 0.04 |
| $CoCl_2.6H_2O$ | 0.015 | Folic acid | 0.4 |

| Constituents | Concentrations (mg/l) | Constituents | Concentrations (mg/l) |
|---|---|---|---|
| $H_3BO_3$ | 2.0 | Glycine | 1.4 |
| $KH_2PO_4$ | 68.0 | Glucose | 99,000.0 |
| KI | 0.25 | Mannitol | 8,19,000.0 |
| $KNO_3$ | 1480.0 | Meso-inositol | 10.0 |
| $MgSO_4.7H_2O$ | 984.0 | NAA | 1.52 |
| $MnSO_4.H_2O$ | 5.0 | Nicotinic acid | 2.0 |
| $Na_2EDTA$ | 37.0 | Pyridoxine HCl | 0.7 |
| $Na_2MoO_4.2H_2O$ | 0.1 | Sucrose | 1,70,000.0 |
| $NH_4NO_3$ | 1444.0 | Thiamine HCl | 4.0 |
| $ZnSO_4.7H_2O$ | 1.5 | pH | 5.8 |

### *Murashige and Skoog Medium*

For protoplast culture, use MS culture medium as described for tissue culture supplemented with 13% mannitol (w/v).

**TABLE 25.7:** Modified B5 Medium (Kartha et al., 1974)

| Constituents | Concentrations | Constituents | Concentrations |
|---|---|---|---|
| + Sorbitol | 4.55% w/v | $CaCl_2.2H_2O$ | 0.0875% |
| Mannitol | 4.55% w/v | 2,4-D | $2.3 \times 10^{-6}$M |
| d-glucose | 0.25% w/v | BAP | $10^{-6}$M |
| D-ribose | 0.125% w/v | NAA | $10^{-6}$M |
| N-Z anime | 0.015% w/v | Calculated molarity | 0.52 |

### Protoplast Culture Media PC II Group

Media calculated under PC I Group with minor changes. The changes are: the concentration of mannitol/sorbitol present in media individually or in combination reduced by half. Concentration of agar in the medium is 0.8%.

### Protoplast Culture Media PC III Group

Media described PC I Group with following minor changes:

1. Mannitol/sorbitol/D-glucose/D-ribose individually or in combination are omitted in the media.
2. The sucrose concentration in all media is adjusted to 2% w/v.
3. PC III Group media may be prepared with 1.6% or 0.8% agar depending on the mode of protoplast culture.

Composition of the digestion mixture, wash medium and regeneration medium used for isolation and culturing of protoplasts.

*Digestion Mixture*

Salts and Vitamins of Murashige & Skoog (MS)
Cellulysin ........ 1.0%
Macerase ........ 0.5%
Rhozyme ........ 1.0%
$CaCl_2$ ........ 4.5 mM (0.5 g/l)
Mannitol ........ 0.3 M (54.7 g/l)
Sorbitol ........ 0.3 M (54.7 g/l)
MES ........ 0.3 mM (640 mg/l)
pH ........ 5.7

*Filter Sterilize Washing Medium (MS)*

$CaCl_2$ .......... 4.5 mM
Sorbitol .......... 0.3 M
Mannitol .......... 0.3 M
pH .......... 5.7

| **Autoclave Regeneration Medium (MS)** | | **Vitamins, Sugar and sugar alcohols (pH 5.7)** | | **Sugar and Sugar Alcohols (g/l)** | |
|---|---|---|---|---|---|
| Sucrose | 20 g/l | Nicotinamide | 1.0 | Fructose | 0.125 |
| 2,4-D | 0.5 mg/l | Pyridoxine HCl | 1.0 | Fibose | 0.125 |
| Kinetin | 1.0 mg/l | Thiamine HCl | 1.0 | Xylose | 0.125 |
| Coconut water | 2.0% (v/v) | D-calcium pantothenate | 0.5 | Mannose | 0.125 |
| Inositol | 100 mg/l | Folic acid | 0.2 | Cellobiose | 0.125 |
| Casein hydrolysate | 200 mg/l | p-aminobenzoic acid | 0.01 | Sorbitol | 54.7 |
| Sodium pyruvate | 5.0 mg/l | Choline chloride | 0.5 | Mannitol | 54.7 |
| Citric acid | 10.0 mg/l | Riboflavin | 0.1 | | |
| Malic acid | 10.0 mg/l | Ascorbic acid | 1.0 | | |
| Fumaric acid | 10.0 mg/l | Vitamin B12 | 0.01 | | |

### Agar Embedded Culture

In this method protoplasts may be allowed to regenerate a new cell wall in liquid culture before embedding or may be embedded directly after isolation. The original method of Nagata and Takebe for the isolated protoplasts is mixed with 1.0% agar-culture medium and maintained at 40 to 45° C. Small amount of the agar (liquid) protoplasts mixture is then poured into sterile petriplates.

### Microchambers

This method for the observation of plant cells is quite common. The number of possible designs of chamber is enormous. A chamber is constructed of a small plastic ring with a coverslip on top. The joints of the system are sealed with mineral oil. The simplest form is to hold a droplet between two coverslips with a ring of mineral oil around to form a seal. The mineral oil prevents evaporation of media; this can be a major problem with protoplast cultures as the osmotic concentration of the media rises. Another good sealing substance, which is easy to handle, is clear nail polish.

### Hanging Drop Cultures (HDC) Techniques

To avoid evaporation of the media while observing protoplast, this method is used. Drop can be placed in the depression of a specially prepared hanging drop-slide, which is then undersealed with a coverslip and oil, or the drops are placed on the lid of a petridish. The petridish contains mannitol solution to maintain the humidity. The dish is sealed with parafilm.

### Multidrop Array (MDA) Techniques

This is a refinement and amplification of the hanging drop technique. Through this technique it becomes easier to screen a wide range of nutritional and hormonal factors rapidly (using a small amount of plant material, less than 1g of plant tissue). The system can be described as follows:

1. Prepare a dilution series of two variables A and B, so as to have tubes $A_1$ to

$A_{10}$, $B_1$ to $B_{10}$, likewise 100 possible combinations can be prepared, i.e., $A_1B_1$, $A_2B_2$..... $A_{10}B_{10}$, etc.

2. To these 100 different media (1.0cm$^3$ each) add 0.1cm$^3$ of protoplast suspension prepared in basal media at 20 times the minimum planting density.
3. For each combination, i.e. $A_1B_1$, $A_1B_2$, etc. make, 49 drops + 4 control drops in order on the lid of a petridish and label the dish, i.e. $A_1B_1$.
4. To each drop add 20 µl of another variable media component previously prepared C-C7, 11-17. Thus each dish contains 49 different media combinations.
5. A total of 4900 different media compositions each inoculated with cells have thus been prepared on only 100 petriplates.
6. 20 cm3 of mannitol is added to the base of the dish. The lid with attached droplets is inverted into position and the plates are sealed and incubated.
7. After a suitable incubation period the plates can be easily evaluated microscopically to determine which media combination, has given the most successful growth.
8. The best method for the culture of protoplast is by plating them in a medium containing 1.2% agar. In this condition, protoplasts grow best by good number of divisions and can be handled conveniently. Moreover, it could be observed under the microscope.

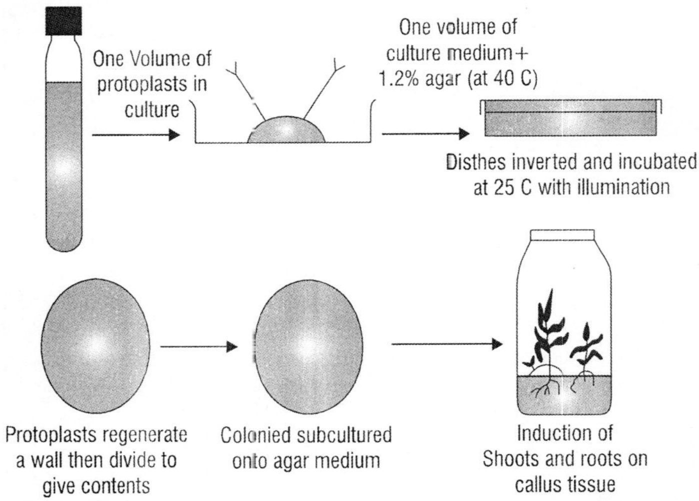


**FIGURE 25.5:** Diagrammatic presentation of culture of protoplast by agar embedding technique in small petridishes.

## GROWTH AND DIVISION OF PROTOPLAST

For the growth and nuclear division, the regeneration of cell wall is not seems to be a prerequisite in protoplast culture. In *Convolvulus* sp. protoplasts undergo one or two nuclear division prior to cytokinesis. The same was observed in *Haplopappus* spp.

It has been observed that for the rapid growth of protoplasts frequent sub culturing in mannitol free medium is necessary. Protoplasts don't grow in large colonies on high osmotic medium. Its growth gradually shows a downward trend, which ultimately inhibited altogether. The colonies, therefore, should be picked up along with small pieces of agar and then transferred on the top of another medium to avoid any damage to protoplasts by rough handling with the forceps.

### Protoplast Culture: Regeneration of Cell Wall

In culture, protoplasts start developing a wall around itself within few hours and it takes only few days to complete the process. Wall materials are progressively deposited at the surface of the plasmalemma. The cellulose is deposited either between the plasmalemma and the multilamellar wall material or directly on the plasmalemma. The nature of biosynthesis of the cell wall depends on the plant material and the system of protoplast culture.

The newly built cell wall can be observed either by plasmolyzing the protoplast by transferring it in a hypertonic solution, or by staining the cell wall with **calcofluor white** fluorescent stain (0.1%). However, electron microscopic studies and freeze etching studies have revealed much about the structure and progressive development of cell wall around the protoplast in culture medium.

1. Observe regularly the regeneration of cell wall, cell division and small callus formation under inverted microscope
2. Examine cell wall formation in protoplasts with a droplet of 0.1% calcofluor white R, American Cyanamid, Bound Brook, N.J. USA, in 0.4 M sorbitol solution on a slide. The cell wall regenerated protoplasts fluoresce
3. Small cluster of calli are observed after 2-3 weeks of culturing protoplasts
4. Subculture the cell clusters on a freshly prepared protoplast culture medium with or without 1/2 the initial mannitol and 0.8-1.6% agar.

### Development of Callus/Whole Plant

Soon after the formation of wall around the protoplasts, the reconstituted cells show considerable increase in size and first divisions usually occur within 7 days. Subsequent divisions give rise to small cell colonies. After 2-3 weeks macroscopic colonies are formed which can be transferred to an osmotic free medium to develop a callus. The callus may be induced to undergo organogenic differentiation, or whole plant regeneration.

## HANDLING OF REGENERATED PLANTLETS

Plantlets regenerated by the process of embryogenesis or organogenesis is then used to transfer to non-sterile conditions. This can be very critical period when the plantlet is removed from sterile controlled environment and transplanted to the real world of agriculture. As the plantlet is moved to non-sterile conditions care must be taken to allow gradual release of these careful controls.

Anatomical analysis of these plantlets has shown that under these conditions often the cuticle (covering the leaf surfaces) is extremely thin and the root hairs are poorly developed. If these plantlets are suddenly move to a situation where they are subjected to water stress they would be unable to control water balance adequately and will die. Both the plantlets should first allow developing a good root system before transferring them in vivo conditions and care should be taken for least possible damage of the roots. If the plants have been grown on agar-solidified medium, this can be removed by gentle washing in a basin of warm water.

These plantlets are then carefully taken out from the test tubes or flasks and adhered agar medium is removed from them. The in vitro plants with roots are then carefully planted in small pots. The care is to be taken that roots should make quick and good contact with the potting mix because the plant may suffer water stress. The planting mix should be well balanced for water retention and sterile. Small, sterile peat pots (Jiffy pots) are excellent for this purpose.

## PROTOPLAST FUSION

Plant protoplasts represent the finest single cell system that could offer exciting possibilities in the fields of somatic cell genetics and crop improvement. The isolated protoplasts are surrounded by plasma membrane only. The lake of the cell wall allows the plasma membrane of two or more protoplasts to come into intimate contact. The important aspect has been that incompatibility barriers do not exist during the cell fusion process at interspecific, intergeneric, or even interkingdom levels. Under certain conditions they will stick together rather like two soap bubbles. Later, if they are given an appropriate stimulus they will fuse together forming a single mass surrounded by a common membrane. Kuster was able to see occasional spontaneous fusion of protoplasts. However, with the development of enzymic methods for producing large numbers of isolated protoplasts, great interest has now been developed in the use of protoplast fusion (**somatic hybridization**) as a possible tool of plant breeding. Many people believe that protoplast fusion offers a useful tool to plant breeders to make crosses between sexually incompatible species for transfer of nuclear or cytoplasmic characters. Protoplast fusion can be used to make crosses within species (interspecific), between species (interspecific), within genera (intrageneric) and between genera (intergeneric).

Protoplasmic fusion permits the formation of cytoplasm containing organelle of 2 different parents like chloroplast of one parent and mitochondria of the other, but coexistence of a particular organelle like chloroplast of both the parents in a same cytoplasm usually does not persist due to sorting during mitotic divisions. Protoplasmic fusion is helpful in several graminae, solanaceae and cruciferae members to overcome sexual barriers, to produce plants with resistance to diseases or tolerance to selective herbicides, to produce somatic hybrids having efficient organelle system and widen to genetic base. Protoplast system can be transformed by incorporation of pure DNA or calcium phosphate-DNA coprecipitate or DNA encapsulated in liposomes or by direct microinjection of DNA. The DNA uptake can be further enhanced under high pH or by heat shock treatment and / or by addition of polyethylene glycol, poly –1 ornithine or polyvinyl alcohol.

By inactivation of nucleus of one parent and subsequent fusion with normal protoplasts or by fusion of normal and subprotoplasts or microprotoplasts or by fusion of normal and subprotoplasts or microprotoplasts or by using mutants and anti metabolites it is possible to produce \`cybrids‘ having nuclear genes of one parent and cytoplasmic genes of other parent. Cytoplasmic organelle like chloroplasts and mitochondria possess DNA, which control the expression of male sterility. Thus the production of cybrids will be helpful in seed propagated annual ornamentals to exploit hybrid vigour by heterosis breeding. It may also be possible to achieve postponement of pollen induced senescence in flowers.

### General Steps of Protoplast Fusion

An efficient method employing polyethylene glycol (PEG) has found wide application in experiments on protoplast fusion. Many fusion inducing agents such as sodium nitrate, high $Ca^{++}$, high pH 10.5, high temperature (37°C) and polyvinyl alcohol have been used for fusion of protoplasts.

### Procedure

Place a drop (1 drop = 50 µl) of each protoplast suspension in sterile petri dish; gently shake the plate to ensure proper mixing of suspensions from callus or suspension culture with leaf protoplast. The concentration of each species should be about 1 x $10^5$/ml before mixing.

1. Wait for 5 min to allow the protoplasts to settle to the bottom of petri dish
2. Slowly add 300-450 µl PEG solution to the edge of the protoplast suspension, then in the centre of the protoplast, and wait for 15 min
3. Add drop by drop 1 ml of 0.7 M mannitol solution to dilute the PEG solution
4. Raise one side of the plate and wash the protoplast clumps adhering to the plastic surface of plate up to with 9 ml of 0.7 M mannitol
5. Remove the PEG and mannitol residues from petri dish. Add a few drop of mannitol solution to the fused cells
6. After 5 min observe the fused product on the inverted microscope. The process of adhesion or agglutination of protoplasts in various combinations can be seen.
7. Study the fused and unfused protoplasts for the synthesis of cell wall, first cell division, formation of homokaryons/heterokaryons, frequency of multinucleate homokaryons from heterokaryons and mitotic index, cytologically examination under microscope.

The technique of protoplasts fusion has developed new vistas for the plant breeders where conventional methods failed or the plant species are incompatible. Protoplasts fusion results in somatic hybrids, the phenomenon is known as somatic hybridization. Protoplast fusion may be spontaneous or induced.

The protoplasts fusion may be of three kinds:

1. Spontaneous fusion
2. Mechanical fusion
3. Induced fusion

### Spontaneous Fusion

Only interspecific protoplasts have been found very often fusing through their plasmodesmata during enzyme treatment. Nearby protoplastsin a Petri dish have been observed to form large multinucleate (coenocytic) mass. Protoplasts of young leaves are prone to fuse spontaneously with one another.

### Mechanical Fusion

The giant protoplasts of *Acetabularia* have been fused mechanically) This kind of fusion is not dependent upon the presence of fusion-inducing agent. However, in this procedure protoplasts are likely to get injury.

### Induced Fusion

This kind of fusion can be done with the help of (a) $NaNO_3$ treatment, (b) High pH/$Ca^{++}$ treatment, (c) Polyethylene glycol (PEG) treatment, (d) Electro fusion. This method can bring together both intra- and interspecific protoplasts. For this, some inducing agent is necessary. The inducing agent is called **fusogens**. In animals **Sendai virus** plays this role but in plants both physical as well as chemical methods are used.

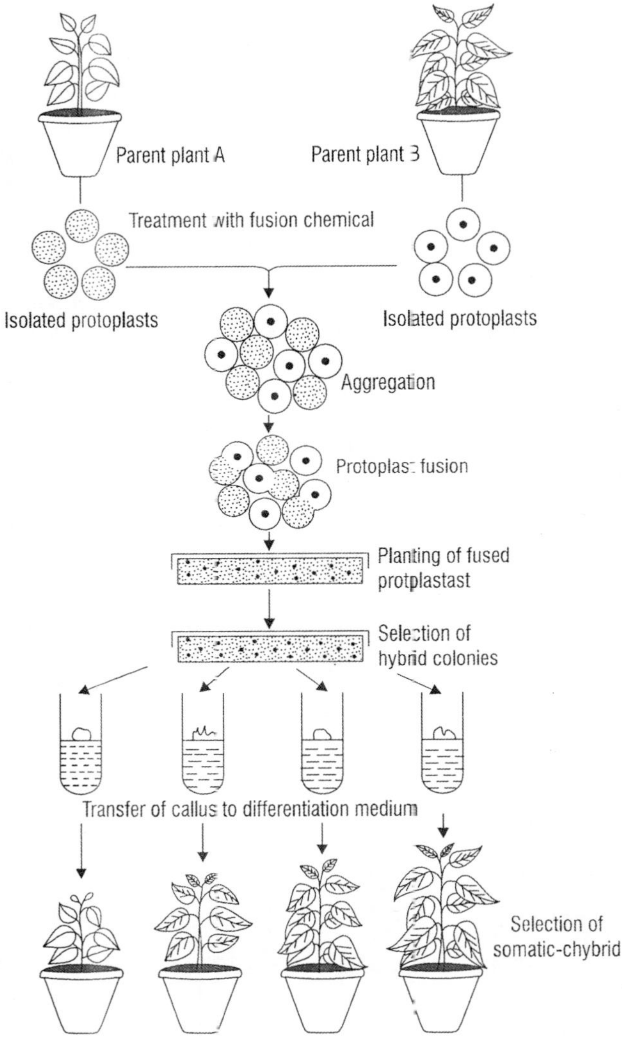


**FIGURE 25.6:** Diagrammatic presentation of fusion of protoplast from two different plant species and the regeneration of somatic hybrid plants.

Several chemicals like sodium nitrate, PEG, poly-L-ornithine, poly-D-lysine, cytocholasin B, Protamine sulphate, lysozyme, glycerol, dimethyl sulphoxide, proteins and calcium ions at high pH have been used as fusogens. Kuster (1909) for the first time used sodium nitrate to plasmolysed onion protoplasts, which underwent fusion on, deplasmolysis. Thereafter, Power et al. (1970) employed the same technique and successfully fused root tip protoplasts of maize and oat.

Keller and Melchers (1973) employed calcium ions at high pH (0.05 M CaCl $2H_2O$ in 0.4 M mannitol at pH 10.5) for fusion of tobacco protoplasts. They spinned the isolated protoplast in the fusion inducing solutions for 3 minutes at 50g and then kept the tubes in water bath at 37° C for 40-50 minutes nearly 20-50% protoplasts undergone fusion process.

**PROTOPLAST FUSION**

**Protoplasts (Leaf)** ⇓ **Protoplasts** (Cell suspension culture)

⇓

? **Combine**

⇓ PEG High $Ca^{2+}$ and

⇓ pH (10.5)

⇓ PEG Dilution

**Fusion**

⇓ Culture Medium

**Mitosis**

⇓

**Hybrids**

**FIGURE 25.7:** Presentation of stages in protoplast fusion and production of hybrids. 1 = Parental protoplasts, 2 = Aggulation adhesion.

The one most widely used fusogens is polyethylene glycol. The effects of PEG are not specific and it promotes the aggregation and fusion of protoplasts from the same or different species or intergeneric protoplasts.

Zimmermann has shown that if protoplasts are placed into a small culture cell containing electrodes and a potential difference is applied, and then the protoplasts will lie between the electrodes. Protoplasts fusion can be induced after applying extremely short, square wave electrical stocks. This method has become very popular for its highly controllable fusions property. It is possible to fuse two single protoplasts.

## PROTOPLAST CULTURE: REGENERATION OF PLANTS

The idea that cell contains all the information necessary for the regeneration of a whole organism was given in the cell theory, proposed by Schwann. At that time, however, the technology did not exist to study regeneration of plants from any single cell other than a zygote, and even this developmental process takes place within plant body and not as an isolated single cell in vitro.

It has been a long practice to regenerate vegetative plants by taking stem cuttings. Haberlandt successfully regenerated whole plant from small meristematic pieces; however, he could not obtain any growth of a single cell system because of its inability to regenerate plants from single cell. This was not due to genetic reason, but due to lack of adequate nutrients in the medium. He further suggested the use of embryo sac fluids in the medium for such culture.

Protoplasts in culture regenerate cell wall around themselves to reconstitute cells. These cells in proper medium by repeated cell divisions form callus. Callus tissue by manipulation of the nutritional and physiological conditions can be induced to regenerate into the morphogenic form of plant.

About 100 plant species have successfully been regenerated from single cell or protoplast. Regeneration of plants from isolated cells or protoplasts is by one of two following methods:

1. By Organogenesis
2. By Embryogenesis

## Organogenesis

Organogenesis takes place from a callus and not directly from a single cell. When isolated protoplasts are put into culture under appropriate conditions of plant inoculum, medium and environmental factors they go through a set series of events.

1. Wall regeneration
2. Early mitotic division and callus formation
3. Organogenesis

## Embryogenesis

Plant tissue in vitro can induce to form somatic haploid embryos. Steward and coworkers first observed the phenomenon of somatic embryogenesis in carrot. Somatic embryos can be induced in cultural conditions from three different sources:

1. vegetative cells of mature plants;
2. reproductive tissues other than the zygote; and
3. hypocotyls and cotyledons of embryos

Embryoids are initiated in callus from small superficial clumps of cells associated with dense cytoplasm and large starch grains. The developing embryoids in vitro then pass through a sequence of growth stages that is exactly the same as seen in the development of a seed embryo.

Like cells, protoplasts are genetically unstable in vitro. Somaclonal variations have also been detected in the plants regenerated from protoplasts. Chromosomal variation has also been observed in plants regenerated from protoplasts. Shepard and his associates conducted detailed studies of plant regeneration from protoplasts of Russet Burbank potatoes. Significant variations were detected for growth habit, tuber colour, and fruit production and tuber uniformity. Several of these traits are considered to be improved over the widely grown commercial variety. They further reported somaclones with resistance to early blight (*Alternaria solani*) and late blight (*Phytophthora infestans*). Our knowledge is insufficient to understand the mechanism of such instability in culture.

In *Nicotiana tabacum*, protoplast cultures were regenerated into cells and callus tissues. These calli were then transferred to fresh medium supplemented with IAA (4 mg/l) and cytokinin (2 mg/l). Calli differentiated into shoots and occasionally in roots within 3 to 4 weeks. The plantlets are then subsequently transplanted to pots.

## Handling of Regenerated Plantlets

Plantlets regenerated by the process of embryogenesis or organogenesis is then used to transfer to non-sterile conditions. This can be very critical period when the plantlet is removed from sterile controlled environment and transplanted to the real world of agriculture. As the plantlet is moved to non-sterile conditions care must be taken to allow gradual release of these careful controls.

Anatomical analysis of these plantlets has shown that under these conditions often the cuticle (covering the leaf surfaces), is extremely thin and the root hairs are poorly developed. If these plantlets are suddenly move to a situation where they are subjected to water stress they would be unable to control their water balance adequately and will die. Both the plantlets should first allow developing a good root system before transferring them in vivo conditions and care should be taken for least possible damage of the roots. If the plants have been grown on agar-solidified medium, this can be removed by gentle washing in a basin of warm water.

These plantlets are then carefully taken out from the test tubes or flasks and adhered agar medium is removed from them. The in vitro plants with roots are then carefully planted in small pots. The care is to be taken that roots should make quick and good contact with the potting mix because the plant may suffer water stress. The planting mix should be well balanced for water retention and sterile. Small, sterile peat pots (Jiffy pots) are excellent for this purpose.

## PROTOPLAST CULTURE: UPTAKE OF FOREIGN MATERIALS

Isolated protoplast of plant cells have the remarkable property of taking up micro- as well as macromolecules like ferritin, polystyrene, latex, protein, DNA, chromosomes, whole nuclei, chloroplasts and even virus and bacteria. This unique property of protoplast can be utilized for the incorporation of desirable traits into plants to get what we call *transferred* or *noval varieties* from wild ones.

### Incorporation of Foreign DNA

Unlike transformation of bacteria by exogenous DNA, transgenosis in cells or protoplasts of higher plants are now possible and have been tried successfully in some plant species. Doy and his associates transformed haploid tomato callus culture by lambda phage. The transformed higher plant cells can grow on a medium containing galactose and lactose as the only carbon source that is considered to be due to phage DNA incorporation. This kind of transformation phenomenon is termed as **transgenosis**. These studies have given a major break through in agricultural biotechnology.

### Incorporation of Nuclei

Intra- and interspecific nuclear incorporation into protoplasts of *Nicotiana tabacum*, *Petunia hybrida* and *Zea mays* have been observed. These exogenous nuclei enter the protoplasts without any damage to plasmalemma. However, the role of these incorporated nuclei in the metabolism of protoplasts in culture is yet to be confirmed.

### Incorporation of Chloroplasts

In addition to nuclear material and nuclei as a whole protoplast can take up certain cell organelles like chloroplasts, mitochondria etc. Transplantation of chloroplasts into albino plant species could solve a great physiological deficiency. Potrykus (1973), Bonner and Eriksson (1974) have reported the uptake of chloroplasts by albino *Petunia* and carrot protoplasts. Nass (1969) and Gills and Sarafis (1971) have incorporated spinach chloroplast into animal cell cultures. They have also reported that this chloroplast survives and remain metabolically active in animal cells. Transplantation of chloroplast in plant having insufficient or deficient photosynthetic system will have far- reaching implications in plant improvement programme.

Several methods for the induction of chloroplast into plant protoplasts have been reported, all of them depending either on some form of osmotic effect or on hypothesized changes in the charge pattern on the plasmalemma.

### Incorporation of Cyanobacteria

Experiments have been performed to incorporate certain cyanobacteria or blue-green algae (*Gleocapsa*, *Anacystis*, etc.) in protoplast of higher plants. The method is used to co-incubate the algal preparation with the isolated protoplast preparation in the presence of 25% polyethylene glycol, when high Ca concentration is added the protoplasts begin to engulf the algal cells. The method of uptake has been shown to be by invagination of the plasma membrane.

### Incorporation of Bacteria

Incorporation of nitrogen fixing bacteria (*Rhizobium*, *Azotobacter*) blue green algae and *nif*-genes into non-leguminous plants specially cereals is a important task in the hands of geneticists and agronomists. Davey and Cocking (1972) introduced a symbiotic nitrogen fixing bacterium (*Rhizobium*) in legume chloroplasts. These chloroplasts containing nitrogen-fixing bacteria could be fused with non-legume chloroplast to impart the capability of nitrogen fixation in non-leguminous cash crops. *Agrobacterium tumefaciens* have been adapted to become efficient and reliable gene vectors for genetic engineering of dicotyledonous plants. *Chimeric*-genes have been constructed which upon transfer into differentiated plant cells allow the identification of DNA sequences involved in the regulation of gene expression. Recently, direct DNA transfer and expression of a bacterial gene in protoplasts of *Triticum monococcum* and *Nicotiana tabacum* have been reported; and in the case of tobacco transformed protoplasts were regenerated into plants which obeyed Mendelian law of inheritance.

### Incorporation of Virus

Virus, an obligate parasite in nature enters the plant cells, leaving behind its protein coat only in those plant cells whose cells wall is damaged. The virus genome (RNA) enters the protoplasts of the cell and then multiplies in cytoplasm. Takebe and Otsukic (1974) reported that protoplasts can be infected by one or more than one virus. Cocking (1966) observed the uptake of TMV by tomato fruit protoplast.

The inclusion of these micro- and mega molecules in plant protoplasts has opened a new path to solve the genetic deficiency in economically important plants through biotechnology.

## INCORPORATION OF NON-BIOLOGICAL MATERIALS

Incorporation of non-biological materials have been tried such as polystylene in to plant protoplasts. These types of studies have proved useful as a method of determining the maximum size of an object that can be engulfed by endocytosis. However, no direct biological value can be seen to these studies at the present time.

## SELECTION OF FUSION HYBRIDS

### Visual Selection

This method can be used to select fusion products of protoplasts that have distinct physical characters. For example fusion between leaf mesophyll protoplasts of green colour and a cultured cell that has an anthocyanins containing red protoplast. A single fusion product between the two will have both chloroplasts and an anthocyanin vacuole together with two nuclei. The same method can be applied to fusions between leaf mesophyll and hyaline cultured cell protoplasts. Obviously, the drawback with this method is that it is very laborious and only a very limited number of fusion products can be selected.

### Fluorescent Labels

1. In this method fluorescent-labelled dyes are used to detect fusion products.
2. If the two original protoplast cultures are pre-incubated for 12-15 hours, one in octadeconyl aminofluorescein and the other in octadecyl palamine B each group of protoplasts takes on a specific fluorescence colour. The dyes are non-toxic and do not affect viability, wall regeneration or growth.
3. After fusion of the protoplasts fusion products may be identified by their fluorescence characteristics under a fluorescence microscope.

### Fluorescence Activated Cells Sorting

Protoplasts are fluorescently labelled as already described. After fusion three types of fluorescence can be found; two original fluorescence of the two parental populations and the fluorescence of the fused hybrids. A machine now exists to separate these populations. A droplet containing a single cell is held between two fluorescence detectors. If it is one of the two parent populations the machine charges two electrical plates, as the droplet falls it is deflected by the electrical charge. If the fluorescence is of the hybrid, no electrical charge is applied to the plates and the droplet fall vertically. Thus the protoplasts are automatically separated into the parent types and the fusion products. The fusion products can then be plated into an appropriate culture system.

### Nutritional Selection

In some cases, Carlson et al. were able to separate fusion products by their ability to grow on a media that would not permit growth of the parent protoplast lines. For example hybrid of *Nicotiana glauca* and *N. langsdorfii* could be cultured in auxin free culture medium in contrast to its parents protoplast.

### Drug Sensitivity and Resistance

Power and co-workers used sensitivity to actinomycin-D to select fusion products between *Petunia hybrida* and *Petunia parodii*. The two parental species are sensitive to actinomycin-D; however, the hybrid is resistant. Again it would be of great benefit to the field of somatic cell hybridization if a wider range of resistance mutants could be isolated. Some research groups are now attempting to produce large numbers of resistant mutants for use in protoplast fusion studies.

## SOMATIC CELL HYBRIDIZATION OR CYBRID OR CYTOPLAST

Success in protoplast fusion between diversified plant species gave great impetus to geneticists and plant breeders though the technique has some shortcomings. Nevertheless, it has stimulated research activities in many laboratories around the world. When two protoplasts fuse along with their nuclei true hybrid is formed. However, often the two nuclei will survive independently in the mixed cytoplasms, forming a heterokaryon. Chromosome loss can occur from one or both of the nuclei; if all the nucleus of one parent is lost, what is left is one nucleus in a mixture of both cytoplasms. This product is called a **cytoplast** or

**cybrids**. Cybrids can also be prepared by fusing a normal protoplast with an enucleated protoplast. Presently, researchers are studying the use of cybrids to transfer resistance to some herbicides. This resistance character is carried on the chloroplast genome. Transfer of the mitochondrial genome can control transfer of cytoplasmic male sterility. The study of cybrids is also important for investigation of possible recombination within these extra-chromosomal genomes.

The scenario presented in this chapter gives an overall view of protoplast culture. The technique has unraveled the complexities of protoplast culture and its application to agricultural biotechnology. The protoplast technique is used in studies of plant cell genetics especially in somatic embryogenesis and transfer of genetic information by DNA uptake and organelle implantation. Though tissue culturists have developed many plantlets and new variants through this technique, care must be taken to achieve cost-effective results so that the in vitro technique of protoplast culture can be adapted for commercial exploitation of agricultural crops in general and plantation crops in particular.

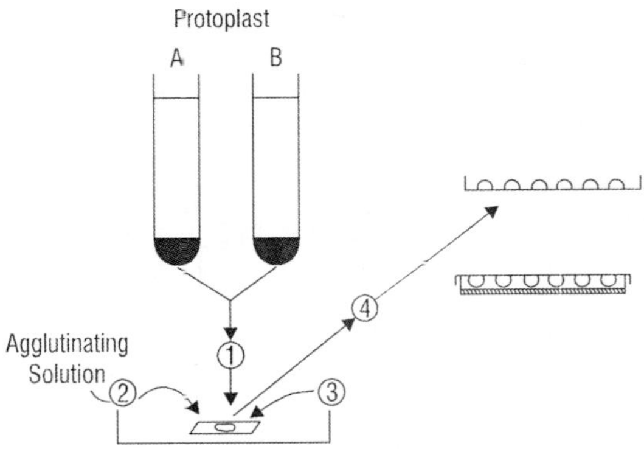


**FIGURE 25.8:** Typical protocol for protoplast fusion of potato using PEG. (1) Dilute protoplast mixture to $10^4$ cells per milliliter in 0.5 M glucose, 3 mM $CaCl_2$, 0.7 mM $KH_2PO_4$, pH 5.7; (2) add 3 to 200 µl volumes of agglutinating solution (0.09 mM PEG 6000, 10 mM $CaCl_2$, 0.7 mM $KH_2PO_4$, pH 5.5); (3) add five 100-µl volumes of diluting solution (50 mM $CaCl_2$, 0.5 M glucose, 50 mM glycine buffer, pH 10.0); (4) wash cells with culture medium and culture in hanging drop suspensions.

## PROTOPLASTS FOR ISOLATION OF CELL COMPONENTS

The danger of erroneous interpretation of protoplast metabolism as truly representing intact cell processes is hardly encountered when the aim is merely the isolation of cell components. The objective of isolating such components as plasmalemma, nuclei, and chloroplasts from protoplasts, rather than from cells or intact tissues, is in most cases to obtain them by convenient methods, which will cause the least destruction. Therefore, if the functionality of these components is of primary importance, the evaluation of the rate and quality of damage (e.g. in nuclei) as well as their unquestionable identification and purity (e.g. of plasmalemma) are important considerations. Most of the problems involved in the isolation of cell components from protoplasts are specific to plant cells because of the latter's rigid cell wall and their specific organelles, e.g. chloroplasts and vacuoles. These problems are very rare in animal cells but are common to plant cells.

# 26. DNA Isolation and Sequencing

Recombinant DNA technology deals with isolation and manipulation of genes. Before nucleic acids can be cut or otherwise manipulated, they must be either isolated and purified or artificially synthesized. The isolated genes are also sometimes sequenced for a better understanding of its structure that needs to be manipulated. In this chapter we will discuss the methods available for isolation, sequencing and synthesis of genes.

## Isolation of Total Cellular DNA

Shear forces very easily damage DNA; even rapid stirring of solution can break even with high DNA molecular weight into much shorter fragments. Consequently, DNA is recovered from cells by gentlest possible method of rupture, in the presence of EDTA to chalate the $Mg^{2+}$ ions needed for DNase activity. Cell walls, if present, are digested enzymatically (e.g. lysozyme treatment of bacteria) and the cell membrane is solubilized using detergent. Cell disruption is performed at 4°C using autoclaved glassware and solutions to destroy DNase activity. After release of nucleic acids from the cells, RNA can be removed by treatment with ribonuclease (RNase). The other major contaminant, protein, is removed by shaking the solution gently with water-saturated phenol, or with phenol/chloroform mixture, either of which will denature protein but not nucleic acids. Centrifugation of the emulsion formed by this mixture produces a lower, organic phase, separated from the upper aqueous phase by an interface of denatured proteins. The aqueous solution is recovered and deproteinised repeatedly, until no matter is seen at interface. Finally, the deproteinised DNA preparation is mixed with ethanol and allowed to precipitate out in freezer. After centrifugation, the DNA pallet is redissolved in a buffer containing EDTA for protection against DNases and this solution can be stored at 4°C for at least a month. DNA solutions can be stored frozen, but repeated freezing and thawing tends to damage long molecules by shearing; so preparation in frequent use is normally stored at 4°C. This procedure is suitable for total cellular DNA. If the DNA from specific organelle or viral particle is needed, it is best to isolate the organelle or virus before extracting its DNA.

## Nucleic Acid Hybridization

If a double-stranded DNA molecule is exposed to high temperature, or to very alkaline conditions, then the two strands will come apart. The molecule becomes denatured. The temperature at which denaturation occurs is termed as melting temperature or $T_m$. If the denatured DNA is returned to a temperature below its $T_m$, or to neutral pH when alkali was used to denature it, each strand will, after a time, find its complementary strand. The two strands will 'zipper' back together to re-form a double-stranded DNA molecule. This ability of complementary sequences to anneal, or hybridize, to one another is called nucleic acid hybridization. This technique helps in determining the gene structure and in identifying molecules, which contain same sequences of nucleotides. In a complex mixture of nucleic acid molecules, nucleic acid hybridization technique helps in separation of complementary sequence.

When employed analytically, hybridization is normally performed using one labelled sequence, termed the probe, and an unlabelled sequence called the target. **Probe** is a short synthetic oligo deoxyribonucleotide, which is complementary to target DNA sequences. Target DNA may be obtained from total cellular DNA preparations. The probe is labelled by incorporation of either radioactively labelled nucleotides or with some chemicals (described later). The probe is the known, pure species in the hybridization and the target is the unknown species to be identified.

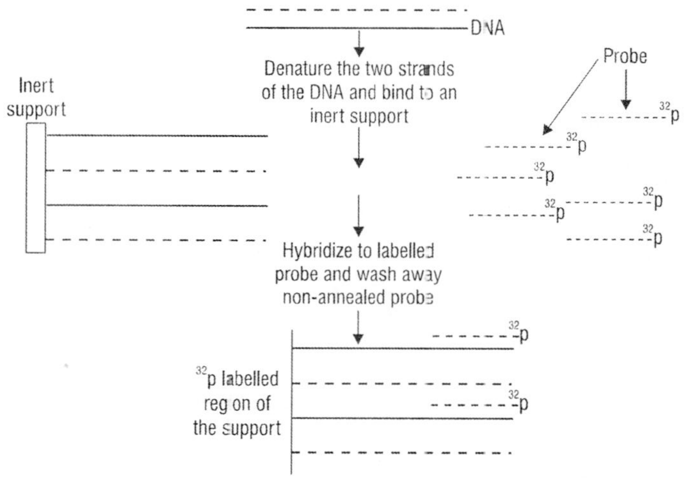


**FIGURE 26.1:** The principle of hybridization using an inert support. The target DNA is most often localized to one part of the support and, in most cases, the aim of the hybridization experiment is to identify this region. The probe is prelabelled by some method, most often by the incorporation of radioactivity. After hybridization, the annealed probe is detected by autoradiography.

The target will most often form part of a mixture of unrelated nucleic acid sequences. The probe is usually added in considerable excess over the target, and its concentration determines the rate of reaction. Hybridization reaction time is extended to ensure complete reaction. The target and probe can be hybridized to each other in solution, or alternatively the target can be immobilized on an inert support such as nitrocellulose. When the target is immobilized the rate of annealing is greatly reduced.

The rate of hybridization depends upon the DNA concentration, the concentration of salt in the solution and the temperature. Hybridization reactions are normally performed at 10-20°C below the $T_m$ of the duplex. Under any given conditions of ionic strength and temperature, the $T_m$ of a duplex also depends on the base composition of the annealed partners.

Hybridization involving RNA is performed in the presence of formamide, as RNA cannot withstand higher temperatures. Formamide disrupts hydrogen bonding and thus reduces the effective $T_m$ of the duplex, allowing annealing to occur at a lower temperature (e.g. 37°C).

## METHODS FOR LABELLING NUCLEIC ACIDS

For identification of hybridized nucleic acid duplexes, labelling of probe or target is necessary. As probe is short synthetic oligonucleotide it can be easily labelled. A $^{32}P$ isotopic label can be added to the 5' hydroxyl group using polynucleotide kinase to a probe. Alternatively, labelled nucleotides can be incorporated at the 3' end using terminal transferase. Because label is added at only one place within the oligonucleotide, such end-labelled probes contain relatively little radioactivity per unit weight of DNA. This low specific activity limits their sensitivity as probes. If a high specific activity probe is required, then some method of uniform labelling must be used. Here the probe sequence is copied along part or all of its length using DNA polymerase and labelled nucleoside triphosphate.

## METHODS OF LABELLING NUCLEIC ACID AND PROBES

There are five basic methods for labelling nucleic acids. These are: nick translation, primer extension, methods based on RNA polymerase, end labelling methods and direct labelling methods.

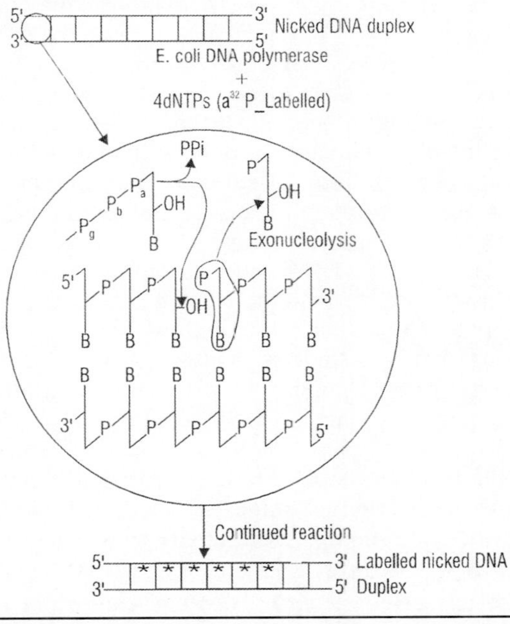


**FIGURE 26.2:** $^{32}P$-labelling of duplex DNA by nick translation. Asterisks indicate radiolabelled phosphate groups.

### Nick Translation

This is done by making single-strand cuts ("nicks") in the double-stranded DNA molecule by brief exposure to a dilute solution of an endonuclease (usually deoxy ribonuclease I of *E.coli*). DNA polymerase I is then used in the presence of at least one radioactive precursor (e.g., Â-$^{32}P$-dATP) to "translate" the nick along the molecule in the 5' to 3' direction. This occurs by the concerted action of the 5'→3' polymerase and the 5'→3' exonuclease activities built into DNA polymerase I. The 5'→3' polymerase adds nucleotides to the 3'-OH terminus at each nick, while the

5'-phosphate terminus at the nick. The net result is that a radioactive strand replaces a nonradioactive strand of DNA. The DNA is then denatured and used as a radioactive probe in hybridization experiments (Southern blots, northern blots, etc.).

Nick translation can be used with a variety of labels to generate probes suitable for most hybridization applications. It is also appropriate for the generation of biotinylated probes (described later).

A major advantage of nick translation over other uniform labelling methods is its flexibility with respect to probe size, specific activity and concentration. It is particularly suitable for the production of large quantities of probes for use in multiple hybridization and/or where a high probe concentration is required.

## Primer Extension Method

Primers are synthetic oligodeoxyribonucleotides, which are complementary to specific regions of known vector DNA. Some commonly used primers are M13 sequencing primer, pBR322 primer etc. The 3' termini of these primers serve as initiation site for template dependent DNA synthesis by enzymes like DNA polymerase I.

DNA polymerase works by extending a short double-stranded region made by annealing an oligonucleotide primer to the single-stranded template. Thus this method of uniform labelling requires a primer, which matches the probe sequence. Radiolabelling of primers can be done with two methods.

If the probe sequence is not known then random oligonucleotide labelling can be used (**FIGURE 26.3**). It is often in the case when natural cellular DNA is used. These primers are made by adding a mixture of all four bases at each step in the chemical synthesis reaction. The DNA is denatured and the two complementary strands are copied in the presence of labelled primers as well as, nucleoside triphosphates. The polymerase used is Klenow fragment derived from DNA polymerase-I of *E.coli*. Chance homology ensures that these primers anneal to the separated DNA strands at many points along their length, thus providing a base for polymerase to initiate DNA synthesis. This is only one of several uniform labelling methods.

The second method uses a unique primer to restrict labelling to a particular sequence of interest. In the primer extension method it is essential to use a polymerase lacking a 5'→3' exonuclease activity otherwise degradation of the primer will occur. Both the Klenow fragment of *E.coli* DNA polymerase I, which lacks the 5'→3' exonuclease activity and reverse transcriptase have been used successfully.

An advantage of random priming method is that is may be used to label relatively impure DNA. In primer extension method, there is not synthesis of DNA and labeled nucleotides are incorporated in concentrations. It is an ideal method for situations where high specific activity and low probe concentration are frequently employed.

## Methods based on RNA Polymerases

RNA polymerases catalyze the synthesis of RNA from nucleoside triphosphates using a DNA template. Thus they can incorporate labelled ribonucleotides into RNA during transcription if such labelled nucleotides are provided to it. If a specific site of a vector or DNA is transcribed in such way, RNA probes (or transcripts) of defined length and sequence can be obtained.

Initially RNA probes were obtained from RNA polymerase from *E.coli*. This enzyme when used *in vitro* exhibited very little template and promoter specificity and therefore produces transcripts, which have been initiated and terminated at random. Today, RNA polymerases from a number of phages have been used, which possess a high degree of specificity. Thus transcription can be limited to sequences cloned downstream of an appropriate promoter.

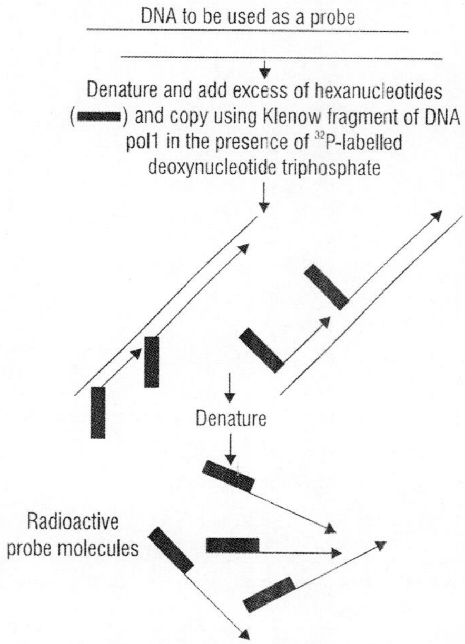


**FIGURE 26.3:** The principle of random primed (oligo-) labelling. The DNA to be used as a probe is denatured by heating and mixed with hexanucleotides of random sequence which then act as primers for DNA synthesis.

RNA probe's specific activity is determined solely by the specific activity of labelled nucleotide and moreover such probes are needed not to be denatured before hybridization. These probes are more sensitive than nick translated probes because of avoidance of probe reannealing during hybridization (as RNA-RNA hybrids are more stable than RNA-DNA hybrids)). A further advantage of RNA probes is that non-specifically bound probe can be removed with RNase A. RNase A is very specific for single stranded RNA.

The only difficulty with RNA polymerase-based-labelling is that a specific promoter is read by RNA polymerases before transcripting the downstream. This promoter has to be subcloned and inserted into the vector. However, once an insert has been cloned into such a vector, DNA probes may still be prepared using nick translation or random primer labelling.

## END-LABELLING OF NUCLEIC ACIDS

A wide variety of techniques are available for introducing label at either the 3' or 5' ends of linear DNA or RNA. Usually only a single label is introduced at the terminus. Nucleic acid can be 5' end labelled using T4 polynucleotide kinase. The enzyme catalyzes the transfer of the γ-phosphate of ribonucleotides 5'-triphosphate donor to the 5'hydroxyl group of a poly- or oligo-nucleotide or even to nucleoside-3-triphosphate. Radiolabelled phosphate group is donated by [$\gamma^{32}$-P] ATP to DNA or RNA containing a 5'-hydroxyl terminus. This is termed as forward reaction.

If 5'-phosphate group is present in DNA or RNA, then it is removed with alkaline phosphatase. This reaction is driven by excess ADP, which causes the enzyme to transfer the terminal 5'-phosphate from DNA to ADP. This is known as exchange reaction. The DNA is rephosphorylated by transfer of labeled γ-phosphate from [$\gamma^{32}$-P] ATP.

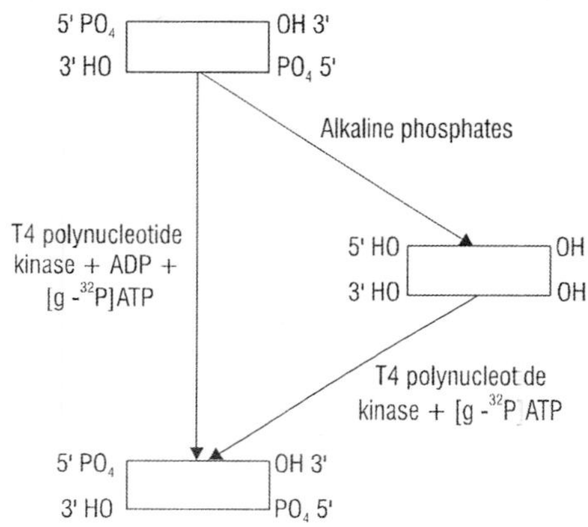


**FIGURE 26.4:** The 5'-end labelling reaction using T4 polynucleotide kinase.

The major advantages of 5'-end labelling are:

- Both DNA and RNA can be labelled.
- Oligonucleotide can be conveniently labelled e.g. for detection of mutants following site-directed mutagenesis.
- Location of labelled group is known.
- Very small fragments can be labelled.
- Restriction digest fragment can be labelled.

DNA fragments also can be labelled at their 3'end by using terminal deoxynucleotidyl transferase, however, this method is not of wide use.

## CHOICE OF LABEL

There are two technical parameters, resolution and sensitivity, which determines the success of probe application. High degree of resolution is required to known the relative position of a nucleic acid fragment. High sensitivity is necessary because sequence of interest may be present at low abundance. Other factors are probed stability, safety and ease of use. Broadly labels can be categorized into radioactive and non-radioactive types (Biotin labelled probes, Digoxigenin labelled probes, Alternatives to biotin and digoxigenin labeling). The major disadvantage of biotin, and of other nonradioactive labels, is low sensitivity in comparison to radioactive labels.

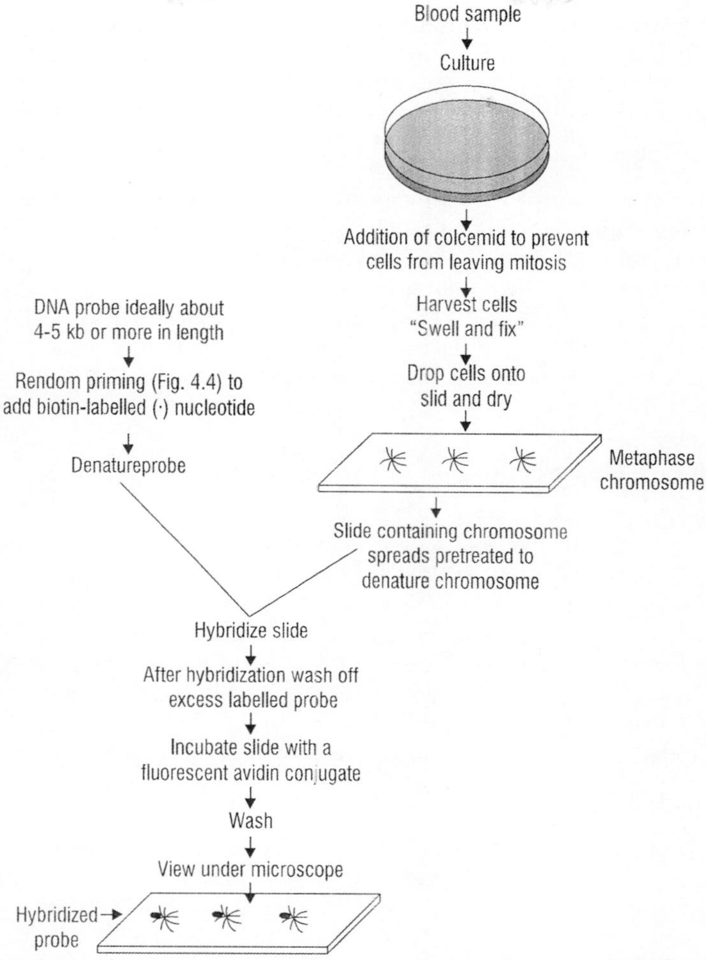


**FIGURE 26.5:** FISH is used where gene mapping is done with in-situ hybridization. It uses a non-radioactive labelling system involving biotin attached. When viewed under special filters the specific sites of DNA probe can be seen on a chromosome.

## MAPPING GENES ON SPECIFIC CHROMOSOMES

One of the most important applications of nucleic acid hybridization is construction of genetic maps. This helps in determining the chromosomal location of individual genes. There are three commonly used methods that differ in the degree of resolution they afford viz somatic cell hybrids, *In situ*, hybridization and genetic linkage mapping.

### In-situ Hybridization

In this technique a labelled DNA probe is hybridized to metaphase chromosome. The position of hybridization is visualized at the light microscope level. Metaphase chromosomes are liberated from cells and bound to the surface of a cover slip. The DNA within the chromosomes is denatured in such a way that the individual chromosomes remain visually identifiable. The probe is labelled, either by incorporation of a $^{3}$H-nucleotide or non-isotopically. The chromosomes are incubated with the labelled probe for many hours and, after hybridization, the excess unbound probe is washed away and the annealed probe detected.

### Transposon Tagging

Isolating a eukaryotic gene by transposon tagging involves inserting into target DNA, fractionating fragments of the latter on southern blots (described later in this chapter) and hybridizing the transposon with a probe. Cleaving out the transposon, together with flanking DNA, reveals the adjoining sequences of a gene. The probes are made for these adjoining sequences to find the fragment having gene of interest. Presence of transposon is confirmed by altered phenotype. This method was first employed to isolate *white* locus in *Drosophila*. The mutants with *white- apricot* eyes harboured a *copia* retrotransposon as well as accompanying *white* locus.

### Genetic Linkage Mapping

This form of mapping is used to construct genetic linkage maps of individual chromosomes and also to localize genes involved in genetic disease. Genetic linkage analysis relies on an ability to estimate the frequency of crossing over (recombination) that occurs between homologous chromosomes during meiosis. The closer two genetic markers are to each other then the lower is the chance that they will be separated during meiosis. The statistical probability that two markers are linked is measured as the logarithm of the odds (Lod). A Lod score of 3 is usually taken as the threshold of significance, while a Lod score below -2 suggests that the two loci are not linked. Lods between -2 and +3 are not significant.

## GENOMIC LIBRARY

The complete set of cloned fragments of DNA of an organism is known as genomic library. This is also called gene library. Such a library contains all or most of the genes present in the cell or the organism. It is difficult to identify a particular gene from an organism and then clone it. Hence the organism's entire genome as DNA fragments is cloned and the clone containing the sequence of interest is identified. Cloning experiments can be divided into two types (1) those where the cloned DNA segment is of unknown composition and function and, (2) those where the cloned DNA segment is of known composition and function. In the first approach the entire cellular DNA has to be fragmented into clonable size by either digestion with restriction endonucleases or, mechanical sheering.

**Colony hybridization:** Once a genomic library or cDNA library is available, gene sequence can be isolated by colony hybridization techniques. It involves replica plating the colonies formed by transformed cell into nitrocellulose filter and their *in situ* hybridization with a radioactive probe. This method was first developed by Hogness and Grunestein (1975).

**Plaque hybridization:** This technique is applicable where a chimeric phage particles is carrying the gene of interest. In such a situation, a bacterial lawn is infected with a mixture of chimeric phage particles (i.e. the library) and a large number of plaques develop overnight. These plaques can be treated just like the colonies in colony hybridization to identify and isolate the chimeric phage particle carrying the gene of interest. Two new techniques have shown their potential in gene isolation and characterization. They are chromosome walking and chromosome jumping. Let us have a small account on both of these techniques.

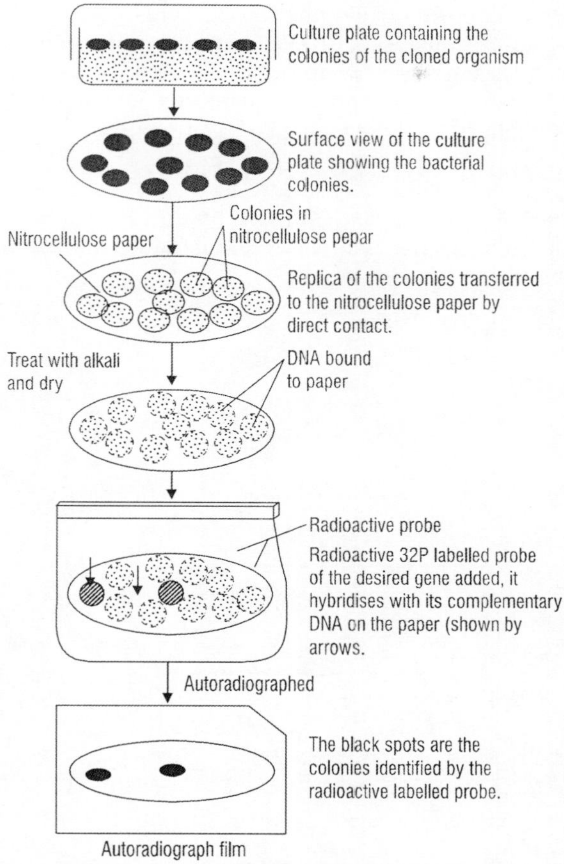


**FIGURE 26.6:** Colony hybridization to identify the clones containing the DNA of interest.

**Chromosome walking:** Identification of fragments with overlapping sequences may be a key to the reconstruction or characterization of large chromosome regions. This is achieved by the technique popularly described as chromosome walking.

**Chromosome jumping:** In order to bring the molecular marker close to the gene of interest, *chromosome-jumping* approach has been utilized. The technique of chromosome jumping helps in narrowing the gap between the gene and available molecular markers. After several cycles of chromosome jumping followed by cloning the regions proximal to the gene, it will be possible to approach very close to the desired gene capable approached and cloned. The gene for **cystic fibrosis** has recently been isolated using this technique.

## CLEAVING DNA WITH RESTRICTION ENZYMES

After locating gene of interest in colony or on chromosome, the DNA is extracted and cleaved with enzymes. Most restriction enzymes cleave DNA molecules in a site-specific manner. The genomic DNA from an individual organism is digested with either a single or more restriction enzymes. This results in fragmentation of DNA molecules with variation in length. These fragments are known as restriction fragment length polymorphisms or RFLPs. Since restriction enzymes cut DNA in a sequence specific manner, every homologous DNA molecule will be cut at exactly the same sites. This allows isolating large quantities of specific DNA fragments for sub cloning and isolating the DNA sequence of interest.

## Gel Electrophoresis

Nucleic acid molecules can be distinguished by differences in their chain length, and study of gene organization and expression includes nucleic acid hybridization with a chain length measurement.

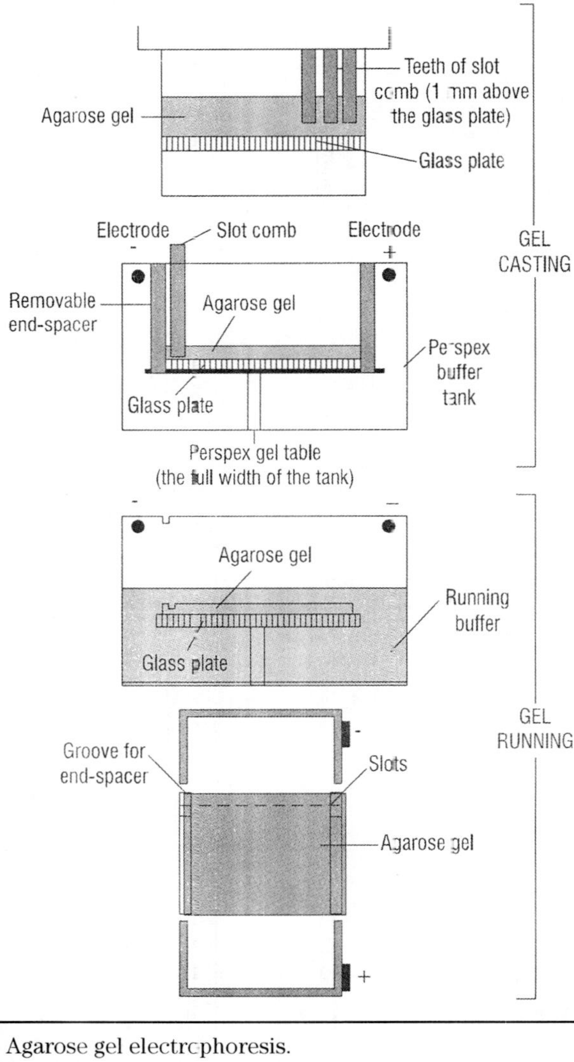


**FIGURE 26.7:** Agarose gel electrophoresis.

DNA and RNA have a net negative charge. At neutral pH therefore, these molecules migrate towards the positive terminal when placed in an electric field.

This process of electrophoresis is performed in an agarose or polyacrylamide gel. Nucleic acids are loaded into slots in the gel and allowed to migrate towards the positive terminal. The pores in the gel act to sieve the molecules. So that the mobility of a particular nucleic acid species depends on its length. All the molecules of a particular size move at approximately the same rate through the gel, forming a band, which gradually increases in width during electrophoresis because of diffusion. Running DNA molecules of known size in a parallel track on the gel, and determining their migration position relative to the species of unknown size can estimate the size of the molecules.

Polyacrylamide has a smaller average pore size than agarose and so is effective at separating molecules of 10-1000 nucleotides in length (1000 nucleotides is termed 1 kilobase or 1 kb), with very high resolution. The larger pores in agarose allow it to resolve much bigger molecules, up to 100 kb in length. However, the separated molecules migrate as a wider band so that the resolution (i.e. the ability to separate molecules which differ only slightly in size) is lower with agarose than with polyacrylamide. In pulsed-field gel electrophoresis (PFGE) the electrical field across the gel is repeatedly switched back and forth between two or more directions. This allows very long DNA molecules, several million base pairs in length, to be separated, since very big molecules change direction less rapidly as they move through the agarose gel.

After gel electrophoresis, the nucleic acid molecules must be detected. They can be visualized directly by including ethidium bromide in the polyacrylamide or agarose gel. This dye binds to double-stranded nucleic acids by intercalating between adjacent base pairs, and emits a red fluorescence when exposed to UV radiation. As little as 5-10 ng of DNA can be detected. If the nucleic acid molecules are radioactively labelled, then individual resolved species could be detected by autoradiography.

## BLOTTING TECHNIQUES

### Analysis of DNA by Southern Blotting

In 1975, E. M. Southern invented a procedure to identify the locations of genes and other DNA sequences on restriction fragments separated by gel electrophoresis. The essential feature of this technique is the transfer of the DNA molecules separated by gel electrophoresis to a nitrocellulose or nylon membrane.

The DNA is denatured either prior to or during transfer by placing the gel in an alkaline solution. After transfer is complete, the DNA is immobilized on the membrane by drying or UV-induced cross-linking to the filter. A radioactive DNA (probe) containing the sequence of interest is then hybridized or annealed with the immobilized DNA on the membrane. The probe will anneal to form a double helix only with complementary DNA molecules on the membrane. Nonannealed probe is then washed off the membrane, and the washed membrane is exposed to X-ray film that detects the presence of the radioactivity in the bound probe. After the autoradiogram is developed, the dark bands show the position(s) of DNA sequences that have hybridized with the probe.

The ability to transfer DNA restriction fragments or other DNA molecules that have been separated by gel electrophoresis to nitrocellulose or nylon membranes for hybridization studies and other types of analyses has proven to be extremely useful. Such transfers of DNA to membranes are called Southern blots after E. M.

Southern, which developed the technique.

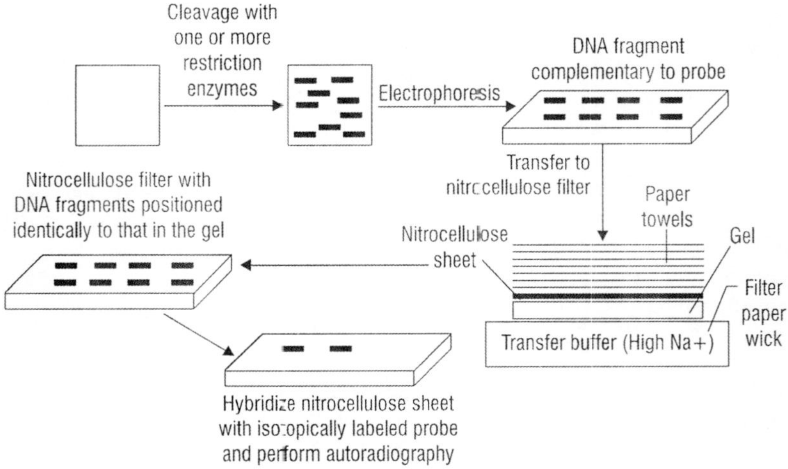


**FIGURE 26.8:** Southern blotting method of analyzing DNA segments that share homology with a nucleic acid probe.

## Analysis of RNAs by Northern Blot Hybridizations

As DNA molecules can be transferred from agarose gels to nitrocellulose or nylon membranes for hybridization studies, RNA molecules can also be separated by agarose gel electrophoresis similarly and transferred for analyses analyzed. Such RNA transfers are used routinely in molecular genetics laboratories and are called *northern blots*, as it is the mirror image of the Southern blotting technique.

The northern blot procedure is essentially identical to that used for Southern blot transfers however, degradation by RNA molecules are more sensitive to degradation by RNases. Thus, gloves must be worn at all times to prevent contamination of solutions with RNases on one's fingers. In addition, all glassware must be baked overnight to inactivate RNases, and all chemical reagents must be baked or be highly purified to ensure that they are RNase-free. RNase molecules are extremely stable enzymes; they are not inactivated by boiling or other treatments that would totally destroy most enzymes. Most RNA molecules contain considerable secondary structure. Therefore, they must be kept denatured during electrophoresis to separate them on the basis of size. Denaturation is accomplished by adding formaldehyde or some other chemical denaturant to the loading buffer and the electrophoresis running buffer. After transfer to an appropriate membrane, the RNA blot is hybridized to either RNA or DNA probes just as with Southern blots. Now an alternative to nitrocellulose, DBM or diazobenzyloxymethyl paper is used. It may be reused several times for different probes, without appreciable loss of nucleic acid signal.

Northern blot hybridizations are extremely useful in studies of gene expression. They can be used to determine whether a particular gene is transcribed in all tissues of an organism or only certain tissues. They can also be used to study the temporal pattern of expression of individual genes during growth and development. Major disadvantage with northern blot hybridizations is that it only measures the accumulation of RNA transcripts. But explains nothing about why the observed accumulation occurs.

## Analysis of Proteins by Western Blot Techniques

Polyacrylamide gel electrophoresis has been used as an important tool for the separation and characterization of proteins.

Since many functional proteins are composed of two or more subunits, individual polypeptides are separated by electrophoresis in the presence of the detergent sodium dodecyl sulfate (SDS), which denatures the proteins.

After electrophoresis, the proteins are commonly detected by treating with Coomassie blue or silver stain. However, the separated polypeptides in the gel can also be transferred to a nitrocellulose membrane, and individual proteins can be detected by using specific antibodies. This transfer of proteins from acrylamide gels to nitrocellulose membranes is called western blotting and is performed by using an electric current-so-called *electroblotting*. After transfer, a specific protein of interest is identified by placing the membrane with the immobilized proteins in a solution containing an antibody to the protein. Nonbound antibodies are then washed off the membrane, and the presence of the initial ("primary") antibody is detected by placing the membrane in a solution containing a "secondary" antibody. This "secondary" antibody reacts with immunoglobulins (the group of protein containing all antibodies) in general. The secondary antibody is conjugated to either a radioactive isotope (permitting autoradiography) or an enzyme that produces a visible product when the proper substrate is added. The western blot procedure is a very powerful tool for identifying and characterizing specific gene-products.

### *Dot Blots and Slot Blots*

Another variety of Southern or Northern blots is dot blots or slot blots. In dot blots, spotting them adjacent to each other on a nitrocellulose membrane tests cloned or purified extract of DNAs. DNA blots thus produced are immobilized and denatured so that they are bound on the membrane as single stranded DNA blots. The membrane is then hybridized with radioactively labelled probe (DNA or RNA). The dot representing sequence related to probe would light up by autoradiography. The intensity of the dot will indicate the relative concentration of a sequence related to the probe in the DNA sample used for a particular dot. No steps involving digestion with enzyme, gel electrophoresis or transfer from gel to membrane is needed in this technique. Dot blot are useful in screening viral genomic RNA, RNA transcripts and especially in samples where restriction sites are not considered.

Sometimes, an apparatus is used for placing spots on the membrane, through slots made in this equipment. The spots made thus are in the form of oblong slots rather than round blots. These slots are used just like dot blots and are described as slot blots.

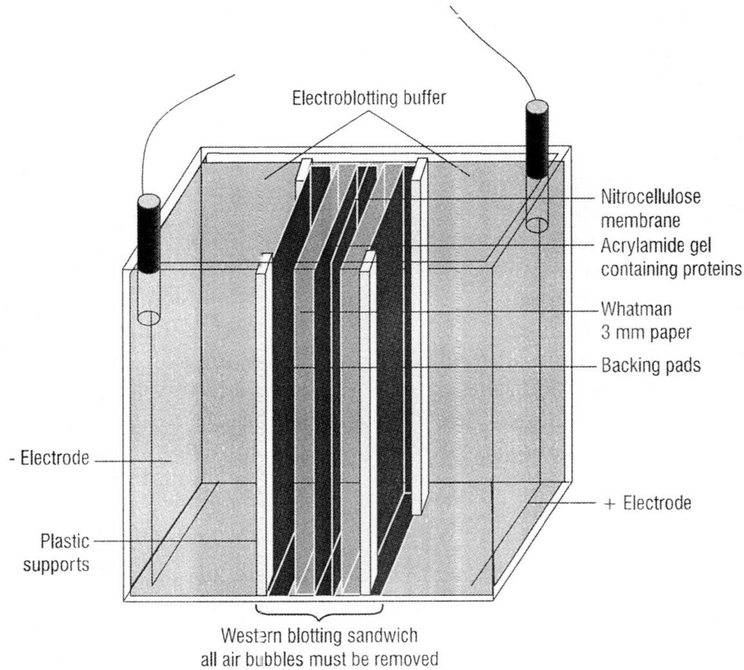


**FIGURE 26.9:** Diagram of a typical Western blotting or electro blotting apparatus. An electric current is used to transfer the proteins from the gel to the nitrocellulose membrane placed next to it in the blotting sandwich. All other components of the sandwich function to provide gentle but firm support; tight contact between the gel and the membrane is essential for good transfer. All bubbles must be carefully removed from the sandwich to assure uniform flow of current and complete transfer of all proteins.

## DETECTION OF RFLPS

A RFLP can be demonstrated using the following steps:

- Extract and purify DNA from several individuals, which may differ in some respects among themselves.
- Digest DNA samples with a restriction endonuclease.
- Subject each DNA digest to gel electrophoresis separately, but on the same gel slab.
- The DNA fragments of different lengths resulting from digestion can be separated by gel electrophoresis, but the differences in the distribution patterns of fragments in several DNA samples due to digestion by several endonucleases can not be detected directly. This is because the number of fragments is large and the range in size is rather continuous, such that they form a continuous smear on the gel. The following steps therefore, follow electrophoresis.
- Fragments are transferred from gel to nitrocellulose filters using the technique of **Southern blotting**.
- The filters with small fragments, fully or partially homologous to the probe, can be detected by autoradiography after hybridization.

With a sufficient number of digests, it is possible to deduce the origin of each band on the gel, that is, which enzyme or combinations of enzymes have generated it. This yields a restriction enzyme cleavage map or restriction map. Obtaining such a map is very often essential first step in analyzing a piece of DNA of unknown structure. Another method is by labelling with probe followed by autoradiography. Since DNA clones having repetitive DNA may hybridize with many fragments, they will give a smear on autoradiography. Therefore, unique DNA sequences are generally used as probes for detecting RFLPs. The DNA fragments hybridizing with the probe, can be visualized by autoradiography, and will be called RFLPs at the phenotypic level. RFLPs can be studied in a set of related species, using a random or a specific DNA probe. The similarities and differences be used to infer phylogenetic relationships. This has been done in a number of cases, both in plants and in animals. RFLPs have also been used to prepare chromosome mops in humans mice, fruitfly, maize, tomato, lettuce and rice.

## DNA SEQUENCING

Restriction enzyme mapping and Southern blotting help establish the gross organization of a piece of DNA, and allow the approximate position of sequences of interest to be determined. The next step in analysis is normally to determine the nucleotide sequence.

### Methods for Analyzing DNA Sequence

Our present ability to sequence essentially any DNA molecule is the result of four major developments. (1) The most important breakthrough was the discovery of restriction enzymes and their use in preparing homogeneous samples of specific segments of chromosomes. (2) A second major advance was the improvement of gel electrophoresis procedures to the point where DNA fragments that different in length by a single nucleotide could be resolved. (3) The development of gene cloning techniques greatly facilitated the preparation of large quantities of a particular gene or DNA sequence of interest for use in sequencing studies. (4) Finally, two different approaches to the rapid sequencing of DNA molecules (i.e. by chemical degradation and by chain termination) were developed in the period from 1974 to 1977.

### Sequencing by Chemical Degradation

Determination of DNA sequence depends upon the resolution obtained from polyacylamide gels, which are run in the presence of denaturant e.g. urea. If two single-stranded DNA molecules, identical in sequence over most of their length but differing in just one base on one of the two molecules, are subjected to electrophoresis they will be separated. In the chemical degradation procedure a DNA restriction fragment is isotopically labelled with P at either its 5' or its 3' termini and each of the two strands is separated and isolated. The DNA is then partially modified with chemical reagents specific for the different bases, and cleaved at the modified nucleotides. This generates a set of molecules differing in length but with the same isotopically labelled terminus. These fragments are subjected to denaturing gel electrophoresis in parallel slots of high-resolution gel, and the sequence of the DNA can be deduced from the resulting autoradiogram.

## Sequencing by Chain Termination

In practice the chemical degradation procedure has been replaced by chain termination method. In this procedure DNA is synthesized rather than degraded, but the principle is similar to the chemical method. DNA polymerases function by adding a primer, a short synthetic oligonucleotide, to pre-existent DNA strand, to provide a fixed 5' terminus for the growing chain.

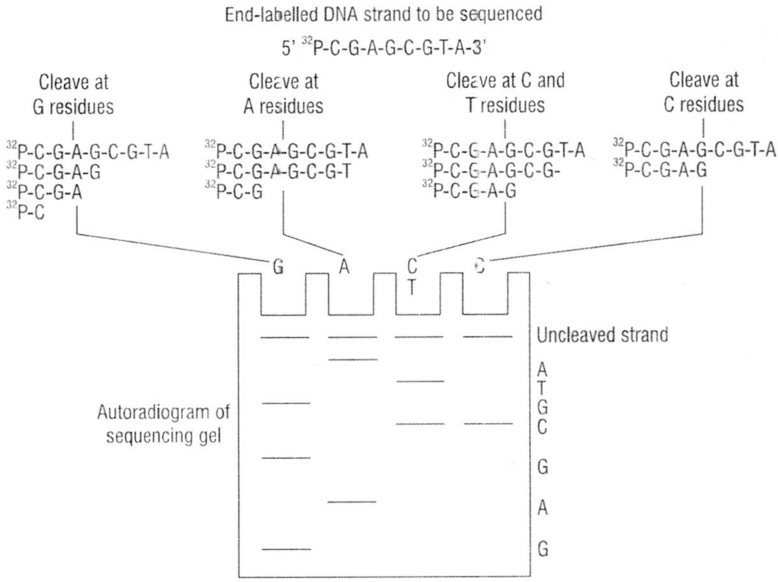


**FIGURE 26.10:** DNA sequence analysis by the chemical degradation (Maxam and Gilbert) technique.

The reaction is performed in the presence of chain terminators. These nucleotide derivatives act as blocking group at their 3' hydroxyl terming and once there are incorporated into the growing chain, they prevent further addition of nucleotides. For example, if dideoxy GTP (**FIGURE 26.12**), is mixed with unmodified GTP and the other three nucleoside triphosphates, it terminates at a subset of G residues in the newly synthesized strand (**FIGURE 26.12**). Four parallel reactions are run; with dideoxy radiolabelled nucleoside triphosphates of all four kinds are included. The nucleotide sequence of the template strand can be deduced by comparison of the pattern of autoradiographic bands in the four lanes of the gel.

A suitable template for single-strand DNA sequence analysis can be created by inserting the double-stranded template DNA into the genome of a bacterial virus, which naturally packages only one of the two strands into viral particles. Another method is denaturing of double-stranded DNA template by treatment with alkali. An excess of primer is added before re-annealing occurs, so that primer will compete with the complementary strand for annealing to the template. This is a preferable method over insertion of DNA into viral vector.

DNA sequence analysis using chain terminators is best performed using a single-stranded template. This allows the primer to be free accessible to the template.

The major disadvantage of double-stranded DNA sequence analysis is the presence of competing complementary strand. This leads to a higher level of sequencing artifacts like presence of more than one band. In such situation it may be impossible to decide which is the correct residue at that position. Artefacts of this kind are not restricted to double-strand DNA sequence analysis; some level of artefacts is a universal problem in sequence analysis.

The DNA sample then re-sequenced using another primer, or by using the other strand (i.e. the strand which is complementary to that used in the original analysis) as template.

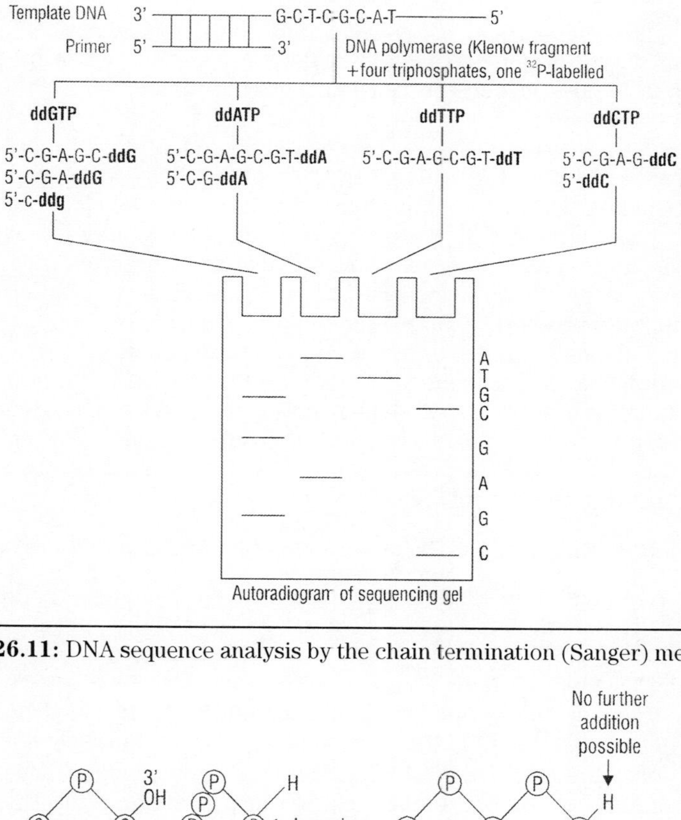


**FIGURE 26.11:** DNA sequence analysis by the chain termination (Sanger) method.

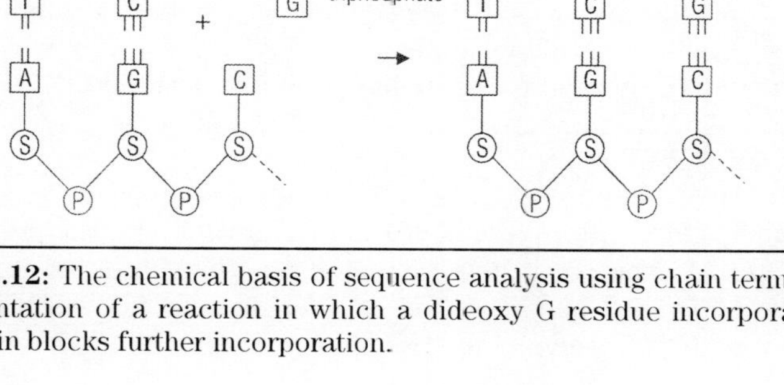


**FIGURE 26.12:** The chemical basis of sequence analysis using chain terminators. This is a representation of a reaction in which a dideoxy G residue incorporated into the growing chain blocks further incorporation.

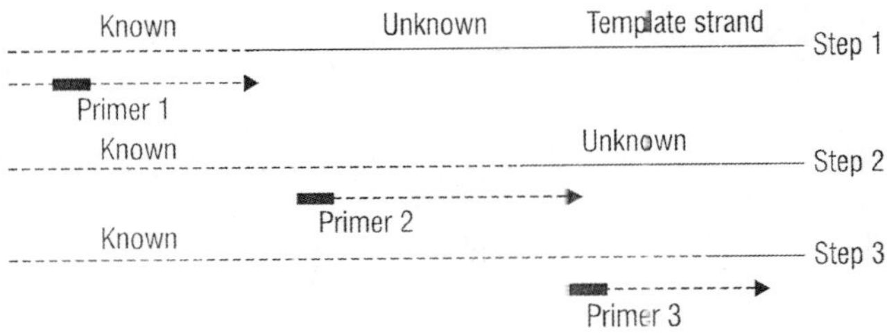


**FIGURE 26.13:** DNA sequence analysis by *primer walking.*

Even when a single-stranded DNA molecules with more than 500 nucleotides, clearly it is not possible on a polyacrylamide gel to resolve when the molecules are larger than about the nucleotides. This sets the upper limit to the length of sequence to be determined. One way to extend from a known into an unknown region of sequence is to synthesize an oligonucleotide primer, which is complementary to a tract near the end of the region whose sequence has already been established. In this way it is possible to 'walk' along a sequence, at each step synthesizing a new primer, which is used to extend the region of known sequence (Figure 5.14). Alternatively, it is possible to delete sequences enzymatically from the template at a position downstream of the annealing site of the primer, to generate a deletion series. The nucleotide sequence of a set of such molecules can be used to deduce the sequence of the intact template.

By determining the sequence of both strands of a DNA molecule, and confirming that the two deduced sequences are perfectly complementary, it is possible to detect and eliminate artefacts generated in the reactions or during electrophoresis. When this check is performed, DNA sequence can be established with a very high degree of accuracy.

### *Direct DNA Sequencing using PCR (also called Ligation Mediated PCR=LMPCR)*

Polymerase chain reaction (PCR) has also been used for sequencing the amplified DNA product. This method of DNA sequencing is faster and more reliable and can utilize either the whole genomic DNA or cloned fragments for sequencing a particular DNA segment. The DNA sequencing using PCR involves two steps: (i) generation of sequencing templates (double stranded or single stranded) using PCR and (ii) sequencing of PCR products eight with the thermolabile DNA polymerase or with the thermostable Taq DNA polymerase. Thus the DNA sequencing method using PCR eliminates the need of cloning the DNA in single stranded DNA phage vector i.e. M13. Although double stranded DNA product of PCR can be utilized for sequencing, this may cause difficulty due to reassociation after denaturation thus preventing the sequencing primer from annealing to its complementary sequence to allow extension. To reduce this problem, either a variant of the standard method for sequencing double stranded DNA is employed or single stranded DNA templates are produced following **asymmetric PCR** earlier discussed in Chapter 4.

A number of thermolabile DNA polymerases have been used for sequencing of *in vitro* amplified DNA. Alternatively, **Taq polymerase** (thermostable enzyme) used for PCR can also be used for sequencing reaction. In either case, Sanger's synthetic method involving incorporation of dideozynucleotide (ddNTP) for chain termination is used for sequencing. However Maxam and Gilbert's chemical degradation method can also be used.

In Sanger's method, as usual, four mixtures are prepared, each using one of the four ddNTP. The sequencing primer is labelled with $^{32}P$ and the mixtures with amplified DNA, Taq polymerase and appropriate buffer are incubated at 70°C for 5 min. The reaction is stopped by addition of formamide stop solution in each tube and mixtures are run on polyacrylamide sequencing gel to obtain the ladders, which can be read by computer or manually. This method using PCR helped in automation of DNA sequencing.

### Automated DNA Sequence Analysis

Using the chain termination method, DNA sequencing can be mechanized to make the process rapid. In a conventional sequence reaction described above four separate single-strand extension reactions are performed, with each containing a different chain-terminating dideoxynucleoside triphosphate. The radiolabelled reaction products are separated on four separate tracks of a polyacrylamide gel. In machine-based sequencing a single reaction is run, with each of the four-dideoxynucleoside triphosphates containing a different fluorescent label.

When excited by an argon laser, the fluorescent labels emit light at a different wavelength. The same chain-termination reactions are carried out; however, it is no longer necessary to run the products in separate lanes of a gel. The reaction products are mixed and run in a single lane in a gel and as the different length fragments pass the laser each fluorescent label gives a unique signal, which is stored in the computer.

Machine-based DNA sequencing is beginning to change the speed and scope of sequencing projects dramatically. The first complete genomic sequences were those of simple viruses such as the SV40 tumor virus, containing around 500 based of DNA. Recently, the DNA sequence of over 106 000 bases of DNA from human chromosome 19 was determined in a few months using DNA sequencing machines. The sequence of a significant fraction of the *E.coli* genome has been established, the entire sequence of yeast chromosome 3 is known, and there are plans eventually to determine the sequence of the entire human genome.

## MECHANICAL SHEARING OF DNA

In addition to digestion DNA with restriction endonucleases to produce discrete fragments, there are a variety of treatments, which result in non-specific breakage. Non-specific endonucleases and chemical degradation can be used but the only method that has been much applied to gene manipulation involves mechanical shearing.

Shear forces in solution cannot easily break the long, thin threads, which constitute duplex DNA. Intense sonication with ultrasound can reduce the length to about 300 nucleotide pairs. More controlled shearing can be achieved by high-speed stirring in a blender. Breakage occurs essentially at random with respect to DNA sequence. The termini consist of short, single-stranded regions, which may have to be taken into account in subsequent joining procedures.

## TRANSCRIPT MAPPING TECHNIQUES

Analyzing the nucleotide sequence of a piece of DNA is a necessary preliminary to understanding its precise organization and function, but is not normally sufficient for these purposes. Because eukaryotic gene transcripts are subjected to a complex series of post-transcriptional processing events, the structure of a gene can be understood only by correlating the organization of the genomic DNA with the organization of its mRNA transcript. This has required the establishment of sophisticated cloning and mapping techniques, which can, when used in conjunction, give a complete description of the structure of a gene.

It is possible to isolate, by gene cloning, the double-stranded DNA copies of individual mRNA sequences. These cDNA clones will normally contain a DNA copy of the poly (A) sequence at the 3' terminus. Hence, by determining which strand of the cDNA clone contains a poly (A) tract at one end, it is possible to identify the coding strand of the cDNA clone. Provided a sufficient length of sequence has been established to make their presence statistically likely (in practice 300-500 nucleotides), termination codons will occur in two of the three potential coding frames, ruling them out as translational open reading frames (ORFs).

Sometimes this level of analysis, that is, identification of the reading frame and deduction of the encoded protein sequence, will be all that is required. However, it is often desirable to determine the structure of the gene itself. It will have been isolated as part of a genomic clone. This will contain the gene and a variable amount of flanking DNA depending upon the method of cloning used.

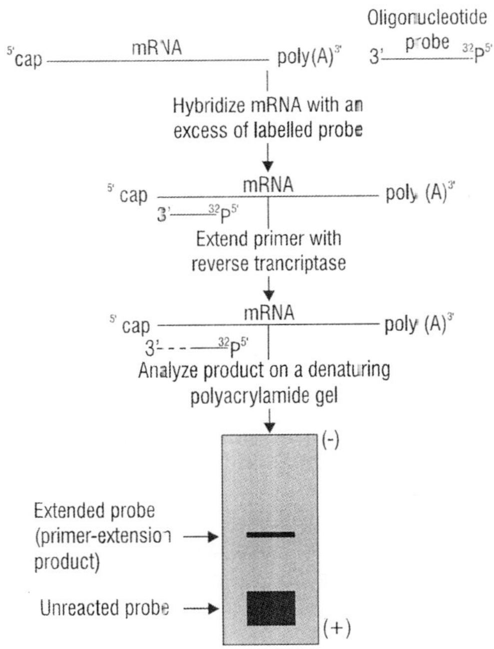


**FIGURE 26.14:** Identification of the 5' end of an mRNA by primer extension.

By correlating the sequence of the gene with the sequence of the mRNA deduced from the cDNA clone, it is possible to identify the site of poly (A) addition and the positions of introns. However, technical limitations in the cDNA cloning procedure ensure that the cDNA clone will not normally contain sequences derived from the extreme 5' terminus of the mRNA. There are two potential methods to determine the position of the 5' end, primer extension and S1 mapping.

## Primer Extension

In primer extension, a short oligonucleotide complementary to a sequence tract near to the 5' end of the mRNA is annealed to the mRNA. This is then used to prime synthesis with the enzyme reverse transcriptase (**FIGURE 26.14**). This enzyme produces a DNA copy of the mRNA (a cDNA), which extends to the 5' terminus of the mRNA. By correlating the length of the cDNA primer extension product with the sequence of the genomic DNA, it is possible to deduce the start site of transcription.

## S1 Mapping

The alternative technique is called S1 mapping (**FIGURE 26.15**). An isotopically labelled probe is generated from the genomic clone. This probe is complementary to the mRNA over a region encompassing part or all of the first exon, and extending upstream of the region where the cap site is expected to be located. The cap site may have already been putatively localized using primer extension.

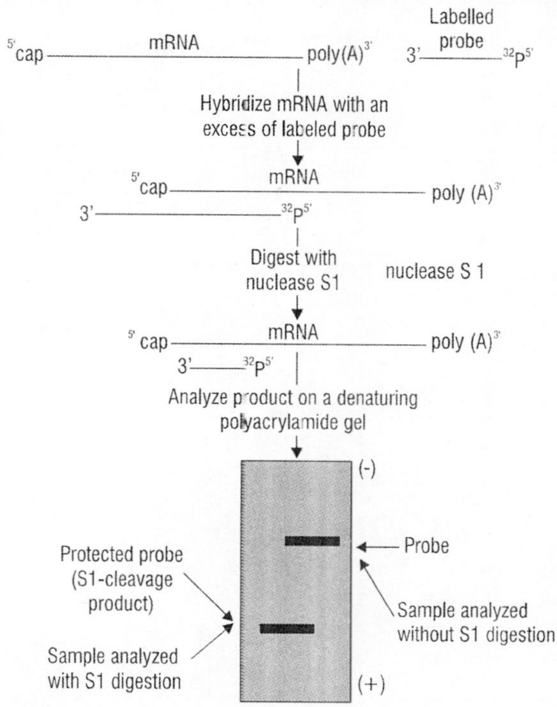


**FIGURE 26.15:** Identification of the 5' end an mRNA by S1 mapping.

The probe is annealed to the mRNA and digested with an enzyme, which destroys single-stranded nucleic acids but leaves double-stranded molecules intact. In early experiments an enzyme called nuclease S1 was generally used, and the name S1 mapping has been retained, even though other nucleases are now sometimes used. During digestion, the 3' proximal region of the probe, which has no complement in the mRNA molecule and which therefore remains single-stranded, is degraded. A molecule is generated which is the complement of the region between the cap site and the position at which the 5' terminus of the probe annealed to the mRNA. Again, by determining the length of this molecule it is possible to infer the position of the cap site.

These two techniques are often used in parallel to determine the cap site. They are both subject to artefacts, but of different kinds. Therefore if both methods agree, it greatly increases confidence in the assignment of the cap site. If the hybridization is performed with an excess of probe, the amount of primer extension or nuclease-resistant product is proportional to the amount of primer extension or nuclease-resistant product is proportional to the amount of complementary mRNA. Both techniques can therefore be used an alternative to Northern transfer to quantitated mRNA levels.

## SYNTHESIS OF cDNA FROM mRNA

Complementary DNA can be prepared by isolating mRNAs from tissues which are actively synthesizing proteins, like roots and leaves in plants, ovaries or recticulocytes in mammals etc.

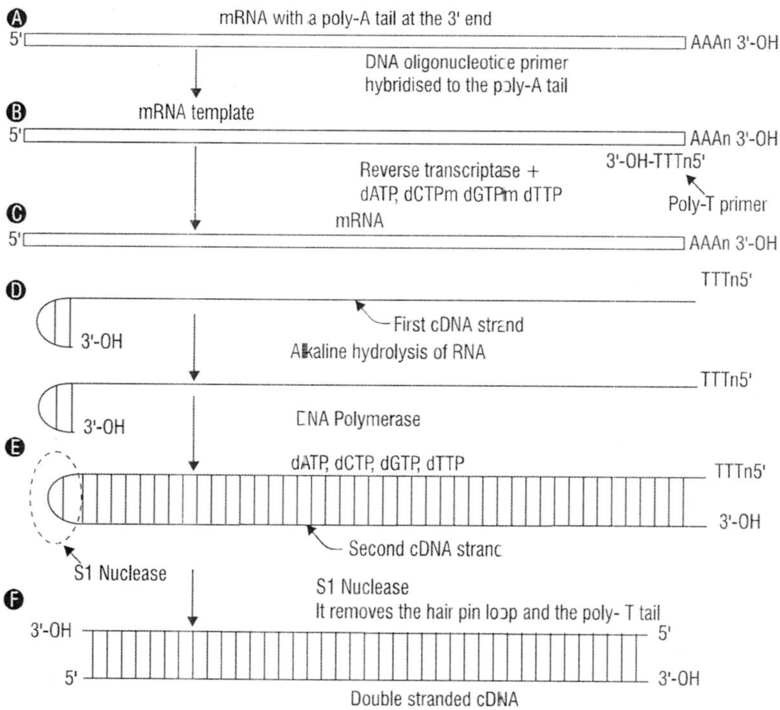


**FIGURE 26.16:** Synthesis of double stranded cDNA from an mRNA.

Isolated and purified mRNA can then be directly used to synthesize complementary DNA strand using the enzyme reverse transcriptase. This results in a DNA/RNA hybrid. The RNA part of the hybrid is then removed with alkali. Thus a single-stranded DNA is generated which can act as the template for the synthesis of its complementary strand in the presence of DNA polymerase. The double stranded DNA molecule thus formed is called complementary DNA (cDNA), as it is synthesized complementary to the mRNA. In the prokaryotes the double stranded cDNA will be identical to the gene (DNA) from which the mRNA is formed. But, most of the eukaryotic genes contain intervening sequences (introns), which interrupt the coding sequences. Hence, the cDNA of the eukaryotic cells and their native DNA are not collinear.

As bacterial genes do not have the introns - exon organization found in the eukaryotic genes they find the expression of eukaryotic genes in them (bacterial) difficult. Therefore one has to depend upon cDNA of eukaryotes, which does not have introns, for cloning.

## Preparation of Double-Stranded cDNA

The mRNA that is specific for a particular kind of protein is isolated and purified. All naturally occurring mRNAs of eukaryotes have about 200 adenylate residues at their 3'-termini. This is called polyadenylate tail. The enzyme reverse transcriptase can mediate the synthesis of a new DNA strand only when it has a growing point in the form of a pre-existing primer, which is base paired with the mRNA template and has a free 3' OH group. Such a primer is provided by hybridizing a short complementary oligo (dT) molecule with the poly adenylate (poly-A) sequence.

The 3'OH end of the poly-T primer is made double stranded by reverse transcriptase forming RNA-DNA hybrid. The RNA strand of this RNA - cDNA hybrid is digested by alkaline hydrolysis (DNA is resistant to alkali). At the 3'-OH end of the newly formed single stranded cDNA, a hairpin is formed which is stabilized by base - pairing. This hairpin, which has a 3' -OH free end, acts as a self-priming structure. Forming second strand of the cDNA in the presence of the fragment of the *E.coli* DNA polymerase-1 (Kornberg enzyme) and the four deoxynucleotidyl triphosphates.

The enzyme reverse transcriptase or T4DNA polymerase can also be used for the synthesis of the second strand of cDNA. The synthesized double stranded DNA is treated with the single-strand specific nuclease S1. This enzyme digests the curved region of the hairpin and any polynucleotide tails. Thus, finally a double stranded cDNA molecule is formed. This DNA can now be inserted into a suitable vector and can be further cloned in the desired host.

Complementary DNA of a given mRNA is inserted into a suitable vector can then be cloned in a host. The cDNA is blunt ended. It has no cohesive ends that are complementary to the cohesive ends of any of the vectors. This is overcome by homopolymer tailing. Once inside the host cell, the cDNA in the hybrid will express the gene(s) it carries.

## cDNA Library

Complementary DNAs (cDNAs) of the genome of an organism can be synthesized and cloned to generate a cDNA library. Thus, cDNA library is a collection of cDNA clones of an organism. The starting material for the synthesis of cDNA is the mRNA. Hence the resulting cDNAs will represent the mRNAs present in the cells from where these mRNAs were isolated. In the case of multicellular organisms there is specialization of individual cells. For example, a mammal is made up of a large number of cell types: brain cells, mammary cells, blood cells, liver cells, cells of the pancreas, nerve cells etc. Each type of cell will contain the same complement of genes. However, only a particular set of genes will be expressed in a given type of cell. Other genes will remain silent. Only those genes that are expressed are transcribed into mRNA. Hence the cDNA clones of this type of cell will be only a selection of the total number of genes in a cell. Hence for such organisms cDNA clones of different cell are prepared.

## SEARCHING FOR GENE USING COMPUTERS

Although the series of analyses described above are necessary ultimately to establish the organization and mode of expression of a particular gene, less laborious means using computer programs can seek genes. These search for regions of a DNA sequence which look like coding regions flanked by the signals that direct gene transcription and processing of their RNA products. This procedure is of greatest value when large regions of genomic DNA for which cDNA clones are not available are being analyzed. It is thus an essential part of the Human Genome Project.

One of the principles, that of searching for long ORFs (= open reading frames), has already been described. Although the sequence (the Kozak sequence) that precedes the ATG initiation codon is only weakly conserved between different genes it can provide a helpful guides as to the start site of translation. The start site of transcription often lies downstream of a TATA box, and it is also possible to search for the consensus binding sites of other transcription factors.

RNA processing signals can also be used to help establish the likely gene organization. The sequence that signals polyadenylation is well conserved, and splice junctions have a consensus sequence comprising an almost perfectly conserved core (5' GU---AG 3'), with moderately well conserved flanking sequences.

Identification genes by analysis of DNA sequences are an evolving science. The structures of new genes are constantly adding and the sophistication of the search programs is increasing. The databanks exists which contain many thousands of gene sequences e.g. Gene bank and the EMBL. Although some scientists initially submit their gene sequence data to one rather than other, the two organizations that compile the databases then exchange data.

# 27. Gene Cloning Vector/Plasmid

Genetic engineering can be defined as transfer of DNA between hosts (or species) by *in vitro* enzymatic manipulations. This implies that the DNA to be transferred will be duplicated in the new host. Since most DNA fragments are incapable of self-replication in *E.coli* or any other host cell, an additional segment of DNA, capable of autonomous replication, must be linked to the fragment to be cloned. This autonomously replicating fragment is the **molecular cloning vector** and, by definition, plays a central role in recombinant DNA technology. Most cloning vectors were originally derived from naturally occurring extrachromosomal elements such as bacteriophage and plasmids.

Stanley Cohen and co-workers first reported the use of bacterial plasmids as molecular cloning vectors. Since that initial report thousands of cloning vectors have been constructed, and their versatility in terms of cloning sites, host-range, and function appears to be limited only by the ingenuity and imagination of the inventor. Whereas insertion of foreign DNA fragments into pSC101 was limited to a single restriction enzyme cleavage site and to *E.coli* as a host, most vectors currently in use carry multiple restriction-enzyme cleavage sites and have been modified for use with most common laboratory microorganisms. Many vectors now include dual origins of replication that enable them to be shuttled between hosts as dissimilar as *E. coli* and human cells.

Specialized vectors are now available that enable the investigator to; (1) identify and isolate regulatory DNA sequences such as promoters and terminators; (2) identify open translational reading frames; (3) overproduce useful RNAs and proteins; and (4) determine the nucleotide sequence of genes and segments of DNA. Primarily, a cloning vector is necessary as carriers of the cloned gene. Without a vector, a DNA molecule introduced into a cell would be diluted out by cell division and eventually lost.

There are a number of features common to all cloning vectors as:

- The vector must be able to replicate in the host cell. Thus, it must have a replication origin. Replication origin is a sequence element, which is recognized by the host cell's replication machinery, and hence the vector can be amplified in the host cell.
- Vector must contain a selectable marker for identification and isolation of subpopulation of bacteria containing vector. Actually, efficiency of transformation is very low and only or minority of cells in the population takes up and retains the exogenous DNA. Marker in vector helps in identifying transformed bacterial colony.
- Most commonly used vectors contain a short DNA sequence with many closely placed restriction enzyme cleavage sites. It is known as multicloning site (MCS) or polylinkers. These MCS can also be chemically synthesized in a vector and are very helpful during foreign gene insertion. Any two sites within an MCS can be simultaneously cleaved with the help of range of restriction enzymes and that too without excising the vector sequences.

This chapter deals with major types of cloning vectors.

## PLASMIDS

A *plasmid* is a *replicon* (unit of genetic material capable of independent replication) that is *stably inherited* (maintained without specific selection) *in an extrachromosomal state.* Most of the plasmids are not required for the survival of in, which they reside. In many cases, however, they are essential under certain environment, such as in the presence of an antibiotic such as chloramphenicol, ampicillin, tetracycline, kannamycin etc. In *Rhizobium leguminosorum*, the genes for nitrogen fixation and nodule formations are located in plasmid.

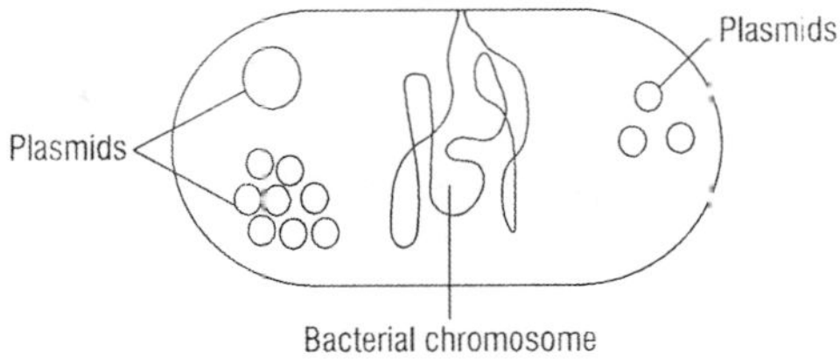


**FIGURE 27.1:** Bacterium showing chromosome and plasmids.

### Replication of Plasmids

The independent replication of plasmids is due to the presence of certain sequences acting as the origin of replication. The smaller plasmids use the DNA replicative enzymes of the host cells. Larger plasmids carry genes that code for special enzymes necessary for their replication. Under certain conditions, some plasmids may integrate into the bacterial chromosome. They are called episomes or integrative plasmids. At this stage they replicate along with the bacterial chromosome.

### Size of Plasmids

The size of the plasmid varies from less than 1.0 kb to more than 200 kb. Smaller plasmids are much desirable for gene cloning experiments. Larger plasmids are less in number whereas smaller ones are more in number.

**TABLE 27.1:** Various sizes of some plasmids

| Plasmids | Size (Nucleotide length in kb) | Organism |
|---|---|---|
| pBR345 | 0.7 | *E.coli* |
| pBR 322 | 4.363 | *E.coli* |
| Col E1 | 6.36 | *E.coli* |
| F | 95 | *E.coli* |
| TOL | 117 | *Pseudomonas putida* |
| pTiAch 5 | 213 | *Agrobacterium tumefaciens* |

### Copy Number

The number of molecules of a plasmid found in a single bacterial cell is termed as copy number. It ranges from 1 to more than 50 per cell but this number is specific for a given plasmid residing in bacterial cell.

## Amplification of the Plasmid

The copy number of a plasmid can be increased manifold. This process is called **amplification** of the plasmid DNA. This property is exploited to increase the plasmid copy number. When the bacterial culture is at its exponential phase chloramphenicol is added to the medium. This arrest chromosomal DNA replication (and cell division). The culture is then incubated for another 12 hrs for the plasmid molecules to replicate. This increases the number of plasmids per cell. Sometimes the plasmid copy numbers may reach to several thousand.

## Types of Plasmids

Plasmids can be categorized on the basis of number of copies per cell as,

- Relaxed plasmids, which are normally maintained at multiple copies per cell, and
- Stringent plasmids, which have a limited number of copies per cell.

Plasmids with larger copy number are more useful for gene cloning experiments.

Plasmids can also be classified as conjugative and non-conjugative plasmids, depending on whether or not they carry a set of transfer gene called the ***tra*** genes. These ***tra*** genes promote bacterial conjugation. Generally conjugative plasmids are of high molecular weight and are present as 1-3 copies per cell, whereas non-conjugative plasmids have low molecular weight and are present in multiple copies, i.e., 20-25 copies per cell.

## Isolation of Plasmid DNA

Plasmids, which are very common in bacteria, are not so common in eukaryotic organisms. The best-known eukaryotic plasmid is the 2μ circles that occur in many strains of the Yeast, *Saccharomyces cerevisiae*.

Gene cloning and related experiment require pure plasmid DNA. Various methods have been used for the purification of plasmids. Usually these methods involve following main steps:

- Growth of bacterial and plasmid amplification.
- Breaking the bacterial cells to release their contents.
- Treatment of bacterial cell extracts to remove all components except the DNA.
- Separation of Plasmid DNA from the chromosomal DNA.

The bacterial strain containing required plasmid is grown usually on LB medium (Luria Bertani medium). LB medium contains trypton (10 g/l), yeast extract (5 g/l) and sodium chloride (10 g/l) and pH of the medium is adjusted to 7.5. Late long cultures are then transferred to LB medium containing appropriate antibiotic with small amount of chloramphenicol for 12-16 hours incubation.

Isolation of DNA from such bacteria culture can be done by very gentle cell lysis, using lysozyme and then detergent, followed by clearing the lysate by centrifugation. Centrifugation sediments the high molecular weight DNA (predominantly chromosomal) and cell debris, leaving the small plasmid molecules and RNA in supernatant. Undamaged plasmid is particularly compact, since it is super coiled as a consequence of having slightly too few turns of double helix per unit length. Further purification of the plasmid can be achieved by

**cesium chloride density gradient ultracentrifugation** of the nucleic acid preparation in presences of ethidium bromide. Ethidium bromide causes unwinding of DNA as it binds to it, simultaneously producing a decrease in its buoyant density. Since supercoiled plasmid DNA can unwind to only a very limited extent, it will not bind the dye and have a higher density than other types of DNA (latter binds to dye). Because of this density difference, plasmid DNA can be separated from other DNA by isopycnic ultracentrifuge

## Criteria for Plasmid Cloning

The following criteria are used to design an ideal cloning vector:

- A plasmid vector should be small and should contribute as little as possible to the overall size of the recombinant molecule. Low molecular weight (small) plasmids are easy to handle, present in multiple copies and do not get damaged during isolation.
- It should be well characterized with respect to gene number, location and should contain single cleavage sites for a large number of restriction endonucleases located in one or the other genetic marker. During insertion of a foreign DNA into plasmid vector, latter should be linearized (thus single cleavage site should be present). Presence of more than one site will break the plasmid into two or more fragment, which is avoided.
- The plasmid should have the ability to confer readily selectable markers (phenotypic trait) on the host cells that can be used to distinguish the transformed cells from the non-transformed ones.
- The plasmids should be easily propagated in the desired host so that the number of recombinant DNA molecule can be amplified.
- It should have a large number of copies.
- A plasmid vector should have an additional genetic marker that can be inactivated by insertion of the foreign DNA, so that the inactivated gene may help in distinguishing the cells harbouring recombinant molecules on the basis of a readily altered phenotype.

These properties provide maximum flexibility in terms of cloning different types of restriction fragments. In addition to these properties one should also keep in mind following characters when designing future molecular cloning vectors.

- The vector should use, whenever possible, dominant genetic markers that can be expressed in a wide range of hosts. $C_m^R$ (dominant genetic marker) do not restrict the vector to a particular genetic background. Convenient and quantitative assays for all vector markers should also be available.
- Partitioning sequences such as *par* or *cer* should also be incorporated into the vector to ensure efficient segregation of plasmids to the daughter cells.
- Incorporating a mechanism for controlling vector copy number will greatly increase the versatility of the vector. Plasmid vectors in which the RNAI (inhibitor) transcript is placed under the control of an inducible promoter are available. With these vectors, plasmid copy number can be lowered when RNAI is induced.
- Since a number of recombinant DNA techniques require single strands of DNA, a vector with single-strand synthesis capability is desirable. This can be most easily achieved by incorporating the fd, f1, or M13 origin of replication

along with the vector origin.

- Because of its simplicity, a positive (direct) selection scheme for recombinant transformants is more desirable than the commonly used insertional inactivation method. The latter process usually requires the screening of large numbers of clones to identify a few recombinant molecules.

## CLONING VECTORS BASED ON BACTERIAL PLASMIDS

Bacteria are the common source of plasmids. For the sole purpose of cloning experiments, new plasmids with specific features are constructed *in vitro*. Such artificial plasmids are superior and efficient. The most popular plasmid vector thus obtained is pBR 322.

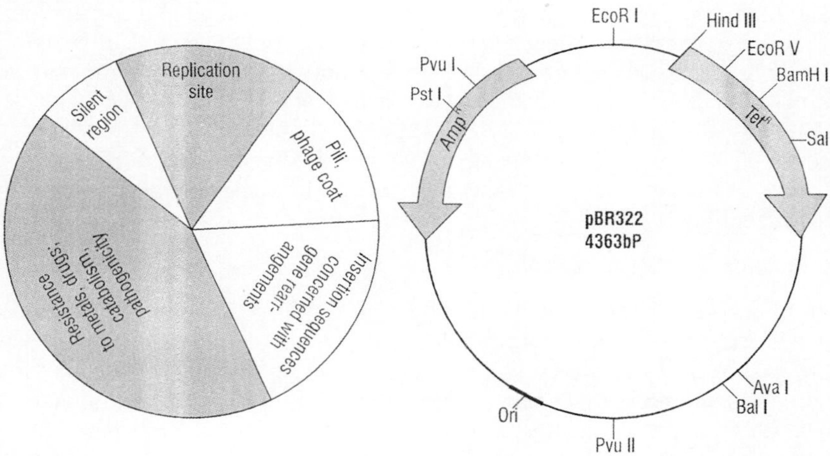


**FIGURE 27.2:** The plasmid pBR322 amp$^R$ and tet$^R$ are genes for resistance to ampicillin and tetracycline, respectively; ori is the origin of DNA replication. Some unique sites of cleavage by a selection of restriction endonucleases are indicated (e.g. *EcoR* I, *Cla* I).

The plasmid pBR322 was constructed by using three different naturally occurring plasmids. The ampicillin resistance gene was derived from RSF 2124. The tetracycline resistance gene was taken from pSC 101. The origin of replication was obtained from pMB1, which is related to the plasmid ColE.

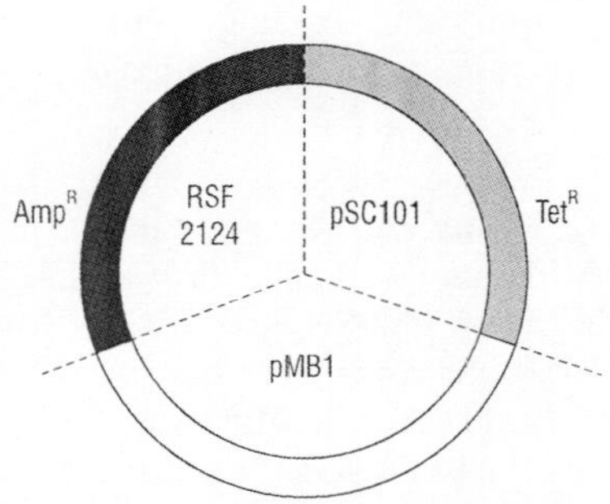


**FIGURE 27.3:** Diagram showing the origin of the plasmid pBR 322.

**COL E1 Plasmid DNA:** Bacterial strain containing Col E, plasmid produce Colicin E1, a protein that inhibits the growth of other *E.coli* strains. Col E1 plasmid also imparts immunity to low level of colicin E1 for strains carrying the plasmid. Challenging stains with Colicin E1 can test colicin immunity:

**Col E 1 Amp Plasmid DNA:** This strain carries col E1 Amp and produces colicin. It has colicin immunity and is ampicillin resistant.

**pBR 325 Plasmid DNA:** This is a derivative of pBR 322 which confers resistance to ampicillin, tetracycline and chloramphenicol.

**pMB9 Plasmid DNA:** Confers tetra cycline resistance and colicin immunity.

**pTZ Plasmids:** A versatile cloning vector containing F1 and pBR322 origins of replication allowing replication of the episomal DNA in ds or ss form. For ss replication, helper phage M13 K07 is required. pTZ also feature the multiple cloning sites from pUC 18 or pUC 19 to facilitate cloning. The ß-galactosidase gene provides easy identification of recombinant colonies by color detection when employing X-Gal (5-bromo, 4-chloro- 3-indoyl -ßD-galactopyranosidase) in the growth medium. The T7 promoter, located within ß-galactosidase gene, serves as a template for *in vivo* or *in vitro* synthesis of specific RNA transcripts from cloned DNA. The *in vitro* synthesized RNA may be used as hybridization probes. Ampicillin resistance may be utilized for additional selection for plasmid containing colonies.

## Bacteriophage Vectors for *E. coli*

The cloning of single genes is usually best carried out using plasmids, since the insert will rarely be larger than about 2 kb. But, for cloning of larger pieces of DNA (e.g. during gene library construction), these plasmids are not suitable as larger inserts increase the plasmid size, making the transformation inefficient. Large molecules can be injected in host bacterial cell by viral particles (bacteriophages). Commonly used bacteriophages are M13, f1, fd and λ phage.

**Lambda replacement vectors:** Lambda replacement vectors contain a restriction site for phage propagation in suitable bacterial host. Remaining part of lambda genome is removed and is replaced by foreign DNA. Ligation is performed at a ratio of arms to target that favours the formation of very long concatemers, in which vector and target molecules are interspersed.

**Lambda Insertion Vectors:** When cloning into an insertion vector, the phage DNA is cleaved with a restriction enzyme that cuts it only once, and the target is inserted into this site. No phage DNA is removed therefore; much lower size of DNA can be inserted. In commonly used vector λ gt 10 the E col RII cloning site is in a gene which is deleterious to phage replication in certain host strains. This allows selection against non-recombinant phage.

## M13 Bacteriophage

M13 is a filamentous bacteriophage of *E.coli*. It is 870 nm long and 6 nm wide. It has a protein coat (capsid) made up of three kinds of capsomeres. These filamentous single-stranded DNA phages infect cells by adsorbing to and entering through F pili; thus, these phages only infect $F^+$ or Hfr cells, not $F^-$ cells. Lastly, these phages do not lyse the host cells (releasing progeny phage plus bacterial DNA from the ruptured cells) like phage λ during the lytic cycle. Instead, the progeny viruses are extruded through the layers of the cell membrane and cell wall without major interference with cell growth. Infected cells will continue to grow and extrude thousands of progeny virus particles, each containing a single-stranded genome to the medium. Because the virus particles are very small relative to the size of the host bacteria, the host cells can be removed by low-speed centrifugation. The virus particles can be collected from the supernatant suspension by high-speed centrifugation, and their single-stranded DNA molecules can be isolated by simple phenol-chloroform extractions. The same DNA strand of the virus is always packaged; it is called the "+" strand (its complement is the – strand). The packaged + strand has the same *sense* as the mRNA; its nucleotide triplets correspond to the mRNA codons, but with T in place of U.

### Cosmids

Cosmids are hybrid DNA molecules and can live a dual life. Their plasmid part enables them to replicate as it has origin of replication. Plasmid part also helps in selection due to the presence of marker gene. Cleavage site is located in the marker gene. Their lambda part (*cos* sequences) allows them to be packaged in a phage coat and to be transduced to a recipient by the lambda infection machinery. It has no genes for viral proteins; therefore, viral particles are not formed in the host. Host cell lysis is also absent. The vector cosmids ligate passenger DNA to increase the size of packaging and thus becoming stably packaged. This requirement is the basis for the size selectivity used for cloning.

Two cosmids are described below:

**Cosmid pHC79:** It is especially suitable for the cloning of large DNA fragments (upto 40 kb) in connection with an *in vitro* packaging system. It is 6.5 kb in size and contains part of the pBR 322 sequence and a fragment of L-phage DNA containing the cos region. The cos region is required for packaging into phage heads. The pBR 322 segments have the genes for Ampicillin and Tetracycline resistance.

**Cosmid pJB8:** This cosmid is 5.4 kb in size. It has ampicillin resistant genes of plasmid pBR 322 and the cos site derived from L -phage DNA.

## VECTORS FOR PLANT CELLS

Several vector systems are used for the introduction of foreign genes into plant cells. Among them, the Ti-plasmid of *Agrobacterium tumefaciens* and a few plant viruses such as Caulimoviruses and Geminiviruses are the most important ones. The role of Ti-plasmid of *Agrobacterium tumefaciens* has been discussed in chapter 5. Cauliflower Mosaic Virus (CaMV), one of the Caulimoviruses is discussed in this section.

Caulimoviruses are a group of plant viruses, which have double stranded DNA. These have a very narrow host range. One such virus is Cauliflower Mosaic Virus (CaMV). This virus infects plants of the family Cruciferae. The Caulimoviruses are isometric particles and are 50 nm in diameter. The cells of the infected plants accumulate virus containing proteinaceous inclusion bodies (viroplasms). These regions are regarded as the sites of virus assembly.

### Shuttle Vectors

Certain vectors can replicate in different host systems for example, in *E.coli* and in yeast. Such vectors are called shuttle vectors. For this purpose shuttle vectors carry different origins of replication, which are characteristic for different host systems.

## Yeast Episomal Plasmids (YEps)

Yeast episomal plasmids (YEps) are shuttle vectors. They can replicate in *E.coli* and also in yeast. pJDB219 is an example. It is an artificial plasmid and has the following segments.

- the 2 μm circular plasmid of yeast.
- the entire pBR 322 sequence.
- the selectable LEU2 gene of the yeast chromosome. This will also carry an origin of replication.

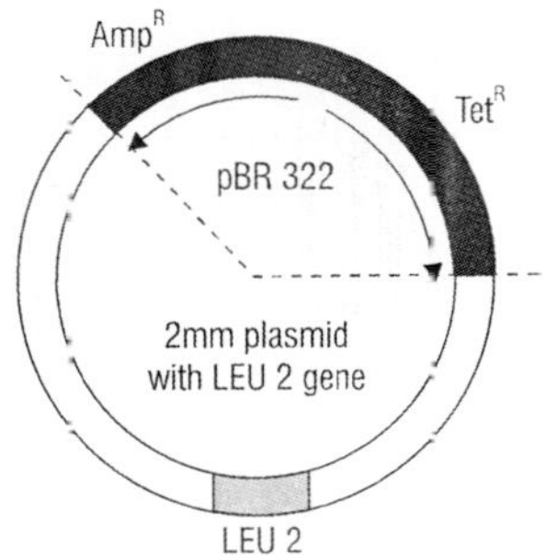


**FIGURE 27.4:** Structure of a yeast episomal plasmid (YEp), pJDB219 (after T.A. Brown, 1990. Gene cloning: An introduction).

## YAC Vectors

Large DNA segments can be cloned in yeast artificial chromosomes (YACs). YACs are linear DNA segments that contain all the molecular components required for replication in yeast: in a replication origin known as an autonomously replicating sequence (ARE), ii. a centromere and iii. the telomeres (the ends of linear chromosomes). They are also called minichromosomes. DNAs of several hundred kb (200-400 kb) can be introduced into YACs and successfully cloned. Example pYAC2.

## Expression Vectors

The primary function of a vector in DNA Recombinant Technology (DNART) is to carry a gene of interest into a recipient cell. When such a transfer is successful, the foreign gene should be expressed in the recipient cell. Some times, the foreign gene in the recipient cell may not be expressed. This may be due to the following reasons.

- Transcription of the gene did not occur due to the absence of an effective promoter.
- A terminator sequence to terminate transcription may be absent.
- The initiation codon for the particular protein is absent.
- The mRNA lacks a suitable Shine-Dalgarno sequence (in the case of prokaryotes).

Hence the vectors are constructed in such a way that they contain suitable expression signals. Such vectors are called expression vectors. A foreign gene carried by an expression vector into the recipient cell will have complete expression, i.e., the gene will be transcribed into mRNA and then translated into the protein (gene product). Several expression vectors are constructed to suit the needs.

There are two main strategies for the construction of expression vectors:

- Vectors, which can allows the synthesis of a fusion protein (hybrid protein) having vector and insertion sequences.
- Vectors that can synthesize pure proteins exclusively encoded by the inserted gene (foreign gene).

### Vectors for the Synthesis of Fusion Proteins

The foreign gene inserted into a vector gene in such a way that the reading frame of the vector gene is conserved. During transcription the hybrid mRNA is formed. The promoter of the vector gene initiates transcription. The hybrid mRNA is translated to yield a hybrid protein consisting of vector and foreign proteins. Translation begins from the start codon of the vector gene and ends at the stop codon of the foreign gene. The fusion protein is then cleaved to release the foreign protein.

### Vectors for the Synthesis of Pure Unfused Foreign Protein

There are vectors that can produce only the proteins encoded by the foreign gene. Such vectors will be constructed by linking a suitable and strong prokaryotic promoter, a bacterial Shine-Dalgarmo sequence and the start codon infront of a desired structural gene. This will allow the synthesis of the corresponding gene product as a pure unfused protein.

pSV240 is an example for this kind of expression vector. This vector is designed to express SV40t - antigen gene. This vector is derived from pBR 322 to which the lac promoter, the S/D sequence and the start codon are linked in front of the SV40 -t antigen gene.

The lac promoter allows transcription of SV40-t antigen gene. The mRNA thus produced has the S/D sequences, the start codon AUG for translation and the codons for SV40 t- antigen. This mRNA is translated into t-antigen.

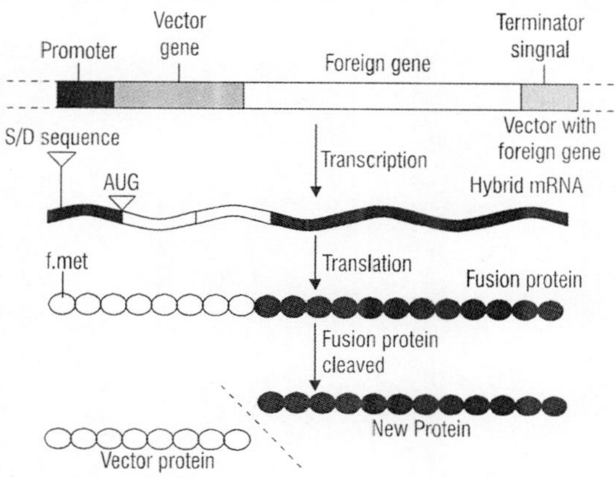


**FIGURE 27.5:** Expression vector for the synthesis of fusion protein.

From the above two examples, it is clear that a foreign gene introduced into a host genome should have the following components for proper expression.

- A suitable vector
- A terminator sequence to terminate transcription.
- Sequences to give Shine-Dalgarno sequence on the mRNA (if the gene is to be cloned in bacteria).
- The start codon to initiate translation.
- A stop codon to terminate translation.

## GENE CARTRIDGES

Purified DNA fragments, which may be a control region, a gene for a protein or another sequence, are readily available. Such fragments are called gene cartridges. These genetic fragments are very useful for plasmid construction. Some gene cartridges and their parent plasmids are given in the following Table.

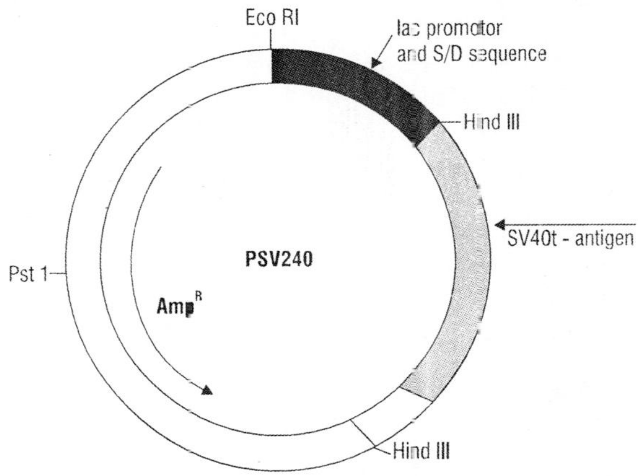


**FIGURE 27.6:** Expression vector for SV40 t-antigen. Restriction sites of a few endonucleases are shown.

# 28. Methods of Gene Transfer

## BACTERIA AND GENE TRANSFER

Agricultural scientists made a similar discovery about natural gene transfers in plants in 1907. Erwin Smith and Charles Townsend discovered that plant traits could be changed in nature in much the same way bacteria are altered by the activity of viruses. The scientists discovered this when they observed the formation of a plant tumor called a crown gall. Crown galls form on a plant at the site of a fresh wound when the wound is infected by a common soil bacterium. Although at the time they didn't understand the genetic science involved, the researchers were observing the activity of *Agrobacterium tumefaciens*, which is capable of inserting its genes into plants. The foreign genes can then *hijack* the genetic code of the plant, making it a better host for the bacteria. What Smith and Townsend did come to realize was that the cell mechanisms of all living things, whether plant or animal, are fundamentally the same. This discovery later made it possible to apply the process of gene transfer, first observed in the plants changed by *A. tumefaciens*, to the modification and improvement of many plants useful in agriculture. In time, this was accomplished by finding a way to delete the genes that cause the crown gall disease and substitute other genes that produce desirable traits. Scientists found that they could harness this unusual ability of *A. tumefaciens* to carry out gene transfers in the lab.

### How Bacteria are used to Modify Plants

Once scientists have discovered the genes responsible for the trait they want to introduce into another plant, they face several difficult problems in transforming the target plant.

- **First** they must accumulate enough pure DNA needed to genetically alter a number of target plant cells.
- A **second** problem is finding a way to deliver genes through the walls of the cell and into the nucleus. After all, the cell is designed by nature to keep foreign material out.

Several ways have been found to solve these problems, but one of the most useful methods harnesses bacteria to both increases the amount of DNA as well as to *deliver* genes through cell wall and into the nucleus.

### How do they do it?

Scientists have discovered that a number of enzymes (as described previously), complex organic substances originating in living cells, can be used as tools to *cut* genes from plants. Some 300-400 enzymes are now available for use in biotechnology. Different enzymes are used to cut and isolate specific genes. Once scientists have discovered and isolated the genes that control a desired trait in a plant (such as resistance to a disease), they have to find a way to replicate or clone enough exact copies of the genes to transform a large number of cells from the target plants.

Their first step is to take the genes responsible for a desired trait from a plant and insert them, using enzymes, into a plasmid taken from a common type of bacterium called *E. coli*. Plasmids (as discussed previously) are very important in biotechnology because they are the basis of gene cloning. Once a gene is inserted into a plasmid, the plasmid can be reintroduced into the bacterial cell. When the altered bacteria are grown in a culture (a medium that promotes rapid growth), the plasmid containing the desired gene is copied or cloned in every cell. Because bacteria grow quickly, this method makes it possible to make millions of copies of the desired gene in a short period. Using enzymes, the plant genes are now removed from the E. coli plasmids and inserted into the plasmid of A. tumefaciens, the soil microorganism that can move foreign genes into plant cells.

Finally, the altered *Agrobacterium tumefaciens* is mixed with the target plant cells. The bacteria ferry the engineered plasmid into the target plant cells. The "foreign" plant gene is then incorporated into the plant cell's DNA. New plants can be grown directly from these modified cells in a special growth chamber. Some of the plants grown from these cells will have the trait specified by the new gene. By growing the plants, scientists can now test to see which of the plants have been genetically transformed.

Three gene-transfer techniques have been developed to such a stage of reproducibility and practical applicability that they can be recommended for future practical work. These are *Agrobacterium-mediated* gene transfer, *ballistic* gene transfer, and *direct* gene transfer to *protoplasts*. These techniques are described in some detail in the following chapters. These are additional technique that have yielded confirmed transgenic plants and that show some promise for future development, such as *microtargeting*, *microinjection*, *electroporation* into tissues, and *whiskers*. These techniques are described to a smaller extent and the literature citations will support further reading There is a further group of techniques for which unequivocal evidence for the production of transgenic plants is still missing such as the pollen tube-pathway.

## *AGROBACTERIUM* MEDIATED GENE TRANSFER

For genetic transformation in plants *Agrobacterium tumefaciens* (a soil bacterium) is used. It transfers T-DNA into plant genome. Foreign genes are inserted in T-DNA through *Ti*-plasmid, co-transferred and integrated in the host genome. Only some sequences of T-DNA in *cis*-position are required for its transfer. By deleting T-DNA encoded genes and adding selectable markers a *disarmed plasmid vector* is prepared without disturbing endogenous balance. Disarmed strains with *Ti*-plasmid can be used to produce transgenic plants.

### Plant-transformation Vectors

These vectors are designed on the basis of naturally occurring gene-transfer mechanisms. They are widely employed in plant molecular biology and in the genetic engineering of plants (Fraley et al., 1985, 1986; Hiel et al, 1997; Tyagi et al., 1999). Chilton (1993) described these vectors as as *poor man's vector*. They are used for gene delivery to higher plants.

#### *Advantages of Vector*

- They are convenient and efficiency.
- They allow the transfer of large segments of DNA with minimal

rearrangements.

- They are the integration of small numbers of copies of genes into plant chromosomes (Hamilton et al., 1996; Klee et al., 1987; Zambryski, 1992).
- Foreign genes delivered by this method are usually transmitted to progeny plants in Mendelian manners.
- In the *Ti*-plasmid, the genes responsible for the transfer of T-DNA are clustered together in a region. The transfer process is active even when the virulence genes and the T-DNA are located on separate replicon in an *A. tumefaciens* cell.
- In the widely used binary system, a *Ti*-plasmid services as helper; providing the virulence functions, and artificial T-DNA, which contains a selectable marker gene and other genes of interest, is placed on a second, small plasmid, which is often referred to as binary vector.
- In dicotyledonous plants the systems of *Agrobacterium* mediated transformation have been well established for many. However, monocotyledonous plants (cereal grasses) were originally outside, the host range of *A. tumefaciens*.

## GENE CLONING

Gene cloning is a technique, where an isolated single copy of a gene is cloned to obtain its indefinite identical copies. In gene cloning, we introduce a vector containing the desired gene into the host cell. Cloning can be done in bacterial host (e.g. *E.coli*) or in eukaryotes (e.g. yeast). Basically, cloning involves four steps:

- The vector DNA is cleaved with one or more restriction enzymes;
- The DNA to be cloned, the target or insert is joined to the vector, generating a recombinant molecule;
- The recombinant DNA molecules are introduced into the host bacterial cell. The host cell is said to be *transformed* by the introduction of such vector molecule;
- Transformed colony is selected and amplified.

Another method of amplifying the desired gene is PCR, which offers this possibility in test tube. The two methods PCR and gene cloning are complementary techniques. Molecular cloning of a PCR product is now commonly used to isolate and manipulate genes. In this chapter, we will discuss gene cloning in *E.coli* and yeast, as well as PCR.

### Creating Recombinant DNA

Inserting a foreign DNA segment into the DNA molecule of a vector produces Recombinant DNA molecules. This involves joining of the broken ends of vector DNA with the two ends of sequence to be cloned. These recombinant DNA molecules (also called as rDNA molecule) are subsequently cloned. There are three methods of joining fragments *in vitro*. The first uses ability of DNA ligase to join the cohesive ends produced by certain restriction enzymes. The second depends on the ability of T4 ligase to form phosphodiester bonds between blunt-ended fragments. The third utilizes the enzyme terminal deoxynucleotidyl transferase to synthesize homopolymeric tails at the ends of fragments. The three enzymes and their mechanism of joining DNA are separately dealt below.

## DNA Ligase

The sticky ends created by restriction enzyme have the nick after a few base pairs in opposite strands. DNA ligase repairs these nicks and forms intact duplex. The optimum temperature for ligation of nicked DNA is 37°C, but at this temperature the hydrogen-bonded joint between the sticky ends is unstable. Thus temperature is kept in range of 4 to 15°C.

This technique has the advantage of regenerating two restriction sites in recombinant DNA. Thus, the foreign DNA segment can be retrieved from cloned DNA copies by cleaving again with the same enzyme. The disadvantages within this technique include:

- The cohesive ends of foreign DNA may join end-to-end and self-circularize before getting inserted in vector.
- The recognition site, particularly in sequence to be cloned, may not lie at a convenient position so that sometimes only a part of gene is inserted.

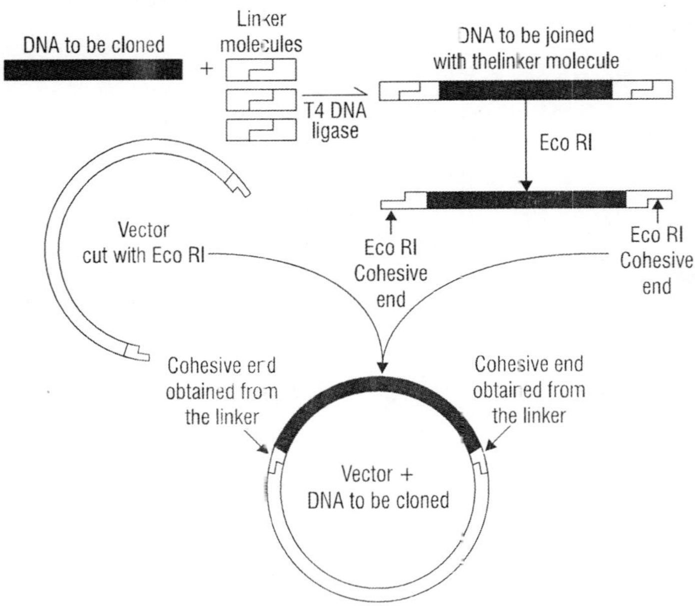


**FIGURE 28.1:** Linker molecules containing Eco RI cleavage site are joined by T4 DNA ligase to both the ends of blunt-ended DNA-to-be-cloned. Cutting with Eco RI then generates cohesive ends. This DNA is then inserted into a vector molecule cut by the same restriction enzyme.

## T4 DNA ligase

The *E.coli* DNA ligase cannot catalyse ligation of blunt ends except under special reaction conditions of macromolecular crowding. Such blunt end ligation can be done with T4 DNA ligase. T4 DNA ligase can join the broken blunt ends of two DNA molecules irrespective of the sequences. The disadvantage of this technique is that any two broken ends may join including those belonging to same DNA molecule. However, blunt-ended ligation is a useful way of joining together DNA fragments, which have not been produced by the same restriction enzyme and therefore probably have incompatible sticky ends. These ends are removed prior to ligation using the enzyme S1 nuclease, which digests single stranded DNA.

Blunt ligation is most usefully applied where blunt-ended fragments are joined via *linker* molecules (described later). The linker molecule can be ligated to both ends of foreign DNA to be cloned, and then treated with restriction enzyme to produce sticky ended fragments. The vector molecule is also cut with same restriction enzyme. Finally, the foreign DNA is ligated to vector via linker created sticky ends. These linkers also help in excising and recovering the foreign DNA (gene of interest) after cloning and amplification in host bacterium. Here a brief account on various types of linkers is given.

### *Linkers*

They are self-complementary synthetic oligonucleotides, which contain restriction endonuclease recognition sites. They may be used to clone DNA fragment into plasmids or other vectors.

These blunt ended linkers are first phosphorylated using polynucleotide kinase and labelled or unlabelled ATP molecules. These are then ligated to the blunt ended DNA fragment using T4 DNA ligase. The DNA fragment is finally treated with restriction enzyme to generate sticky ends and ligated into vector, which has been cut by the same enzyme.

#### Stop Linker

They are 5'-phosphorylated double-stranded synthetic oligo-deoxynucleotides, which can be cloned into an expression vector to ensure transcription termination. Such linkers contain stop codons in each reading frame from both directions. Each stop linker contains an SspI restriction site, which can be used to screen clones for linker insertion.

#### Adaptors

They are linkers, which contain more than one restriction enzyme recognition sites. They are used to insert new restriction sites into cloning vectors or to join restriction fragments to vectors with incompatible cohesive ends.

#### Double-linkers

In some plasmid vectors like pUC8, duplex cDNA molecules can be cloned. This is done by double-linker approach, in which different linker molecules are added to the opposite ends of the cDNA. This is an advantageous method as:

- Orientation of inserted DNA is fixed.
- Vector reclosure without insertion of foreign DNA can be done.
- Efficient for cloning of cDNA obtained from rare mRNA molecules.
- Better than homopolymer where expression from vector-borne promoter is desirable.

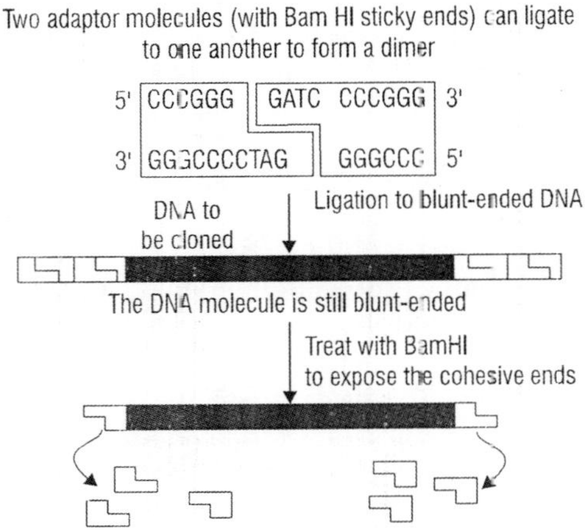


**FIGURE 28.2:** Diagram to explain the problems associated with the use of adoptor molecules. A. Two adaptor molecules with Bam HI cohesive ends have ligated to each other to form a dimer. B. Such dimer of the adoptor molecules can ligate to a blunt-ended DNA molecule. The DNA molecule thus obtained is still blunt-ended. C. The cohesive ends are recreated by treatment with Bam HI.

### *Terminal Transferase*

Besides adding cohesive ends or linkers, blunt ended DNA fragment and vector can also be ligated. This technique makes use of producing and annealing complementary homopolymer sequences. An enzyme, terminal transferase (purified from calf-thymus), is used to synthesize homopolymeric extensions. Using precursor dGTP, poly dG is added at both the 3'cut ends of vectors while poly dC is added to 5'cut end of foreign DNA. The vector and clone can then be joined by annealing the poly dG with poly dC tails and then ligating them using DNA ligase enzyme. Using Pst I restriction enzyme can retrieve the cloned sequence. The method works out the chances of reannealing between the two cut ends of same DNA molecule or of two similar DNA molecules, thus is advantageous.

## CLONING IN BACTERIA AND EUKARYOTES

### Cloning in Bacteria

After ligation of target into vector, the recombinant DNA is multiplied in a host. Various bacterial strains are available as host. Most common prokaryotic host is *E.coli*, which is also termed as **workhouse of genetic engineering**. In this technique the recombinant DNA is subjected to *in vitro* packaging and used to transform bacteria for its amplification. The host bacteria are modified and selected on the basis of high transformation efficiency. The type of bacterial host is selected on the basis of plasmid vector. Various bacterial hosts commonly employed are:

- Gram-positive bacteria e.g. *E.coli*: host for plasmids only.
- Gram-negative bacteria e.g. *Bacillus subtilis:* host for plasmid, phages and cosmid vectors.

**Chimeric DNA** is inserted into bacterial cells to create transformed bacterial cells. Colonies of transformed cells are selected and used for multiplication in suspension cultures. The recombinant DNA is retrieved and isolated from the host cell when required.

### Cloning in Eukaryotes

*E.coli* is workhouse of genetic engineering but the chromosome structure and gene expression in prokaryotes and eukaryotes differ radically. Therefore, *E. coli* offers limitation to its use in cloning eukaryotic genes. The split genes cannot be correctly expressed in prokaryotic host as latter do not have introns and thus splicing of mRNA cannot take place. Therefore, eukaryotic cells are used to clone and express eukaryotic genes.

Among eukaryotes, DNA cloning has been done in yeast, mouse and some plant species. Yeast has following attractions as host for genetic engineering:

- Short doubling time (90 minutes).
- Can be maintained in either a haploid or diploid state.
- Can be grown on simple defined media and auxotrophs can be selected.
- The wall can be removed and foreign DNA entry can be facilitated with PEG and $CaCl_2$ mixture.
- Confers extra-chromosomal replication of chimeric DNA just as plasmids in *E.coli*.

The simplest kind of yeast vector contains a selectable marker such as TRP1, joined to plasmid vector. TRP1 encodes an enzyme in the tryptophan biosynthetic pathway and when introduced into $TRP^-$ cell by DNA transformation, it renders the cell capable of growth in absence of added tryptophan. Vector used is a hybrid or shuttle type. When transformed into yeast, such a vector integrates into the yeast genome by recombination and makes the transformed yeast selectable on the basis of growth in absence of added tryptophan.

Yeast is becoming an important organism for the commercial production of medically important proteins such as viral vaccines. It is of great value as an organism for basic research, for producing heterologous proteins and as a vehicle for cloning large segments of DNA. For some purpose, however, it is essential to be able to transform mammalian cells. While yeast and mammals share many basic properties, there are important different cells in the precise mechanics of gene expression. Mammalian RNA transcripts are not spliced in the yeast cell, nor the mammalian proteins are glycosylated correctly.

In animal cells like mouse cells, special animal viruses are used as cloning vehicle. Simian virus 40 (SV40) with foreign DNA can be transcribed and translated in mouse kidney. Retroviral vectors are used for transformation of mammalian cells. These animal viruses comprise single stranded RNA genome, gene of interest and G418R gene for resistance against protein synthesis inhibitor G418.

## AMPLIFICATION OF DNAS BY POLYMERASE CHAIN REACTION (PCR)

The major scientific development of year 1989, as named by prestigious journal 'Science' was **Polymerase Chain Reaction** (PCR). Kary Mullin is the inventor of PCR, who works in Cetus Corporation, a biotech company, of United States. Since its introduction, the polymerase chain reaction has revolutionized the method of DNA analysis in both research and clinical laboratories.

The PCR is an extremely powerful new procedure that allows one to make million fold copies of a selected DNA sequence in a genome. PCR can be used to clone a given DNA sequence *in vitro* without using living cells during the cloning process. However, the procedure can only be applied when the nucleotide sequence of at least one short DNA segment on each side of the 'region of interest' is known. The PCR procedure makes use of synthetic oligonucleotides complementary to the known sequences spanning the region of interest to fetch *enzymatic amplification of this segment of DNA in the test tube.*

PCR may eventually replace gene-cloning technique. This technology is much easier and requires much smaller quantity of DNA (in ng) whereas for cloning, DNA needed in microgram (μg) quantities.

## Procedure

The PCR procedure involves three steps, each repeated many times to produce cycles of amplification.

- **Step 1:** The genomic DNA containing the sequence to be amplified is denatured by heating.
- **Step 2:** The denatured DNA is annealed to an excess of the synthetic oligonucleotide primers.
- **Step 3:** DNA polymerase is used to replicate the DNA segment between the sites complementary to the oligonucleotide primers. The primer provides the free 3'-OH required for covalent extension; the denatured genomic DNA provides the required template function. These three steps (denaturation, primer binding and DNA synthesis) represents a single PCR cycle.

The products of the first cycle are replicated for further amplification. This reaction can be performed many times to supply unlimited copies of amplified DNA. Repetition is made possible by regular denaturation of freshly synthesized double stranded DNA molecules by heating it to 90-98°C.

At this temperature, the two strands separate. Once the double stranded DNA is made single stranded, it is mixed with two primers and is cooled to 40-60°C.

This allows the primers (which are in excess) to bind to their complementary strands through renaturation followed by replication with DNA polymerase. The cycle is repeated many times until the desired level of amplification is achieved.

### *Applications of PCR*

During the last few years, with the improvement of PCR protocols and due to the availability of automatic thermal cycles commercially, the applications of PCR have increased manifold. Some of the applications are listed here.

### *Study of DNA Polymorphism using PCR*

DNA polymorphism is amplified using PCR and can be used for study. Loci either with known DNA sequence or at random sites both can be amplified. PCR can be used to study RFLPs as well as RAPDs.

Loci with poly (A), poly (G), poly (TC) and poly (TG) etc., known as short tandem repeats or simple sequence repeats (SSR) can also be amplified with PCR. Some of the these SSR loci such (CA), or (GT)n occur in the human genome with many copies. SSRs can be utilized in preparing saturated molecular maps.

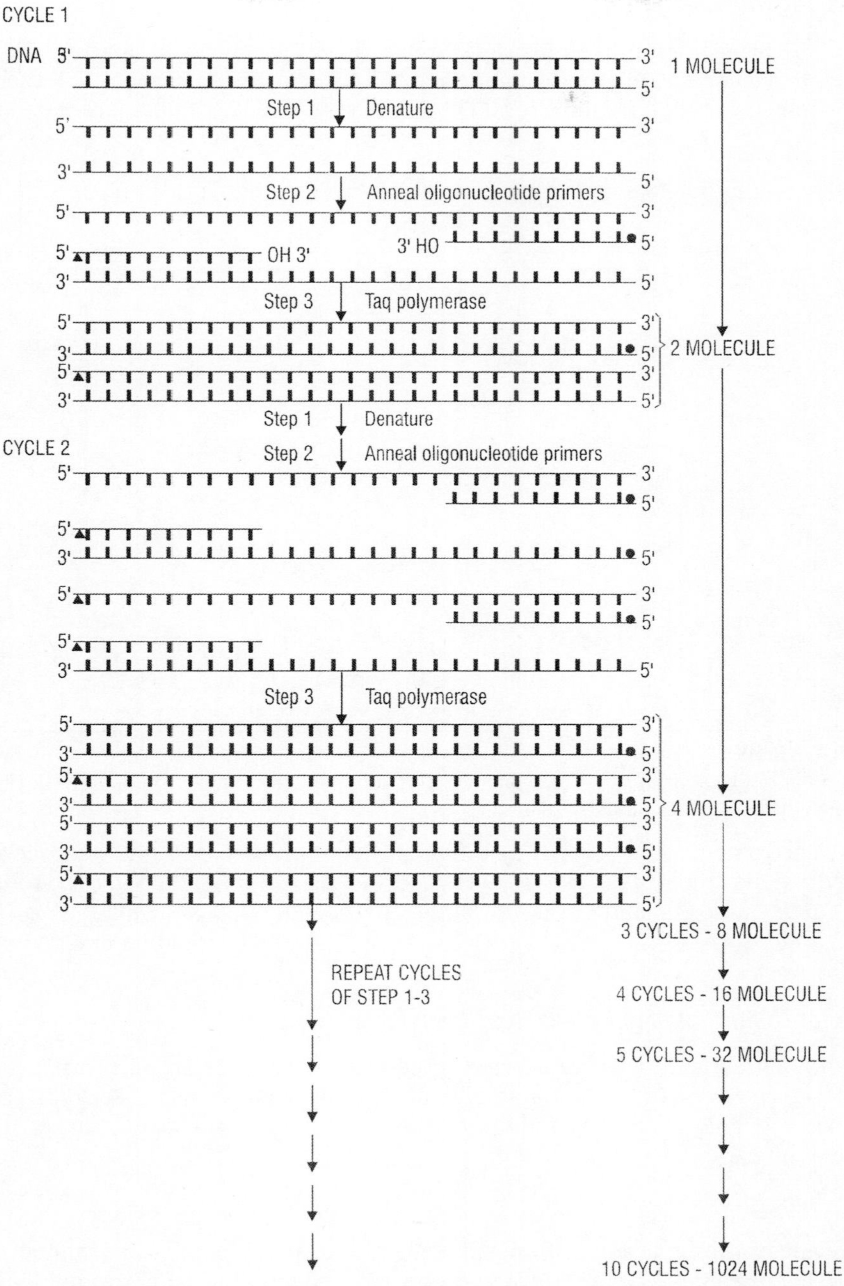


**FIGURE 28.3:** Diagram illustrating the use of PCR to amplify DNA molecules *in vitro*. Each cycle of amplification involves three steps: (1) denaturation of the genomic DNA being analyzed, (2) annealing of the denatured DNA to chemically synthesized oligonucleotide primers with sequences complementary to sites on opposite sides of the DNA region of interest, and (3) enzymatic replication of the region of interest by Taq polymerase.

### Gene Tagging using PCR

PCR has also been utilized for developing molecular markers closely linked to specific genes of economic importance. For instance in tomato, 144 random primers were used to produce 625 PCR products from a set of near isogenic lines.

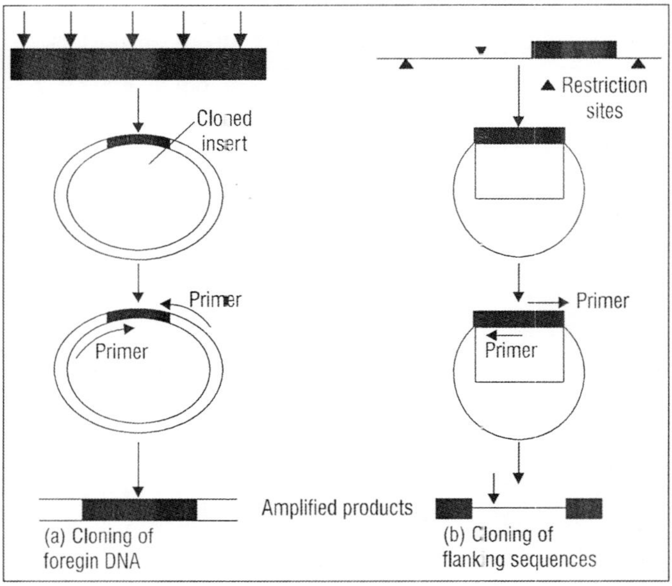


**FIGURE 28.4:** Inverse PCR (a) amplification of genes using flanking vector sequences as primers; (b) amplification of sequences flanking the gene, using sequences at the two ends of the gene as primers.

Near isogenic lines (= NIL's) differ only for the presence and absence of a specific gene for *Pseudomonas* resistance. Atleast three of these PCR products were present in one and not in the other of a pair of NIL's showing linkage with *Pseudomonas* resistance gene (Pto). These PCR products could be mapped close to Pto.

### PCR for Confirming the Presence of Transferred Gene

When a chimeric gene is transferred to cultured cells or organisms, primers can be designed to conduct PCR for amplification of the gene sequence so transferred.

This technique allows the confirmation of transfer and maintenance of the gene of interest.

### DNA Fingerprinting using PCR

DNA fingerprinting is use in forensic medicine for identification of criminals and disputed parentage. Some well-characterized sequences of microsatellites are being used for designing primers. PCR allows amplification of DNA from individual hairs, stains of blood or seminal fluid having partially degraded DNA. Characterization of individuals with such small amount of DNA could not be done earlier.

## *PCR in Gene Therapy*

In gene therapy experiments, the transfer and presence of a marker gene for neomycin resistance (NeoR) could be detected in the blood of patients, even after 60 days, although the frequency of marked lymphocytes gradually decreased necessitating a second transfusion of such marked cells. Similarly, ADA (adenosine deaminase deficiency) gene was transferred for therapeutic purposes and PCR was used to detect the presence of this gene. Thus PCR proves to be of immense help in monitoring a gene in gene therapy experiments.

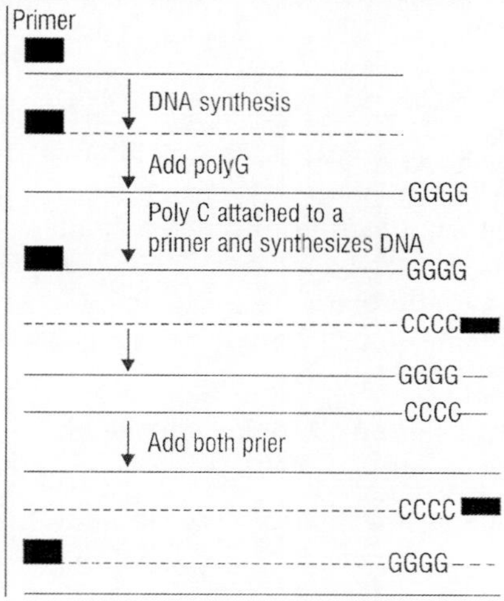


**FIGURE 28.5:** Anchored PCR, using a single primer in the first cycle.

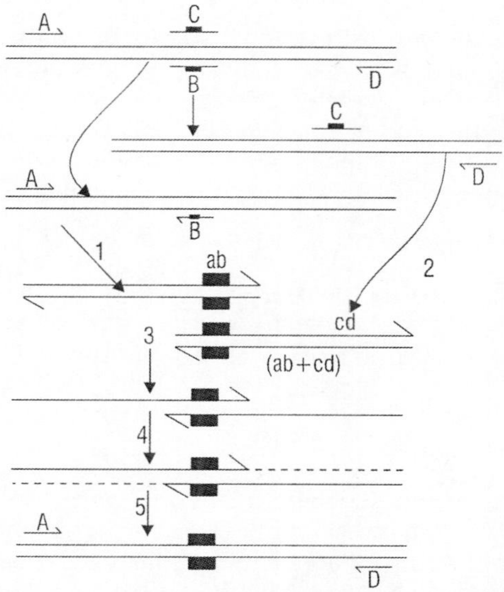


**FIGURE 28.6:** Introduction of mutation at an internal site using overlap extension (two pairs of primers A & B and C & D are used. B and C represent internal sites and overlap each other).

PCR technologies thus provide shortcuts for many cloning and sequencing applications. These procedures permit scientists to obtain definitive structural data on genes and DNA sequences when very small amounts of DNA are available. One important application involves the diagnosis of inherited human disease, especially in cases of prenatal diagnosis, where very limited amounts of fetal DNA are available. A second major application occurs in forensic cases involving individual identities. No criteria can provide more definitive evidence of identity than DNA sequences; after all, other phenotypic traits are controlled by the expression of these DNA isolated from small blood samples, sperm, or even individual human hairs. Thus, PCR DNA printing experiments are likely to play a major role in legal cases involving disputed identity.

## GENE TRANSFER TECHNOLOGY

Gene transfer is of basic as well as applied interest. It enables to study the structure and function of isolated genes via modification, reintroduction and subsequent analysis of their function in transgenic plants, cell lines or even cell populations. Besides these, transfer of specifically designed hybrid genes to plant genomes creates novel varieties with designed characters.

Gene transfer in plants can be accomplished by sexual and asexual methods. Asexual methods include:

- Non-specific gene transfer by protoplast techniques.
- Specific gene transfer methods.

In animals, gene transfer is done using egg and cultured stem cells or by specific gene transfer.

### Sexual Method

#### *Gene Transfer using Pollen or Pollen Tube*

Pollen grains are treated with heavy dose of X-ray, which pulverizes chromosomes. The pollen, if able to pollinate, assures incorporation of changed chromosomes into daughter progeny. It is assumed that DNA can be taken up by the nuclei or reach the zygote through the pollen tube pathway. Both these approaches have been tried, though several problems exist in this method like presence of cell wall, nucleases, and callose plugs in pollen tube etc.

#### *Incubation of Dry Seeds, Embryos, Tissues or Cells in DNA*

This method has been tried in many cases and expression of defined genes has been witnessed. In all the cases, plant cell walls work as efficient barriers, thus DNA uptake can be enhanced by PEG or electroporation or by any other device.

### Asexual Methods

#### *Non Specific Gene transfer*

Fusion of protoplasts to form somatic hybrid and partial hybrids are the known methods. The most direct method of gene transfer to plant cells is the uptake of plasmid DNA but at present its application relies on the availability of isolated genes. Agronomic traits are not coded for by single genes but rather by families of interacting genes. Thus, there is an interest in techniques, which allow the transfer of large blocks of intact DNA encoding Polygenic traits.

Protoplast membranes may be induced to fuse with nucleated or mitotically inactivated protoplasts to form hybrids; they may be induced to take up isolated nuclei or chromosomes; or they may be fused with bacterial spheroplasts, or liposomes.

Protoplast fusion may be carried out with three different methods; by using chemicals, by an electric field or by using a U.V. laser microbeam. Electrofusion is reported to be free of cytotoxicity, associated with chemical fusagen such as polyethylene glycol (PEG) or polyvinyl alcohol (PVA). Electrofusion appears superior to laser fusion, as former is commercially available and affordable for a typical biological laboratory while this is not presently true for the laser microbeam. Furthermore, the number of hybrids that can be produced in a given time is much higher using electrofusion.

Foreign DNA can be directly introduced into mammalian cells. Firstly, DNA is precipitated with $CaCl_2$ and then this precipitate is mixed with the cells to be transformed. The cells will take up the DNA. Inside the cell, the new DNA fragments get ligated to give a concatemers, which is then integrated into the nuclear DNA as one single piece. The integration occurs at random sites. In order to select the transformed cells several marker genes are also introduced along with the genes to be cloned. The marker genes will be linked with the genes of interest prior to integration into the host genome. One such marker is the neomycin resistance gene, which would normally inhibit protein synthesis.

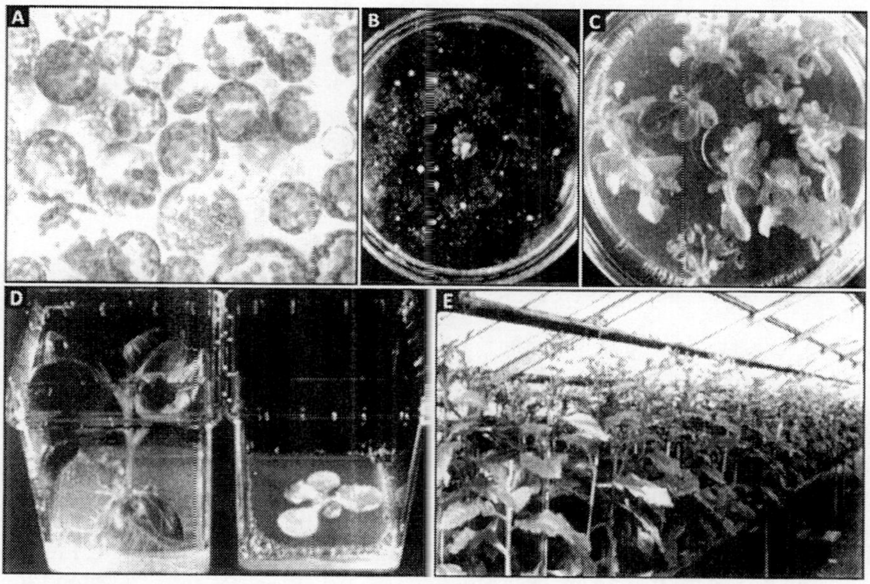

**FIGURE 28.7:** Direct transfer to leaf protoplasts and regeneration of transgenic plants in tobacco (*Nicotiana tabacum*). (A) Freshly isolated leaf protoplasts of tobacco treated with polyethylene glycol and plasmid DNA bearing a chimeric *npt2* gene as selectable marker; (B) selection of kannamycin-resistant transformed colonies from protoplasts in bead-type culture; (C) shoot differentiation from tobacco calli obtained from transformed protoplasts; )D) rooting of transformed (left) and untransformed (right) shoot from protoplasts on selective medium; (E) fertile transgenic tobacco plants obtained from protoplasts.

### *Specific Gene Transfer*

Specific gene transfer in plants is mostly done with *Agrobacterium* species having natural biological delivery system. There are several other techniques developed recently for introducing cloned genes into plant cells. These are categorized as artificial delivery system transferring gene directly. Some examples of artificial delivery systems are direct DNA uptake, electroporation, fusion with liposomes etc. Unlike with *Agrobacterium* species mediated gene transfer (where cells with intact cell walls and tissue explants can be used); artificial delivery systems can be applied to protoplasts only.

The basic pre-requisites for successful gene transfer are:

- selectable marker gene, which could be expected to yield a clear phenotype in transformed cell clones;
- expression signals, which are independent of physiological state of host cell;
- precise linkage of the protein-coding region of the selectable gene to the expression signals. This will guarantee the expression of hybrid gene in the host genome;
- DNA stretches on both sides of the hybrid gene for its integration into the host genome without disrupting the gene construct;
- an absolutely tight selection protocol for presumptive transformants; and
- where of artificial delivery systems are utilized, the protoplast system taken should have routine high plating efficiency and plant regeneration capacity.

Different methods for specific gene transfer are discussed in this chapter.

## BIOLOGICAL DELIVERY SYSTEM

To date, *Agrobacterium. tumefaciens* mediated transformations have been in wider practice and a much broader range of transgenic plants have been produced with this technique. It is a method of choice for routine production of transgenic species since it can transform regenerable tissues with relative ease.

### Agrobacterium tumefaciens: An Important Vector for Genetic Engineering

*Agrobacterium tumefaciens* is the most important tool in the genetic engineering of plants (the Ti plasmid of the soil bacterium *A. tumefaciens*). This bacterium is the causative agent of crown gall disease of dicotyledonous plants only (unable to develop gall in monocotyledonous species).

The name refers to the galls or tumours that often form at the crown (junction between the root and the stem) of infected plants. Since the crown of the plant is usually located at the soil surface, this is where a plant is most likely to be wounded (e.g., due to a soil abrasion from the plant blowing in a strong wind) and infected by a soil blowing in a strong wind) and infected by a soil bacterium such as *A. tumefaciens*. However, *A. tumefaciens* can infect a plant and induce a tumour at any wound site.

In the normal disease process the bacterium enters the plant at the site of a wound. The infection often occurs at the crown, but it may involve an upper part of the stem, the petiole, or the leaf. The bacterium attaches to the wall of a cell and transforms the cell. Infected cells proliferate at the site of infection, forming a tumor or gall. The tumor can be fatal to young plants if it grows big enough to crush the vascular system, although often the tumor is merely debilitating and unsightly. Transformed cells also have the ability to grow in culture on a medium that lacks the hormones auxin and cytokinin. Finally, transformed cells produce opines, such as octopine and nopaline. These are chemicals that are formed from two amino acids. Argenine and alanine form octopine, and argenine and glutamine produce nopaline. Octopine and nopaline are not found in normal plant cells.

### *Infection of Agrobacterium tumefaciens*

During the infection of a wound site by *A. tumefaciens*, two key following events occur:

- The plant cells begin to proliferate and form tumours.
- They begin to synthesize an *arginine* derivative called *opine.*

The synthesized opine is usually either nopaline or octopine depending on the *A. tumefaciens* strain involved.

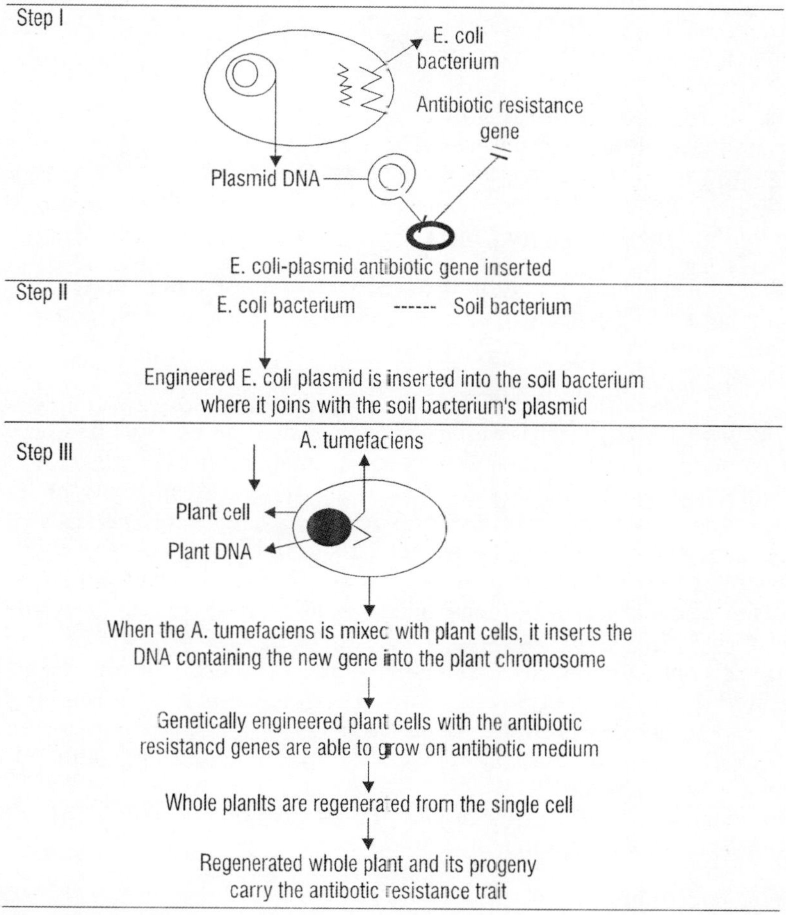


**FIGURE 28.8:** Various steps (I-III) used in plant genetic engineering system.

These opine are catabolized and used as energy sources by the infecting bacteria. Interestingly, *A. tumefaciens* strains that induce the synthesis of nopaline can grow on nopaline, but not on octopine, and vice versa. Clearly, an interesting interrelationship has evolved between *A. tumefaciens* strains and their plant hosts.

*Agrobacterium tumefaciens* is able to divert the metabolic resources of the host plant to the synthesis of opines, which are of no apparent benefit to the plant, but which provide sustenance to the bacterium.

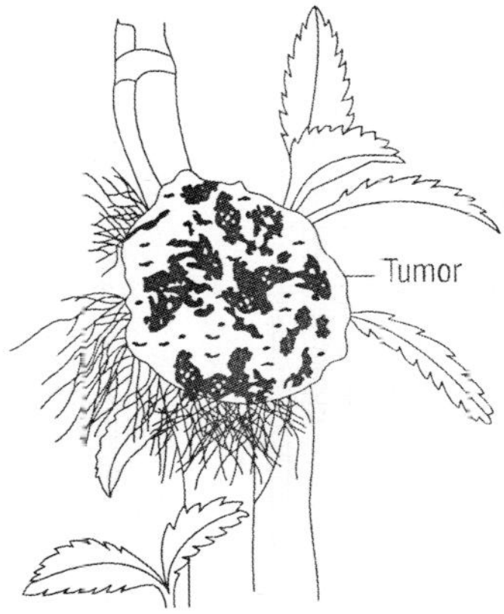


**FIGURE 28.9:** A plant tumor induced by Agrobacterium tumefaciens.

### Genetic Information Needed for Crown Gall Development (Ti Plasmid)

The ability of *A. tumefaciens* to induce crown gall disease in plants is controlled by genetic information carried on a large (about 200,000 nucleotide-pairs) plasmid called the Ti plasmid for its tumor-inducing capacity.

### Tumour-inducing Principle

Armin Braun at Rockefeller University explained that the bacteria had provided a tumour-inducing principle. The nature of this principle remained obscure until the 1970s, when Jeff Schell and his co-workers at the State University of Ghent in Belgium found that virulent bacteria all contain a large circular DNA plasmid. When strains of bacteria lose the plasmid, they lose the ability to infect plants. When bacteria regain a plasmid, they regain virulence. Further experiments by Rob Schilperoort's group in Leiden, Holland, demonstrated that many of the genes that control

Tumour characteristics (morphology, opine synthesis) lie on the plasmid. Finally, Mary-Dell Chilton et al. at the University of Washington in Seattle showed that a segment of the plasmid, the T (transferred) DNA, integrates into the genome of transformed cells. The plasmid (Ti plasmid) is thus identified as the tumour-inducing principle.

Two components, the T-DNA and the vir region, of the Ti plasmid are essential for the transformation of plant cells.

### T-DNA (Transferred DNA)

During the transformation process, the T-DNA (for Transferred DNA) is excised from the Ti plasmid, transferred to a plant cell, and integrated (covalently inserted) into the DNA of the plant cell. The available data indicate that integration of the T-DNA occurs at random chromosomal sites; moreover, in some cases, multiple T-DNA integration events occur in the same cell. In nopaline-type Ti plasmids, the T-DNA is a 23,000-nucleotide-pair segment that carries 13 known genes. In octopine-type Ti plasmids, there are two separate T-DNA segments.

Some of the genes on the T-DNA segment of the Ti plasmid encode enzymes that catalyze the synthesis of phytohormones (the auxin-IAA and the cytokinin isopentenyl adenosine). These phytohormones are responsible for the tumorous growth of cells in crown galls. The T-DNA region is bordered by 25-nucleotide pair imperfect repeats, which are required in cis for T-DNA excision and transfer. The deletion of either border sequence completely blocks the transfer of T-DNA to plant cells.

### *Vir Region (For Virulence)*

The *vir* (for virulence) region of the Ti plasmid contains the genes required for the T-DNA transfer process. These genes encode the DNA processing enzymes required for excision, transfer, and integration of the T-DNA segment. *vir* genes can supply the functions needed for T-DNA transfer when located either *cis* or *trans* to the T-DNA. The *vir* genes are normally expressed at very low levels in *A. tumefaciens* cells growing in soil. However, exposure of the bacteria to plant cells or exudates from plant cells induces increased levels of expression of the operons containing the vir genes. Surprisingly, this induction process is very slow for bacteria, taking 10 to 15 hours to reach maximum levels of expression. The phenolic compounds like acetosyringone have been purified from plant exudates and shown to act as inducers of the *vir* operons. Thus, the induction process can now be studied *in vitro* by using these purified inducer molecules.

### Disarmed Ti Vectors

When the T-DNA region of the Ti plasmid of *Agrobacterium tumefaciens* is transferred to plant cells and becomes integrated in plant chromosomes, the potential use of *A. tumefaciens* in plant genetic engineering seems to be obvious. The foreign genes can be introduced into the T-DNA region and, hopefully, these genes can be transferred to the plant with the rest of the T-DNA segment. In fact, this works very well. All the available data indicate that any DNA inserted any where between the T-DNA border sequences is transferred to plant cells and integrates in plant chromosomes with the rest of the T-DNA. The problem is that the transformed plant cells that have received the T-DNA from a wild-type Ti plasmid lose their normal control of cell division and form tumours. This oncogenicity of T-DNA renders wild type Ti plasmids incompatible with the goals of most gene-transfer experiments.

### Vectors

A shuttle vector is commonly used, which can replicate in *E.coli* (where recombinant manipulations can be easily handled), and then transferred into *A. tumefaciens* in preparation for transfer into plants. There are two basic modes of maintenance of the shuttle vectors in *Agrobacterium*:

- either by integration into Ti plasmid by recombination at a region of DNA homology
- or, by autonomous replication in trans to the Ti plasmid.

The former type of vector is referred to as cis or integrating vector while the latter is called a trans or binary vector.

A selectable markers is required to recognize the transformed cells and on other hand, suppress the growth of wild type cells. This helps in identification of transformed cells, which are in small proportion.

**Limitations:** *A. tumefaciens* based gene transfer is hampered by host range limitations, which are especially disturbing when aiming at gene transfer into cereals or other graminaceous monocots.

### Artificial Delivery Systems

Escape from host range limitations can be made by transferring the genes *directly* without using any biological vectors, into plant cells. Cell wall constitutes an effective barrier for DNA molecules to pass; naked plant cells, the 'protoplasts', have been found ideal experimental system to study vector less gene transfer. Various artificial delivery systems are given here.

## GENE TRANSFER USING PARTICLE BOMBARDMENT TECHNIQUE

The first genetically engineered plants were produced in middle 1980s (Horsch et al. 1985) and developed interest of genetic engineers in the production of transgenic plants for both basic and applied work. The major efforts were in the development and refinement of methods for transgenic plant production and also in the isolation and characterization of useful gene for introduction into plants. Through gene transfer, plants have already been produced that contain genes for disease and insect resistance, and modified fruit and grain quality. Particle bombardment genes transfer (or biolistics) is a method of transferring genes. Biolistics has been used to transform a large number of different plant species (reviewed by Christou 1994), as well as some animals, fungi, and bacteria.

The basic principle of particle bombardment for gene transfer is to use high velocity microprojectile to incorporate the introduced genes. The outer cell layers are penetrated in order to introduce genetic material into living cells.

### History

Biolistics (a combination of *biological* and *ballistics*), is a method in which high-velocity microprojectiles are used to deliver exogenous genetic material into plant tissue. It was developed, and first described by John Sanford in collaboration with Ted Klein, Ed Wolf, and Nelson Allen (Sanford et al. 1987). The first used penetration particles were onion tissue, where a monolayer of large epidermal cells was used as the target tissue. By observing cytoplasmic streaming in these cells, this tissue could be easily evaluated for viability following bombardment. Onion epidermal cells, transformed with the chloramphenicol acetyl transferase (CAT) gene were the first transiently transformed cells produced using this method.

The first stably transformed plants obtained using biolistics were reported by Christou et al. (1980) and Klein et al. (1988) using soybean and tobacco respectively. Since these initial reports of successful transformation, a broad range of species has been transformed including bacteria (Smith et al. 1992), fungi (Toffaletti et al. 1993) and even animals et al. 1988). In addition, organelles such as chloroplasts (Boynton et al. 1988) and mitochondria (Johnston et al. 1988) have also been transformed using Biolistics.

This method of particle acceleration has been described as the *particle gun, the gene gun, the bioblaster*, and the *microprojectile bombardment method.* It initially christened the biolistics method by its inventor (Sanford, 1988). Biolistics is a technique, which utilizes instrumentation to accelerate DNA coated microprojectiles into cells, past the cell wall and cell membrane. The microprojectile is small enough (0.5-5μm) to enter the plant cell without too much damage, yet large enough to have the mass to penetrate the cell wall and carry an appropriate amount of DNA on its surface into the interior of the plant cell. The important aspects of biolistics are:

- It is a direct gene transfer method.
- It does not require biological vector.
- It has been used successfully to transform cells that were once impossible to transform by other means.
- DNA delivery appears to be genotype and tissue independent.
- With this technique, the microprojectile is shot through the physical barriers of the cell.
- There are no biological limitations to DNA delivery.
- Transformation via the biolistics method is simple and safe.

## Applications of Biolistic

- The utility of the Biolistic method is broad and the technology can be used for basic studies on transformation.
- Modifications of gene expression.
- Understanding essential components of genes.
- The production of transgenic plants of economic value.

## Particle Gun Design

Several particle guns have been designed in various laboratories. The basis of all of these designs is to coat DNA onto small dense particles and accelerate the particles towards a target tissue. The particles usually consist of either gold or tungsten.

**Gold particles** are chemically inert. They do not produce cytotoxic effects. They are more uniform in size than tungsten particles. However, the cost of gold particles may limit their accessibility and the more affordable tungsten particles are sufficient for most studies.

**Tungsten particles** are somewhat phytotoxic (RUSSELL et al. 1992a) and tend to be more variable in size than gold particles. Ideally the particles used for bombardment should have good initial affinity for the DNA, yet freely release it once in the cytoplasm or nucleus of the target cell.

## Preparation of DNA-coated Microprojectiles

- Wash gold or tungsten particles and mixed them with plasmid DNA.
- The DNA is bound on the particles using either ethanol or $CaCl_2$ precipitation methods.
- Spermidine may be added to the mixture, possibly protecting the DNA from degradation and/or altering its conformation.
- After precipitation, the particles may be washed, resuspended and either dried or stored on ice as an aqueous suspension until use.

### *Use of Macrocarrier*

Many of the particle guns utilize a macrocarrier. They supports or carries the particles and gun is accelerated along with the particles towards the target. The macrocarrier is usually retained by a stopping plate or screen before it collides with the target, whereas the particles continue along their course. In most cases, the particles are accelerated under partial vacuum in a vacuum chamber to reduce air drag.

Modifying the intensity of the explosive burst controls the penetration of particle. It is done by changing the distance that the particles must travel to reach the target tissue or by using different sized particles.

### Gene Gun

The original gene gun described in the literature used DNA-coated tungsten particles in aqueous slurry loaded on the end of a small plastic bullet, which served as the macrocarrier. The plastic macrocarrier was placed into a 0.22 caliber barrel in front of a gunpowder cartridge. Once the cartridge was fired, the plastic bullet was propelled towards a solid stopping plate containing a small opening in the center. The microprojectiles continued through the hole towards the target tissue while the macrocarrier was retained. Both the stopping plate and the macrocarrier were discarded afterwards. The initial blast from the gunpowder cartridge necessitated frequent cleaning of the chamber and the target variation in results were common with this device, this first unit was used successfully by a number of different laboratories.

### Modified Bombardment Devices

### *Helium-Modified Bombardment Device*

The first important modification of the original device included the substitution of a helium blast for the gunpowder discharge (Sanford et al., 1991). Helium pressure continually builds in a reservoir and is released using *rupture discs.* The release of helium produces a shockwave, which travels to a second disc or macrocarrier. It holds the DNA-coated particles. The macrocarrier carries the DNA coated particles a short distance into a stopping screen. It retains the macrocarrier. The microprojectiles continue traveling, ultimately penetrating the target tissue which is held in a partial vacuum. The PDS-1000/He is the most commonly used device.

### *Accel Particle Gun*

This gun uses a high voltage electrical discharge to vaporize a water droplet, which produces a controlled shock wave (Mccabe and Christou, 1993). The device is constructed so that the initial shock wave is reflected to produce secondary shock waves. This in turn accelerates a Mylar sheet coated with particles. The Mylar sheet is accelerated towards a retaining screen stops the carrier and allows the particles to continue onto the target tissue. Adjusting the intensity of the voltage passing through the water droplet accurately modifies particle speed, and the resulting depth of particle penetration. It is the most efficient particle bombardment device in current. However, it is not commercially available.

FIGURE 28.10: PDS 1000 particle gun.

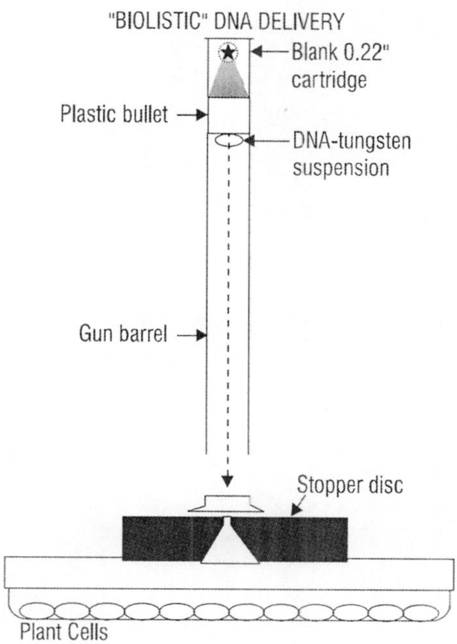


FIGURE 28.11: The gene gun.

### *Particle Inflow Gun*

It is a costly device due to manufacturing expenses. This created a need to develop more affordable, simple, yet safe and effective devices such as the **Particle Inflow Gun** or PIG. With this device, a solenoid controlled by a timer relay is used to generate a burst of low-pressure helium. The particles, supported by a screen in a reusable syringe filter unit, are accelerated directly in a stream of helium without the need for a macrocarrier. The target tissue is held in a chamber under a partial vacuum. To minimize the localized impact of the particles and to aid in dispersal, a nylon mesh baffle may be placed over the tissue. For this device there is no need to purchase macrocarriers or rupture membranes. It is less cleanup and minimal down time between shots.

### *Microtargeting Device*

This device accelerates small amounts of a DNA/particle mixture in a focused stream of high-pressure nitrogen. It was originally developed to allow precise delivery of particles to the shoot meristem. DNA is not precipitated on the gold particles but is used as a DNA/particle mixture. If the shoot meristem could rapidly give transgenic progeny.

### *Helios Gene Gun*

In these guns targeted tissue held in a vacuum, and apparatuses are not very portable. The Helios Gene Gun is a hand-held device, which uses low-pressure helium to accelerate DNA, coated particles from the interior of a small plastic cartridge towards the target tissue. A spacer at the tip of the gun maintains optimal target distance and minimizes cell damage by venting the helium gas away from the tissue. An evacuated chamber could not easily use them to transform target tissue that was sensitive to those conditions created. The Helios Gene Gun can accommodate up to 12 loaded cartridges at once thus allowing for multiple firing of the device before reloading is necessary. The greatest utility of this device seems to be *in situ* transformation of animal cell and tissues (Sundaram et al., 1996).

## Factors Affecting Particle Bombardment-Mediated Transformation

- Particle bombardment is a physical process of DNA introduction.
- After the DNA-coated particles are introduced, there are numerous factors that influence gene expression and the subsequent recovery of transgenic plants.

## Components of the Introduced Plasmid DNA

The introduced plasmid DNA(s) typically contains the necessary plasmid backbone (allows growth and selection in the bacterial host), plant marker genes and the gene(s) of interest.

The *plant marker genes* can be selectable markers (which allow growth of transformed plant cells in the presence of a selective agent) and/or scorable markers (reporter genes). The genes themselves are composed of the promoter, coding region and terminator. Introns, enhancers and attachment regions are defined regions of additional DNA that can be used to enhance, modulate or stabilize gene expression.

### *Selectable Marker Genes*

The selection of proper marker genes of choice is critical for the successful recovery of stably transformed tissues. The selectable marker allows the transformed cells to proliferate in the presence of a selective agent while nontransformed cells either do not grow or multiply at a much-reduced rate. Selectable marker genes may be either antibiotic or herbicide resistance genes. The resistance genes can act either by modifying the selective agent itself to inactivate it or by producing an insensitive from of the normal target of the selective agent.

### Selectable Marker Gene: NPTII

- It is most commonly used selectable marker gene.
- It imparts resistance to the antibiotic kannamycin.
- The NPTII gene product is neomycin phosphorylation, which inactivates the antibiotic by phosphorylation.
- Other antibiotics that are commonly used with selectable marker genes in plants include hygromycin, phleomycin and gentamycin.

### Selectable Marker Gene: Bar Gene

- It is most commonly used herbicide resistance marker gene.
- It is typically used with species of the Gramineae.
- The *bar* gene encodes phosphinothricin acetyl transferase, which inactivates phosphinothricin herbicides by acetylation.
- The *bar* gene can be used with bialaphos of glufosinate; again, selection of the proper herbicide must be empirically determined as with antibiotics.
- Other commonly used herbicide resistance genes impart resistance to glyphosate, sulfonyl ureas and dalapon.

### Reporter Genes

- β-glucuronidase (GUS) is the most widely used reporter gene for plant transformation.
- The GUS gene has proven to be very useful in monitoring transient expression in a number of species.
- Plant tissues could be evaluated at various times following bombardment by reporter genes whose expression can be monitored without causing tissue damage. Such a system could greatly reduce the effort and time involved in selection of stable transformants.
- The firefly **luciferase gene** is a nontoxic reporter gene that has been successfully used to monitor gene expression in plants. After activity bombarded tissues are treated with luciferin (substrate) solution, luciferase activity is documented using either X-ray film.
- Luciferase activity can also be assayed using extracts of plant tissue that are incubated with the luciferin substrate.
- The detection techniques for luciferase severely limit the flexibility of using this reporter gene.

### Introns

- The introns are present between the promoter and coding region of a gene.
- They significantly affect gene expression. Monocotyledonous species, in particular the Graminae, show significant enhancement of gene expression when using introns.
- Introns may act by stabilizing mRNAs, thus leading to greater amounts of translation products.

- Introduction of different introns between the cauliflower mosaic virus 35S promoter and GUS gene caused significant enhancement of transient gene expression in maize. The degree of expression varied depending on the intron.

### Promoter Analysis

The use of reporter genes such as GUS or GFP under control of strong promoters (e.g., CaMV 35S) allows for optimization of transformation efficiency of crop plants. Particle bombardment also provides a rapid and convenient means for evaluation of promoter strength, analysis of the function of different regions of promoters, and for examination of tissue specificity of gene expression. The promoter, either in its entirety or with various deletions/mutations, is fused to a reporter gene or introduced into the target tissue. This approach has successfully been used in a number of different species.

### Codon Usage

The ability to introduce genes into plants from nonplant sources had led to the realization that genes from different organisms can be processed very differently with respect to their expression when compared with have native plant genes (Diehn, et al., 1996). Unaltered DNA from nonplant sources may contain DNAs that are recognized by plants as signals for splicing or polyadenylation. Changes for codon usage resulted in a 300-fold increase in expression levels of a synthetic Bt gene.

### Target Tissue

A variety of different plant tissues have been used as targets for particle bombardment-mediated transformation. Selection of the appropriate target tissue is dependent on the nature of the research. For rapid gene expression studies, various plasmid constructions can be introduced into different tissues and transient expression can be quickly analyzed to assess promoter activity without the production of stably transformed plants (IIDA et al. 1995). Almost any tissue can be used for transient expression studies, as long the cell wall is penetrable by the DNA-coated particles. The choice of the appropriate target tissue, physiological state of the plant material, and pre- and post bombardment treatments are critical to the successful regeneration of stable transformants.

### Embryogenic Cultures

Embryogenic cell cultures have been very useful for the production of stably transformed plants. Particle bombardment of embryogenic maize and wheat cultures was used to generate the first transgenic plants of these species. Embryogenic cultures provide a reliable source of tissue for optimization of bombardment conditions and can be used for year-round production of transgenic plants.

### Shoot Apical Meristems

Shoot meristem cells are most effective target for production of stable transgenics. Theoretically, introduction of DNA into would result in the production of chimeric plants, where the transformed cells would directly give rise to germ-line tissue and the introduced DNA would be passed on to the progeny. This approach could potentially avoid or reduce tissue culture manipulations but there are many problems with shoot meristem bombardment.

In sunflower, bombarded shoot apices of particles alone (no DNA) was done and then inoculated the tissue with *Agrobacterium*. In this case, particle bombardment was used as a method to wound the target tissue for transformation via *Agrobacterium* rather than for direct DNA introduction itself. Although the resultant plants were chimeric for the introduced gene(s), it is important to note that many of the transformation events were germ-line and transgenic progeny were recovered.

### *Other Target Tissues*

Particle bombardment can be used to introduce DNA into any target tissue that is penetrable by the DNA-coated particles. The other tissues that have been subjected to particle bombardment include leaves, pollen, styles, cereal aleurone cells and tassel primordial.

## Tissue Treatment

One treatment that greatly influences gene expression is osmotic conditioning. Osmotic conditioning can be both pre- and post-bombardment. Tissues are subjected to either partial drying in a laminar flow hood or cultured in a medium containing an osmotic agent such as mannitol and sorbitol at relatively high concentration (~0.4 M). Osmotic treatment can enhances both transient expressions as well as significantly increase stable transformation. The osmotica may act by causing plasmolysis of the target cells that lessens or eliminates extrusion of the protoplasm in cells that are penetrated by particles.

Other media additives are silver thiosulfate along with mannitol. They increase GUS expression in bombarded scutellar tissue of wheat, presumably by reducing the negative effects of wound ethylene. Inclusion of abssisic acid with an osmoticum (myo-inositol) increased transient GUS expression 1 day after bombardment of *Picea* embryogenic cultures.

## The Fate of the Introduced DNA(s)

Although particle bombardment has been used for DNA delivery to plant cell for some time, it is still not clear what happens to the majority of DNA that is introduced. Transient expression studies indicate that large numbers of cells can receive the particles resulting in approximately 10,000 GUS-positive cells per bombardment. In cells that transiently express the introduced DNA, particles are most often seen either in or directly adjacent to the nucleus.

### *Recombination*

Plasmid DNA which is introduced into plant cells via particle bombardment undergoes recombination with other plasmid molecules as well as with the host chromosomal DNA. Recombination occurs whenever new gene arrangements are formed through exchange, elimination or insertion of DNA. Plasmid DNA can either insert cleanly into plant chromosomal DNA as a single entity or large linkages of plasmids can form and subsequently integrate (the full range of intermediate combinations is possible). Integration of a large chain of plasmids can result in a single insertion event with a high copy number of introduced plasmid DNA. With transformed event tissue obtained via particle bombardment, a single integration event rarely results in the introduction of a single plasmid. When large numbers of plasmids integrate into the same site, the plasmids may recombine with each other prior to integration into genomic DNA.

Although the exact fate of DNA introduced via particle bombardment is not clear, it appears that plasmid DNA molecules rearrange or recombine with each other to yield linkages of both whole plasmids and plasmid fragments.

### Position Effect

The expression of the introduced gene(s) is dependent on the nature of the DNA as well as the final physical arrangement of the foreign DNA in the plant genome. The influence that adjacent, native plant DNA has on expression of the introduced DNA is called *position effect.*

If the foreign DNA integrates in a region of DNA that is active, the introduced gene may be expressed at relatively high levels. If the introduction occurs in a region that is not transcriptionally active, gene expression may be reduced.

Although the T-DNA from *Agrobacterium* is generally inserted into transcriptionally active regions, DNA that is introduced via particle bombardment show no known preference for insertion sites. Position effects control the variation in gene expression in different transformed clones. Methylation of introduced DNA has been shown to occur in cases where the introduced gene shows minimal activity. However, the precise association of methylation with position effects is still not clear.

### Co-suppression

A reduction in expression of a gene resulting from the introduction of multiple copies of the gene is called *co-suppression.* Co-suppression is useful in some situations when a reduction in expression of a native gene is desirable. Unfortunately, co-suppression is also observed as an undesirable product of particle bombardment-mediated transformation when multiple copies of the same gene are introduced.

### Scaffold Attachment Regions

To stabilize gene expression in transgene plant tissue, scaffold attachment regions (SARs; also known as matrix attachment/associated regions, MARs) have shown some utility. SARs are regions of DNA, which were isolated based on their ability to bind to the nuclear scaffold. For transformation, the SARs are ligated to the flanking regions of the gene of interest. They apparently act by stabilizing or normalizing gene expression rather than enhancing overall activity. SARs from yeast and tobacco have helped to stabilize gene expression in plants while a human and soybean SAR were nonfunctional. Particle bombardment-mediated transformation efficiency could potentially be increased using SARs as SAR activity was found in Transformation Booster Sequence isolated from petunia. SARs appear to be especially useful in cases of high copy number DNA introductions, which would normally result in highly variable gene expression profiles.

### Targeted Recombination Systems in Plants

In order to have better control on the expression of the introduced DNA(s), it would be beneficial to predict or *direct the site integration.* This can be accomplished through either *agrolistic transformation* or homologous recombination

**Agrolistic Transformation or Homologous Recombination:** For agrolistic transformation, plant tissues are bombarded with plasmids that contain some of the DNA excision components from the *Agrobacterium* gene transfer system. Approximately 20% of agrolistically transformed plant cells contained precisely defined DNA inserts that would be expected for *Agrobacterium*-mediated transformation.

For targeting using homologous recombination, genes could potentially be placed in certain areas of the genome based on sequence homology between the introduced and the genomic DNA. Although targeted or homologous recombination is fairly well characterized in some organisms, it has not been well studied in higher plants. Very little is known about native plant recombinases with most recombination studies in plants relying on information and genes that are available from nonplant systems.

**Homologous Recombination Systems:** Two homologous recombination systems have been evaluated in plants.

- *Cre-lox system* from bacteriophage P1
- *FLP/FRT system* from yeast

In these systems, a target site containing specific DNA sequences must be introduced specific recombinase, a second piece of DNA using transformation. Following activation with a specific recombinase, a second piece of DNA replaces the target DNA through homologous recombination.

## GENE TRANSFER BY MICROINJECTION TECHNIQUE

Intranuclear microinjection is a direct physical approach that overcomes many of the biological and other obstacles. Originally developed for animal cells, it is now applied to plant cells also. The microinjection technique uses a fine capillary needle to deliver DNA precisely into cells (even nuclei of cells) in such a way that the cells are not killed but survive the treatment and can develop further into clones.

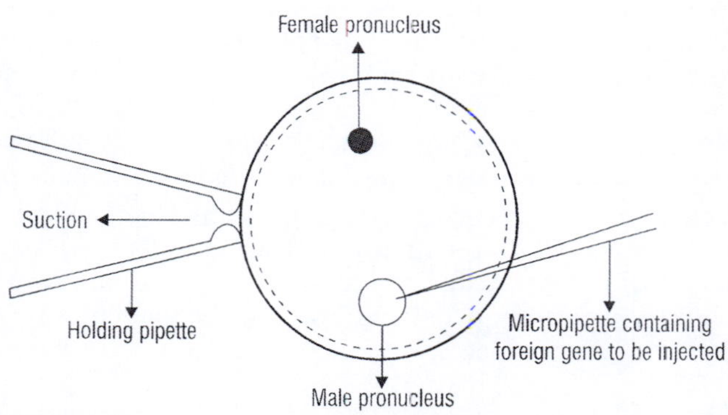


**FIGURE 28.12:** Microinjection system.

For transformation, microinjection systems are designed for speed and operational simplicity as these factors govern the total number of cells that can manipulate. To facilitate rapid injection rates, protoplasts with partially reformed cell walls can be attached to solid support (glass coverslips or slides etc.) with poly-L-lysine or artificially bound substrate without damaging the cells. After one day in culture, the cell walls of protoplasts are firm enough to provide support and protection yet they can be easily penetrated with micropipettes with 0.5-10 $\mu$m diameter tip). Each protoplast is microinjected, segregated and transferred to optimal condition to obtain transformed cell line. An alternative method employs a blunt holding pipette to hold each protoplast under suction and segregating it directly into micro drop culture after microinjection.

Several advantages are associated with the microinjection technique. First of all, the amount of DNA delivered per cell is not limited by the technique and can be optimized. This improves the chance for integrative transformation. Secondly, the delivery is precise, even into the nuclei of the target cell, again increasing the chance for integrative transformation. Thirdly, the small structures can be injected containing only a few cells and with high regeneration potential, for example, microspore derived and zygotic proembryos. Lastly, it is a direct physical approach and therefore host-range independent.

However, there are some major drawbacks like, only one cell is targeted per injection and the handling requires specialized skill and instrumentation. If hundreds of injections can be performed per working day, low transformation rate is yet not a serious disadvantage.

## Applications of Microinjection Technique

Different applications of microinjection technique in plant science have been reported. Some experiments dealt with the introduction of fluorescent dyes into differentiated cells to study intracellular transport. Efforts have been made to microinject microorganisms into suspension cells. Attempts to transfer chromosomes or even cell organelles have been made. A common feature to these experiments is low numbers of manipulated cells.

DNA has also been microinjected in mice and fish. In mice, foreign DNA is injected into the mole pronucleus, residing in fertilized ovule. The injected zygotes are cultured *in vitro* to blastocysts and then placed inside a foster mother. This technique has been used to introduce rat growth hormone gene. Transgenic fishes have been obtained in many species like carp, catfish, gold fish, salmon, Tilapia, rainbow trout and Zebrafish. As the male pronucleus is difficult to locate in fishzygote, the DNA is injected into cytoplasm of fertilized ovule within first few hours of after fertilization. As the fishes undergo external fertilization, transfer of embryos into foster mothers is not needed. Transformed embryos are instead hatched in trays. Various gene transferred into fish are human or rat growth hormone, chicken delta-crystalive protein, *E.coli* β-galactosidase gene, *E.coli* neomycin resistance gene, *E.coli* hygromycin resistance gene, winter flounder anti freeze protein (AFT) gene, and, cDNA of rainbow trout growth hormone. Cytoplasmic microinjection is a tedious and time-consuming procedure and is unsuitable for mass gene transfer.

## The Technique

The macroinjection technique involves the use of injection needles with diameters far wider than cell diameters for delivering DNA into wound sites within tissues. However, at the same time it destroys those cells, which are in direct contact with the injected DNA. To integrate, the DNA must gain entry to intact cells adjacent to the wound site.

There are a variety of problems associated with experiments that require DNA passage through single, double (in the case of pollen) or even many cell walls. The uptake of DNA across cell walls is a very rare event. Majority of DNA molecules applied get attached to cell walls before they have a chance to enter the cell. Penetration is further reduced by subsequent presence of cell wall. Moreover, DNA does not normally move from cell to cell. Thus, recombination occurs at relatively low frequency by this method. Microinjection has been mostly applied to ovule, embryos and other tissues

### *Lipofection*

The use of liposomes for transformation of transfection system is called lipofection. Liposomes, as defined by Bongham, in 1965, are special type of lamellar phase in which water is self contained and the lipid molecules are disposed in bimolecular layers attached by their non-polar interfaces.

Liposomes possess many properties akin to those of biological membranes. They are easy to manipulate, their lipid contents can be varied at will and many substances may be trapped inside the interlamellar spaces. They are found to be ideal carrier systems and are now used in genetic engineering to deliver foreign DNA molecules into a desired cell.

DNA or RNA, encapsulated in liposomes, can be transferred into plant protoplasts by either direct fusion with the naked plasma membrane or through endocytosis of the liposome. Although it is efficient in terms of DNA encapsulation and transfer, the method once again requires protoplasts and its application is limited to species for which protocols are available for protoplast regeneration. Foreign macromolecules, for example impermeant fluorescent dyes, RNA and DNA, can be introduced into walled plant cells by an alternative strategy which involves encapsulation of the desired probe into liposome that can be pressure-injected into the vacuole. There are two advantages of this method. First of all, the liposome protects the DNA from being damaged by the acidic pH and proteases present in the vacuole. Secondly, when liposomes are constructed from material that is compatible with the lipids of the tonoplast, liposome-tonoplast fusion occurs (Madore *et al.*, 1986) and so the encapsulated DNA should be successfully delivered into the cytoplasm.

Liposomes would appear to offer the following advantages as vectors for gene delivery into plant cells:

- Enhanced delivery of encapsulated DNA by membrane fusion.
- Protection of nucleic acids from nuclease activity.
- Delivery into a variety of cell types besides protoplasts entry through plasmodesmata.
- Delivery of intact small organelles.
- Protect the genetic material from immunogenic reaction (liposomes are non-immunogenic).
- Composition of lipid can be manipulated to confer specific properties to liposomes.

The use of liposome for delivery of gene was first described for delivering polio virus RNA. Later liposomes were used in combination with polyethylene glycol for delivery of plasmid DNA into tobacco protoplast.

## SILICON CARBIDE FIBER-VORTEX (SILICON WHISKERS

Silicon carbide forms microscopically small needles (silicon whiskers) that can be used to perforate plant cell walls and plasmalemma of suspension cells or protoplasts. Advantages of the silicon carbide fiber-mediated transformation method over other procedures include the ability to transform walled cells (thus avoiding protoplast isolation), relative ease of the procedure, and very low equipment costs.

Technically, the method is very simple: a mix of whiskers and plant cells in a small test tube is shaken. This method involves a mixture of plasmid DNA encoding a selectable or screenable marker gene, silicon carbide fibers, and tissue of the species to be transformed in an Eppendorf tube. Mixture is vortexed to facilitate cellular penetration by enhancing frequent and forceful intercellular collisions. DNA deliver in this system is presumably due to cell wall penetration by DNA coated silicon carbide fibers. The silicon carbide fibers apparently act as microinjection needles facilitating DNA delivery.

Only Eppendorf tubes and a vortex is sufficient equipment. This opens the way for DNA diffuse into the cells (Kaeppler et al., 1992). Although, details of the mechanisms by which silicon carbide wiskers work is unclear, stable transformation with maize has been recently published (Frame et al., 1995). The overall efficiency of the wiskers approach is between one-fifth and one-tenth that of conventional microprojectile shooting, but is advantage would be its *low tech* character, dispensing with any sophisticated or expensive instrumentation. However, currently, only maize has been transformed and a highly regenerating system was used. It remains an open question whether the whiskers approach will be useful in organized tissue.

A procedure has been recently developed for the injection of DNA into *Zea mays* and *Nicotiana tabacum* tissue cultures using silicon carbide fibres.

## SONICATION

The latest method developed uses ultrasound to facilitate the uptake of nucleic acids into plant cells. By this technique, plant protoplasts and cells in suspension and intact pieces of plant tissue can readily be transformed. The protoplasts, suspension cells or small pieces of tissue are suspended in a few ml of sonication medium in a microcentrifuge tube. Plasmid DNA (and possibly carrier DNA) is then added and after rapid mixing the samples is ready for sonication.

Ordinary machines deliver the pulses of ultrasound. The microtip (normally tapered) is immersed 2-3 mm into the suspension and pulses of selected intensity and duration are delivered. The cells are finally transferred to fresh growth medium.

There are two possibilities for permeabilization of cell and uptake of DNA. Firstly the violent collapse of activation bubbles, generating high-pressure and high-temperature shock waves could possibly cause localized rupture of the plasmalemma and lead to uptake of exogenous solute. This may be followed by re-establishment of membrane integrity. The second hypothetical possibility may be the existence of a critical hydrostatic pressure at which the intrinsic membrane potential is sufficiently high to induce mechanical breakdown of the membrane (60-100 MPa) in a hyperbaric chamber. Consequently it is possible that either the high oscillating pressure generated by the ultrasonic field and/or the high-pressure shock waves originating from collapse cavitations could produce such higher hydrostatic pressures that reversible membrane breakdown would occur.

This method has been applied for transferring plasmid DNA into protoplasts, suspension cells and tissues of sugar beet and potato.

## ELECTROPORATION

This method uses an electrical pulse to introduce DNA into cells. By exposure to brief electric shock, the cells become temporarily permeable to DNA. If the required DNA fragment is kept in direct contact with the protoplast membrane, the entry of DNA molecules into the cells is facilitated. Successful DNA transfer through this technique has been shown in bacterial, yeast, fungal, plant and mammalian cells. Among plants, successful gene transfer using electroporation method has been reported in protoplasts of tobacco, petunia, maize, rice, wheat and sorghum. In carrot, maize and tobacco protoplasts, bacterial CATgene with different plant promoters has been introduced and efficiently expressed. Electroporation is applicable to both monocot and dicot protoplasts.

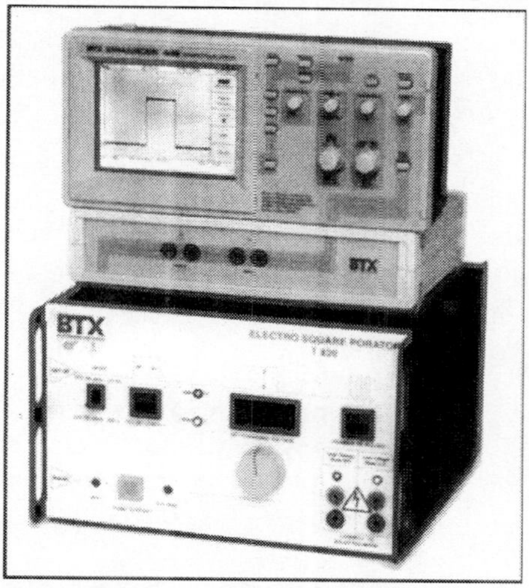


**Fig 28.13:** Electroporation apparatus

It has been recently hypothesized that the mechanism of DNA entry into the cell during electroporation may be either due to unspecific membrane process, or it may be specifically mediated by permease. The activation of permease may be due to high electric field, and thus, may lead to thinning of membrane. Hole may be created as a result and stabilized by a favourable dipole interaction with the electric field. Various advantages of electroporation have been listed. It is a convenient, simple, fast and efficient method applicable to a wide range of plant protoplasts and involves low cell toxicity.

The main disadvantage of electroporation is its requirement for protoplasts and therefore a system where regeneration of these protoplasts to calli or whole plants has been established. Consequently, it may not be applicable to some plant systems. Recently an electroporation system using organized and easily regenerable tissues such as immature maize embryos has been developed.

## POLY-CATION MEDIATED DNA UPTAKE

Direct DNA uptake by protoplasts can be stimulated by using hydrophilic, long chain polycations such as polyethylene glycol (PEG), poly L-lysine, poly L-ornithine or DEAE-dextran with or without metal ions such as $Zn^{2+}$, $Li^{+}$, $Cs^{+}$, $Rb^{+}$, $K^{+}$, $Na^{+}$, $Ca^{2+}$ or $Mg^{2+}$. The technique is efficient and can be applied virtually to every protoplast system and for transfer of even small quantities of DNA.

Polyethylene glycol is an extremely hydrophilic molecule and it removes 'free water' when placed in a solution. Free water indicates such molecules, which interact with charged (usually ionic) molecules soluble in water.

Thus, under high PEG concentration (15-25%), ionic macromolecules such as DNA can no longer stay in solution and get precipitated. The protoplast membrane as well as DNA is normally negatively charged, and therefore, charge repulsion is obvious. This repulsion limits the interaction between DNA and protoplasts. At very high concentration, PEG seems to minimize charged repulsion effects thus stimulating DNA uptake by endocytosis without any gross damage to protoplasts. Studies have been conducted with *Petunia*, *Nicotiana*, rice, maize etc. In these experiments, PEG was used in combination with pure Ti plasmid or calcium phosphate precipitated Ti plasmid mix with a carrier DNA. Mechanism of cross membrane transport in presence of PEG is still to confirm.

DNA-uptake may be enhanced as a result of the interaction of the polycations at one or more of the following levels:

- Protection of the exogenous DNA against nuclease degradation.
- Increase in the permeability of the plasma membrane.
- DNA transformation by using the polycation "polybrene" (hexamidethrine bromide) is new, although similar techniques have been used for several years in the transformation of animal cells. The major advantages of the polybrene technique are the following:
- It is less toxic than other polycations
- The high transformation efficiency allows very small quantities of plasmid DNA to be used, thus it is relatively easy to introduce only one sequence of a genome. Even with quite high plasmid concentrations only one or a few copies of the plasmid are integrated and concatamers of integrated plasmid are rarely observed.
- Since no carrier DNA is used to integrate plasmid DNA into the host genome thus, the sequences surrounding the site of integration can be analysed.
- The transformation frequency appears to be independent of nature of plasmid (supercoiled or linear).

The polybrene method can be used for both stable and transient expression.

## CA-DNA CO-PRECIPITATION METHOD

The basis of the Ca-DNA technique is the formation of a co-precipitate of plasmid DNA with calcium phosphate, formed by adding $CaCl_2$ to a solution of DNA in phosphate buffer. On contact with protoplasts, the co-precipitate is transferred across the plasma membrane in a calcium requiring process; and in order to encourage the endocytotic uptake of the precipitate, the protoplast/Ca-DNA complex is treated with polyvinyl alcohol and high pH calcium. This method was first used for plant transformation by Hain *et al.*, (1985) using a tobacco leaf protoplast system.

## ULTRASONICATION

Gene transfer by ultrasonication is the latest and simple method, where, plant protoplasts, cells in suspension and intact pieces of plant tissues can be transformed. The target cells are suspended in sonication medium in a microcentrifuge tube. Carrier DNA is added and microtip (generally tapered) is immersed in the sonication medium. The ordinary homogenizing machines deliver pulse of selected intensity and duration. Transformed cells are finally transferred for selection.

There are two hypotheses for permeabilization of membrane by sonication method. First assumes that reversible localized rupture of plasmalemma results in is due to violent collapse of activation bubbles, which results in generating high-pressure and temperature shock waves to plasmalemma. The second hypothesis predicts the existence of critical hydrostatic pressure responsible for reversible mechanical breakdown of membrane.

This method has been used in tobacco, wheat and sugarbeet. In tobacco, transgenic plants were obtained at 22% frequency with ultrasonication. Ultrasound transformation is unique among existing methods because of its versatility as plant protoplasts, cell in suspension and intact pieces of plant tissue can be readily transformed. This feature, along with the fact that the equipment for ultrasound transformation is simple, cheap and multifunctional should allow for substantial future applications.

## UV LASER MICROBEAM

A new method of introducing exogenous gene materials into cells is laser microbeam irradiation. DNA can be incorporated into whole plant cells and has been successfully applied to tobacco. DNA enters the cell through a self-healing hole made with a laser microbeam.

There are several advantages to this method. A laser beam can be used inside of tissues, cells and organelles, thus increasing the rate of transformation and the speed of operation. Wide controllability of the laser beam intensity, flexibility in optical manipulation wide choice for laser wavelength and pulse width is other advantages. It is also applicable to cells with hard walls. Its wide application in the field of cell biology is the ability to transfer cytoplasmic material. Finally, future versions of this system are expected to be more automated which should greatly increase output as well as the success rate.

This technique can be used for genetic manipulation of plant tissue, individual cells, protoplasts and organelles. Furthermore, it permits selective fusion in of part of a subcellular organelle present inside a single living cell.

A number of techniques exist today for transferring genes into plant protoplasts and cells. Most of these techniques use protoplasts because the cell wall is not sufficiently permeabilize for DNA uptake. As protoplasts of many species are often difficult to regenerate efficiently into fertile plants it cannot be applied to any system. These techniques can be preferably used as an analytical tool, for example, in testing the function of gene constructs.

Moreover, most of the systems require sophisticated and costly equipment. In addition, it does not allow the combination of both accurate quantitative estimates of gene expression and efficient generation of transgenic plants.

However, successful gene transfer has been reported in few systems using above techniques, and much research in the context is still in progress.

## GENETIC TRANSFORMATION IN MONOCOTYLEDONOUS PLANTS

The initial development of system for genetic transformation of monocotyledonous was slow. Formation of tumors and production of opines in certain monocotyledons suggested that at least some monocotyledonous species might be susceptible to infection by *A. tumefaciens*. *Agrobacterium tumefaciens* (*Ti*-plasmid).

### Factors involved in *Agrobacterium*-Mediated Transfer of Genes to Monocots

- This work is mainly done on cereals (particularly rice).
- Induction of *vir* genes, one of the most commonly used techniques in the transformation is the addition of phenolic compounds. Such as acetosyringene to co-culture or bacterial cultures.
- The cell division should be active.
- Bacterial strains and vectors
- Composition and quality of the culture medium.
- High osmotic pressure.
- Host genotypes, and
- Acidic pH.
- Selectable marker
- Temperature of the room below 28° C.

## PROBLEMS INVOLVED IN GENE TRANSFER

### Expression System

Adequate expression of gene is essential with high expression at right time and in a particular tissue. For high expression strong promoters like 35-S promoter of Cauliflower Mosaic Virus (CaMV) is used which has a wide host range. More specific promoters are required for tissue specific attachment regions (SAR's) are used to increase expression. SAR's act by forming loops in chromatin structure thereby reducing gene silencing in cells having limited number of transgenes. But high amount of m-RNA may not have corresponding amount of product.

### Cellular Localization

Abiotic stress causes injury to membrane (by formation of reactive oxygen intermediates) affecting membrane bound process finally influencing metabolism of plant. It is desirable to locate membrane modifying, osmoprotectant producing or radical scavenging enzymes to defined organelles. Transit peptides are required to ensure correct localization of enzymes expressed. Typical transit sequence used to direct polypeptides to chloroplast is from small subunit of Rubisco.

### Proteolysis, Protein Folding and Prosthetic Group Acquisition

Short polypeptides after expression may undergo proteolytic degradation. Processing and folding of proteins by co-expression of chaperon proteins in transgenic plants especially in case of large enzymes can avoid this. Oxygen radical metals like, Fe, Zn, Cu, Mn, Therefore, the metal ions should be present in adequate amount.

### Precursor Availability

Amount of precursor has a great effect on product formation. Precursor can be extracted from environment or synthesized in cell from existing metabolic pathways.

## Inhibitory Environments

Intracellular condition in transgenic plants (pH, temperature, ionic strength) affects foreign proteins. It is usually operated during varied environmental conditions. Therefore, the introduced gene should be taken from closely related plant.

## Side Reaction of New Compounds

Endogenous enzyme activities may deplete the product pool or form toxic compounds. Therefore, analysis of transgenic plants using NMR is important for detection of any metabolic changes.

# 29. Biotechnology Production of Secondary Plant Metabolites

Secondary metabolites, including antibiotics, are produced in nature and serve survival functions for the organisms producing them. The antibiotics are a heterogeneous group, the functions of some being related to and others being unrelated to their antimicrobial activities. Secondary metabolites appear to serve (i) as competitive weapons used against other bacteria and fungi, amoebae, plants, insects, and large animals; (ii) as metal-transporting agents; as agents of (iii) plant-microbe relations, (iv) nematode-microbe symbiosis, and (v) insect-microbe symbiosis; (vi) as sexual hormones; and (vii) as differentiation effectors. Although antibiotics do not appear to be obligatory for sporulation, some secondary metabolites (including antibiotics) stimulate spore formation and inhibit or stimulate germination. Formation of secondary metabolites and spores may be regulated in a similar way. This could serve to insure secondary metabolites production during sporulation for the following possible reasons: to slow down germination of spores until a less competitive environment and more favorable growth condition appear, to protect the dormant or initiated spores from consumption by protozoa, or to cleanse the immediate environment of competing microorganisms during germination.

Higher plants are valuable and the most important producers of natural products including food, fiber, wood and oils. However, several plants are rich in secondary metabolites, which are potential source of drugs, fragrances, pigments, food flavours and essential oils. Biosynthesis of metabolites although, controlled genetically, is also affected strongly by environmental factors and soil conditions. UNESCO (1960) has published a survey report with details of old world medicinal plants growing in different arid zones belonging to different families, which are most adapted to dry conditions for their growth and production of secondary metabolites.

It was in 1950s that the potentialities of the isolated cells of higher plants in producing the useful metabolites were recognized. At that time the technique of tissue culture was not very elaborate and thus could not be taken up for general use. Since then, technique of tissue culture has been developed for large scale cultivation of plant cells and production of metabolites has created keen interest in the researchers. The metabolites including, proteins, flavonoids, antimicrobials, pyrethrins, rotenoids, alkaloids, steroids, and glycosides have been detected from the tissues developed *in vitro*.

Concerns for health and well-being have made plants collected for drugs, colors, species, and fragrances a focal point of curiosity since the dawn of history. The arduous route from there to analysis and production of plant drugs is marked by medieval herbals, by the discovery of morphine in opium (Serturner, 1783-1841), and synthesis of acetylsalicylic acid (Aspirin/BAYER-Leverkusen) by H. Dreser in 1893. Biotechnology-assisted production of drugs from plants is roots in physiological plant anatomy (Heberlandt, 1854-1945), *in vitro* culture of plant cells (White, 1939; Gautheret, 1939), phytochemistry (Mothes, 1900-1983), the design of bioreactors (Tulecke and Nicell, 1959), and plant transformation (Horsch et al., 1985). Today, recognition of ecologic implications mandates that research and development (R&D) personnel better understand the synthesis and function of plant metabolites and guarantee supply while guarding natural resources. The recent paclitaxel (Taxol) supply crisis (Cragg et al., 1983) may underscore academic and public demand.

Primary metabolites in plant (i.e., proteins, oils, and starches) maintain essential physiological processes. Secondary metabolites, such as alkaloids, terpenoid, flavonoids, and a host of glycosides, mediate the relation between plants and their environment. Specifically, they constitute a chemical response to pollinators and distributors, to competitors and herbivores, and to pathogens and stress. Manipulation of secondary metabolites, therefore, not only may change the quality and quantity of products, but also may significantly change plant interaction with the environment. Low-alkaloid lupine mutants, for example, are said to have been widely appreciated by rabbits. Modern zero-glucosinolate rapeseed plants may or may not affect flea bettle populations.

The 20,000 plus organic compounds listed as secondary metabolites show an enormous diversity and, thus, allowed chemical description to complement conventional classification of plants by morphological features.

Research and development for the production of secondary metabolites have been directed primarily at cell technologies. Arguments in support of plant cell cultures stress the advantages of year-round availability of plant material, process isolation, and magnification-accentuation of chemical reactions under growth control.

The secondary metabolites have high economical and pharmacological importance and the industries are deeply interested in large variety of chemical substances being produced by plants due to their lesser toxicity. Though these substances are generally extracted from plant parts, the plant tissue culture technique has widened the scope and opened new vistas for the production of secondary metabolites.

Recently, Mitsui Petrochemical Industries (Tokyo) has achieved success in producing *shikonin* commercially in 750-lit fermenters from *Lithospermum* sp. plant cells. Shikonin is a valuable antibacterial dye. This is the only successful attempt so far achieved in the world by Japan. Thus the chances of exploitation of tissue culture technique for large-scale production of useful metabolites are bright.

In India, the beginning of this work goes back to 1964 when Mitra and Kaul at NBRI (National Botanical Research Institute, Lucknow) showed the production of *reserpine* from *Rauwolfia serpentina* tissue culture. Later on work on various metabolites was carried out in other laboratories as well in India.

## PRODUCTION OF SECONDARY METABOLITES IN PLANTS

The production of secondary metabolites in plants can be understood as the result of plant development, which differentiation in metabolism as well as in morphogenesis (Constabel, 1988). Or, the production of secondary metabolites is the expression of a plant genome under developmental control. On the other hand, secondary metabolites accidentally appear as a result of stress and disease. Such metabolites may be synthesized *de novo*, (i.e., by gene activation) and are referred to as *phytoalexins*. Offen metabolites produced in response to stress and disease result from stimulation of enzymes active regular metabolism and are referred to as *phytoanticipins* (van Etten et al., 1994). Both concepts invite genetic manipulation to better understand the biology of secondary metabolites in plants and to increase yields.

Today, the concept of secondary metabolite products as differentiation products is supported by the following four arguments:

- During the stationary phase of plant cell cultures
- Under hormonal control
- During maturation of cells to tracheids, giant cells, laticifers. or idioblasts
- During organ development (i.e., growth of roots and shoots)

## STAGES OF SECONDARY METABOLITES PRODUCTION

Extensive studies have been made in various laboratories worldwide to know when plant synthesizes secondary metabolites. Methodologies have been designed to know the stage for enhancement of synthesis and increase yields of secondary metabolites in cell cultures. Following are the phases/stages when plant synthesizes secondary metabolites *in vitro*.

### Stationary Phase

Kinetics of growth and secondary metabolite synthesis and accumulation has long and repeatedly been shown to co-occur with the stationary phase of cell cultures.

### *Hormonal Control*

Hormonal control of metabolite formation and accumulation has been well established for many systems. As a rule, deceleration of growth because of deletion or reduction of auxins in culture media prompts the appearance of products such as pigments or alkaloids. Cytokinins, on the other hand, have been demonstrated to stimulate production, as shown for nicotine in cell cultures of *Nicotiana tabacum* (Tabata et al., 1971); for indole alkaloids in cultures of *Catharanthus roseus* (Decendit et al., 1992); or for anthocyanins (Seitz and Hinderer, 1988) catechins, proanthocyanidins, and lignins (Zaprometov, 1988).

Growth regulators affect growth and synthesis of secondary metabolites of cultured cells. Brain and Lockwood reported that the dose and nature of hormone supplementation and the culture-age have very marked effect on steroidal levels while studying the cultured tissue of *Trigonella foenumgracum*. According to them certain combinations of auxins and cytokinins had synergistic effect while other had antagonistic influence on steroidal synthesis. The nicotine synthesis in the tobacco cell cultures is strongly inhibited by 2,4-D, whereas it is promoted by kinetin. The effect of growth regulators on secondary metabolism varies greatly depending upon kinds of metabolites.

Tabata and his associates observed that the synthesis of shikonin is inhibited both by 2,4-D and NAA but not affected at all by IAA using the *Lithospermum* sp. cell cultures. In *Morinda* cultures, Zenk et al. (1975) observed that formation of anthraquinones takes place in presence of NAA but not in presence of 2,4-D. However, the synthesis of anthraquinones in *Cassia tora* remained ineffective by 2,4-D. Brain (1974) observed stimulatory effect of 2,4-D in L-DOPA synthesis. Although some promising information have already been obtained, more researches are needed for improving biosynthetic rate of secondary metabolites in *in vitro* using biotechnological advances.

## *Precursors and Secondary Metabolites Production*

When precursors or intermediate precursors are fed in culture medium they increase the synthesis of various photochemical in *in vitro*. They are considered to be the limiting factors such as, cinnamic acid or phenyl alanine increases the synthesis of flavonoids and tropic acid for tropane alkaloid. High doses are proved to be lethal to tissue growth.

Shikonin production increases three-folds when *Lithospermum* sp. cultured cells are fed with L-phenylalanine, however, *p*-hydroxybenzoic acid are found ineffective. When *Datura* sp. cell suspension cultures are supplemented with hydroquinone, in traces, the arbutin synthesis increases considerably.

From economic point of view it is necessary that the precursor must be of low price than that of the ultimate product. Ogutuga and Northcote used inexpensive ammonia as nitrogen source and increased caffeine production nearly four times in tea suspension cultures.

### Cell Maturation

In *Papaver bracteatum* cell cultures, deprivation of hormones led to organogenesis (differentiation) and subsequently to an elevation of the morphinan alkaloid concentration. Concurrent microscopic analysis clearly revealed a correlation between the appearance of thebaine and laticifer-like (Kutchan et al., 1983). Supplementary experimentation of morphinan alkaloids. Interestingly, it was not a precondition for the accumulation of related benzophenanthridine alkaloids (sanguinarine), or their mutual precursor, dopamine. The relation between cell morphology and metabolite synthesis and accumulation again came to the fore when cells of protoplast-derived *Catharanthus roseus* clones were investigated. In each clone, most of the cells maintained only one of two shapes, spherical or cylindrical. The cell aspect ratios (length/width) for most clones was stable for more than 2 years of subculture. Remarkably, the production of indole alkaloids ajmalicine and catharanthine was significantly greater when the cell aspect ratio was more than 2.8.

The degree of differentiation of a given tissue certainly makes for differences in the spectrum of products. The appearance of triterpenoids as a function of the degree of tissue differentiation is not surprising. The correlation of accumulation of laticifer-specific triterpenes in cell cultures of *Asclepias syriaca* and *Taraxacum officinale* and the occurrence of proper laticifer cell in tissues and regenerants demonstrates the point. Regenerants (somaclonal variants) without latex, as observed once with *Papaver somniferum* cultures.

### Organ Development

The competency of *in vitro* cultures for metabolite synthesis and accumulation, once roots or shoots are formed, has been established rather solidly. The ultimate in morphological differentiation for accumulation of secondary metabolites is the formation of trichomes in shoot cultures. Their structure can be rather complex because they provide effective compartmentalization of products that are generally toxic. Hairy roots and shoots present the most definite answer to metabolite production by organized tissues and organs. These materials are known to grow profusely while retaining histological integrity. For example, poppy plants (*Papaver* spp.) readily respond to *in vitro* culture and may synthesize and accumulate benzophenanthridines such as sanguinarine. Recently, co cultivation of *P. somniferum* hypocotyl cells with *Agrobacterium rhizogenes* gave rise to transformed cell cultures and, subsequently, to adventitious shoots. These accumulated tenfold higher level of morphinan alkaloids, specifically of thebaine and codeine, than controls. Conversion of these zero-morphine shoots to codeine plants is intriguing (Yoshimatsu and Shimomura, 1992).

### *Growth Factor*

Biosynthetic activity of cells in a batch culture depends on cell growth and substrate utilization. Not much is known about the correlation between the rate of secondary metabolite formation and age of individual cell in culture. However, production-growth pattern could be categorize in three major types:

- **First type:** Product-production precedes parallel with cell growth for example anthraquinones, morindone, nicotine, tropane alkaloids, etc.
- **Second type:** Product-production is delayed until cell growth declines or stops, for example polyphenol, shikonin, etc.
- **Third type:** Product-production declines as the cell growth increases for example, diosgenin, ascorbic acid, etc.

The growth phase of cells in cultures in many cases such as shikonin, anthocyanin and certain phenols production can be altered by auxin level in growth medium.

### Morphogenic Differentiation

In nature, certain compounds are synthesized and stored up only in some specific plant parts such as essential oils in certain sex glands or duct; tropane alkaloids in roots of tobacco; and latex in laticifers ducts. Cells in suspension cultures cannot synthesize such compounds reported in intact plants but if organogenesis is induced in cell cultures, these are synthesized *in vitro*. In *Scopolia parviflora* suspension culture, root initiation coupled with normal production of tropane alkaloids; alkaloid content increases many folds when organogenesis is induced in suspension culture of *D. innoxia*.

Shikonin, a derivative, found localized in cork cells only, has been produced exceptionally in suspension cultures. This finding has opened the path for *in vitro* production of certain monotrepenes, α-pinene, linalool, etc. without organogenesis in future.

### *Somaclonal Variation and Biosynthetic Activity*

Cells grown in cultures for a long period undergo extensive genetic changes such as polyploidy, aneuploidy, Anaphase Bridge, fragmentation of chromosomes and gene mutation. Biosynthetic ability of cultured cells may be improved selecting high producing clones or by the artificial induction of genetic mutation.

The chemical mutagens tested for the induction of genetic variations in carrot cells are N-methyl-N'-nitro-nitrosoquanidine. As a result, carotenoid yield in some improved clones increased three times in comparision to its wild types. Similarly, variants of sugarcane were isolated which showed appriciable increase in sucrose content in addition to disease resistance against Fiji virus, downy mildew, eye spot disease and smut. In some cases, the somaclones of sugarcane are unstable, while in other cases somaclones have been found stable when grown in four locations over five years. Somaclonal variation has also been reported in plants regenerated from protoplasts. Cellular variation, is therefore, another factor which can regulate secondary metabolism and has potential use in improving biosynthetic capabilities of culture strains. It is theoretically possible that the frequency of obtaining mutant cells may be markedly enhanced by mutagen treatment of haploid cells. However, mutation research on plant cell culture should be one of our major concerns for the near future.

### *Effects of Nutrients on Production of Secondary Metabolites*

The addition of nutrients, growth hormones, vitamins, etc. in culture media is primarily aimed to increase cell growth in cultured condition. It has been reported that certain nutrients in culture media increase some secondary metabolites while others show an inhibitory effect.

Addition of sucrose in culture media above its ordinary level increases shikonin accumulation in cultured cells, lower concentration of sugar increases production of ubiquinone-10 in tobacco cell culture. Carbon-nitrogen ratio (C:N) plays a vital role in the increase production of catechol tannins in *Scynamore* sp. suspension culture. Unfortunately, not much study has been made focused in this field.

### *Environmental Control on Production of Secondary Metabolites*

Light is considered as an important environmental factor, which controls the synthesis of most of the secondary metabolites *in vitro*. Appreciable quantitative changes have been observed in volatile oil content in cell cultures of *Ruta graveolens* when grown in light and dark separately.

Using Parsley cell culture, Hahlbrock et al. (1971) noted a marked increased in the accumulation of flavone and flavonol glycosides in illuminated cultured cells, especially with ultraviolet light. Many workers have also reported stimulatory effects of light on the production of medicinal compounds. Increased yield of polyphenols, plastoquinones, flavonoids and carotenoids has been observed when cultures were grown under light.

Inhibitory effects of light have also been observed specially with white and blue light on shikonin derivative formation when cell cultures of *Lithospermum* sp. However, in other cases no significant changes in the content of secondary metabolites could be observed when cell cultures were grown in light or dark.

## Stress Products

Secondary metabolite production has been viewed as the result of differentiation, exclusively, until the concept of chemical defense gained ground, and elicitors were employed for metabolite production *in vitro*. In a wider context, the occurrence of some metabolites was seen and interpreted as response to stress. The following are stress factors influencing the production of secondary metabolites in plant cell, tissue, and organ cultures.

- Exposure to osmotica
- Exposure to UV light
- Exposure to elicitors

### *Osmotic Stress*

The relation between osmotic stress caused by high levels of sucrose in nutrient media and production of secondary metabolites has been known for some time, and corroborative observations have been added ever since. Interestingly, products now include not only anthocyanins and alkaloids, but also terpenoids. For example, increasing the carbohydrate concentration to 6% (3% sucrose plus 3% mannitol) and adding 25 μM abscisic acid significantly increased the accumulation of monoterpenes in shoot cultures of *Lavandula latifolia*, primary of cineole.

### *Ultraviolet Radiation*

The response of plants in high altitudes to UV irradiation by anthocyanin formation is a common observation. Pigmentation in cell cultures of parsley (*Petroselinum hortense*) on treatment with UV light has not only improved conditions for metabolite production, but also has significantly affected plant molecular biology in our understanding of the induction, signal transduction, and gene expression in the biosynthesis of plant metabolites.

### *Biotic and Abiotic Stress*

Cell cultures have been exposed to very different kinds of stress, biotic and abiotic. Most radical was the use of fungal homogenates and extracts in nutrient media as elicitors of alkaloid synthesis and accumulation in plants and plant cell cultures. When applied to poppy (*P. somniferum*) cell cultures, not morphinan as expected, but saguinarine production was the result. Shortly thereafter in periwinkle (*Catharanthus roseus*) cell cultures, fungal homogenates led to the formation of catharanthine. The list of materials successfully treated with fungal elicitors for production of secondary metabolites has grown rapidly and conditions to do so. Investigations of elicitor-induced production of benzophenanthridine alkaloids in *Sanguinaria canadensis* cell cultures were extended to include the effect of pretreatment with a specific calcium chelatant and a calcium channel inhibitor. As a result, elicitor-induced alkaloid accumulation was seen as mediated by calcium and possibly calmodulin and/or protein kinase C. Product formation through fungal elicitors was stimulated by heavy-metal ions such vanadyl ions.

### Genetic Engineering Products

The manipulation of metabolism in plants through DNA technologies for enhanced production of secondary metabolites is in part curiosity-driven. Finally, this technology allows introduction of experimentation into biochemistry and phytochemistry, where previously the *modus operandi* was description. In part, recombinant DNA technologies are also pulled by the resource and processing industries owing to prospects of marketable and profitable botanicals. Here, the concept is that nature did not maximize product formation in plants, that value may be added through genetic engineering.

Generally, the strategy is aimed at enabling transgenic plants to synthesize new and novel products by

- Completion of partial pathways
- Amplification of regular pathways
- Blockage of competing pathways
- Interdiction of regular pathways
- Revising metabolic regulation
- Minimizing response cascades

### *Insertion of Heterologous Genes*

Gene insertion for synthesis of enzymes that will upgrade natural products has been accomplished. Specifically, a hydroxylase gene from *Hyoscyamus niger* was introduced into hyoscyamine-rich *Atropa belladonna*, with the result that transgenic material now produced the more valuable scopolamine (Yun et al., 1992). Interestingly, one single copy of the transgene was sufficient to create an all-scopolamine-type shoot. In the future, conventional breeding efforts for including metabolically engineered plants, as may well complement improved botanicals demonstrated here.

### *Amplification of Pathways*

The processes secondary metabolites can be enhanced by genetic modification of cells and plants. Such amplification would *simply* require activation of rate-limiting enzymes by insertion of homo- or heterologous genes, thereby removing *bottle-necks*. Apart from pigment synthesis in petunia, the isolation, characterization, and heterologous (sense-) expression of the tryptophan decarboxylase and strictosidine synthase genes for monoterpene indole alkaloid synthesis *Apocynaceae* and of the berberine bridge enzyme in *Papaveraceae*, which catalyzes the conversion of (*S*)-reticuline to (*S*)-scoulerine in the pathway to benzophenanthridine alkaloids (sanguinarine), are significant recent steps toward the goal of genetically improving the quality and quantity to secondary metabolites. Yields, however, are not yet overwhelming.

Attempts to enhance the production of nicotine in root cultures of *Nicotiana rustica* by over-expressing a heterologous ornithine decarboxylase gene amounted to a twofold increase of putrescine and, consequently, a twofold increase in nicotine.

### *Competitive Pathway Blockade*

Glucosinolates accumulate in the leaves and seeds of *Cruciferae* and so in rapeseed plants. They give an unpleasant teste to meal and cause thyroid disorders in livestock. The strategy to reduce glucosinolates in rapeeed crops would be to insert genes encoding enzymes that complete without harm for a common precursor, here a sulfur compound (i.e., flavonol-3-sulfate). So far, expression in host material of a sulfotransferase gene has been low. An alternative suggested strategy would be to produce rapeseed with tryptophan decarboxylase transgenes to direct tryptophan away from the synthesis of indole glucosinolates and to increasing the tryptamine pool, and indeed, transgenic rapeseed plants responded with higher tryptamine and dramatically lower glucosinolate levels, only 3% of that found in nontransformed controls.

### *Use of Antisense Genes*

Experimentation based on the insertion of antisense genes of rate-limiting enzymes of the phenylpropanoid pathway (i.e., phenylalanine ammonia-lyase and chalcone synthase) led to curtailment of transcription processes and inhibition of flower pigmentation in transgenic petunia and tobacco. The result has had a strong influence on attempts manipulate metabolite reduction in botanicals, as demanded by industry: lignin in wood and glucosinolates in rapeseed meal.

Reduction of the amount of lignin in cell walls of trees could assist the pulp and paper industry, and its reduction in herbs could benefit the forage industry.

The omnipresence of lignin in higher plants has not made it any easier to elucidate its analysis, synthesis, polymerization, and deposition. The identification of rate-limiting enzymes in lignin synthesis has been slow, three enzyme systems have now been isolated and characterized. For one, an *O*-methyl-transferase antisense gene from aspen has been expressed in tobacco. The expected suppression of precursor (i.e., ferulic and sinapic acid synthesis) did materialize and led to lignin reduction.

Antisense technology has also been applied to suppress the activity of two ayringyl alcohol (i.e., cinnamyl-CoA-reductase and cinnamyl alcohol dehrydogenase; CAD) in tobacco. As a result, CAD activity was reduced; however, judged to be more accessible to chemical extraction, however, and thus may well affect the pulp industry. Interestingly, staining sections of tobacco with phloroglucinol·HCl made possible a quick first differentiation of transformed and control material (Halpin et al., 1994). Lignin polymerization is catalyzed by peroxidase. However, transgenic tobacco plants with suppressed levels of this enzyme did not show changes in the amount of lignin in tissues.

### *Metabolic Expression Revision*

Much attention has been paid first to genes that encode enzymes, which catalyze steps along the biosynthetic pathway of secondary metabolites. In analogy to studies of flower pigmentation in petunia, such experimentation will have to be followed by studies of regulatory genes that control the flux of compounds along a pathway for expression, at specific sites, for specific patterns. For anthocyanin synthesis in petunia flowers the dihydroflavonol 4-reductase gene has been characterized as such. It will be interesting to see to what extent organized and unorganized tissue respond to changes designed for control of product synthesis and accumulation, by regulatory genes.

A most promising step has been accomplished with identification and characterization of genes that encode the major latex proteins in poppy and, thus, will permit us to target these cell components for genetic manipulations much more specifically than before.

### *Minimization of Response Cascades*

The cascade of events following insertion of a heterologous gene may bring surprises rather than the expected outcome. It will be important to minimize unwanted response cascades (Bailey, 1991). Transgenic plants with lower than normal phenylalanine ammonia lyase activity did show corresponding lignin reduction. However, they also exhibited several unusual and aberrant phenotypes, including stunted growth, altered leaf shape, reduced pollen viability, and altered flower morphology. Over expression of genes encoding enzymes of anthocyanin synthesis in petunia did not always demonstrate only the complexity of gene expression. The basic cause may be gene-gene interaction, interference of RNA strands with transcription process, and methylation.

One handicap lies in the employment of constructs that are chimeric (i.e., the gene of interest has been aligned with a precursor derived from a heterologous donor, for example, a virus, as CaMV S35). For better specificity and expression, a homogenomic promoter sequence should be used. Significant progress has been made when the gene for a major latex protein of poppy was analyzed and characterized. The promoter sequence contained may well be suited to direct gene activation in laticifers for increase or interdiction of morphinan production, without a concomitant cascade of undersirbale phenotypes. Besides, it can be expected that a zero-morphin poppy would benefit agriculture in the future as did the Bronowski, zero-glucosinolate rapeseed variety in the past.

## USES OF TISSUE CULTURE TECHNIQUES FOR PRODUCTION OF SECONDARY PLANT METABOLITES

Plant cells in culture offer many advantages over intact plants for secondary metabolites production and their biosynthetic studies. These are as follows:

- Plant cells are relatively easy to grow and can be kept under strict controlled nutritional and environmental conditions. Hence the uncertainties of climate and soil can be avoided.
- Cells are cultured aseptically, devoid of many microorganisms or insects etc.
- Suspension cultures, offers a very effective way of incorporating precursors which are often difficult to administer to the plant growing in nature.
- The technology is now available for the relatively large scale production of plant suspensions, in batch cultures, closed continuous cultures system, open continuous culture system etc. and may also eventually provides an efficient means of producing commercially important plant products.

However, in order to realize the industrial applications of plant cell culture for medicinal compounds production, it is essential to satisfy the following conditions as minimum requirements:

- The rate of cell growth and biosynthesis should be high enough to give a good production of the final product in a short period of time.
- The cultured cells should be genetically stable to give a constant yield of the product.
- The metabolites should be accumulated in cells without being catabolized rapidly or preferably, they should be realized into the liquid medium.
- Production cost including the culture medium, precursor and chemical extraction should be low enough to be profitable to the industries.

As far as the first requirement is concerned experiments with various plant suspension cultures showed that the growth rate can be accelerated considerably by improving cultural conditions and selective breeding of cultured cells. Frozen storage technique (cryopreservation) may be used for the second requirement as suggested by Street (1975). For the above third condition, as plant cells generally tend to accumulate their secondary metabolites in vacuoles or the cytoplasm, it would be desirable to device a method for altering the permeability of plasma membrane. A surface-active agent was used successfully by Tanaka et al. (1974). The forth requirement is related to the social demand of the product and production cost. The production cost depends upon the total cost of carbon source, cost of electricity necessary for aeration, stirring, temperature regulation etc. Therefore, it is advisable to use molasses, starch and alcohols as carbon source. Secondly, the introduction of continuous or semi-continuous culture in place of the batch culture system can reduce the cost for sterilization of the fermenter and the necessity for propagation of cells from the original stock.

## Applications of New Culture Methods for the Producton of Secondary Metabolites

Recently, two new techniques which have enhanced the accumulation of secondary metabolites (phytochemical) in plant tissue culture have been developed: (i) Hairy root culture, and (ii) Elicitation of product accumulation. The simplicity of the procedures and outstanding results made the techniques most useful. They have drawn attention of most of the tissue culturists.

### Hairy Root Culture

Plants, mostly, growing in tropical areas are the major source of many of the fine chemicals (secondary metabolites). Their structural complexity makes their chemical synthesis uneconomic. Therefore, the biosynthesis of these plant products using plant cells in culture has long been accepted as a worthwhile objective. The production of red naphthoquinone, an anti-inflammatory drug, shikonin by suspension cultures of *Lithospermum erythrorhizon* is the important landmark in this direction.

Plant cell cultures proved to be more complex and very often the rate of production of desired compounds is also very low. Low productivity in cell suspension culture to some extent over comes by:

- By manipulation in the composition of the medium.
- By altering physical environment of cells, and
- By screening for clones of high producing cells.

The ability of selecting high-producing cells from cell suspension cultures is not an easy task. To maintain elevated productivity, it is necessary to screen repeatedly the desired clones. This inherent instability is associated with the changes at the genomic level, both inter- and intrachromosomal.

Recently, highly productive and stable hairy root culture has been obtained by the genetic transformation of plant tissue by the pathogenic soil bacterium, *Agrobacterium rhizogenes*. The infection of dicotyledonous plants by *A. rhizogenes* causes roots to proliferate rapidly at the infection site. This phenotypic change results from the insertion into the plant genome of t-DNA (transfer DNA), carried on the bacterial Ri-plasmid, coding for auxin synthesis and other rhizogenic functions.

The hairy roots can be removed from the parent plant and can be cultured indefinitely in simple defined media free of auxins and phytohormones. In this respect they differ from untransformed root cultures, which are often dependent upon added auxins and phytohormones. Hairy root cultures are potentially applicable to the production of all **root-derived secondary** metabolites from dicotyledonous plants.

Many of the root synthesized compounds including tropane alkaloids, atropine and hyoscyamine, steroidal precursors such as solasidine, and *Cathranthus* alkaloids are of sufficient high value (1000 US $/kg) to justify the exploitation of hairy root culture for their commercial production.

#### *Establishment of Hairy Root Culture*

For the establishment of hairy root cultures, the explant material is inoculated with a suspension of *A. rhizogenes* generated by growing bacteria in YMB medium for two days at 25° C with gyrator shaking, pelleting by centrigugation (5 x $10^3$ rpm, 20 min) and resuspending the bacteria in YMB medium to form a fix suspension (approx. $10^{10}$ viable bacteria/ml). Transformation may be induced on aseptic plants grown from seeds, or on detached leaves, leaf-discs, petiols or stem segments from green-house plants following sterilization of the excised tissue with 10% (v/v) Domestos (Lever Bros.) for 20 min. Scratching the midrib of a leaf or the stem of a plantlet, with a needle of a hypodermic containing the thick bacterial suspension allows inoculation with small droplets containing *Agrobacterium rhizogenes*.

In some species, hairy roots may appear directly at the site of inoculation while in others a callus will form initially and hairy roots appear subsequently from it. In both cases hairy root appears within one to four weeks. Including 10M acetylsyringone in the medium in which the bacteria are suspended might increase infection in resistant plant species.

*Solanum nigrum* protoplasts have been cultured with *A. rhizogenes* and obtained transformed protoplast, which subsequently developed calli. The calli spontaneously formed hairy roots when cultured on hormone-free media.

In most of the plant species, establishment of hairy root cultures in suspension is brought about by excising the transformed roots from the stock plant and placing them in growth medium with 30 g/L sucrose as the energy source. However, some species are more difficult to establish in independent culture. Various techniques, including suspending roots on the explant, until they are established, buffering the pH of the medium and/or using half or quarter strength, high phosphate or nitrate, or the addition of exogenous hormones, have been found to cause extensive callus formation.

Cultures may be cleared of bacteria by several passages in medium containing 500 mg/L ampicillin, or 500 mg/l ampicillin + 200 mg/l cephalosporin C.

### *Characteristics of Hairy Root Cultures*

When normal roots of dicotyledonous plants are infected with *A. rhizogenes* it is transformed into hairy roots, which are characterized by:

- High degree of lateral branching.
- A profusion of root hairs.
- Absence of geotropism.

Profuse branching, resulting in the formation of many meristems, is one of the reasons, why such roots have the property of high growth-rate in culture. Both growth rate and extent of branching, are highly inter-related characteristics, vary between species and with the culture conditions. In *Nicotiana rustica* and *Datura stramonium* growth of cultures is influenced by the initial pH of the culture medium. In *D. stramonium* the density of branching per unit length of primary root is also markedly affected by the ionic strength of the medium.

The property of hairy root culture, which is most important from their commercial exploitation point of view is their high level production of secondary metabolites in absence of hormones in growth medium. Betalain synthesis in *Beta vulgaris* and hyoscyamine synthesis in *D. stramonium* increases two-folds and five-folds respectively in hairy root cultures.

Another key feature of hairy root culture is their productivity which is stable over many generations. In *Hyscyamus muticus* and *N. rustica* hairy root cultures remained genetically constant for at least one year. While in untransformed root cultures both genetic and chromosomal abnormalities are frequently observed. In hairy roots, the chromosome numbers resembles normal roots.

### *Genetics of Hairy Root Formation*

The infection of plant root tissue with *A. rhizogenes* causes to pieces of t-DNA to be inserted into plant genome. The integration of foreign genome though is random, it alters the auxin metabolism of the transformed plant tissues in such a way that:

- Hairy root morphology is expressed.
- Amino acid metabolism alters to produce opines, and
- Secondary metabolites synthesis is enhanced.

The synthesis of the opines (manopine or agropine) is the firm indication that hairy roots are indeed transformed. The strains of *Agrobacterium rhizogenes* which are used for transformation have good virulence. These are identified to have following plasmids, pRiA4, pRiHR1, pRi1855 and pRi8196.

### *Manipulation of Hairy Root Cultures*

Chemical or biochemical manipulation of hairy root cultures can enhance synthesis of biochemicals and/or a noval or uncharacteristic plant products. For this purpose natural precursors or matabolite analogues are used. When nicotinic acid was fed with *Nicotiana rustica* cultures it showed enhanced production of nicotine and anatabine. However, feeding of cadaverine to *Datura stramonium* cultures produces noval, as yet, to be identified alkaloids.

### *Hairy Root Cultures: Improvement of Secondary Metabolites Formation*

Every plant species has its biosynthetic capabilities depending upon its genetic makeup and environmental conditions of its habitat. Therefore, some species are considered as high yielding for a particular fine chemical while others are made to be high-yielding by selective breeding programmes. For the production of a fine chemical, only high-yielding varieties are selected. In cultures these varieties are not genetically stable and thus their product formation is not always constant. Hairy roots have stable productive capabilities similar to the plants from, which they were derived as these remain genetically stable for a long time.

As plasmid (gene) transfer is involved in the formation of hairy roots, the system is highly amenable to manipulate at the genetic level. Recently foreign genes have also been inserted into plants during transformation with *Agrobacterium rhizogene* by either insertion into the Ri-plasmid. tDNA or in binary co-transformation using disarmed vectors derived from *Agrobacterium tumefaciens* (pBiN19). In the later case tDNA from both plasmids is inserted into the plant genome. Thus the manipulation of secondary metabolite by altering the expression of key genes in the pathway is now a reality in present context.

In presence of high levels of exogenous auxins, hairy root cultures spontaneously form callus from which suspension cultures can be obtained. However, such transformed suspension cultures are returned to hormone-free media resulting in roots regeneration. Therefore, commercial exploitation of hairy roots for secondary metabolite production in the high producing strains has two following advantages:

- Regeneration of improved varieties of plants for growth in the fields, and
- the exploitation of the secondary metabolite formation of the roots in bioreactor system.

A significant progress to both these ends is being made. However, no definite procedure of cell selection or genetic manipulation has been developed to improve valuable metabolite synthesis in a plant.

## Elicitation of Product Accumulation

Elicitors strictly speaking are compounds of biological origin involved in plant-microbe interaction. In the context of product accumulation by plant cells in suspension cultures elicitors are mediator compounds or stress agents. There are two kinds of elicitors, these are: (1) Abiotic elicitors, (2) Biotic elicitors

Abiotic elicitors are physical or chemical in nature e.g., ultra violet radiations, alkalinity, osmotic pressure or heavy metal ions etc. Biotic elicitors are complex culture homogenates of fungal or bacterial origin or fractions thereof. Both pathogenic (*Phytophthora*, *Botrytis*, *Verticillium* etc.) as well as non-pathogenic (*Aspergillus*, *Micromucor*, *Rhodotorula* etc.) microbes have been employed for this purpose. The chemical nature of the elicitors has identified as oligosaccharids, polysaccharids, glycoproteins and low molecular weight compounds like archidonic acid.

The products, which accumulate in plant cell cultures due to elicitation, may be antimicrobial in nature but they should not be confused with phytoalexins (a term used in plant pathology for pathogen-induced chemical defence in host plant), unless there is sufficient proof that the source plant responds to pathogens with rapid accumulation of the same product. Therefore, a new term has been coined for those compounds, which in cell cultures are induceable by way of elicitation - **Elicitation product** or **Elicitation metabolite**.

Elicitors can be regarded as substitute of production media (that is optimum cultural conditions). Optimum employment of elicitors depends upon factors like: 1. Elicitor specificity, 2. Elicitor concentration, 3. Duration of elicitor contact, 4. Characters of cell line (clones), 5. Time course of elicitation, 6. Growth stage of culture, 7. Growth regulation and 8. Nutrient composition

A variety of elicitor preparations are generally tried for the yield and quality of a secondary metabolite for a cell line. For example *Botrytis* sp. preparation proved best with *Papaver sominiferus* (popy). The procedure followed in this experiment by Eilert et al. (1985) is as above.

1 $cm^2$ mycelium of *Botrytis* sp.

↓

Cultured on 100 ml B5 medium without hormone

↓

for 7 days at room temperature

↓

Homogenated entire culture

↓

Sterlized by autoclave

↓

Add 1-5 ml culture homogenate in culture media for

↓

*Papaver sominiferus* (popy)

↓

Cell culture responded in 2-6 hr different cell lines responded differently to added elicitor

↓

After 10 hr to 5 days maximum product accumulation was found even in those cell lines which never synthesized it from last several years

**TABLE 29.1:** Some elicitor-induced products accumulation (Adopted from the abstracts of VI Congress of International Association of Plant Tissue Culture held at Minneapolis, 1986).

| Elicitor | Preparation | Cell culture | Product |
|---|---|---|---|
| *Aspergilus niger* | Homogenate | *Cinchona ledgeriana, Rubia tinctoria, Morinda citrifolia* | Anthraquinones |
| *Botrytis* sp. | Homogenate | *Papaver somniferus* | Sanguinarine |
| *Chrysosporium palmorum* | Filtrate and extracts | *Catharanthus roseus* | Tryptamine, Ajmalicine, Catharanthine |
| *Dendrythion* sp. | Extract | *Papaver somniferus, P. bacteatum* | Sanguinarine |
| *Eurotium rubrum* | filtrate and extract | *Catharanthus roseus* | Tryptamine, Ajmalicine, Catharanthine |
| *Micromucor isabellina* | Filtrate and extract | *C .roseus* | Tryptamine, Ajmalicine, Catharanthine |
| *Nigeran* sp. | Homogenate | *Solanum melongena* | Polyacetylenes |
| | | *Vigna angularis* | Isoflavones |
| *Pythium aphanidermatum* | Filtrate and extract | *C. roseus* | Tryptamine, Ajmalicine |
| *Pythium aphanidermatum* | Homogenate | *C. roseus* | Strictosidine, Ajmalicine, Taberosonine, Lachnuricine |
| Yeast | Carbohydrate preparation | *Glycine max* | Glyceolline |

Elicitation of cell in suspension culture may react in the following ways:

- In a given cell line different products may show highest levels of accumulation at different times.
- Product accumulation may be observed in cell lines which never synthesized it from last several years.
- Elicitation will not cause an additive effect when applied to cells in production media, but may shorten the culture period required for maximum product accumulation.
- Product accumulation due to elicitation has also been observed in growth media. Such occurrence may be due to excretion or leakage caused by cell breakdown.

Elicitation of secondary product accumulation in plant cell cultures would appear to develop into a powerful tool for biotechnologists and biochemists.

## BIOTRANSFORMATION: PRODUCTION OF PHRMACEUTICAL COMPOUNDS

One of the most promising fields in the biotechnological application of plant cell culture is the biotransformation of some less important substrates to medicinally useful products. The abiotic synthesis of these fine chemicals is very difficult and uneconomical also. Their synthesis by plant cell culture method is easier than by microorganisms. The importance of biotransformation technique has, therefore, very recently been realized by several biotechnologists. Many workers have undertaken detailed studies on biotransformation of steroids such as sex hormones and cardenolides in recent years.

Biotransformations in plant tissue cultures to transform steroids in a manner similar to that have widely been used with microbial system is very promising field. Stohs and Staba first time studied the biotransformation of steroids in which digitoxigenin was shown to convert (during a 7-day incubation with cultures of *D. lanata* and *D. purpurea*) in to a keddepositive compound, 3-dehydrodigitoxigenin. Furuya and his associates reported biotransformation of digitoxin to gitoxin purpurea glycoside-A and purpurea glycoside-B from suspension culture of *D. purpurea*. Reinhard (1974) has studied the transformation capacity of *D. lanata* suspension culture transforming digitoxin to purpurea glycoside-A, digoxin desacetyl lanatoside-C, lanatoside-A, and lanatoside-C. Thus suspension culture of *D. lanata* are able to hydroxylate - the steroid nucleus of digitoxin at C-12, to link glucose to the carbohydrate side chain and to acetylate digitoxose.

Rainhard reported that both digitoxin and β-methyl digitoxin can be converted into medicinally more important cardiotonics by the specific hydroxylation at position 2 of the steroidal skeleton by suspension culture of *Digitalis lanata*. Feeding experiments showed that cultured cells of *Datura innoxia* possess a remarkably high capability for glycosylation of hydroquinone to form arbutin used as diuretic and urinary antiseptic.

It has been observed that anaerobic condition of cell physiology favours the biotransformation rate. Barz reported that anaerobic conditions which inhibits degrative reactions and expidite glycosylation reaction of phenolic compounds nearly six-fold. Isomerization of D-tryptophan into a optically active form, L-tryptophan have been demonstrated into tobacco cell cultures using recemase enzyme.

## Evaluation of Antimicrobial and Antifertility Activity

Screening of various cultures and extracts by bioassay for definite pharmacological activities may be an effective method for exploiting substances produced by cultured cells. Khanna screened aqueous and/or other tissue extracts of several plant species viz., *Atropa belladonna*, *Argemone mexicana*, *Artemisia scaparia*, *Agave weightii*, *Datura metel*, *D. tetula*, *Glycine max*, *Hyoscyamus niger*, *Lycopersicon esculentum*, *Solanum nigrum*, etc. and found antimicrobial activity against either Gram-positive or Gram-negative bacteria or both. She also isolated anit-microbial principles from tissues of these plant species. Quercetin obtained from *Agave weightii*, *Crotalaria burhia*, *C. juncea*, *D. metel*, *D. tatula*, *L. esculantum*, *Papaver rhoeas*. Whereas isorhamnetin isolated from tissue of *A. maxicana* was active against Gram-negative bacteria. She isolated kaempferol, a very strong flavonoid, from the tissue of *A. wightii*, *Pisum sativum*, *L. esculantum*, etc. and was found active against both bacteria and fungal human pathogen - *Candida albicans*. A-lignan and phyllemblin, isolated from tissue of *Emblica officinalis* also possess antibacterial and antifungal properties. She further extracted and screened kaempferol from various plant tissues pharmacologically against white male rats and observed antifertility activity (71 per cent) which had no side effects when tested by conventional methods. Nishi and Mitsuoka (1975) found that aqueous extracts of callus and cell cultures of

*Isodon japonies* inhibit the secretion of gastric juice in rats and promote the healing of acetic acid induced peptic ulcers on oral administration.

## Biological Control on Production of Secondary Metabolites

Active production of pharmacologically important compounds by plant cell cultures is mainly dependent upon following biological factors:

**TABLE 29.2:** Some secondary metabolites detected in plant tissue culture.

| Secondary Plant Products | Type of culture(s) | Stock plant |
|---|---|---|
| **Flavones and Flavonoids** | | |
| Quercetin | (C) | *Agave wightii* |
| | | *Datura metel* |
| | | *D. tatula* |
| | | *Lycopersicon esculantum* |
| | | *Solanum eviculare* |
| | | *Petroselinum hortense* |
| | (SC) | *Crotolaria juncea* |
| | | *Artemisia scaparia* |
| Apigenin | (C) | *Trigonella corniculata* |
| | | *Solanum dulcamara* |
| | (SC) | *Glycine max* |
| Luteolin | (C) | *Datura pinnata* |
| | (SC) | *Papaver hortense* |
| Chrysoerion | (SC) | *P. hortense* |
| Scopolatin | (SC) | *S. luteum* |
| Negretin | (SC) | *Solanum tuberosum* |
| Isorhamnetin | (SC) | *Argemone maxicana* |
| | | *P. hortense* |
| Kaempferol | (C) | *A. wightii* |
| | (C) | *C. cheiri* |
| | (C) | *P. sativum* |
| | (C) | *L. esculentum* |
| | (C) | *S. vaccaria* |
| | (C) | *S. dulcamara* |
| | (C) | *A. scoparia* |
| | (C) | *A. sativum* |
| | (C) | *G. max* |
| | (C) | *D. lablab* |
| Phyllemblin | (C) | *E. officinalis* |
| Artimissinine | (C) | *A. scaparia* |
| Vitexin | (C) | *V. decurrens* |
| Sinensetin | (C) | *Citrus aurantium* |
| Nobeletin | (C) | *C. aurantium* |
| **Chalcones and Deoxyflavones** | | |
| Daidzein | (C and& SC) | *P. aureus* |
| | (C and SC) | *G. max* |
| 2',4,4'-trihydroxy chalcone | (C) | *Phaseolus aureus* |
| **Coumestanes and Coumarinochromans** | | |
| Pisatin | (C) | *Pisum sativum* |

| **Secondary Plant Products** | **Type of culture(s)** | **Stock plant** |
|---|---|---|
| Soyagol | (C & SC) | *P. aureus* |
| Coumestrol | (C & SC) | *P. aureus* |
| **Anthocyanins** | | |
| Cyanidin | (C) | *H. gracilis* |
| | (C) | *L. usitatisimum* |
| | (C) | *Dimorphothica auriculata* |
| Dephinidin | (C) | *D. auriculata* |
| **Anthraquinones** | | |
| Rhein | (C) | *Cassia anqustifolia* |
| Digitoluutein | (C) | *Digitalis lanata* |
| 4-hydroxydigitolutein | (C) | *D. lanata* |
| 3-methyl purpurin | (C) | *D. lanata* |
| Alizartin | (C) | *Morinda citrifolia* |
| **Tannins** | | |
| Catechin | (C) | *Camellia sinensis* |
| | (SC) | Paul's Sacrlet Rose |
| Epicatechin | (C) | *C. sinensis* |
| **Naphthoquinones** | | |
| Plumbagin | (C) | *Plumbago zeylanica* |
| **Sesquiterpenes** | | |
| Linderane | (C) | *Lindera strychinifolia* |
| Lindenenol | (C) | *L. strychinifolia* |
| Cryophyllene | (C) | *L. strychinifolia* |
| -bisabolene | (SC) | *Andrographis paniculata* |
| **Sterols and Triterpenes** | | |
| Beta-sitosterol | (C) | *N. tabacum* |
| Stigma sterol | (C) | *Dioscorea tokora* |
| Campesterol | (C) | *Withania somnifera* |
| | (C) | Paul's Scarlet Rose |
| | (C) | *Tylophora indica* |
| | (C) | *Helianthus annus* |
| | (C) | *D. metel* |
| | (C) | *T. foenumgracum* |
| Lanosterol | (C) | *S. nigrum* |
| | (C) | *S. elaegnifolium* |
| | (C) | *S. xanthocarpum* |
| | (C) | *S. carota* |
| | (C) | *D. pinnata* |
| Cholesterol | (C & SC) | *H. annus* |
| | (C & SC) | *S. indicum* |
| Cycloartinol | (C & SC) | *N. tabacum* |
| Isofucosterol | (C & SC) | *H. annuus* |
| Obtusifoliol | (C & SC) | *N. tabacum* |
| Beta-amyrin | (C) | Paul's Scarlet Rose |
| | (C) | *T. indica* |
| **Steroidal Alkaloids** | | |
| Solasonine | (C) | *S. xanthocarpum* |

| **Secondary Plant Products** | **Type of culture(s)** | **Stock plant** |
|---|---|---|
| Tomatin | (C) | *L. esculentum* |
| **Carotenoids** | | |
| Voilaxanthin | (C) | Paul's Scarlet Rose |
| Zeaxanthin | (C) | *Ruta graveolens* |
| Neoxanthin | | |
| Beta-carotene | (C) | *R. graveolens* |
| **Lutein** | | |
| Lutein-5,6 epoxide | | |
| Antheraxanthin | | |
| Sapogenins | | |
| Diosgenin | (C) | *S. nigrum* |
| | (C) | *S. xanthocarpum* |
| | (C) | *S. aviculare* |
| | (C) | *S. dulcamara* |
| Gitogenin | (C) | *A. wightii* |
| Hecogenin | | |
| Tigogenin | (SC) | *S. nigrum* |
| Smilagenin | (C) | *Yucca aloifolia* |
| **Sarsasapogenin** | | |
| Tiggenin | | |
| Hecogenin | | |
| Chlorogenin | | |

(C) Callus, (SC) = Suspension callus.

APPENDIX 1

# Physical Quantities and Preparation of Solutions

## INTRODUCTION

For a researcher, the use of physical quantities involved in various functions and detailed knowledge about the nature of various forces in action are necessary. Physical quantities and their units provide an idea of how they operate to create the various interactions. For the study of such matter the knowledge of mathematics, physics and chemistry is involved. All physical quantities are to be assigned with units to make them meaningful. For example, the distance between two points, it is not appropriate to say the answer as 2.5; a unit must be specified here. The answer in this case might be 2.5 m; many such different systems of units are in use. Thus 2.5 m is the same length as 98.38 inches, 2.73 yards and 0.00155 mile. All physical quantity has a dimension, for example the unit of measurement should be specifically as metre or inch or foot or mile. The dimension of any physical quantity can defined in terms of six **primary dimensions**; length mass, time, electric current, temperature and light intensity. For example, the dimension of area i.e., $(length)^2$ abbreviated to $l^2$ and the force is (mass x $length/time^2$) abbreviated $mlt^2$. Correspondingly, each primary dimension has a primary unit. It is an accepted standard unit measure of that dimension.

**TABLE 1:** Base units

| Unit | Name | Symbol |
|---|---|---|
| Length | metre | m |
| Mass | kilogram | Kg |
| Time | second | s |
| Electric current | ampere | A |
| Temperature | kelvin | K |
| Light intensity | candela | cd |

## PRIMARY AND SECONDARY UNITS

Different systems of units are used to measure physical quantities and may lead to considerable confusion. In an effort to have a standard set of units acceptable to all disciplines, an International Commission has proposed a system known as SI (**Systeme Internationale Units**). These are now widely accepted and used by many scientific bodies. All other units are derived from these. Of the sex primary units, we can have a good idea of the magnitude of 1 m, 1 kg. Following is the list of units and dimensions. These unites can be used from some other scales and may be related to the new scale you may have been using before.

- **Measurement of Length:** For accurate measurement of length, a laboratory standard is **cathetometre**. This is a graduated metal ruler with a vernier scale and an attached travelling microscope. Cross hairs in the microscope eye piece are used to align the microscope whose position is then read on the calibrated rule. Accuracies of 0.001 mm are possible.

- **Mass:** For mass a good set of balance weight will be accurate enough for most measurements.
- **Time:** Clocks can be readily calibrated against broadcast time and used for time intervals greater than one second. Short time intervals 1 second can be standardised against the 50 Hz mains alternating voltage using an oscilloscope.
- **Voltage:** It can be measured by comparison in the standard cell and many viltmeters contain a calibration cell.
- **Resistance:** It is measured by comparison with known resistors using a conductivity bridge.
- **Temperature** (°C) can be measured with an accuracy of 0.1 in a good mercury glass thermometer. Absolute accuracy to better than this is difficult, although estimation of temperature difference can be made to 0.0001°C.

**TABLE 2:** Units and Dimension

| Variable | Name | Abbreviation | SI units | Dimension |
|---|---|---|---|---|
| Acceleration | | | m/s$^2$ | lt$^{-2}$ |
| Area | | | m$^2$ | l$^2$ |
| Capacitance | farad | F | AS/V | |
| Charge | Coulomb | C | As | |
| Current | ampere | A | A | |
| Density | | | Kg/m$^3$ | ml$^{-3}$ |
| E.M.F. | Volt | V | W/A | |
| Electric field | | | V/m=N/C | |
| Force | Newton | N | kgm/S$^2$ | mlt$^{-2}$ |
| Frequency | hertz | Hz | s$^{-1}$ | t$^{-1}$ |
| Length | metre | m | m | 1 |
| Light intensity | Candela | cd | cd | |
| Magnetic | | N/(Am) Weber/m$^2$ | | |
| Mass | kilogram | Kg | Kg | m |
| Moment of interia | | | Kg/m$^2$ | ml$^{-2}$ |
| Power | watt | W | J/S | ml$^2$t$^{-3}$ |
| Pressure | | | n/m$^2$ | ml$^{-2}$t$^{-2}$ |
| Radioactivity curie | | Ci | s$^{-1}$ | |
| Resistance | Ohm | | V/A | |
| Surface tension | | | N/m, J/m$^2$ | mt$^{-2}$ |
| Time | Second | s | s | t |
| Torque | | | Mm | ml$^{-2}$t$^{-2}$ |
| Velocity | | | m/s | lt$^{-1}$ |
| Viscosity | | | Ns/m$^2$ | ml$^{-1}$t$^{-1}$ |
| Volume | | | m$^3$ | l$^3$ |
| Work, energy | joule | J | Nm | ml$^2$t$^{-2}$ |

Secondary set of units can be defined in terms of primary set. For example, the unit of force (Newton) is defined as that force required to give a mass of 1 Kg and acceleration of 1 m/s. Its primary units are thus $Kgm/ls^2$ but this is abbreviated N or Newton. Units named often famous scientists are given in capital letters.

## Changing Units

As there as different kinds of units, it is necessary to change from one set of units to another. Let us taken an example, the process of converting 10 miles/hour into feet/second.

$$\text{Velocity} = 10\ \frac{\text{Mile}}{\text{Hour}}$$

The procedure is first to write down the relationship between the individual units of the same dimension in the expression. In this example, we should have the conversion from miles to feet (length) and from hours into seconds (time).

$$\text{One mile} = 5280 \text{ feet, and}$$
$$1 \text{ hour} = 3600 \text{ seconds,}$$

Thus, $$1 = \frac{500 \text{ feet}}{1 \text{ mile}} \text{ and } \frac{1 \text{ hour}}{3600 \text{ sec.}} = 1$$

Since these expressions are both equal to units, we can multiply the original expression by both without changing its numerical value. Velocity = 10 miles/hour = 5280 feet/1 mile = 1 hour/3600 seconds; 10 x 5280/3600 = 14.66 feet/second.

## Multiples and Submultiples of Units

The value of a physical quantity remains greater or much smaller than the value of 1 unit as defined in Table 3. For example, it is inconvenient to say the distance between two towns in metre, since the numbers are very large. Instead we can use a multiple of basic units, in this case kilometers, where kilo stands for $10^3$ (1000). SI units use the scale of multiples. Some useful physical constants are given in Table 3.4.

**TABLE 3:** The scale of multiple and submultiples of units

| Abbreviation | Prefix | Factor | |
|---|---|---|---|
| T | tera | $10^{12}$ | 1000 000 000 000 |
| G | giga | $10^9$ | 1000 000 000 |
| M | mega | $10^6$ | 1000 000 |
| K | kilo | $10^3$ | 1000 |
| H | hecto | $10^2$ | 100 |
| D | deca | $10^1$ | 10 |
| D | deci | $10^{-1}$ | 0.1 |
| C | centi | $10^{-2}$ | 0.01 |
| m | milli | $10^{-3}$ | 0.001 |
| μ | micro | $10^{-6}$ | 0.000 001 |
| N | nano | $10^{-9}$ | 0.000 000 001 |
| P | pico | $10^{-12}$ | 0.000 000 000 001 |
| F | femto | $10^{15}$ | 0.000 000 000 000 001 |
| A | atto | $10^{-18}$ | 0.000 000 000 000 000 001 |

mega, kilo, centi, milli, micro, etc; are commonly met with in biology.

**TABLE 4:** Some physical constants

| Parameter | Value | Unit in abbreviator |
|---|---|---|
| Atmospheric pressure (atm) | $1{,}01 \times 10^5$ | $N/m^{-2}$ |
| Acceleration due to gravity(g) | 9.81 | $m/S^{-2}$ |
| Force to hold up a 1 kg mass | 9.81 | N |
| Gravitational constant(G) | $6.67 \times 10^{-11}$ | $Nm^2\ Kg^{-2}$ |
| Gas constant(R) | 8.32 | $K\ JK^{-1}\ mole^{-1}$ |
| Avogadro's no.(N) | $6.02 \times 10^{23}$ | $mole^{-1}$ |
| Planck's constant(h) | $6.63 \times 10^{-34}$ | Je |
| Boltzmann's constant(k) | $1.38 \times 10^{23}$ | $JK^{-1}$ |
| Elementary charge(e) | $1.60 \times 10^{-19}$ | C |
| Velocity of light(c) | $3 \times 10^8$ | $m/s^{-1}$ |
| Faraday(F) | $09.65 \times 1^4$ | $C\ mol^{-1}$ |
| Density of water | $10^3$ | $Kg/m^{-3}$ |
| Density of air | 1.3 | $kg/m^{-3}$ |
| Absolute Zero | | 0°K=273.15°C |
| Electron Volt | $1.60 \times 10^{-19}$ | J |
| Year | $3.16 \times 10^7$ | S |
| Radius of earth | 6400 | km |
| Radius of moon | 1740 | km |
| Mass of earth | $6 \times 10^{24}$ | Kg |
| Mass of moon | $7.4 \times 10^{22}$ | Kg |
| Stefan's constant | $5.69 \times 10^{-8}$ | $Wm^{-2}K^{-4}$ |
| Latent heat of vaporization of water | $2.256 \times 10^6$ | JKg |
| Surface tension of water (2°C) | $7.3 \times 10^{-2}$ | $Nm^{-1}$ |

## ACCURATE MEASUREMENTS

As all measurements are liable to errors, the sources of these errors should be taken into considerations. Errors may be caused by faulty instruments or carelessness. Wrong use of experiments can result errors. Even modern equipments can provide wrong results if the instruction manual of the instrument is not followed correctly. Such mistakes can be corrected with practice. The random errors are unpredictable. Random errors are usually corrected by taking a large number of measurements and calculating the average value. Usually triplicate experiments are sufficient if there is good agreement between them. The degree of agreement between experiments is called **precision**. The precision does not mean accuracy; measurements may be highly precise but inaccurate due to faulty instrument or technique.

While working with biological material various factors should be taken into consideration. A physical quantity such as refractive index of a liquid may be obtained with correct figure; but for biochemical measurements, it is difficult to obtain a correct value, but there is a range of normal values. This means that if an animal is healthy, the value (serum chloride) may be in the normal range. If the same animal is administered with a drug or hormone, the value may be varied. Larger deviations are usually result of disease or extreme stress.

## TYPES AND PREPARATION OF SOLUTIONS

Solutions play a large role in many scientific investigations; therefore, it is necessary for the scientist (or prospective scientist) to become familiar with what solutions are and with some of the techniques used in preparing them. In laboratory block you will use several different solutions- some of them in varying concentrations. The term concentration means the amount of one substance that is contained in a definite amount of another substance. There are numerous ways of expressing concentration, for example;

**Percent strength:** The number of parts by weight of one substance that is contained in 100 parts by weight of solution.

**Parts (grams, micrograms, milliliters, etc.) per liter:** The weight or volume of one substance contained in one litre of solution.

**Molarity**: The number of moles of one substance contained in one litre of solution. Mole means the molecular weight of a substance expressed in grams. Litre is a unit of volume in the metric system. It is the volume occupied by one cubic decimeter of a substance. Your teacher will show you a container which measures this volume.

The terms solute and solvent are used to designate the components of a solution. The solute is the substance that is dissolved. The solvent is the liquid in which the solute is dissolved. A solution is the product obtained when a solute is dissolved in a solvent. The following conversion factors will aid you in understanding the discussion of dilution which is used to obtain weights of chemicals that are too small to be weighed on the balances in your laboratory:

- 1 gram = 1,000 milligrams = 1,000,000 micrograms
- 1 milligram = 1,000 micrograms
- 1 litre = 1,000 millilitres

One has to prepare the concentration as per requirement of the experiment. It is essential note that the concentration of solution used fo any treatment should be **physiologically acceptable**, that it should not be **toxic** to the plant material.

## Preparation of Solutions of Low Concentrations

Sometimes we need extremely small quantities of some of the chemicals- many of these below the weighing capacities of the balances at your disposal. Therefore it will be necessary for you to start with a volume of solution containing an amount of the chemical which you can weigh. Assume that you have prepared 1 liter of a uniform solution that contains 1 gram of the chemical you are using. Since 1,000 milliliters of this solution contains 1,000 milligrams of the chemical, you may conclude that 1 milliliter contains 1 milligram. Even smaller weights of the chemical can be obtained easily by preparing more dilute solutions. For example, if we place 1 milliliter of the above solution- thus 1 milligram of the chemical- in 1 liter of water: we then have 1,000 milliliters of solution containing 1,000 micrograms of the chemical. We can now obtain 1 micrograms of the chemical by using 1 milliliter of this solution.

A stock solution is a relatively concentrated solution that is used in preparing larger volumes of relatively dilute solution. For preparing 1ppm or $mg/l^{1-3}$ solution dissolve 1 mg in one lit water. or if you need 1000 ppm dissolve 1 g in one lit of water. If you have to prepare various dilutions (e.g., 25 ppm, 50 ppm 100 ppm, 250 ppm, amd 500 ppm) from the stock solution of 1000 ppm concentration the following procedure will be followd:

**TABLE 5:**

| Concentration of Stock Solution | Add Distilled water (ml) | Total volume (ml) | Obtained Concentration (ppm) |
|---|---|---|---|
| Take 20 ml of Stock Solution of 1000 ppm add | 20 | 40 | 500 |
| Take 20 ml of Stock Solution of 500 ppm add | 20 | 40 | 250 |
| Take 20 ml of Stock Solution of 250 ppm add | 20 | 40 | 125 |
| Take 2.5 ml of Stock Solution of 1000 ppm add | 22.5 | 25 | 100 |
| Take 20 ml of Stock Solution of 100 ppm add | 20 | 40 | 50 |
| Take 20 ml of Stock Solution of 50 ppm add | 20 | 40 | 25 |
| Take 4 ml of Stock Solution of 100 ppm add | 36 | 40 | 10 |
| Take 20 ml of Stock Solution of 10 ppm add | 20 | 40 | 5 |
| Take 20 ml of Stock Solution of 5 ppm add | 20 | 40 | 2.5 |
| Take 2 ml of Stock solution of 10 ppm | 18 | 20 | 1 |

Similarly you can prepare the various concentration of molar solution by using above procedure.

## STERILIZING SOLUTIONS

In biological work it is often necessary to use sterile solutions. This is particularly essential in solutions which are to be injected into living tissue. Two common methods of sterilizing solutions are autoclaving and pasteurizing.

## AUTOCLAVING

This is a method of sterilizing objects by steam under pressure. Prepare the solution and place it in the container in which it is to be maintained. Place the cap or stopper on the container loosely so that steam can escape, and place the container of solution in an autoclave or pressure cooker. Bring the temperature of the autoclave up to 120°C and maintain it for 20 minutes. If a pressure cooker is used, a pressure of 15 pounds will produce approximately this temperature. After 20 minutes remove the heat source and allow the apparatus to cool. When the pressure gauge returns to 5 pounds, you may open the steam valve and allow the remainder of the steam to escape. Allowing the pressure to go down suddenly may cause the caps or stoeps to blow off; so note carefully the above instructions. Remove the solution from the autoclave or pressure cooker and tighten the caps or stoppers immediately.

## PASTEURIZING

This technique is especially useful for solutions that are to be injected. Prepare the solution and place it in a sterile serum bottle or vial. Cap the bottle with a sterile rubber cap of appropriate size. (You can sterilize the bottle and cool satisfactorily by immersing them in 70% alcohol). If the rubber cap is new, prick a fine hole in it with a sterile needle, so that the expanding air can escape when the vial is heated. Place the sealed container of solution in a little water on a heat source, and bring the temperature to about 60°C. Maintain approximately this temperature for 30 minutes. Then remove the container from the water, cool it and place it in the refrigerator or deep freeze until needed.

# APPENDIX 2

# Laboratory Help

## WEIGHING

Many different kinds of balances are used in chemistry laboratories, including (a) triple-beam balances, which require manual adjustment of weights on each of three balance beams; (b) top-loading automatic balances; and (c) analytical balances, which are accurate to closer than 1 mg. Strictly speaking, such balances measure mass rather than weight; however, the word weight is commonly used for mass when this will not cause error or confusion. Your instructor will explain or demonstrate the operation of the balances used in your laboratory.

### Weighing Solids

Most solids can be weighed in glass containers (such as vials or beakers) or on special glossy weighing papers. Filter paper and other absorbent papers should not be used for accurate weighings, since a few particles will always remain in the fibers of the paper. If you are weighing an indeterminate amount of a solid (such as the product of a reaction), its container or the weighing paper should first be weighed separately and the mass recorded (unless the balance has a taring system). Then the solid is added, the total mass is recorded, and the mass of the container or weighing paper is subtracted. If the balance has a taring system, the mass readout scale is set to zero (using the taring control) while the container or weighing paper is on the balance pan. Then the solid is added, and its mass is read directly from the scale.

If you are measuring out a specified quantity of a solid, the expected total mass of the solid plus its container should be calculated and the balance should be set to approximately the value. Then the solid is added in small portions, using a clean scoop or spatula, until the desired reading is attained. If the balance has a taring system, its readout scale is zeroed with the container on the pan and is then set to the desired mass of the solid.

Most products obtained from a preparation are transferred to vials or other small containers, which should be weighed empty and then reweighed after the product has been added. As a general practice, the container should be weighed with its cap and label on and this tare weight recorded. Then the mass of the contents at any given time can be obtained by subtracting the tare weight from the total mass of container and contents.

*Note: Allow a moistened label time to dry before weighing a labelled container.*

### Weighing Liquids

The mass of an indeterminate quantity of a liquid is measured as described above for a solid. A tared container should be used, and it should be kept stoppered during the weighing to avoid loss by evaporation. To weigh out a specified quantity of a liquid from a reagent bottle, you should first measure out the approximate quantity of the liquid by volume and then weigh that quantity accurately in a closed container. For example, if 9.0 g of chloroform (d = 1.5 g/ml) is required for a synthesis, a little more than 6 ml (9.0 g/1.5 g/ml) of the liquid is measured into a graduated cylinder and transferred to a tared vial. If the measured weight is not close enough to that required, liquid can be weight is not close enough to that required, liquid can be added or removed with a clean dropper. If the liquid is provided in dropper bottles, it can be transferred directly from the bottle to the container on the balance pan (*Care: Do not get any thing on the balance)* without pre-measuring.

*Note: Never withdraw liquid directly from a reagent bottle with your dropper or pipet- you may contaminate the liquid.*

Calculate the volume from: V = mass/density
Clean up any spills in the vicinity of the balance immediately.

## MEASURING VOLUME

A given volume of a liquid can be measured using either a graduated cylinder, a pipet, or a syringe, depending upon the quantity and accuracy required. Burets and volumetric flasks are also used to measure liquid volumes accurately. Their use is discussed in most general and analytical chemistry laboratory manuals.

### Graduated Cylinders

Graduated cylinders are not highly accurate, but they are adequate for measuring specified quantities of solvents and wash liquids as well as liquid reactants that are present in excess. The liquid volume should always be read from the bottom of the liquid meniscus.

### Pipets

Graduated or volumetric pipets can be used to accurately measure relatively small quantities of a liquid. Suction is required to draw the liquid into a pipet; however, using mouth suction is unwise because of the danger of drawing toxic or corrosive liquids into the mouth. An ordinary ear syringe works quite well as a pipetting bulb. Another convenient pipetting-bulb assembly.

- The top end of the pipet is inserted into the pinch-cock valve.
- The pinchcock is opened by pinching it at the glass bead and the bulb squeezed to eject the air.
- The pipet tip is placed in the liquid and the pinch cock is squeezed open to fill the pipet to just above the calibration mark.
- The bulb is removed from the narrow end of the dropper and the pinchcock is carefully opened until the liquid falls to the calibration mark.
- The liquid is then delivered into another container by opening or detaching the pinchcock valve.

Most volumetric pipets are calibrated to deliver (TD) a given volume, meaning that the measured liquid is allowed to drain out by gravity, leaving a small amount of liquid in the bottom of the pipet. This liquid is not removed, since it is accounted for in the calibration. Graduated pipets are generally filled to the top (zero) calibration mark and then drained into a separate container until the calibration mark for the desired volume is reached. The remaining liquid is either discarded or returned to its original container. The maximum indicated capacity of some graduated pipets is delivered by draining to a given calibration mark and of others by draining completely. It is important not to confuse the two, since draining the first type completely will deliver a greater volume than the indicated capacity of the pipet.

### Syringes

Syringes are most often used for the precise measurement and delivery of very small volumes of liquid, as in gas-chromatographic analysis <OP-32>. A syringe is filled by placing the needle in the liquid and slowly pulling out the plunger until the barrel contains a little more than the required volume of liquid. Then the syringe is held with the needle pointed up and the plunger is pushed in to eject the excess pointed up and the plunger is pushed in to eject the excess sample. Excess liquid is wiped off the needle with a tissue.

Syringes should be cleaned immediately after use by rinsing them several times with a volatile solvent, then removing the plunger and letting the barrel dry. Microsyringes can be dried rapidly by aspiration. The needle is inserted carefully through the dropper bulb and the aspirator is turned on for a minute or so. The pinchclamp is then opened to release the vacuum, the aspirator is turned off, and the syringe is removed.

## HEATING MANTLES

A heating mantle is perhaps the most satisfactory laboratory device yet developed for heating over a wide temperature range. Unlike a Bunsen burner, a heating mantle can be controlled precisely. Unfortunately, most heating mantles are costly, take a long time to warm up, and accommodate only a narrow range of flask sizes. It is also difficult to monitor the operating temperature of a heating mantle. Despite these disadvantages, heating mantles are highly recommended for most heating operations.

A heating mantle must be used in conjunction with a variable transformer or a time-cycling heat control to regulate the heat output. Since the temperature of the mantle itself can be measured only by a thermocouple, it is difficult or inconvenient to set a heating mantle for operation at a particular temperature. A mantle will generally not be at thermal equilibrium with the contents of a flask, so if the flask is not filled to a level near the top of the mantle, the part of the flask above the liquid level will be hotter than its contents and can cause decomposition of materials splashed onto it. When possible, the mantle should have a well of nearly the same diameter as the flask being heated. Some kinds of all-purpose mantles are intended for operation with a range of flask sizes; however, heating efficiency is reduced and the chance of superheating is increased when a small flask is heated in a large mantle.

The mantle is mounted on a lab jack, ring, set of wood blocks, or some other support so that it can be lowered and removed quickly if the rate of heating becomes too rapid. The flask is clamped in place so that it is in direct contact with the heating well, and the heating control dial is adjusted until the desired rate of heating is attained. Because a heating mantle responds slowly to changes in the control setting, it is easy to overshoot the desired temperature by turning the control too high at the start. If this occurs, the mantle should be lowered so that it is no longer in contact with the flask. The voltage input should then be reduced and the mantle allowed to cool down. Further adjustment may be required to maintain heating at the desired rate.

## MIXING

Reaction mixtures are frequently stirred, shaken, or otherwise agitated to promote efficient heat transfer, improve contact between the components of a heterogeneous mixture, or mix in a reactant that is being added during the course of a reaction. If the reaction is being carried out in an Erlenmeyer flask, this can be accomplished by manual shaking and swirling, or by using a stirring rod. If the apparatus is not too unwidely and the reaction time is comparatively short, ground glass assemblies can sometimes be manually shaken for adequate mixing. This is most easily done by clamping the assembly securely to the ringstand and carefully sliding the base of the ringstand back and forth. But when more efficient and convenient mixing is required, particularly over a long period of time, it is necessary to use some kind of magnetic or mechanical stirring device.

### Magnetic Stirring

A magnetic stirrer consists of an enclosed unit containing a motor attached to a magnet, underneath a platform. As the magnet inside the unit rotates, it can in turn rotate a teflon- or glass-covered stirring bar inside a container placed on (or above) the platform. The rate of stirring is controlled by a dial on the stirring unit. Since no moving parts extend outside this container, a magnetically-stirred reaction assembly can be completely enclosed if necessary.

Magnetic stirrers can be used with heating mantles or heating baths that are not constructed of ferrous metal. They work particularly well with oil baths, since they can be used to stir the oil and a reaction mixture simultaneously. The reaction flask must be positioned close enough to the bottom of the oil bath to allow sufficient transfer of magnetic torque from motor to stirring bar. When a copper or aluminium steam bath is used for heating, the flask should be clamped inside the rings, close to the bottom of the steam bath. Some hot plates have an integral magnetic stirrer, and these units can be used (with a heating bath) to simultaneously heat and stir a reaction mixture.

When magnetic stirring is used during a reaction, the heat source is set directly on the stirring unit and a stir bar is placed in the reaction flask in place of boiling chips. The stirring motor should be started and cooling water for the reflux condenser turned on (if applicable) before heating is begun.

### Mechanical Stirring

Mechanical stirring utilizes a stirring motor connected to a paddle or agitator by means of a shaft extending through the neck of the reaction vessel. A glass sleeve or bearing is used to align the shaft, which is ordinarily made of glass to reduce the likelihood of contamination.

Mechanical stirrers can exert more torque than magnetic stirrers, and are preferred when viscous liquids or large quantities of suspended solids must be stirred. A variety of stirring paddles made of teflon, glass, and chemically resistant wire are available.

## ELECTROPHORESIS

Any charged particle suspended between the poles of an electrical field tends to travel toward the pole that bears the charge opposite to its own. The rate at which it travels is conditioned by a number of factors, including the characteristics of the particle, the properties of the electrical field, and environmental factors, such as temperature and the nature of the suspending medium. The mobility of a particle is approximately proportional to its charge: mass ratio. Thus, an oxalate ion with two charges and a formula weight of 88.1 (charge/mass = 0.0227) would be expected to move more rapidly than a stearate ion (1/283.5 = 0.0035). Unfortunately this relationship is complicated by such factors as the molecular volume of the migrant, coordination of the migrant with molecules of solvent, and interference with migration by the supporting medium factors such as these make it impossible, with our present knowledge, to make accurate quantitative predictions of electrophoretic mobilities unless experimental data are available. It is true, nevertheless, that when a solution containing substances with different charge: mass ratios is acted upon by an electrical field, the components tend to separate by migrating at different rates. The word electrophoresis will be used to mean any application of this principle without regard to whether the substances are colloidal or ionic, and without considering whether the purpose of the application is preparation, purification, or measurement.

### Electrophoretic Mobilities

The rate at which a particle moves under a controlled set of circumstances is reproducible, making it possible to calculate how far it will travel during an electrophoresis, once the necessary data have been accumulated. Let it be emphasized that mobilities can be established solely by experimentation, and that they are reproducible only when all conditions are controlled. Only voltage gradient and time of migration can be treated as variables if mobility calculations are to be valid. Variations in pH, temperature, ionic strength, medium, and the like, have not successfully been taken into account in mobility calculations. Knowing the mobilities of the components of a mixture enables one to predict the positions of the components after arbitrary time intervals or in response to varying field strengths. This is useful for locating and identifying components after separations have been obtained and for calculating the time necessary to effect complete separations.

Conventionally, mobility is defined as the distance a particle will travel in a unit of time per unit of strength of an electrical field. Distance of travel is customarily stated in centimetres and time in second; field strength is expressed as the voltage gradient, in volts per centimetre, along the electrophoretic bed. From this it follows that the dimensions of mobility, u, are cm/sec divided by volts/cm

which simplifies to: $u = cm^2/volts \times sec$

If this formula for mobility seems strange remember that it does not describe velocity but is instead a factor intended for use in calculating velocity under defined conditions in response to any given voltage gradient.

## INSTRUMENTATION

The area upon which electrophoretic separations occur, called the bed, can be composed of any of a number of materials including gels, films, and powders. It is moistened with an electrolyte solution (usually a buffer). The ends of the bed are immersed in more of the electrolyte contained in two chambers designed to hold electrodes that are connected to a dc power supply. Provision is made for adjusting the electrolyte in the electrode chambers to equal levels so that siphoning action does not occur through the bed. The entire apparatus, excluding the power supply, is enclosed in an airtight chamber to prevent excessive evaporation of buffer. When a spot of sample mixture is applied to the bed and the power is turned on, those components of the mixture whose particles are charged will migrate toward the electrode having the opposite polarity.

# APPENDIX 3

# pH and Buffers

The term pH was introduced by a Danish biochemist, Soren Sorensen in 1909. The term pH may be defined as the negative logarithm of hydrogen ion concentration (moles/litre). In other words, pH is the negative logarithm of the hydrogen ion activity or function. pH may be expressed as pH = log $1/aH^+$ = -log $aH^+$; $aH^+$ is the activity of $H^+$.

The quantity termed pH was originally defined in a fundamental way. Srensen intended that the pH should be a means of expressing the hydrogen ion concentration of a particular medium, thus pH = -log $c_H$. The acidic or basic nature of a solution is expressed in terms of hydrogen ion concentration (pH). For example, $H^+$ concentration of 1N HCl is 1 and 1N NaOH is $10^{-14}$ moles/litre. Srensen proposed that experimental pH numbers be determined by measuring the emf ($E$) of a cell consisting of a hydrogen electrode and a calomel reference electrode separated by a liquid junction of concentrated potassium chloride solution. Essentially the same procedure is in use today, except that the versatile glass electrode has largely replaced the hydrogen electrode as the hydrogen ion indicator.

## INSTRUMENTATION

The great majority of pH measurements are made by electrometric procedures. The solution whose pH is to be determined is placed in a cell assembly consisting of two suitable electrodes. The voltage of this cell is then measured, usually by an instrument called a pH meter. This voltage or electromotive force (emf) is converted to a pH number by a formula relating it to the emf furnished by a standard reference solution of assigned pH. The definition of pH, electrodes for pH cells, and pH instrumentation will be discussed in that order.

With the passage of time it became evident that the emf of the Srensen cell did not respond in a simple way to hydrogen ion concentration, and thus, the pH numbers obtained from this cell were not an exact measure of -log $c_{\text{-H}}$ in the unknown solution. Indeed, they bore no exact simple relationship to either the concentration or activity of the hydrogen ions in these mistures, but were influenced to some extent by other ions present. This experimental defect has its origin in the fact that all of the ions in the solution diffuse independently across the interface between the solution and the potassium chloride bridge, and, therefore, contribute to the liquid-junction potential at that point.

## HYDROGEN ION RESPONSIVE ELECTRODES

The hydrogen gas electrode is the primary standard for pH measurements. The purity of the hydrogen ion response of other, secondary electrodes is affirmed by comparison of their changes of potential upon change of pH with that displayed by the hydrogen electrode. Most pH measurements in practice, however, are made with the glass electrode, the applicability of which is far broader than that of the hydrogen electrode. In specialized situations, the quinhydrone and antimony electrodes also are useful indicators of hydrogen ion.

**Glass Electrode.** The glass electrode is highly versatile and is by far the most useful of the hydrogen ion responsive electrodes for practical measurements of pH. It has a very fast, almost instantaneous, response in buffered solutions, although equilibrium is achieved more slowly in unbuffered media. It is the only one of the common hydrogen ion responsive electrodes that is substantially unaffected by the presence of oxidizing agents. Furthermore, it responds well in most mixed solvents containing at least a few per cent of water, and even in many completely anhydrous media without pronounced dehydrating properties. In addition, it responds in a nearly theoretical manner to deuterium ion in heavy water.

The glass electrode consists of a very thin bulb (about 0.1 mm thick). The bulb is made of high conductivity glass which is sensitive to pH change. The glass electrode consists of silver chloride, silver and 0.1 N HCl. Its tip is covered by a special glass surface. It allows only hydrogen ions. Each of these B components exist in ionized state.

$$Ag \longrightarrow AG^{+} + e^{-}$$

$$AgCl \longrightarrow Ag^{+} + Cl^{-}$$

$$HCl \longrightarrow H^{+} + Cl$$

Different type of glass electrodes are available. Reasonably rugged glass electrodes which operate from 0-80°C and upto pH of about 10. In more recently, combination electrodes (glass and reference electrodes) can be used. Combination electrodes may be used to determine the pH of small volume of solutions. The glass electrode can be used for almost any solution except that with is very acidic or basic.

The faults of the glass electrode are few. Among the most prominent may be mentioned the high resistance of the glass membrane, which commonly runs from 10 to 500 megohms. Measurement of the emf of pH cells utilizing a glass electrode is, therefore, not as simple as the measurement of low-resistance cells and, until recently, not as accurate. Most glass electrodes available today respond in a nearly theoretical manner to changes of pH in the range 0 to 12 or even higher. There is, however, some impairment of the response at very low and at very high pH. These departues are known as the *acid error* and the *alkaline error*, respectively.

**Calomel electrode.:** The other electrode is a calomel electrode. Here also, a series of ionizations take place resulting in the generation of electron but not hydrogen ion.

If these two electrodes (glass and calomel electrode) are connected, the electrons will move from the more positive electrode to other. The electrodes are exposed to solution containing high concentration will not respond directly.

In pH meter, the current is fed into a calibrated dial. In such a way that the dial reading gives the pH of the solution.

Commercial pH meters may be classified broadly into two types: (1) the *null-detector type* and (2) the *direct-reading type*. The pH meter consists of two electrodes namely, (1) glass electrode and (2) calomel electrode. Currently, advanced pH meters have been evolved which have degital measurement.

However, the basic principles are the same.

## ACIDS AND BASES

Acids and bases have developed by Bronsted and Lowry. An acid is defined as a substance that yields $H^+$ ions. A base or alkali forms OH ions. Acids are proton donors and bases are proton acceptors. Some of the salts also yield $H^+$ or $OH^-$ in aqueous solutions. Depending upon the relative concentration of $H^+$ or $OH^-$ they are acidic, basic or neutral. Some example of acids and bases are given in the following Table.

**TABLE 6:**

| Acid | | Base |
|---|---|---|
| $H_3O^+$ | = | $H^+ + H_2O$ |
| $H_2SO_4$ | = | $H^+ + HSO^-_4$ |
| $H_2CO_3$ | = | $H^+ + HCO^-_3$ |
| $HCO_3$ | = | $H^+ + CO^-_3$ |
| $HSO_4$ | = | $H^+ + SO^-_4$ |
| $H_2O$ | = | $H^+ + H_2O$ |
| HOH | = | $H^+ + OH^-$ |
| HCl | = | $H^+ + Cl^-$ |
| $NH_3$ | = | $H^+ + NH_2^-$ |
| $NH^+_4$ | = | $H^+ + NH_3$ |
| HCN | = | $H^+ + CN^-$ |
| $CH_3COOH$ | = | $H^+ + CO_3COO$ |

Acids dissociated to yield $H^+$ ions and ions or molecules capable of combining with or bases. For example, $H_2SO_4$ gives $H^+$ and $HSO_4$. In which $H_2SO^-_4$ is the acid $HSO_4$ the base. The base corresponding to HCl is the acid, Cl.$HSO_4$ is the base form by the dissociation of $H_2SO_4$. However, $HSO_4$ is also the acid corresponding to the base $SO_4$. According to this law, an acid dissociates into a proton and a base, while a base combines with a proton to yield and acid. This phenomenon is called as "conjugate base". For example, the $CN^-$ ions is the conjugate base to the acid HCN.$NH_3$ is the base conjugate to the acid $NH^+_4$.

### Strength

The strength of an acid or base is determined by the degree of its ionization in aqueous solution. The weakest acid has the strongest conjugate base. The strongest acid has the weakest conjugate base. For example, HCN is a weak acid because of its strong conjugate base $CN^-$. It combines firmly with protons, whereas HCl is a strong acid. Because its conjugate base $Cl^-$ combines loosely with protons.

**Acid** is a substance that can act both as an acid and a **base** is said to be amphoteric or ampholytes. When an acid is mixed with a base both $H^+$ and $OH^-$ are produced in the medium. They combine to form natural water molecules. If the concentration of $H^+$ is equal to the concentration of $OH^-$ mixture is neither acid nor base. This phenomenon is called **neutralization.**

### Acid Ionization

The ionization of an acid is explained by the following equation:

$$HA \longrightarrow H^{+} + A^{-}$$

It is considered that $H^+$ ions combine with the solvent molecules to yield **hydronium ions**. For example, in water the $H^+$ ion combines with $H_2O$ to form $H_3O$. This is called hydronium cf hydroxonium or oxonium ion. In glacial acetic acid as solvent, the $H^+$ ion forms the acetonium ion $(H.CH_3COOH)^+$. In alcohol it forms the ion (H; $C_2H_5OH$). when liquid ammonia is the solvent, the $H^+$ ion becomes ammoniated to yield $NH^+_4$. Both liquid HCl and $H_2O$ are poor conductors of the electric current and little ionized. Suppose, if they are mixed, the solutions is a good conductor. This is explained by the following reactions:

$$HCl + HOH \longrightarrow H_3O^+ + Cl^-$$

The dissociation of acetic acid in water,

$$CH_3\,COOH + HOH \longrightarrow H_3O^+CH_3COO^-$$

The reaction of acids with water.

$$HA + HOH \longrightarrow H_3O^+\,A^-$$

- The strength of an acid is determined by the following:
- the dielectric constant of the solvent,
- the strength of its conjugate base, and
- the basic strength of the solvent. Solvents of high dielectric constant greatly reduce the attraction between oppositely charged particle dissolved in them.

The strength of the conjugate base of an acid is dependent upon and controlled by the structure of the acid.

**TABLE 7:** pH Scale

| $H^+$ | 1 | $10^{-1}$ | $10^{-2}$ | $10^{-3}$ | $10^{-4}$ | $10^{-5}$ | $10^{-6}$ | $10^{-7}$ | $10^{-8}$ | $10^{-9}$ | $10^{-10}$ | $10^{-11}$ | $10^{-12}$ | $10^{-13}$ | $10^{-14}$ |
|---|---|---|---|---|---|---|---|---|---|---|---|---|---|---|---|
| pH | 0 | 1 | 2 | 3 | 4 | 5 | 6 | 7 | 8 | 9 | 10 | 11 | 12 | 13 | 14 |
| **Acid** | | | | | | | **Neutral** | | | | **Basic** | | | | |

The pH scale ranges between 0 and 14. When pH 0, it indicaes almost pure acid solution. pH 14 denotes alkali solution. An increase $H^+$ concentration will mean decrease of negative power. The pH of acid solution will be less than 7. If the pH is above 7, the solution is alkaline. Neutral pH is 7, acid pH 0-7 and alkaline pH is 7-14. pH values of some biological materials is as follows:

**TABLE 8:** pH Values of some Materials

| **Materials** | **pH value** |
|---|---|
| Blood serum | 7.35-7.45 |
| Cerebrospinal fluid | 7.4 |
| Distilled Water | about 5.5 |
| Egg white | 8.0 |
| Faeces | 7.4 |
| Gastric juice | 7.5-8.0 |
| Grape juice | 3.2 |
| Intestinal juice | 7.4-8.5 |
| Milk | 6.2-7.5 |

| Materials | pH value |
|---|---|
| Orange juice | 2.6-4.4 |
| Pancreatic juice | 7.0-8.0 |
| Saliva | about 0.9 |
| Seawater | 8.0 |
| Tears | 6.6-6.9 |
| Tomatoes | 4.3 |
| Urine | 7.0-7.5 |

**TABLE 9:** pH values of standard solutions

| Normality | HCl | $CH_3COOH$ | NaOH | $NH_3$ |
|---|---|---|---|---|
| 1.0 | 0.10 | 5.37 | 14.05 | 11.77 |
| 0.1 | 1.07 | 2.87 | 13.07 | 11.27 |
| 0.01 | 2.02 | 3.37 | 12.12 | 10.27 |
| 0.001 | 3.01 | 3.87 | 11.13 | 10.27 |
| 0.0001 | 4.01 | 4.65 | 9.45 | 9.25 |

## BUFFERS

A buffer solution is resistant to changes in pH on the addition of small amount of acid or alkali. This resistance is called the **buffer action**. The magnitude of the buffer action is called the buffer capacity (β). Buffers consist of a mixture of a weak acid and its base (acetic acid and sodium acetate), a weak base and its salt (ammonium hydroxide and ammonium chloride). A salt of weak acid and a weak base (ammonium acetate) also has a buffer action.

### Common Buffers

**Phosphate Buffer**: This buffer readily forms heavy metals. Phosphate buffer plays an important role in several biochemical reactions, where it can act as an inhibitor of metabolites and activator. The buffering capacity above 7.5 is poor.

**Bicarbonate buffer**: Bicarbonate buffer liberates $CO_2$ and therefore maintained in an atmosphere of $CO_2$. Bicarbonate has a pKa of 6.1. The buffering capacity pH 7.4 is poor.

**Zwitterionic buffer**: These buffers were introduced by Good *et al.* (1966). They contain both '+' and '-' groups. Zwitterionic buffers are widely used in biological studies e.g. HEPES. Its dissociation constant is pH 7.6.

**Tris buffer**: It can used with heavy metals, but it also acts as an inhibitor in biochemical reactions. It has a high lipid solubility. A tris buffer of pH 7.8 at room temperature has a pH of 8.4 at 4°C. It has a poor buffering capacity below 7.5.

**Buffers of blood plasma**: The bicarbonate-carbonic acid buffer system is of more important in maintaining the pH of blood. The ratio $HCO/HCO_3$ in blood plasma is 20/1 and pH 7.4.

**Buffer of red blood cells**: The red cells or erythrocytes of blood are heavily buffered. Because blood contains large amounts of protein haemoglobin. The $HCO_3$ / $H_2CO_3$ buffer system is also present in the cells. These buffer systems are efficient in buffering the acids produced in the body; some acids are being formed in the tissue cells. Such as phosphoric, acetoacetic, lactic, ß-hydroxybutyric and sulphuric acids. These acids react with the various buffer systems of the body. They are converted into salts which are then changed into other substances. In the disease diabetes mellitus, 100 or more grams of ß-hydroxybutyric acid and acetoacetic acid per day may be produced. In order to prevent fatal acidosis, the acidity of these acids must be neutralized by the buffer systems of the body fluids and tissues. The $HCO_3$ / $H_2CO_3$ buffer system is most efficient in this process. Here the acetic acid decreases the concentration of $HCO_3$ and increases the $H_2CO_3$ in $HCO_3/H_2CO_3$ buffer system. The pH changes according to the change in ratio of $HCO_3$ to $H_2CO_3$.

**Buffer of tissue fluids and tissues**: The buffer system present in lymph and spinal fluid is $HCO_3/H_2CO_3$ system. Tissues have much buffer capacity which is mainly to protein buffers and $HCO_3/H_2CO_3$ buffer system. Buffers composed of various of various organic acids and salts. They play a major role in regulating the ph of body fluids. In the case of buffer systems of the body fluids and tissues, the pH ranges of efficient action for the buffers vary somewhat and overlap. Thus tending to provide more efficient buffering than would be attained by having all of the buffer capacity present in one buffer system.

## Mechanism of Buffer Action

The mechanism of buffer action can well be explained by taking the buffer pair of acetic acid and sodium acetate. Acetic acid feebly ionises to yield acetate ion and hydrogen ion.

$$CH_3COOH \rightarrow (CH_3COO)^- + H^+$$

Sodium acetate being a salt ionises considerably and forms a high concentration of acetate ions.

$$CH_3COONa \rightarrow (CH_3COO)^- + Na^+$$

If a strongly ionising acid like HCl tending to decrease the pH is now added to the buffer, the acetate ions available combine with the hydrogen ion entering the buffer. It converts it into the weakly ionising acetic acid, and the influence of added hydrogen ion in decreasing pH is overcome by salt of buffer.

$$H + (CH_3COO)^- \rightarrow CH_3COOH$$

## Applications of Buffer

- Buffer solutions are used in many biochemical studies where the pH needs to be accurately controlled.
- Buffers play an important role of *in vitro* studies.
- Buffers are able to prevent large changes in pH of most biological organisms.
- The cytoplasmic fluids consist of dissolved proteins, inorganic salts and organic substrates resist excessive changes in pH.
- The blood plasma is a highly effective buffer solution. It is almost ideally designed to keep the pH of blood within 0.2 pH unit of 7.2-7.3.
- Enzymes are responsible for the catalysis of reactions in which metabolites catalytic action at some definite pH.
- Buffer systems provide protection to cells and tissues against sudden change in pH.
- Buffer systems in the organisms help in carrying on most of the biochemical reactions in a narrow pH range of 6-8. For example, blood maintains its constant pH of about 7.4. Despite the fact that it carries of large number and variety of chemicals.

# APPENDIX 4

# Some Interesting World Wide Web Sites

## SEQUENCE DATABASE SEARCHING

Database searching can be carried out to search for sequences similar to the sequence that interests you. The databases can also be utilized to find out the sequence of specific genes which are utilized for vector construction. Any database sequence and its associated documentation can be retrieved and written into a personal file or a personal data library. The National Center for Biotechnology Information (NCBI) in the NIH (National Institute of Health web (www) and others can be accessed through the Genetics Computer Group (GCG). Programs like the BLAST (Basic Local Alignment Search Tool), Entrez and NLM PubMed (MEDLINE) can be used to search databases maintained at the NCBI over the internet.), Bethesda, Maryland, USA maintain several databases. There are several programs that can be used to search these databases and some of these can be searched through the world wide ograms like FastA, TFastA and several others can be used to access the GCG database.

## AVAILABLE DATABASES

***Several databases are maintained throughout the world***: Gen Bank, PIR-Protein, SWISS-PROT, EMBL, GenEMBL (specifying both the GenBank and EMBL) etc. GenBank is the NIH genetic sequence database, which is an annotated collection of all publicly available DNA sequences.GenBank is part of the International Nucleotide Sequence Database Collaboration, which is comprised of the DNA DataBank of Japan (DDBJ), the European Molecular Biology Laboratory (EMBL), and GenBank at NCBI.These three organizations exchange data on a daily basis.

Most journals require submission of sequence information to a database prior to publication so that an accession number may appear in the paper. The methods of submission of sequences vary with different databases. The databases offer services like PubMed MEDLINE, the Entrez search system, BLAST sequence similarity searching and other E-mail servers. In addition to the DNA sequence databases, there are protein sequence databases like the SWISS-PROT and networks which can be used for the prediction of secondary structure of proteins like the Protein Design Group, EMBL, Heidelberg, Europe. HEADLINE can be used to search for research literature on any topic.

General NCBI information can be obtained over the internet (www) at info@ncbi.nlm.nih.gov and the NCBI home page comes up. The Entrez can be clicked on if the need is to find the sequence of a particular gene. Entrez needs information about the name of the gene etc to start the search. The sequence information can then be downloaded as needed. On the other hand if a search needs to be made to find out any similar sequences to a particular protein, the BLAST program is utilized. In this program, the choice can be made to locate protein-protein similarity by specifying blastp and swissprot or Gen EMBL. The similarity sequences in nucleotide databases can also be specified by choosing the options. The information can be downloaded and utilized. The databases of fasta or tfasta can be searched for similarity sequences through the GCG programs. The GCG program is available only on the Wisconsin package. This runs on UNIX and one can logon to the system over the network by using a terminal emulation software eg.: EM340.

## INTERESTING WWW SITES

There are several interesting **World Wide Web** sites that can be explored by a scientific worker in the areas of Molecular Biology and Biotechnology. Some of these are given below.

http:// webcrawler.cs.washington.edu/webcrawler/webquery.html
  [*For general information.*]
http:// www.public.iastate.edu/~pedro/research_tools.html
  [*Molecular biology info.*]
http:// macserver.molbio.gla.ac.ukbest.gdb.org/best.html
  [*List of researchers in Molecular Biology.*]
http:// www.atcg.com/atcg
  [*Info on restriction enzymes.*]
http:// alpha.genebee.msu.su:80/supplier *or*
  http:// web.frontier.net/MEDMarket/indexes/indexmfr.html
  [*Biotech companies and suppliers of reagents and chemicals.*]
http:// www.atcc.org/atcc.html
  [*American type culture collection of plasmids and strains.*]
http:// cgsc.biology.yale.edu/top.html
  [*Database of E. coli strains & genes.*]
http:// Yahoo. Com. Science search.
http://www.biotech.nwu.edu/nsf/well/html
http://www.biotech.nwu.edu/nsf/callahan.html.
http://www.comunidadandina.org.
http://www.biodiv.org/chm/conv/art19.htm.
http://www.telegraph.co.uk.
http://www.bio.indiana.edu/people/terminator.html.
http://www.fao.org/unfao/bodies/coag/coag15/x0074e.htm.
  FAO (1999). Committee on Agriculture – Biotechnology, Rome
http://www.grain.org/publications/reports/special.htm.
  GRAIN (1995). Genetic Action Resources International.
http://www.grain.org/publications/gtbc/issue3.htm.
  GRAIN (1998).
http://www.telegraph.co.uk.
  Alarm over 'Frankenstein' foods.
http://www.ejb.org/content/vol1/issue3/full/7/
  Ortiz, R. (1998). Critical role of plant biotechnology, Electronic Journal of Biotechnology
http://www.rafi.org/genotype/970618quin.html.
  Bolivian farmers demand researchers drop patent on Andean food crop.
http://www.biotech.nwu.edu/nsf/sheldon.html.
  Biotechnology and ethics: a blueprint for the future.
http://www.hc-sc.gc.ca/pmra-arla/qcont-e.html
  Canadian Pest Management Regulatory Agency
http://www.hc-sc.gc.ca/english/

Health Canada Online
http://infosrv1.ctd.ornl.gov/TechResources/Human_Genome/publicat/primer/prim1.html
http://www.agwest.sk.ca/infsept98.html
Marker-Assisted Selection: Fast Track to New Crop Varieties Ag-west Biotech Inc.
http://www.agwest.sk.ca/genetrans.html
Natural Gene Transfers: How Nature Engineers Plants Ag-west Biotech Inc.
http://www.cfia-acia.agr.ca/english/ppc/biotech/bsco.html
Office of Biotechnology, Canadian Food Inspection Agency
http://www.bio.org/whatis/guidecit.html
The 1998-9 Citizens' Guide to Biotechnology (Biotechnology Industry Organization, Washington DC)
http://www.agwest.sk.ca/transgenic_plants.html
Transgenic Plants Ag-west Biotech Inc.
mark@ofrf.org, http://www.ofrf.org/
Organic Farming Research Foundation

**TABLE 10:** Internet Addresses Relevant to Genome Research

| Description | Address |
|---|---|
| **Site relevant to the Human Genome Project** | |
| CEPH (YAC mapping) | http://www.cephb.fr/bio/ceph-genethon-map.html |
| dBEST (EST sequences) | http://www.ncbi.nih.gov/dbEST/lindex.html |
| Genethon (SSLP maps) | http://www.genethon.fr/ |
| Radiation hybrid database | http://www.ebi.ac.uk/RHdb |
| Whitehead Institute (physical mapping) | http./www.genome.mit.edu/ |
| **Sites relevant to other eukaryotic genome projects** | |
| *Microbial Eukaryotes* | |
| Candido albicans | http://glamdring.ucsd.edu/others/dsmith/dictydb.html |
| Dictyostelium discoideum | http://glamdring.ucsd.edu/others/dsmith/dictyab.html |
| Plasmodium falciparum | http://ben.vub.ac.be/malaria/mad.html |
| | http://www.wehi.edu.au/biology/malaria/who.html |
| | http://parasite.arf.efl.edu/malaria.html |
| Saccharomyces cerevisiae | http://genome-www.stanford.edu/Saccharomyces/ |
| | http://www.mips.biochem.mpg.de/ |
| | http://www.sanger.ac.uk/Projects/ |
| Schizosaccharomyces pombe | http://www.mips.biochem.mpg.de/ |
| **Invertebrates** | |
| Bombyx mori | http://www.ab.a.u-tokyo.ac.jp/sericulture/shimada.html |
| Caenorhabditis | http://www.sanger.ac.uk/Projects/C_elegans/ |
| | http://moulon.inra.fr/acedb/acedb.html |
| | http://www.ddbj.nig.ac.jp/htmls/c-elegans/html/CE_INDEX.html |
| Drosophila melanogaster | http://flybase.bio.indiana.edu/ |
| Mosquito | http://klab.agsci.colostate.edu/ |
| **Vertebrates** | |
| Chicken | http://www.ri.bbsrc.ac.uk/chickmap/chickgbase/manager.html |
| Cow | http://locus.jouy.inra.fr/cgi-bin/bovmap/lintro2.pl |
| Dog | http://mendel.berkeley.edu/dog.html |
| Mekada fish | http://nioll.bio.nagoya-u.ac.jp:8000/ |
| Mouse | http://www.genome.wi.mit.edu/cgi-bin/mouse |

| Description | Address |
|---|---|
| | http://www.genethon.fr/ |
| | http://www.informatics.jax.org/ |
| Pig | http://www.ri.bbsrc.as.uk/pigmap/pigbase/pigbase.html |
| Pufferfish | http://fugu.hgmp.mrc.ac.uk |
| Rat | http://ratmap.gen.gu.se/ |
| Sheep | http://dirk.invermay.cri.nz/ |
| Zebrafish | http://zfish.uoregon.edu/ |
| Plants | |
| Arabidopsis thalliang | http://lenti.med.umn.edu/arabidopsis/arab/arab_top_page.html |
| | http://cbil.humgen.upenn edu/-atgc/ATGCUPhtml |
| | http://www.arabidopsis.com/bb/b950928z.html |
| | http://nasc.nott.Ac.uk/JIC-contigs/ |

# Glossary

**β-replicase:** It helps in replication of viral RNA.

**Aberrant:** A growth that deviates from the normal or usual type; as exceptional growths **(aberrations)** which occur in some tissue cultures.

**Abort:** To stop or fail to reach full development (maturity) or to develop imperfectly; as some hybrids are prone to embryo **abortion** or as some attempts to culture a plant may fail (**abortive** attempt).

**Accession**: Plant or seed sample, strain or population held in a genebank or breeding programme for conservation and use.

**Acetosyringone**: One of the phenolic compounds released by wounded plant cells which can stimulate the transcription of the *vir* region of the *Ti* plasmid in the initial stages of transfer of DNA from *Agrobacterium* to the plant cell.

**Activator**: A substance or physical agent that stimulates transcription of a specific gene or operon.

**Active Genepool Collection:** A collection of germplasm used for regereration, multiplication, distribution, chracterization and evaluation. Ideally, germplasm in the active collection should be maintained in sufficient quantity to be available on request. Active collection germplasm is commonly duplicated in a base collection and is often stored under medium to long-term storage.

**Adenosine Triphosphate (ATP):** The general energy donor molecule in all organisms.

**Adsorb:** the adhesion of a liquid, gas or dissolved material to a solid surface, resulting in concentration of the **adsorbate** (the **adsorbed** material) to the **adsorbent** (the **adsorbing** agent, such as activated charcoal). The process is **adsorption**.

**Aerate:** To supply with or mix with air or gas. The process is **aeration**.

**Aerobic or aerobiotic:** Refers to an organism that lives in or a process occurring in the presence of molecular oxygen.

**Agar or agar-agar:** A gelatinous polysaccharide obtained from the red alga Gelidium corneum and from several other red algae. It is a solidifying agent that mixed with nutrient media (0.6-1%), forms a gel for growing tissue cultured plants and for other purposes. It ranges in quality from rlatively inert to very impure (complex, undefined). Its firmness as a gelling agent is affected by medium pH (it is softer when the medium is more dilute). Agar gels melt at about 100°C but solidify at about 44°C.

**Age:** **1.** The period in the life cycle of an organism; the process of growing older, mature. **2.** The state of being old or senescent. **3.** Culture age is a function of the number of subcultures and the time after subculture.

**Agenesis:** The absence of development.

**Agglutinate:** **1.** To cause to unite. adhere; as if glued. **2.** To gather into a clump or mass; as protoplasts and bacteria, in the presence of specific antibodies, tend to stick to one another. The process is **agglutination.**

**Agglutinin:** An antibody capable of clumping bacteria or other cells.

**Agrenet**: The computer communications network established by the Agriculture and Food Research Council (AFRC) for the agriculture and botanical research community in the United Kingdom.

***Agrobacterium tumefaciens***: The bacterium causing crown gall disease of plants, inducing tumors to form. Tumors may be cultured in vitro and are useful in the study of this disease. The Ti plasmid of A. *tumefaciens*is known to cause the disease and a small portion of it is used as a vector inthe genetic modification of higher plant cells. Novel DNA sequences are spliced into the plasmid DNA segment, the DNA is circularized and introduced into the cultured plant cells.

**Agroinfection**: A method for initiating viral infection by inoculating host plants with *Agrobacterium* containing viral-specific DNA within the T-DNA.

**Airlift fermenter**: A cylindrical fermentation vessel in which the cells are mixed by air, which is introduced at the base of the vessel and rises through the column of culture medium. The cell suspension circulates around of the column as a consequence of the gradient of air bubbles in different parts of the reactor.

**Aliquot or aliquot part:** An evenly divided unit, portion or sample (fraction) of the whole.

**Alphanumeric**: The set of symbols used to represent letters of the alphabet and the integers 0-9.

**Ambiguity code**: A character other than the standard codes for representing the DNA and RNA bases (ACGT and U) used to indicate a base with an ambiguous identity.

**Amphidiploid**: hybrids between two species or genera containing the sum of the complete chromosome complements, which can be of different ploidy levels, of each parent.

**Amplification Fragment Length Polymorphism (AFLP)**: A variant DNA amplification product produced by DAF, PCR, or RAPD of different size.

**Anther culture:** Refers to culture of single pollen grains or of the anther, containing the male gametophytes or micropores, with the objective of producing monoploid plants.

**Antibiosis**: The prevention of growth or development of an organism by a substance or another organism.

**Antigen**: A compound that induces the production of antibodies.

**Anti-idiotype antibody**: An antibody that has the properties of an antigen.

**Antisense construct**: A plasmid that contains the coding regions of a gene placed adjacent to a promotor in such an orientation as to cause the gene to be transcribed from the DNA strand complementary to that from which RNA is normally transcribed.

**Antisense DNA**: (1) The sequence of chromosomal DNA that is transcribed. (2) A DNA sequence that is complementary to all or part of a functional RNA (mRNA).

**Antisense RNA**: A RNA sequence that is complementary to all or part functional RNA (mRNA).

**Artificial seed:** Encapsulated or coated somatic embryos (embryoids) that are planted and treated like seed.

**Asepsis**: Without infection or contaminating microorganisms.

**Aseptic technique**: Procedures used to prevent the introduction of fungi, bacteria, viruses, mycoplasma or other microorganisms into cell, tissue and organ cultures.

**Aseptic:** Free of pathogens, contaminants, algae, bacteria, fungi, viruses, etc; absence of all microorganisms. **asepsis** is a fundamental requirment for plant tissue culture (aseptic culture).

**Asparagine (Asp, $C_4H_8N_2O_3$ mw 132.12):** An amino acid occasionally included in plant tissue culture media, as a source of reduced nitrogen.

**Aspirate:** To draw something in or out, up or through using suction or a vacuum; as **aspiration** (vacuum) may be used in the disinfection process to draw disinfectant into the surface layers of plant tissue.

**Assay: 1.** To test or evaluate. **2.** The substance to be analyzed or the process of examining or testing it (chemically or by other means).

**Assembler**: The lowest level of computer programming language that can be read by humans. The operations are close to machine code and are programmed using memonics for the processor instruction set.

**Asymmetric Somatic hybridization/partial genome transfer**: The production of viable and fertile hybrid plant by irradiating protoplasts from one parental line (the donor) in order to pulverize the chromosomes and facilitate the elimination of substantial amount of the donor genome. Thereafter, the irradiated donor genome may be transferred into the genome of the recipient by protoplast fusion which can be carried out using three different methods: chemicals, an electrical field or a UV laser microbeam. Partial genome transfer system has potential use for the transfer of molecularly uncharacterized traits and multigenetic characters.

**Attachment efficiency**: The percentage of cell plated (seeded, inoculated) which attach to the surface of the culture vessel within a specified period of time. The conditions under which such a determination is made, should always be stated.

**Autoclave: 1.** An closed chamber in which to heat substances under pressure to above their boiling points, to sterilize utensils, liquids, glassware, etc., using steam. The routine method uses steam pressure of 103.4 X $10_3$ Pa at 121° C for 15 minutes, or longer for large volumes to reach temperature. **2.** A pressure cooker. These employed in medium and instrument sterilization for plant tissue culture work. Over-sterilization degrades culture media constituents and caramelizes sugars, so is to be avoided. Under-sterilization results in culture media or equipment contamination. Heat labile components of tissue culture media cannot be **autoclaved**. These are commonly filter-sterilized. **3.** To carry out the process of **autoclaving.**

**Autonomous replicating sequence (ARS):** Any cloned DNA sequences that initiates and supports extrachromosomal replication of a DNA molecule in a host cell; often used in yeast cells. Also called autonomous(ly) replicating segment, autonomously replicating sequence.

**Autoradiography**: A technique that captures the image formed in a photographic emulsion as a result of the emission of either light or radioactivity from a labeled component that is placed next to unexposed film.

**Auxin:** One of a large class of plant hormones (phytohormones) allegedly produced in the growing tips of stems and roots. It causes cell enlargement and elongation but in excised, cultured systems they may promote cell division.

**Auxotroph:** A microorganism or plant cell line possessing atypical nutritional requirements for some item(s) (growth factors) in addition to those required by the control or wild type which can synthesize them; as **auxotrophic** cells requiring certain amino acids.

**Axenic culture**: A culture without foreign or undesired life forms. An axenic culture may include the purposeful co-cultivation of different types of cells, tissues or organisms.

**Axenic:** A pure culture of one species. This implies that cultures are free of microorganisms (aseptic or germ free).

**Axis:** The main plant stem.

**B vitamin:** One of a complex of vitamins made in plants and though to be essential for healthy growth. Some are routinely added to plant tissue culture media to promote growth. The latter include thiamine **($B_1$)**, niacin **($B_3$)**, adenine **($B_4$)** and pyridoxine **($B_6$).**

**Bactericide:** A substance or agent that kills **bacteria** (usually rapidly); is **bactericidal**.

**Batch culture:** A cell suspension grown in liquid medium of a set volume. Inocula of successive subcultures are of similar size and cultures contain abou the same cell mass at the end of each passage. Cultures commonly exhibit five distinct phases per passage; a lag phase follow inoculation, then an exponential growth phase, a linear growth phase, next a deceleration phase and finally a stationary phase.

**Bcteriostat:** A chemical or other agent that does not kill but prevents bacterial growth and multiplication; is **bacteriostatic**.

**Bergmann, L. (1960):** He was first to obtain callus by plating cells from suspension cultures onto solid medium.

**Bioassay:** A biological assay performed on living cells or on a living organism; sometimes used to detect minute amounts of substances which influence or are essential to growth.

**Biotechnology:** The industrial use of biological processes; as yeast fermentation for alcohol production or plant cell culture for extraction of secondary products. The scope of biotechnology has been dramatically increased through genetic engineering.

**Biotin ($C_{10}H_{16}O_3N_2S$ mw 244.31):** Part of the vitamin B complex. It is also called vitamin H. It is present in all living cells, bound to polypeptides or proteins, and is of import in fat, protein and carbohydrate metabolism. It is a common addition to plant tissue culture media.

**Bleach:** A fluid, powder or other whitening **(bleaching)** or cleaning agent. Bleach contains calcium hypochlorite or sodium hypchlorite and is a common disinfectant used for cleaning working surfaces, tools and plant materials in plant tissue culture.

**Buffer:** A substance in solution, or system

that resists pH change, (despite addition of some small amount of acid or base) and withstands shock; as **buffered** media can resist pH drift. The degree to which pH change is resisted is a measure of the solution's **buffering capacity.**

**Bung:** A stopper or closure.

**Burgeon:** To flourish, bud, sprout or grow.

**Calcium hypochlorite ($Ca(OCl)_2.4H_2O$ mw 215):** An algicide, bactericide and fungicide capable of disinfecting and bleaching. It is used in dilute solution (5-10% w/v) as a plant tissue disinfectant often when agitating or using a vacuum. Tissue damage occur if contact is prolonged. Thorough washing with sterile water generally follows treatment.

**Callus, pl. *calli* or calluses: 1.** Wound tissue, tissue formed on or below a wounded surface. **2.** Disorganized tumor-like masses of plant cells that form in culture. These proliferate in an irregular tissue mass, and vary widely in texture, appearance and rate of growth; a function of the tissue type (species and explant) and the composition of the medium. the process of callus formation is **callogenesis.**

**Cannula:** A small tube for insertion into tissue; as a cork borer.

**Carbon source:** A source of the non-metallic element carbon (C); as organic substances like sugars taken up and metabolized by plant tissue cultures.

**Casamino acid:** An amino acid obtained through the digestion (acid hydrolysis) of the milk protein **casein**. Protein digest yields **casamino acids** and other substances. These are undefined constituents of some plant tissue culture media.

**Casein hydrolysate (edamin):** A milk protein digest product composed of amino acids (casamino acids) and other substances. This complex (undefined) product is sometimes used as an addition (0.02-2.10%) in nutrient solutions used in plant tissue culture as a non-specific source of organic nitrogen.

**Casein:** the principal protein of milk, a phosphoprotein.

**Caulogenesis:** Shoot formation; as de novo shoot development from callus.

**Cell count:** The number of cells per unit suspension volume or callus weight. Tissue is treated with chromic acid (5-8%) or pectinase (0.25%) for up to 15 minutes followed by mechanical dispersion, then cell numbers are estimated with a hemocytometer (haemocytometer).

**Cell culture:** The culture of single or groups of cells on solid or dispersed in liquid nutrient media (cell suspension cultures) usually following a set of defined growing conditions (protocols).

**Cell hybridization:** Formation of synkaryons, viable cell **hybrids** produced through cell fusion. they are identifiable by their increased chromosome number compared to the parent cells and the possession of characters found in one or another of the parental cells.

**Cell line:** Developmental history or descent, through cell division from a single original cell; as callus may constitute a group of cell, all descendants from a single cell plating. Numerous **cell lines** may be present in a culture. Any deviation in culture technique may favor one cell line over another.

**Cell number:** The absolute number or approximation of the number of cells per unit area of a culture or medium volume.

**Cell selection:** Selection within a group of genetically different cells; involves competition between cells often under some stress. Selection criteria may involve cell viability l a unique phenotype or biochemical activity; or another basis for choice. Select cells or cell lines are generally relocated to fresh medium for continued selection and often are exposed to an increased level of the stress agent. The final objective is usually to regenerate plants from those select cells, in hope that the plants will exhibit the traits selected for at the cellular level.

**Cell suspension:** Cells and small aggregates of cells suspended in a liquid medium; as in cell suspension cultures. Explants, or callus derived from them are transferred to liquid medium and the cultures are then agitated on a mechanical shaker. The ensuing single cells and small cell clusters are used for a number of purposes in plant tissue culture; as in single cell cloning.

**Centrifuge:** An apparatus used for separating particles from suspension using centrifugal force. Balanced tubes containing the suspension are rapidly rotated causing sedimentation of particles to the bottom of the tubes (pelleting) based on their weight and the density of the suspending medium.

**Chelate:** A chemical compound with which metal atoms may be combined in order to prevent them from precipitating out of solution and so being unavailable to plants; as ethylenediamine tetra-acetic acid, disodium salt ($Na_2EDTA$) is complexed to iron (and other metal ions) in nutrient solutions used for plant tissue culture. The process is **chelation.**

**Chemically defined medium:** A nutrient medium for plant tissue culture in which all of the chemical components are fully known and stated. Quality (high purity) chemicals are utilized, and no undefined constituents are included.

**Chemotherapeutant:** A chemical used to pretreat diseased source plants prior to excision or incorporated into media to support some therapeutic objective; as malachite green or virazole (ribavirin) are used to eliminate virus in meristem tip culture. The process is **chemotherapy.**

**Chimera or chimaera:** A plant or tissue composed of more than one kind of genetic tissue. In a periclinal chimera one genotype occurs in a superficial layer, covering a genetically different core. In a mericlinal chimera one genotype occurs in a localized part of the plant and the rest is occupied by another genotype. In a sectorial chimera two genotypes share distinct sectors of the plant. Chimeras can occur naturally; be established artificially by grafting; or developed through somatic mutation by chemicals such as colchicine, or by other means in plant tissue cultures.

**Choline chloride ($C_5H_{14}ClNO$ mw 139.63):** An alkaloid, (B complex vitamin), occasionally added to plant tissue culture media. **Choline,** found in many plant organs, is the basic constituent of lecithin.

**Clonal multilication or clonal propagation:** Vegetative (asexual) propagation from a single cell or plant. See **clone.**

**Clone: 1.** A basic category of cultivar. The original cultivar or variety. Not to be confused with source clone, indicating different originals within the clone. Members have a common origin, are the extension of a single cell (via mitosis) or plant and have been produced by vegetative means only. Genetic uniformity is accepted. A clone may have one or more sources. **2.** While the expectation is that members of a clone are both phenotypically and genotypically identical, this assumption is not always true and members may not be genetically or phenotypically equivalent (varients). the objective of much tissue culture propagation is to propagate very large numbers of selected plants with the same genotype. The process is **cloning.** The prefix meri reflects the explant used to initiate the clone (meristem tip source clone). In a **calliclone** cloning involves the callus stage. **3.** Specific DNA sequences are said to be **cloned** when isolated and propagated.

**Closed continuous culture:** A cell suspension culture with a continuous influx of fresh medium, maintained at constant volume by the efflux of spent medium. All cells are retained within the unit. See **continuous culture.**

**Cocking, E.C. (1960):** Obtained the first higher plant protoplasts using root cells subjected to fungal cellulase and demonstrated new cell wall regeneration on protoplasts from tomato fruit locule tissue.

**Coconut milk or coconut water:** Liquid endosperm from the center of the coconut seed. A complex, undefined addendum of variable quality and effects in some nutrient solutions (2-15%, v/v) for plant tissue culture. It has growth promoting effects and cell division factors. It is replaceable in some cases by cytokinins and/or sugar. It was first used by van Overbeek et al. (1941) to stimulate *Datura* embryo cultures.

**Co-culture:** The joint culture of two or more types of cells; as a plant cell and a microorganism or two types of plant cells; as is done in various dual culture systems or the nurse culture technique.

**Colony: 1.** A group of interdependent cells or organisms. **2.** An aggregate of cells developed from a single cell; as in single cell platings, a clone composed of one cell line.

**Complex substance:** A complicated and undefined substance; as in some addenda to nutrient solutions used in plant tissue culture. Examples include protein hydrolysate, yeast or malt extracts, endosperm from corn or coconuts, juices like oranges or tomato juice and others.

**Contaminant:** An undesirable bacterial, fungal or algal microorganism accidentally introduced to a culture or culture medium. It may overgrow the plant cells or inhibit their growth through release of toxic metabolities. Rigorous exclusion of potential **contaminants** must be practiced in plant tissue culture.

**Contaminate:** To accidentally introduce a substance or organism **(contaminant)** into a medium or culture. The process is **contamination**.

**Continuous culture:** A cell suspension culture with a continuous influx of fresh medium, maintained at constant volume by the efflux of spent medium (closed continuous) or with the efflux of cells and spent medium (open continuous).

**Control:** **1.** The untreated plant or unchanged (standard) protocol or treatment for comparison with the experimental treatment. **2.** To direct or regulate; as in induction of organogenesis in cultures through hormone regulation.

**Controlled environment:** A chamber, room or situation in which the environmental parameters of light, temperature, and relative humidity are controlled. The partial pressure of gases may also be controlled.

**Corn milk:** Corn endosperm tissue. An undefined complex addendum to some plant tissue culture media.

**Criticl concentration:** The chemical concentration above or below which reaction or developmental process will not proceed.

**Crown gall:** A disease of plants in which tumors form. The causal agent is the bacterium, *Agrobacterium tumefaciens*. Gall tissue has been grown in vitro. A tumor inducing portion of the bacterial genome (the Ti plasmid) my be used experimentally as a genetic tool to incorporate (vector) genetic information into plant cells.

**Cryoprotectant:** An agent able to prevent freezing and thawing damage to cells as they are frozen or defrosted. These substances have high water solubility and low toxicity. They are classified either as permeating (glycerol and dimethyl sulfoxide) or non-permeating (sugars, dextran, ethylene glycol, polyvinyl pyrolidone and hydroxyethyl starch) agents.

**Culture alteration:** A persistent change in a cultured cell or tissue anatomy or physiology, including a change in one or more nutritional requirements or a change in its proliferative capacity.

**Culture medium:** A prepared nutrient solution (substrate) that may be chemically defined for growing plant tissues or other organisms in vitro.

**Culture room:** A controlled-environment room (light, temperature, relative humidity, often air conditioned, etc.) for incubation of plant cultures.

**Culture tube:** A test tube used to contain medium and an explant or culture derived from it, usually incubated in a controlled environment.

**Culture vessel:** A container used to hold medium and an explant or culture derived from it, usually incubated in a controlled environment.

**Culture:** **1.** A general term for the **cultivation** of microorganisms, animals, plants or their cells in vivo or in vitro especially in order to improve the breed. **2.** In vitro aseptic cultivation of microorganisms or of cells, tissues, or organs of plants, animals in or on prepared nutrient media under controlled, aseptic conditions.

**Cybrid:** A cell or plant cytoplasmic hybrid (heteroplast) with the nucleus of one and chtoplasmic organelles of another, or of both cells or plants.

**Cysteine hydrochloride ($C_3H_7NO_2S$. HCl mw 157.63):** An occasionl, amino acid (cysteine) salt additive in plant tissue culture media. It is included primarily for its antioxidant properties.

**Cytochimera or chromosomal chimera:** Cells within a tissue possessing different chromosomal numbers, as when a callus consists of diploid and polyploid regions of cells.

**Cytodifferentiation:** A functional and/or morphological differentiation occurring in cells during ontogenesis, affecting their phenotype.

**Cytolysis:** Cell dissolution or disintegration.

**Cytoplasmic hybrid or cybrid:** A cell or plant hybrid (heteroplast) with the nucleus of one and the cytoplasmic organells of another or of both cells or plants.

**Cytoplasmic variant:** A maternally inherited change in a cellular trait.

**Cytotoxic:** A chemical or other agent toxic to cells.

**De novo:** (Latin: *fromthe beginning, anew*). Arising, sometimes spontaneously, from unknown or very simple precursors.

**Decereration phase:** the declining growth rate phase following the linear phase and preceeding the stationary phase in most batch suspention cultures.

**Decontaminate:** To free from **contamination** or surface sterilize.

**Dedifferentiation:** The resumption of meristematic activity by more or less mature cells through a reversal of the process of cell or tissue **differentiation**. Cell division leading to the formation of small, microvacuolate, isodiametric cells with prominent nuclei, that are capable of organized development; as when adventitious buds or roots develop on mature tissue.

**Dehumidifier:** An apparatus that removes moisture from the air.

**Deionized water:** Water that has been passed through an ion exchange device to remove soluble minerals and some organic salts. The process is **deionization.**

**Deletion: 1.** Omission, removal or cancellation of some factor. **2.** The lost portion of a chromosome or a nucleotide sequence in nucleic acid.

**Derepression:** The mechanism by which repression is alleviated; as when a **repressing** metabolite is removed, resulting in the increased level of a protein or enzyme.

**Detergent:** A cleaning agent; one of numerous synthetic cleaning preparations chemically different from soap. **Detergents** are commonly used to prepare work areas for aseptic plant tissue culture manipulations and to remove dirt and microorganisms from plant material prior to explantation.

**Dewar flask:** A double-walled glass flask used to keep liquids at other than ambient temperatures; as a thermos does.

**Direct embryogenesis:** Embryoid formation directly on the surface of zygotic or somatic embryos or on seedlng plant tissues in culture, without an intervening callus phase.

**Direct organogenesis:** Organ formation directly on the surface of relatively large intact explants, without an intervening callus phase.

**Dispense:** To give, deal or porton out; as nutrient medium is portioned into glassware for plant tissue culture.

**DNA ligases:** Enzymes which bind DNA fragments together

**Donor plant (mother plant):** The source that plant used for propagation, whether an explant, graft or cutting. Source plants used for micropropagation are usually poathogen-tested (disease-free).

**Dry ice:** Frozen (solid) carbon dioxide ($CO_2$). It is commonly used as a refrigerant.

**Dual culture:** A culture system that includes plant tissue and one organism (such as a nematode species) or microorrganism (such as a fungus). Dual cultures are used to study host-parasite interactions or in the production of axenic cultures for a variety of purposes. Usually the microorganism selected is an obligate parasite.

**Embryo culture: 1.** Denotes a culture in which the explant was n embryo. **Embryo cultures** have been used to obtain viable offspring from seeds with a tendency for embryo abortion or when viable seed are limited in number. **2.** Cultures in which embryos are induced to form (embryogenesis) whether in suspension or on a variety of explants or cultures on solidified media. More correctly, these nonzygotic or somatic embryos are termed embryoids.

**Enation:** Outgrowth on a plant surface.

**Enhancer element:** These are short regions of DNA which have the effect of increasing the levels of transcription, but unlike promotors, they affect genes from chromosomal loci at a long distance.

**Environmental chamber:** A controlled environment cabinet (incubator) in which temperature, light quality, intensity and duration; and preferably also the relative humidity and airflow are controlled.

**Enzyme:** Any one of a number of spoecilized proteins produced in living cells, that speed the rate of specific chemical reactions, even at very low concentrations (organic catalyst), but is not used up in the reaction.

**Epigenetic gene control:** This is the new concept. It describes that the control of gene expression in higher eukaryotic organisms is related to the methylation of cytosine residues in gene coding sequence.

**Epigenetic gene control:** This is the new concept. It describes that the control of gene expression in higher eukaryotic organisms is related to the methylation of cytosine residues in gene coding sequence.

**Equimolar:** The same amount of solute per litre of solutiion, the same **molar** concentration.

**Established culture: 1.** Achievement of an aseptic viable explant, synonymous with Stage I culture (culture establishment) or transplant, synonymous with Stage IV culture (establishment in soil). **2.** A suspension culture subjected to several passages, adjusted so that cell number per unit time is consstant from subculture to subculture.

**Ethylenediaminetetraacetic acid, disodium salt ($Na_2$EDTA, $C_{10}H_{14}N_2O_8Na_2.2H_2O$ mw 372.25):** A chelating agent which reversibly binds positive ions such as iron, magnesium, etc. It is commonly added to plant tissue culture media to keep iron and other salts available by releasing them slowly into the medium as required.

**Ex vitro:** (Latin; *from glass*). Organisms removed from culture and transplanted, generally to soil or potting mixture.

**Excise:** To cut, sever or otherwise remove or extract an organ or a segment of tissue from a plant or plant part; as the surgical removal of shoot tips. The process is **excision.**

**Explant donor:** The source plant or mother plant from which the explant used to initiate a culture is taken.

**Explant:** The excised plant portion used to initiate a tissue culture. The process of dissection and removal to culture of these small organs or tissue sections is **explantation.** Explant choice, the timing of excision and pretreatment are important determinants of culture success.

**Factor: 1.** A condition, influence or constituent to be taken into account. **2.** A genetic character determinant (gene).

**Filter sterilize:** to sterilize by passing a solution through a porous material capable of separating out suspended microbes or their spores; as heat-labile components of nutrient media are sterilized.

**Fixative:** A compond that stabilizes, sets or **fixes** other compounds or structures securely so that their structural integrity is retained; as the process of **fixation** utilizes chemical agents to permanently prepare cells or tissues for microscopy.

**Floccule:** An aggregation (coalescence) of microorganisms or colloidal particles floating in or on a liquid. **Flocculation** is seen in some contaminated liquid media appearing as a cloud.

**Folic acid ($C_{19}H_{19}N_7O_6$ mw 441.40):** A member of the vitamin B group (Bc). It is also known as vitamin M. It is present in green leaves and has some coenzyme activities. It is occasionaly added to plant tissue culture media.

**Freeze-dry or lyophilize:** To dry in a frozen state under vacuum; as tissues are **freeze-dried** to obtain a dry weight or to preserve them for analysis.

**Fresh weight or wet weight:** The weight of a plant or plant part including the water content.

**Fusogen:** A fusion-inducing agent used for protoplast agglutination in somatic hybridization studies; as is polyethylene glycol.

**Gametoclone:** A plant regenerated from a tissue culture originating from gametic tissue.

**Gelatin:** A glutinous, proteinaceous material produced by boiling, which partially hydrolyzes the collagen, of animal connective tissues. It is sometimes used to **gel** or solidify nutrient solutions for plant tissue culture. Other **gelling** agents, specifically agars, are usually preferred.

**Gelrite:** The brand name of a synthetic (*Pseudomonas*-derived) refined polysaccharide used as a gelling agent and agar substitute.

**Generation time:** The time between successive generations of cells or organisms within a population.

**Genetic engineering or recombinant DNA technology:** Technology involving man-made changes in the genetic constitution of cells (apart from selective breeding). This technology usually employs a vector (such as the Ti plasmid of *Agrobacterium tumefaciens*) for transferring useful genetic information from a donor organism into a cell or organism that does not possess it. Genetic engineering has mny potential uses.

**Genetic selection:** Selection of genes, cells (cell selection), clones, etc., by man within populations or between populations or species. The usual purpose is to alter a specific phenotypic character. Such selection usually results in differential success rates of the various genotypes, reflecting many variables including selection pressure and genetic variability in populations.

**Genetic transformation:** The transfer of extracellular DNA (genetic information) among and between species; as, for example, with the use of bacterial or viral vectors.

**Genetic variation:** Differences in individuals derived from the same genotype in distincton to differences caused by the environment. **Genetic variance** denotes the proportion of pohenotypic variance caused by differences in the genetic make-up of an individual.

**Genome:** The basic haploid chromosome set of an individual; the sum of its **genes.**

**Genotoxic:** Carcinogenic; toxic to the chromosomes.

**Geotaxis:** Plant orientation with respect to gravity.

**Germplasm or germ plasm: 1.** The reproductive body tissues distinct from somatic (nonreproductive tissues). The genetic material, basis of heredity of an organism, passed on through previous generations. **2.** An individual representing a type of species or culture that may bve held in a repository for agronomic, historic (or other) reasons.

**Gibberellic acid (Gibberellin $A_3$ or $GA_3$, $C_{19}H_{22}O_6$ mw 346.37):** One of the **gibberellins,** a group of growth hormones promoting cell division and elongation. The first of the group to be isolated and the most widely used gibberellin in plant tissue culture. Isolated from the fungal pathogen *gibberella fujikuroi.* It dissolves in base (ca. 1M KOH or NaOH).

**Gyrase:** It is Type-II topoisomerase that catalyzes conversion of relaxed DNA to a superhelical form and requires ATP.

**Hemicellulase:** An enzyme from *Aspergillus niger*, available as a commercial preparation, that degrades hemicellulose to galactose.

**HEPA filter:** An acronym for **high efficiency particulate air filter.** A filter capable of screening out particles larger than 0.3 m. They are used in laminar air flow cabinets (hoods) for sterile transferr work.

**Heterograft or xerograft:** An interspecific graft. The likelihood of success of this graft is proportional to the degree of relatedness between the donor and recipient.

**Heteroplasmon or heteroplast:** A cell with a mixture of two types of cytoplasm (cytoplasmic hybrid); cell with foreign organells.

**Hexitol:** A sic carbon sugar alcohol, such as (meso-, myo-, or i-) inositol.

**Hood:** Laminar air flow cabinet.

**Hydrate:** To add or incorporate water. The process is **hydration.**

**Hypertrophy:** An abnormal increase in cell size causing irregular swelling oir growth. Hypertrophy is usually a disease or other stress-induced response.

**Hypochlorite:** A salt of hypochlorous acid **(sodium hypochlorite, potassium hypochlorite** or **calcium hypochlorite).** All are oxidizing agents used for disinfecting and for bleaching.

**In situ:** In the natural, original place or position; as in the location of the explant on the mother plant prior to excision.

**In vitro**: (Latin = *in glass*). Experimentation on organisms or portions thereof in glassware or culture; growing under artificial conditions as in tissue culture. The antonym of in vivo.

**In vivo:** (Latin = *in life*) Experimentation on organisms under natural conditions within intact living organisms. the antonym of in vitro.

**Incubate:** To maintain under conditions favorable for development, often in an **incubator**. The process is **incubation.**

**Incubator:** An apparatus providing controlled environmental conditions, (light, photoperiod, temperature, humidity, etc.) suitable for **(incubating)** plants or plant cultures. This term is sometimes used as as synonym for cabinet (growth cabinet) but usually implies a greater degree of environmental control.

**Indirect embryogenesis:** Embryoid

formation on callus tissues derived from zygotic or somatic emtryos, seedling plant or other tissues in culture. The antonym is direct embryogenesis.

**Indirect organogenesis:** Organ formation on callus tissues derived from explants. the antonym is direct organogenesis.

**Induction media:** **1.** Media that can induce organs or other structures to form. **2.** Media that will cause variation or mutation in the tissues exposed to it.

**Infestation:** Occupation by threatening numbers of insects, mites or potential disease agents; as mite **infestations** in incubators. An **infesting** organism can cause contamination of tissue cultures. This should not be confused with infected, as infection is not necessarily implied.

**Infrared gas analyzer (IRGA):** An instrument for measuring the proportion of a particular gas in a mixture; as changes in $CO_2$ concentration can be monitored to evaluate a plant's photosynthetic or respiratory activity.

**Innoculum:** **1.** Material introduced **(inoculated)** onto or into a host or medium. **2.** Potential inculum implies that it is infective and may by chance result in natural **inoculation** of a host.

**Inoculate:** **1.** To deliberately introduce something into; as **inoculum** is placed into (or onto) medium to initiate a culture. This process is **inoculation,** but should not be confused with contaminate. **2.** Vaccinate.

**Inositol; myo, meso, or i-inositol or hexahydroxycyclohexane ($C_6H_{12}O_6$ mw 180.16):** A widely distributed sterioisomeric sugar alcohol in plants and essential in animals. Inositol is included by many in the B complex vitamin group. It is involved in the synthesis of phospholipids, cell wall pectins and membrane systems in cell cytoplasm. It is commonly added to the nutrient medium used for plant tissue culture (ca. 100 mg/1) for its growth promoting effects.

**Introns:** Sequences in DNAs which are in fact transcribed into mRNAs but are subsequently split out and there fore form no permanent constituents of the mRNAs (only in eucaryotes)

**Irradiance:** The total radiation on an exposed surface.

**Irradiate:** **1.** To illuminate. **2.** To emit or expose to waves of light, heat or nuclear emissions. **3.** To treat by exposure to radiation. The process is **irradiation.**

**Isograft or syngraft:** A graft or transplant among isogenic (genetically identical) individuals; as on the same organism.

**Isolation medium:** A medium suitable for explant survival and development. It may be synonymous with Stage I medium, or contain antioxidant(s), reduced hormone concentration, bacterial indicators or other addenda, and precedes Stage I culture.

**Jiffy pots:** A brand name for peat pots that are sometimes used for ex vitro transplantation.

**Kinases:** They catalyze the transfer of the α-phosphate of a nucleoside 5'-triphosphate to the 5'-OH terminus of a DNA or RNA molecule.

**Kinin:** The original class name for substances promoting cell-division to which the prefix *cyto* has been added **(cytokinins)** to distinguish them from **kinins** in animals systems.

**Laminar air flow cabinet or hood:** A structure in which a uniform flow of filtered air prevents settling of particles in the work area. The air is filtered through prefilter (furnace-filter quality), then a high efficiency particulate air (HEPA) filter that strains out particles greater than 0.3 m. These hoods are commonly employed for aseptic plant tissue culture manipulations and are designed to accommodate one to the several persons, depending on the model.

**Layering:** **1.** Covering stems, runners or stolons with soil causing adventitious roots to form at the nodes enabling propagation by rooted cuttings. This procedure is used commercially to propagate many plants, such as the brambls. **2.** In vitro layering involves the horizontal placement on agar of cultured shoots (with or without leaves) or nodal segments to promote axillary bud proliferation.

**Lecithin:** A naturally occurring, choline-containing phospholipid present in animal and plant tissue. Chemically **lecithins** are similar to fat, but they also contain phosphorus and nitrogen.

**Ligases:** They catalyze formation of phosphodiester linkages between DNA chain segments.

**Ligation:** Direct linkage of nucleic acid fragments

**Limiting requirement or factor:** An environmental variable whose absolute levels at given time limits the growth or other activity of an organism.

**Linear phase:** The constant increase in cell number following the exponential growth phase and proceeding the deceleration phases in most batch suspension cultures.

**Liquid culture:** The culture of plant cells on liquid medium, in suspension or on supports. Cultures are either held steady (stationary culture) or are shaken (agitated culture or shake culture).

**Liquid medium:** Medium not solidified with a gelling agent. **Liquid media** are used for suspension cultures and for a wide range of research purposes. They are also useful for Stage I culture in some micropropagation protocols. Usually a support structure or wick is used to hold the tissue above the nutrient medium.

**Lysis:** Cell rupture or destruction, as through enzymatic action.

**Macerase:** A brand name for pectinase. Pectinase is useful in the isolation of intact cells and protoplasts of higher plants.

**Macerate:** To disintegrate or separate tissues through cutting, soaking, enzymatic or other action, resulting in cell dissociation.

**Magnetic stirrer:** An apparatus used for stirring solutions. A magnetic stir bar is placed at the bottom of the container and is twirled in the solution that requires mixing by an electrically driven, rotating magnet below the platform of the apparatus.

**Malachite green ($C_{22}H_{25}N_2Cl$ mw 364.66):** A chemical used as a seed disinfectant in

**Mannose ($C_6H_{12}O_6$ mw 180.16):** A hexose component of many polysaccharides and mannitol. It is occasionally employed as a carbohydrate source in plant tissue culture media.

**Maternal inheritance:** Inheritance controlled by extrachromosoml (cytoplasmic) hereditary determinants.

**MDA:** microdroplet array technique.

**Medium formulation:** One of many available, usually derived formulas that are used in plant tissue culture. The formulas commonly contain the microelements and microelements (high and low salt formulas are available), some vitamins (B vitamins, inositol), hormones (auxin, cytokinin and sometimes gibberellin), a carbohydrate source (usually sucrose or glucose) and often other substances such as amino acids (the most common addition is glycine) or complex growth factors. Media may be liquid or solidified with agar, the pH is adjusted (ca. 5-6) and the solution is sterilized (usually by filtration or autoclaving). Some formulations are very specific in the kind of explant or plant species that can be maintained some are very general.

**Medium:** The substrate for plant growth; as nutrient solution, soil, sand, etc. This is a general term for the liquid or solidified formulation upon which plant cells tissues or organs develop in plant tissue culture.

**Mericlinal:** A meristem tip source clone. This term can be synonymous with calli clone or may not involve callus.

**Meristem culture:** The culture of meristems; meristematic dome tissue without adjacent leaf primordia or stem tissue. In practice this usually means meristem tip culture. This may more generally imply the culture of meristemoidal regions of plants or of meristematic growth (associated with or sharing the characteristics of the meristem) in culture.

**Meristem tip culture:** Cultures derived from meristem tip explants. They are excised for virus elimination or axillary shoot proliferation purposes, less commonly for callus production.

**Meristem tip:** A common explant comprised of the meristem (meristematic dome) and usually one pair of leaf primordia. Also refers to apical (apical meristem tip) or lateral (lateral or axillary meristem tip) origin. The term meristem tip is often confused with the term shoot tip, which is much larger and usually has more immature leaves and some stem tissue.

**Meristemming:** the utilization of or process of utilizing meristem tips for explants. This term is often used to refer to micropropagation via axillary shoot proliferation.

**Meristemoid:** A cluster of small, isodiametric **meristematic** cells with a **meristem**, or cultured tissue, with the potentialk for developmental (totipotential) growth. These cells have a dense microvacuolated cytoplasm and a high nucleo-cytoplasmic ratio. They may give rise to plant organs (shoots, roots) or entire plantlets in culture.

**Methionine (Met, $C_5H_{11}NO_2S$ mw 149.21):** An amino acid precursor in ethylene synthesis and occasionally added to plant tissue culture media.

**Methyl methanesulfonate (MMS, $C_2H_6SO_3$ mw 110.13):** A frequently used, very potent chamical mutagen which acts by adding methyl groups to guanine and subsequently causes base pairing errors as it binds to adenine.

**Methylases:** They catalyze transfer of methyl groups from donar *S*-adenosyl-L-methionine (AdoMet) to specific bases of acceptor DNA molecules.

**Microclimate or microhabitat:** The climate in the immediate vicinity or surrounding an organism or its parts.

**Microcutting:** A tiny cutting; as an explant such as a meristem tip removed for culture with the use of a dissecting microscope.

**Microdroplet array or multiple drop array (MDA) or hanging droplet technique:** Introduced by Potrykus, Harms and Lorz (1979), this technique is used to evaluate large numbers of media modifications, employing small quantities of medium into which are placed small numbers of cells. Droplets of liquid culture (medium and suspended cells or protoplasts) are arranged on the lid of a petri dish, inverted over the bottom half of the dish containing a solution with a lower osmotic pressure, and the dish is sealed. The cells or protoplasts form a monolayer at the droplet meniscus and can easily be examined.

**Micrograft:** An in vitro graft. This procedure involves placing a meristem tip or shoot tip explant onto a decapitated rootstock, that has been grown aseptically from seed or micropropaged. The purpose may vary, from virus elimination, using meristem tips and virus-free rootstocks, to viral assays or to erxpedite grafting that is usually done in the greenhouse. The explanted scion is placed directly on the cambium of the severed stock or inserted into a T or inverted T-shaped epidermal incision onthe stock. This technique has been used to eliminate viruses, viroids and mycoplasmas from Citrus species.

**Micromole (M):** One millionth (0,000001) of mole ($10^{-6}$ M). The unit of concentration used for microelements, vitamins, hormones and some other organic addenda used in plant tissue culture media.

**Micropropagation:** Refers to propagation in culture by axillary or adventitious means. It is a general term for vegetative (asexual) in vitro propagation. It sometimes refers specifically to axillary bud proliferation.

**Minimum inoculum size:** The smallest inoculum that can be successfully used for subculture.

**Mitotic index (MI):** The ratio of nuclei undergoing mitosis (including prophase) to total nuclei. MI = (Number of nuclei in mitosis/Total number of nuclei examined in the sample) X 100. A MI of 0.3 means that 30% of cells in the population are observed in mitosis.

**Mitotic index (MI):** The ratio of nuclei undergoing mitosis (including prophase) to total nuclei. MI = (Number of nuclei in mitosis/Total number of nuclei examined in the sample) X 100. A MI of 0.3 means that 30% of cells in the population are observed in mitosis.

**Mitotic nondisjunction:** Occurs when sister chromatids fail to migrate to opposite poles of the cell during mitotic anaphase. The result is daughter cells with hyperploid and hypoploid chromosome counts.

**Monoculture:** A one-crop agriculture system. An intensified monoculture could refer to a clonal (one genotype) agricultural system.

**Mononucleotide:** A nucleic acid basic-building-block. Composed of a pentose, a phosphoric acid plus a purine or pyrimidine base.

**MS:** An abbreviation for the **Murashige, T. and F. Skoog (1962)** medium formulation.

**Mutability:** the propensity of an individual's genes or genotype to undergo heritable mutation.

**Mutagen:** A chemical or physical treatment or agent capable of producing genetic mutation. Causes insertion, deletion or alteration of a base or part of the nucleic acid chain (DNA). Artificial mutations can be induced in cells using chemicals or ionizing radiation. The process is **mutagenesis**. Electromagnetic and particle radiation, such as UV, X-rays and b particles all increase mutation frequency. So do chemical mutagens such as ethylmethanesulphonate, acridine, nitrous oxide and many others

Naphthaleneacetic acid: 1-naphthaleneacetic acid.

**Natural complex:** A complicated, non-synthetic, often undefined addendum in plant tissue culture media, such as orange or tomato juice or coconut milk.

**Neoplasm:** Localized cell multiplication or tumor; a collection of cells which have undergone genetic transformation. These cells differ in structure and function from the original cell type.

**Neoteny:** The occurrence (retention) of juvenile characters in the adult state or vice versa.

**Newcommer's fluid:** A fixative used in chromosome analysis composed of 6 parts isopropyl: 3 parts propionic acid: 1 part petroleum ether:1 part acetone: 1 part dioxane (with or without ferric acetate).

**Niacin or nicotinic aid ($C_6H_5NO_2$ mw 123.11):** Vitamin $B_3$ forming part of a respiratory coenzyme. It is a common micronutrient addition to plant tissue culture media. It is sometimes also known as vitamin PP.

**Niacinamide or nicotinamide ($C_6H_6N_2O$ mw 122.12):** An amide of niacin occasionally added to plant tissue culture media.

**Nitsch, J.P. and C. Nitsch (1969):** Developed a now widely used plant tissue culture medium formulation **NN**, for use in anther culture.

**NN:** An abbreviation for the **Nitsch, J.P. and C. Nitsch (1969)** medium formulation.

NOA: (2-naphthalenyloxy) acetic acid.

**Nobecourt, J.R. (1937):** was among the first to describe the prolonged culture of carrot callus.

**Nucellar embryony**: The process by which individual cells of the nucellus give rise to somatic embryos (embryoids). Plants on which this occurs naturally are considered good candidates for tissue culture of somatic embryoids.

**Nucleases:** They catalyze hydrolysis of the phospho-diester bonds of polynucleotides.

**Nucleo-cytoplasmic ratio:** The ratio of cell nuclear to cytoplasmic volume. This ratio is elevated in meristematic cells and low in differentiated cells.

**Nutrient film technique (NFT):** Hydroponic plant growth whereby plant roots are suspended in shallow, slowly circulating nutrient solutions delivered as a continuous film of liquid rather than in bursts of liquid.

**Nutrient gradient:** A diffusion gradient of nutrients and gases is set up in tissues where only a portion of the tissue is in contact with the medium. This is common in callus culture and causes differential growth rates and senescent regions of callus. Similarly, diffusion gradients of nutrients are set up in the medium adjacent to the tissue. Gradients are less likely to form in liquid media.

**Ontogeny:** The developmental (**ontogenetic**) life history of an organism, its development from fertilized egg (zygote) to seed-forming adult (reproductive phase), with emphasis on embryonic development.

**Open continuous culture:** A cell suspension culture with a continuous influx of fresh medium, maintained at constant volume by the efflux of cells and spent medium.

**Operator:** Regulatory section in a nucleic acid with binding capacity for a repressor protein for blocking the synthesis of mRNAs.

**Organ culture:** The growth in aseptic culture of plant organs such as roots or shoots, beginning with organ primordi or segments and maintaining the characteristics of the organ.

**Organized growth:** The in vitro development of organized explants such as meristem tips or shoot tips, floral buds or organ primordia, or their de novo formation from **unorganized** tissues.

**Organogenesis:** The initiation (de novo) and growth of organs (roots and shoots, usually) from cells or tissues; as in organ culture. Organs may form on the surface of explants (direct organogenesis) or upon an intervening callus phase (indirect organogenesis).

**Organoid:** An anomalous organ-like structure formed in culture; as on leaves, roots or callus.

**Ortet:** The original mother plant or donor plant from which vegetatively propagated plants are derived.

**Osmium tetroxide or osmic acid ($OsO_4$ mw 254.20):** A fixing agent used to prepare tissues for electron microscopy.

**Osmolarity:** The total molar concentration of the solutes affecting the **osmotic** potential of a solution or nutrient medium.

**Ovule culture:** A culture derived from an explanted ovule; which may be fertilized in culture (in vitro fertilization). This technique is used to study development of zygotes and young embryos and is sometimes used to rescue embryos susceptible to abortion when embryo culture is not possible.

**Oxygenase:** An enzyme enabling an organism or system to utilize atmospheric oxygen.

**Packed cell volume (PCV):** A quantitative method of estimating cell growth. It is based on the total cell volume in an aliquot of suspension culture. The aliquot is centrifuged for 5 minutes at 200 g after which the packed cell volume is expressed as a percentage of the aliquot volume.

**Panicle culture:** Aseptic culture of grain panicle segments, usually in an effort to induce microspore development.

**Pantothenic acid ($C_9H_{17}NO_5$ mw 219.23):** Also known as vitamin $B_5$ it is of wide occurrence in most plant and animal tissue, is essential for cell growth and is an important coenzyme in fat metabolism. It is added to some plant tissue culture media as the calcium salt.

**Papain:** A water soluble proteolytic enzyme (protease) extracted from papaya fruit and used especially as a meat tenderizer.

**Paper raft technique:** A technique developed by W.H. Muir (1953) to promote development of single cells taken from suspension cultures in which cells are placed onto filter paper squares set on actively growing callus (nurse tissue). Growth factors and nutrients from the callus tissue diffuse through the filter paper, promoting cell growth and development.

**Para-amino-benzoic acid (PABA, $C_7H_7NO_2$ mw 137.12):** An occasional addition (vitamin Bx) to plant tissue culture media.

**Parahormone:** A substance with hormone-like properties that is not a secretory product; such as ethylene or carbon dioxide.

**Parasexual hybridization:** Refers to genetic recombination by means other than through fertilization of germ cells (parasexual) leading to hybrid cells or individuals; as in hybrid cells or plants derived from somatic cell fusion.

**Particle radiation:** Refers to - particles (positively charged) and B-particles (negatively charged), electrons, protons and neutrons. These particles are used to produce mutant cells or organisms in plant tissue culture.

**Parts per million(ppm):** This term has largely been replaced by the equivalent mg/1 of solution or mµl/1 for liquids or gasses.

**Passage number:** The number of subculture intervals. Culture age is a function of passage number and the culture dilution ratio.

**Passage time:** The interval between subcultures or the culture period.

**Pasteur pipette:** A causal agent of a disease; as a **pathogenic** fungus, bacterium or virus.

**PCV:** packed cell volume.

**Pectinase:** An enzyme that degrades **pectin,** the adhesive material that cement cells together. It is used alone or with other enzymes to digest the polygalacturonic acid of plant cell walls (to sugar and galacturonic acid) in protoplast production.

**Peptide:** A compound consisting of two or more amino acids covalently linked. **Peptides** with three or more amino acids are **polypeptides.**

**Periclinal chimera:** Refers to a condition in which geno-typically or cytoplasmically different tissues are arranged in concentric layers.

**Periclinal chimera:** Refers to a condition in which geno-typically or cytoplasmically different tissues are arranged in concentric layers.

**pH:** A measurement of the degree of acidity or alkalinity of a solution. The negative log of the hydrogen ion concentration, in moles perr litre (M/1). On the pH scale from 0-14, there is a 10 fold difference for each unit of change of pH. The lower the value, the more alkaline (more hydroxyl ions it contains). Most culture media are adjusted in the range pH 4-6 using 1M NaOH or HCl.

**Phenocopy:** A non-hereditary phenotypic

change that is environmentally induced, during a limited developmental phase of an organism, that mimics the effect of a known genetic mutation.

**Phenolic oxidation:** The process by which many plant species which contain phenolics blacken through oxidatin when they are wounded; as in vitro, soon after explantation or subculture. This may lead to growth inhibition or, in severe cases, to tissue necrosis and death. Antioxidants are incorporated into the sterilizing solution or isolation medium to prevent or reduce oxidative browning.

**Phenols or phenolics:** A class of organic compounds with hydroxyl group(s) attached to the benzene ring forming esters, ethers and salts. **Phenolic** substances may bleed from newly explanted tissues, oxidizing to form colored compounds visible in nutrient media or on the filter wicks supporting the tissue.

**Phloretic acid:** A phenolic acid formed by the oxidation of a hydroxyl (OH) of 1,3,5-trihydroxybenzene. An occasional constituent of some plant tissue culture media.

**Photoheterotroph:** A photosynthetic organism that requires an organic hydrogen source. The condition is **photoheterotrophy.**

**Photometer:** An instrument for measuring the luminous intensity of light sources.

**Photosynthetically active radiation (PAR):** The photon flux expressed as moles or microeinsteins per meter squared per second ($molm^{-2}$ or $Em^2S^{-1}$) or as wattsper meter squared ($Wm^{-2}$) over the wavelength range of 400-700 nm; the part of the light spectrum which is primarily absorbed by plants and used in pohotosynthesis.

**Phytokinin:** An absolute term for cytokinin.

**Phytotron:** A controlled environment building or chamber for studying plant growth under defined conditions.

**Plant tissue culture:** A general term encompassing the in vitro culture of plant cells, tissues, organs and whole plantlets.

**Plantlet:** A shoot, sometimes exclusively a rooted shoot, growing in culture or derived from culture.

**Plasmid:** Annular nucleic acid in cells outside the central chromosome with a lower molecular weight than the central chromosome; bearer of nonessential additional information, e.g., sex factors (F factors) and resistance factors (R ractors), plasmids and viral nucleic acids with known functions such as resistance or sequences are denoted by trivial abbreviation derived from their function or origin, e.g., SV 40 for simian virus, $pBR^{322}$ for a plasmid imparting resistance to antibiotics.

**Plasmotype:** A cell type displaying features which are expressions of cytoplasmic, rather than nuclear inheritance.

**Pleiotropy:** The condition in which several characteristics are effected by a single gene.

**Pollen culture:** The culture of pollen grains, which germinate in vitro. Such cultures may eventually form monoploid callus, from which shoots or embryoids develop into monoploid plants.

**Poly(A) adding enzymes:** They attach AMP residues, from ATP, to the 3', 5' phosphodiester bond.

**Polyethylene glycol (PEG) or carbowax:** A fusion-inducing agent (fusogen) for agglutinating protoplasts, which is used in somatic hybridization studies. This compound is also sometimes used in media as a non-metabolite osmoticum and is available in various molecular weights, ranging from ca. 200 to 6,000.

**Polygalacturonic acid:** A long chain sugar acid polymer composed of galacturonic acid and hexose subunits.

**Polygenes:** Systems of genes associated with quantitative character variation in which each gene individually effects the phenotype in a minor way.

**Polymerases:** They catalyze addition of nucleoside triphosphates to 3'-OH terminals of single strands of preexiting polynucleotides (primers) with release of pyrophosphate.

**Polypropylene:** A strong, flexible, transparent thermoplastic formed by the polymerization of propylene. It is used in many labware products.

**Population density:** Cell number per unit medium area or medium volume.

**Potassium hypochlorite (KClO mw 90.6):** A water-soluble bleaching or disinfecting agent used in plant tissue culture practices.

**Potato extract:** A common (undefined) organic addendum to plant tissue culture media in monocot and another culture systems.

**Preadaptation:** In possession of advanageous characters enabling an organism to be well suited **(adapted)** to environmental conditions previously unknown to it.

**Precipitin test or microprecipitin test:** A serological assay in which visible particulates **(precipitates)** form when soluble antigen and antibody react. This precipitin reaction detects and identifies antigens.

**Prefilter:** A coarse filter (furnace filter) such as those used in a laminar air flow cabinet to screen out large particles before air is forced through a much finer filter (HEPA filter).

**Premix:** To mix before use; as nutrient mixtures are commercially available as dry powders of preweighed ingredients, which are put into solution when required.

**Primary culture:** **1.** May be synonymous with Stage I culture. **2.** A recent culture not yet subcultured for the first time.

**Probe DNA:** A radioactively lebeled (usually $^{32}P$) DNA molecule used to detect complementary-sequence nucleic acid molecules by molecular hybridization. To localize the probe DNA sequence and reveal the complementary hybridization sequence auto-radiography is often used.

**Promeristem or protomeristem:** The embryonic meristem-containing organ initials or foundation cells.

**Propagule:** The form or portion of an organism used for reproduction or propagation; as new shoots or callus derived from explants are subdivided into **propagules** and re-culturd for further multiplication.

**Protein digest:** The enzymatic hydrolysis of proteins to yield their building block components, amino acids and short chain peptides (chains consisting of two to several amino acids normally without enzymatic function).

**Protein stability:** Protein, in eukaryotic cells have very diverse half-lives ranging from a few minutes to several hours. The degradation of unnecessary proteins is dependent upon the small protein ubiquitin, which when conjugated to proteins leads to their rapid degradation.

**Protoclone:** Distinct phenotypic regenerates from a plant protoplast. A clone initiated from a protoplast or protoplast-fusion product.

**Protocol:** A sequence of activities, techniques or procedures linked to achieve a specific goal; as a schedule for protoplast preparation.

**Protoplast culture:** The isolation and culture of plant protoplasts by mechanical means or by enzymatic digestion of plant tissues or organs, or cultures derived from these. Protoplasts are utilized for selectio or hybridization at the cellular level and for a variety of other purposes.

**Protoplast fusion:** The coalescence of the plasmalemma and cytoplasm of two or more **protoplasts** in contact with one another. Initial adhesion is a random process but coalescence may be promoted in various ways (induced fusion). When adhesion occurs between adjacent protoplasts during enzymatic wall degradation or between freshly isolated protoplasts in the absence of a fusion agent, it is termed spontaneous fusion.

**Pure culture:** Axenic culture.

**Pure line:** All cell or individual members are homozygous for one or more characters or genes and will give rise to more cells or organisms with the character(s) under consideration.

**Quiescent:** Quiet, at rest, not necessarily dormant and having the potential for resumed activity; can apply to cells unlikely to divide, the non-meristematic cells.

**Radiation:** Rays of heat, light or particles in wave form.

**Ramet:** The vegetatively propagated offspring of an ortet. An individual of a clone.

**Reaction norm:** The range of phenotypic responses of a particular genotype in response to the environmental influences.

**Reciprocating shaker:** A plateform shaker with a back and forth action used for agitating culture flasks at variable speeds.

Recombinant DNA technology: genetic engineering.

**Recombinant DNA:** Genetic material with novel gene sequences produced by cross-overs, chromosome reassortment, by other natural means or through genetic engineering.

**Reconstructed cell or recon:** A viable cell hybrid, cybrid or a transformed cell resulting from genetic engineering.

**Reculture or subculture:** The aseptic transfer of a culture or a portion (division when necessary) of that culture (inoculum) to fresh nutrient medium.

**Redifferentiation:** Cell or tissue reversal in differentiation from one type to another type of cell or tissue.

**Rejuvenation:** **1.** Synonymous with dedifferentiation. **2.** Treatment that leads to culture invogoration (such as subculture) or revival (dormancy breaking).

**Repelcote:** The brand name of a material (dimethyl dichloro-silane) used to coat glassware (soda glass); for tissue cultur containers and for other purposes.

**Repression:** Altered gene expression resulting in the failure of a specific protein synthesis.

**Repressor:** Regulatory protein with binding capacity for operators.

**Resistance transfer factor:** A plasmid, present in certain types of bacteria such as E. coli, that can impart resistance to antibiotics in animal exposd to them.

**Resistivity:** The degree of resistance (in ohm-cm or ohm-m) or interference to the flow of an electrical current or the movement of particles from place to place. A measure of water purity; the purer the water, the greater its **resistivity.** The reciprocal of conductivity.

**Restriction endonucleases:** Enzymes, which specifically cleave DNAs at a definite information-bearing sequence of nucleotides and thereby permit an insertion of foreign sequences. Restriction endonucleases are named in genetic engineering by trivial cell, e.g., EcoRI from *Escherichia coli* R Y 13, HindII from *Haemophilus influenza* Hpall from *Haemophilus parainfluenzq* or psti from *providencia stuartii*

**Reverse transcriptase:** It catalyze transcription of RNA nucleotide sequence into complementary DNA (cDNA).

**Reverse transcriptase:** It catalyzes transcription of RNA nucleotide sequence into complementary DNA (cDNA).

**Rhizogenesis:** Root formation and growth; as in root development de novo from callus.

**Rhodamine isothiocyanate or tetramethylrhodamine isothiocyanate ($C_{25}N_3SO_3H_{21}$ mw 536.10):** A dye used to stain plant protoplasts in the identification of fusion products.

**Root culture:** Isolated root tips of apical produce in vitro root systems with indeterminate growth habits. These were among the first kinds of plant tissue cultures (White, 1934) and remain important research tools in the study of developmental phenomena; mycorrhizal; symbiotic and plant-parasitic relationships.

**Rotary shaker:** A platform shaker with a circular motion used for agitating culture flasks at vaariable speeds.

**Rotating culture or rotator:** A wheel-like device for slowly (ca. 1 rpm) rotating and gently agitating cultures, usually in a vertical plane.

**S phase:** The cell cycle phase during which DNA synthesis occurs.

**Schiff's reagent:** A mixture of pararosaniline. HCl (an analine dye) and sodium bisulfite ($NaHSO_3$), used in staining chromosomes and nuclear material.

**Scion or cion:** The portion of a bud or shoot used for grafting onto another plant (the stock).

**Screen house:** A metal or plastic mesh enclosure similar in function to a greenhouse but with more exposure to the elements.

**Scutellum:** That part of the cotyledon in Gramineae (grasses) that absorbs food from the endosperm at germination. An explant source for plant tissue culture.

**Selection culture:** Utilizes difference(s) in environmental conditions or more usually in culture medium composition, such that preferred variant cells or cell lines (presumptive or putative mutants) are favored over other variants or the wild-type.

**Selection pressure:** The measure of the effectiveness of natural or experimental selection in altering the genetic composition of a population.

**Selection unit:** Single cells or small clusters, units of optimum size for isolating and regenerating variants or mutants;l the minimum number of cells effective in the screening process.

**Semicontinuous culture:** The maintenance of cells in a culture vessel in an actively dividing state by periodically draining the medium and adding fresh medium.

**Serial float culture:** Sunderland's (1977-1979) technique of floating anthers on liquid mediumand subculturing them to new medium at several day intervalsan anther dehiscence, pollen release and development occur, increasing anther productivity.

**Shake culture:** An agitated suspension culture. Usually a flask (commonly an Erlenmeyer flask) containing the culture is attached to a horizontal or platform shaker, or agitated with a magnetic stirrer, to provide adequate aeration for cells in the liquid medium.

**Shakeror platform shaker:** A platform fitted with clips for grasping flasks (usually Erlenmeyer flasks), or with surfaces suitable for attaching flasks, with set or variable speed control. It is desirable to adjust the **shaking** speed for gentle, even agitation of suspension cultures.

**Shoot tip graft or micrograft:** The grafting of a very small shoot tip or meristem tip onto a prepared seedling or mcropropagated rootstock in culture. Meristem tip grafting is used for in vitro virus elimination with Citrus and for other plants as an alternative to grafting in the greenhouse.

**Sodium hypochlorite (NaOH mw 74.44):** A frequently used plant tissue sterilant at 0.5-2% w/v. The usual source is dilute (10-21% v/v) commercial laundry bleach, sometimes with surfactant is common. Tissue damage may occur if contact with tissue is prolonged. Thorough washing with sterile water generally follows treatment.

**Somaclone:** A plant regenerated from a tissue culture originating from somatic tissue.

**Somatic cell embryogenesis:** The production of embryos from somatic cells of explants (direct embryogenisis) or by induction on callus formed by explants (indirect embrygenesis). These two processes may not be materially different in results.

**Somatic cell variant or embryoid:** An organized embryonic structure morphologically similar to a zygotic embryo but initiated from somatic (non-zygotic) cells. These develop into plantlets in vitro through developmental processes that are similar to those of zygotic embryos.

**Somatic hybrid:** A cell or plant product of somaic cell fusion; as the result of cell or protoplast fusion and implying genomic integration. The process is **somatic hybridization.**

**Somatic mutation:** Mutation occurring in vegetative cells or tissues.

**Somatic organogenesis:** The production of shoots, roots or other organs on somatic tissues of explants (direct organogenesis) or by induction oncallus formed by explant (indirect organogenesis).

**Spontaneous fusion:** Uninduced protoplast fusion which may occur between freshly isolated protoplasts or following adhesion of adjacent cells during enzymatic cell wall degradation.

**Spontaneous variation:** The variation in plant populations derived fromtissue cultures not exposed to mutagens but occurring as a result of the culture conditions.

**Stages of culture (I-IV): Stage I:** Aseptic explantation or establishment of the explant in culture. **Stage II:** Multiplication of the propagules. **Stage III:** Rooting of the progagules and preparation for transplant to soil. **Soil IV:** Establishment of Stage II or III propagules ex vitro in soil or potting mix.

**Stationary culture:** A non-agitated culture. the antonym is shake culture.

**Stock plant:** The source plant from which cuttings or explants are made. These are usually maintained carefully in an optimum state for (sometimes prolonged) explant use. Preferably they are certified, pathogen-free plants.

**Stock solution:** A solution, usually concentrated (10 to 100 times the final medium concentration), of select medium constituents that re grouped for compatability to avoid precipitation and prepared before hand to save time during medium preparation. Usually they are frozen or stored in the refrigerator and portions are utilized as media are prepared.

**Stock:** **1.** The root and a portion of the stem (rootstock) of a plant to which is grafted a part of the same or another plant (scion). **2.** A group of closely related plants.

**Subculture or passage:** A culture derived from another culture or the aseptic division and transfer of a culture or a portion of that culture (inoculum) to fresh nutrient medium. Sub-culturing is usually done at set time intervals, the length of which is called the subculture interval or passage time.

**Subline:** A cell line regenerated from a unique cell line of a hybrid callus colony.

**Supraoptimum:** An amount (level) greater than required; as an inhibitory concentration of an exogeneous growth factor.

**Surface sterilization:** The removal of plant surface microflora prior to aseptic excision of explants. Surface sterilization is accomplished by immersion of tissue in one of many sterilants, such as calcium hypochlorite or sodium hypochlorite, hydrogen peroxide, mercuric chloride, silver nitrate or bromine water for an empirically determined period of time.

**Suspension culture:** Cells and group of cells (aggregates) dispersed in an aerated, usually agitated, liquid culture medium. These are obtained by adding friable callus to the medium. The plant species, explant type and treatment; the composition of the nutrient medium; and many other features determine the size and nature of the cell aggregates, which may appear different during each culture growth phase. These cultures are used to study cell division, differentiation and metabolism, in secondary product synthesis or form the basis for single cell lines, callus cultures, somatic cell embryogenesis and many other culture purposes.

**Synchronized cells:** Synchronized mitosis in a group of cells in culture by natural or artificial means.

**Synchronous culture:** A plant cell or microbial culture treated in such a way as to have all (or most) cells or individuals in the same stage of development or mitosis. This can be achieved in various ways including via temperature variation and nutrient limitation.

**Ti plasmid:** A portion of the genome of the bacterium *Agrobacterium tumefaciens*, the agent involved in crown gall disease. This plasmid is a useful (experimental) vector for the transfere of genetic information into plant cells.

**Tissue explant:** An excised plant portion of tissue used to initiate a culture.

**Totipotency:** the potential **(totipotential)** or inherent capacity of a plant cell or tissue to develop into (recreate) an entire plant if suitably stimulated. Totipotency implies that all the information necessary for growth and reproduction of the organism is contained in the cell. Although theoretically all plant cells are **totipotent** the meristematic cells are best able to express it.

**Toxic:** Poisonous; as are some chemicals **(toxicants)** or any substances present to excess and detrimental to normal plant function or growth.

**Trace element:** Miroelement.

**Tracer:** A substance that can be followed within a reaction or an organism; as radioactive isotopes and certain dyes.

**Transcription:** Conversion of a DNA (gene) information sequence into a mRNA sequence

**Transduction:** Introduction of foreign genetic material into a cell by means of a virus or, in the case of bacteria, by means of phages

**Transfer room:** A small room, sometimes sterilized internally by a bactericidal lamp (u.v. irradiation) and provided with clean (fltered) air; employd for sterile transfer work.

**Translation control**: In certain cases, after a gene has been transcribed, the mRNA can fold back upon itself to prevent its translation. This process is known as attenuation.

**Tumor inducing principle (TIP):** The plasmid carried by *Agrobacterium tumefaciens*, the crown gall organism. Through incorporation into the host genome the host tissue is transformed into tumor tissue.

**Tyrosine (Tr, $C_9H_{11}NO_3$ mw 181.19):** An amino acid, sometimes used in plant tissue culture media as a source of nitrogen.

**Ultrasonic cleaner:** A device to include high frequency vibration of materials, removing adhering substances from surfaces by mechanical action. This device is useful for cleaning glassware and for disinfecting plant material.

**Ultraviolet light (UV):** Radiation with wavelength (100-400 nm) at the violet end of the visible spectrum.

**Unorganized growth:** In vitro formation of tissues with few differentiated cell types and lacking recognizable structure; as with many calli.

**Unwinding enzymes:** They catalyze separation of complementary strends of DNA; require ATP.

**Valine (Val, $C_5H_{11}NO_2$ mw 117.1:** An amino acid found in seeds and proteins. Occasionally added to plant tissue culture media.

**Vector:** Plasmid or virus that after ligation with a foreign DNA is used to introduce new information into the cell.

**Virus elimination:** Thermotherapy, chemotherapy and meristem or meristem tip culture, used alone or in combination have been employed for the elimination of systemic viruses from plants.

**Virus-tested or virus-free:** A plant that appears healthy and repeatedly tests negatively for the presence of one or more identifiable viruses. Such a plant may then be used as a stock or donor plant (explant source) for propagation purposes, and may be certified as virus tested (certified virus tested). The term virus-free is incorrect in most cases, as such a plant may contain one or more viruses which have not been assayed.

**Vitamin B complex:** Includes a large group of water soluble vitamins that function as coenzymes; thiamine (**$B_1$**), riboflavin or vitamin G (**$B_2$**), niacin or nicotinic acid (**$B_3$**), pantothenic acid (**$B_5$**), pyridoxine (**$B_{12}$**), biotin or vitamin H, folic acid or vitamin M (**BC**), inositl, choline and others.

**Vitamin $B_x$: para-amino-benzoic acid.**

**Vitamin C:** ascorbic acid.

**Vitamin D:** A fat-soluble vitamin composed of a group of related steroids. Present in all plants.

**Vitamin K:** A fat-soluble quione involved in photosynthetic electron transfer in plants and blood clotting in animals.

**w/m$^2$ or watt per square metre:** A common unit of light measurement.

**W/v:** Weight in volume; as the number of grams of constituent in 100 cm$^3$ (mls) solution.

**Water of hydration:** The amount of water chemically bound to a substance. This amount may be variable and must be taken into account when solutions of salts are prepared; as in medium preparation.

**Wound tissue:** See callus.

**Xerograft:** See **heterograft.**

**X-ray or Roentgen ray (r):** Electromagnetic radiation of short wave length produced by high speed electrons impacting on a metal object under vacuum.

**Xylose ($C_5H_{10}O_5$ mw 150.1):** An aldopentose sugar (wood sugar) present in the woody tissues of many plants as polymeric xylan. It is occasionally used as a carbohydrate source in plant tissue culture media.

**Yeast extract:** A complex, undefined, B vitamin-containing addendum to some plant tissue culture media.

**Yield test:** Productivity assessment.